Probiotics and Prebiotics in Food, Nutrition and Health

Probiotics and Prebiotics in Food, Nutrition and Health

Editor

Semih Ötleş

Food Engineering Department
Ege University
Izmir, Bornova, Turkey

CRC Press
Taylor & Francis Group
Boca Raton London New York

CRC Press is an imprint of the
Taylor & Francis Group, an **informa** business
A SCIENCE PUBLISHERS BOOK

CRC Press
Taylor & Francis Group
6000 Broken Sound Parkway NW, Suite 300
Boca Raton, FL 33487-2742

Printed in the United States of America on acid-free paper

International Standard Book Number: 978-1-4665-8623-9 (Hardback)

Visit the Taylor & Francis Web site at
http://www.taylorandfrancis.com

CRC Press Web site at
http://www.crcpress.com

Science Publishers Web site at
http://www.scipub.net

Preface

Consumer interest in the relationship between diet and health has increased the demand for information on functional foods. Functional foods may do more than simply supply the nutrients our body needs for normal biochemical reactions. They contain compounds or ingredients that reduce risk for certain health conditions or promote better health. One type of functional food is probiotics and prebiotics, which have a long history of safe use and have been known for their health benefits for human. Recent years have seen a series of advances in the research and applications of probiotics and prebiotics medicine. The contributors of the book were asked to discuss not only data in available literature, but also the new concepts and possible health effects that are important in their work.

The book comprises 23 chapters including an introductory chapter and four parts (*food aspects, nutritional aspects, health aspects* and *safety aspects*). The part on *food aspects* comprises chapters on sources, production and microencapsulation of probiotics, production of prebiotics, probiotics and prebiotics in pharmaceutics, analysis of probiotics and prebiotics, regulations and guidelines of probiotics and prebiotics. The part of *nutritional aspects* consists of chapters on probiotics and prebiotics in lipid metabolism, in infant nutrition, in elderly nutrition, in animal nutrition and interactions of probiotics and prebiotics with minerals. The part on *health aspects* consists of chapters on probiotics and prebiotics in energy metabolism and obesity, gut microbiota, in immune system protection, in pediatric diarrheal disorders, in gastrointestinal diseases, probiotics and prebiotics and *Helicobacter pylori*, probiotics and prebiotics in infections, in cancer prevention, in allergy and asthma, in Crohn's disease, in genitourinary system health. The part on *safety aspects* comprises chapters on genomics of probiotics and prebiotics, the future of probiotics and prebiotics. I hope the book will be a useful and practical educational and scientific tool among academics, health professionals, students as well as in private companies and governmental institutions worldwide.

As the Editor of this book I would like to express my sincere thanks to all the authors for their excellent contributions to this book.

Semih Ötleş

Contents

1

Probiotics, Prebiotics and Synbiotics: An Introduction

Ravinder Nagpal,[1] Hariom Yadav,[2,*] Manoj Kumar,[3] Shalini Jain,[2] Yuichiro Yamashiro[1] and Francesco Marotta[4,*]

Probiotics: The Historic Milieu

The term probiotic is derived from Greek and means 'for life' as opposed to antibiotics which means 'against life'. The history of probiotics began with the consumption of fermented foods. Consumption of fermented foods was first observed in ancient Greeks and Romans (Gismondo et al. 1999, Guarner et al. 2005). In 1907, Ellie Metchnikoff, a Nobel Prize winner, first proposed the beneficial effects of probiotic microorganisms on human health. Metchnikoff hypothesized that Bulgarians were healthy and lived long because of the consumption of fermented milk products which consists of rod shaped bacteria (*Lactobacillus* spp.). These bacteria affect the gut microflora positively and decrease the microbial toxic activity (Gismondo et al. 1999, Chuayana et al. 2003). The first documentation about health

[1] Division of Laboratories for Probiotic Research Juntendo University Graduate School of Medicine, Tokyo, Japan.
[2] National Institute of Diabetes and Digestive and Kidney Diseases, National Institutes of Health, Bethesda, Maryland, USA.
[3] National Institute of Nutrition, Hyderabad, India.
[4] ReGenera Research Group for Aging-Intervention, Milano, Italy & Hon. Research Professor, Department Human Nutrition, Texas Women University, Denton, USA.
* Corresponding author(s): yadavhariom@gmail.com; fmarchimede@libero.it

promoting effects of fermented milk dates back to a Persian version of the Old Testament (*Genesis* 18:8) that states that 'Abraham owed his longevity to the consumption of sour milk'. Kollath in 1953 and Vergio in 1954 were probably the first to introduce the term 'Probiotic' (Holzapfel and Schillinger 2002). The term 'probiotic' was first used in 1965 by Lilly and Stillwell to describe substances which stimulate the growth of other microorganisms. Since then the word 'probiotic' has been used in different contexts according to its mechanism and the affects on human health. The term, probiotic, as is used today was first used by Parker in 1974. Parker defined 'probiotic' as substances and organisms which contribute to intestinal microbial balance. In 1989, the term was modified further by Fuller. Thus, probiotic is a live microbial supplement which affects the host's health positively by improving its intestinal microbial balance. This definition was broadened by Havenaar and Huis in't Veld in 1992 to include mono or mixed culture of live microorganisms which benefits animals or man by improving the properties of the indigenous microflora (Çakır 2003, Guarner et al. 2005, Sanders 2003). Some authors have interpreted probiotics as *microbial cell preparations or components of microbial cells that have a beneficial effect on the health and well-being of the host.* Bacterial cell-wall components, heat-killed whole cells or metabolites can have a specific probiotic effect, for example, improvement of lactose digestion or treatment of acute or chronic diarrhoea (Ouwehand and Salminen 1998, Romond et al. 1998, Salminen et al. 1999, Simakachorn et al. 2000, Xiao et al. 2002). Through the years, lots of research has been done on probiotics and therefore, many definitions have been suggested. They are listed below in Table 1.

Probiotics offer challenges for industrial applications. The probiotic concept is open to lots of different applications in a large variety of fields relevant for human and animal health. Probiotic products consist of different enzymes, vitamins, capsules or tablets and some fermented foods contain microorganisms which have beneficial effects on the health of the host. They can contain one or several species of probiotic bacteria. Most probiotic products destined for human consumption are in the form of fermented milk or given in powder or tablet forms. These capsules and tablets are not used for medical applications but as health supporting products. The oral consumption of probiotic microorganisms produces a protective effect on the gut flora. Lots of studies suggest that probiotics have beneficial effects on microbial disorders of the gut, but it is really difficult to show the clinical effects of such products. The probiotic preparations used for traveller's diarrhea, antibiotic associated diarrhea and acute diarrhea show that they have a positive therapeutic effect (Gismondo et al. 1999, Çakır 2003, Quwehand et al. 1999).

Table 1. Some proposed definitions of probiotic.

S.No.	Definition	Reference
1	A live microbial supplement which affects host's health positively by improving its intestinal microbial balance	Fuller 1989
2	Living microorganisms, which upon ingestion in certain numbers, exert health benefits beyond inherent basic nutrition	Shaafasma 1996
3	A live microbial food ingredient that is beneficial to health	Salminen et al. 1998
4	A microbial dietary adjuvant that beneficially affects the host physiology by modulating mucosal and systemic immunity, as well as improving nutritional and microbial balance in the intestinal tract	Naidu et al. 1999
5	A preparation of or a product containing viable, defined microorganisms in sufficient numbers, which alter the microflora (by implantation or colonization) in a compartment of the host and by that exert beneficial health effects in this host	Schrezenmeir and de Vrese 2001
6	Live microorganisms which when administered in adequate amounts confer a health benefit on the host is accepted by FAO/WHO (report in October 2001)	Klaenhammer 2000, Sanders 2003, Guarner et al. 2005

Commercial Strains of Probiotics and their Sources

The probiotic bacteria generally belong to the genera Lactobacillus and Bifidobacterium. However, other bacteria and some yeast also have probiotic properties. Common probiotics include: 1) Lactobacilli such as *Lactobacillus acidophilus, L. johnsonii, L. casei, L. delbrueckii* ssp. *bulgaricus, L. reuteri, L. brevis, L. cellobiosus, L. curvatus, L. fermentum, L. plantarum* 2) Gram-positive cocci such as *Lactococcus lactis* ssp. *cremoris, Streptococcus salivarius* ssp. *thermophilus, Enterococcus faecium, S. diaacetylactis, S. intermedius* and 3) Bifidobacteria such as *Bifidobacterium bifidum, B. adolescentis, B. animalis, B. infantis, B. longum, B. thermophilum* (Collins et al. 1998, Gibson 1999, Mercenier et al. 2002). Other microbial species, besides lactic acid bacteria (LAB), like *Bacillus subtilis, Propionibacterium* spp. and yeasts (*Saccharomyces boulardii*) have also been accepted and used as probiotics (Chukeatirote 2002, Jan et al. 2002). The mechanism of the action of probiotics (e.g., bifidobacteria and lactobacilli) relies on their metabolic end products; mainly organic acids which may lower the human gut pH and consequently inhibit the growth of pathogenic microbes. Other factors are occupation of normal colonization sites by probiotics, competition for available nutrients and production of antimicrobial substances. The second generation of probiotics are genetically modified microorganisms that provide the host with some necessary components, e.g., production of immunomodulators (e.g., interleukines) or *Helicobacter pylori* and rotavirus antigens (Mercenier et al. 2004). Probiotic preparations used as food supplements can consist of one single strain (e.g., Yakult, Japan—*L. casei* Sirota) or are mixed cultures of two (e.g., Bacilac, Belgium—*L. acidophilus* plus *L. rhamnosus*) or even more (e.g., food supplement VSL-3, Italy contains 8 LAB species) strains. Table 2 summaries the probiotic bacterial species and the strains primarily used in the food industry.

Table 2. Some commercial probiotic strains.

Strain	Source
L. acidophilus NCFM	Rhodia, Inc. (Madison, Wisconsin, USA)
Lactobacillus brevis KB290	Kagome Co., Ltd. (Tochigi, Japan)
L. acidophilus DDS-1	Nebraska Cultures, Inc. (Lincoln, NE)
L. acidophilus SBT-2062 *B. longum* SBT-2928	Snow Brand Milk Products Co., Ltd. (Tokyo, Japan)
L. acidophilus R0011 *L. rhamnosus* R0052	Institut Rosell (Montreal, Canada)
L. paracasei CRL 431 *B. lactis* Bb-12	Chr. Hansen (Horsholm, Denmark)
L. casei Shirota *B. breve* strain Yakult	Yakult (Tokyo, Japan)
L. casei DN014001 (Immunitas)	Danone Le Plessis-Robinson (Paris, France)
L. fermentum RC-14 *L. rhamnosus* GR-1	Urex Biotech Inc. (London, Ontario, Canada)
Streptococcus thermophilus MN-ZLW-002	Inner Mongolia Mengniu Dairy Industry Co. Ltd., (Hohhot, China)
L. johnsonii La1 (same as Lj1)	Nestlé (Lausanne, Switzerland)
L. plantarum 299V *L. Rhamnosus* 271	Probi AB (Lund, Sweden)
L. reuteri SD2112 (same as MM2)	BioGaia (Raleigh, North California, USA)
L. rhamnosus GG	Valio Dairy (Helsinki, Finland)
L. rhamnosus LB21 *Lactococcus lactis* L1A	Essum AB (Umeå, Sweden)
L. salivarius UCC118	University College (Cork, Ireland)
B. longum BB536	Morinaga Milk Industry Co., Ltd. (Zama-City, Japan)
B. lactis HN019 (DR10)	New Zealand Dairy Board
L. acidophilus LB	Lacteol Laboratory, (Houdan, France)
L. paracasei F19	Arla Dairy (Stockholm, Sweden)
L. crispatus CTV05	Gynelogix, Boulder, Colo.
L. casei DN 114	Danone, Paris, France
S. boulardii	Biocodex Inc. (Seattle, Washington, USA)
L. delbrueckii subsp. *bulgaricus 2038*	Meiji Milk Products (Tokyo, Japan)

Probiotics Attributes/Characteristics

Most probiotics are related to the *Lactobacillus* and *Bifidobacterium* genera (Bezkorovainy et al. 1997, Salminen and von Wright 1998, Sanders 2003, Guarner et al. 2005, Nagpal et al. 2007, 2012a, Kumar et al. 2009a). However, to be considered as probiotics, the different strains should be normal inhabitants of a healthy intestinal tract, survive the upper digestive tract and be capable of surviving and growing in the intestine (acid and bile resistant), safe for human consumption, produce antimicrobial substances like bacteriocins and have the ability of adherence to human intestinal cell lines and colonization (Guarner and Schaafsma 1998, Morelli 2000, Guarner et al. 2005).

Lactic acid bacteria have an established safety record as compared to the probiotics and are rarely involved in disease. The most commonly used probiotics are *Lactobacillus* spp., *Bifidobacterium* spp. and *Lactococcus* spp. and these have been accorded the GRAS (Generally Recognized As Safe) status (Salminen and von Wright 1998).

However, not all the probiotics available on the market have been shown to meet the requirements determined by FAO and WHO experts. These requirements hold particular importance because there is no information available on the risks related to the long-term use of probiotics. Experts have identified 4 kinds of potential adverse effects, i.e., systemic infections, harmful metabolic activities, excessive immune system stimulation in susceptible individuals and transfer of genetic material (De Groote et al. 2005, Herbrecht and Nivoix 2005, Salminen 2004b).

Indeed, as was very recently pointed out in a review (Paone 2012), additional characteristics that probiotics must possess are a demonstrated genetic stability and the capability to not develop antibiotic resistance. Such characteristics are rarely checked for in the currently available formulations. *Lactobacillus* F19 (species *paracasei* subsp. *paracasei* F19) has been developed in accordance with the abovementioned requirements, and in particular is genetically stable (Morelli et al. 2002) and capable of not developing antibiotic resistance (Sullivan et al. 2003, 2004). A worthwhile eco-friendly strategy of probiotic production has been originally developed by Kagome laboratories in Japan. Here culture collections of lactic acid bacteria from Japanese traditional fermented vegetables have been accumulated along the years while preserving the natural season sampling at proper microclimate conditions. Among almost a thousand isolated and tested strains, *Lactobacillus brevis* KB290 has been extensively studied in humans with proven safety and efficacy (Nobuta et al. 2009).

Mechanism of Actions of Probiotics

Probiotic microorganisms are considered to support the host health. However, the support mechanisms have not been explained (Holzapfel et al. 1998). There are studies on how probiotics work. These studies are trying to explain how probiotics could protect the host from intestinal disorders. The study mechanisms are listed below briefly (Rolfe 2000, Çakır 2003, Salminen et al. 1999, Castagliuola et al. 1999, Nagpal et al. 2007, 2010, 2011, 2012a,b, Kumar et al. 2009a,b, 2010, 2011, 2012).

- Production of inhibitory substances: Production of some organic acids, hydrogen peroxide and bacteriocins which are inhibitory to both gram-positive and gram-negative bacteria.
- Blocking of adhesion sites: Probiotics and pathogenic bacteria are in a competition. Probiotics inhibit the pathogens by adhering to the intestinal epithelial surfaces and blocking the adhesion sites.
- Competition for nutrients: Despite of the lack of studies *in vivo*, probiotics have been shown to inhibit the pathogens by consuming the nutrients which pathogens need.
- Stimulating of immunity: Stimulating of specific and nonspecific immunity may be one possible mechanism of probiotics to protect the host from intestinal disease. This mechanism is not well documented, but it is thought that specific cell wall components or cell layers may act as adjuvants and increase humoral immune response.
- Degradation of toxin receptor: Because of the degradation of toxin receptor on the intestinal mucosa, it was shown that *S. boulardii* protects the host against *C. difficile* intestinal disease.

Some other offered mechanisms are suppression of toxin production, reduction of gut pH, attenuation of virulence (Fooks et al. 1999).

Gastrointestinal Microflora Balance and Probiotics

More than 400 bacterial species exist in the human intestinal tract. It is an enormously complex ecosystem that includes both facultative anaerobic and anaerobic microorganisms (Naidu et al. 1999). The numbers of genera is nearly steady, because each of them has its own growth niches (Fooks et al. 1999). The composition of the gut microflora is constant but can be affected by factors such as age, diet, environment, stress and medication. To have a healthy intestine, the balance of the bacteria must be maintained but this is difficult due to the changing life style of the population. Lots of factors may change the balance away from potenially beneficial or health promoting bacteria to potentially harmful or pathogenic microorganisms. This makes the host more susceptible to illnesses. In this case, the prevalence of the beneficial bacteria must be supported. The use of probiotics helps to protect the host from various intestinal diseases and disorders by increasing the number of beneficial bacteria and making the balance steady again (Fooks et al. 1999, Nagpal et al. 2007, 2010, 2012a,b, Kumar et al. 2009a,b, 2010, 2011, 2012). Probiotics are suggested as food to provide for the balance of intestinal flora (Holzapfel et al. 1998). Probiotics have been used for long times in food ingredients for human and animals without any side effects.

Probiotics, naturally found in the mouth, lower intestine and vagina of healthy individuals, help defend the body against invading pathogenic bacteria. Due to the dominance of common antibiotic treatment, many people lack healthy intestinal flora. The composition of the intestinal flora is relatively stable in healthy human beings between harmful and beneficial or natural bacteria. Among the beneficial bacteria are *Lactobacillus* spp. and *Bifidobacterium* spp. which play a useful role in the production of vitamins, organic acids and antimicrobial factors that inhibit pathogens. Any imbalance in the gut microflora leads to dominance of harmful bacteria in the intestinal flora, which affects the production of essential nutrients and increases the level of damaging substances, including carcinogens, putrefactive products and toxins (Mitsuoka 1996, Salminen and Gueimonde 2004). Therefore, to maintain a well-balanced microflora in the gastrointestinal tract, it has been suggested that live bacteria be introduce to stimulate growth of beneficial bacterial population groups which prevent harmful effects and promote beneficial actions of the intestinal microflora (Salminen et al. 1996, Shah 2000). Consuming probiotics with dairy foods buffers the stomach acid and increases the likelihood that the bacteria will survive in the intestine. Dairy products containing probiotics also provide a number of essential nutrients including calcium and protein (Stanton et al. 2001, Nagpal et al. 2011, 2012a).

Site of Action of Probiotics: The Small Intestine

The intestinal epithelium is a highly organized, single-cell layer covering the interface between tissues and the intestinal lumen. This monolayer is mainly constituted of enterocytes, which are the cells responsible for taking up nutrients, Paneth cells, which secrete the mucus bathing the epithelium, and intra-epithelial lymphocytes, which are part of the mucosa-associated lymphoid tissue (MALT). Yet, all epithelial cells arise from common non-differentiated precursors present in the epithelium (Brandtzaeg 1995). This monolayer is constantly being renewed as epithelial cells undergo a lifecycle, which starts deep within the crypts, from where they arise, continue with their differentiation and migration towards the tip of the villi and end with apoptosis and exfoliation (Stadnyk 1994, Turner 2003, Dommett et al. 2005). This cycle takes about 3 to 5 days in humans and allows epithelial self-renewal. Because of this turn over, the gut surface is covered by dead and exfoliating cells, which provide together with the mucus and the nutrients passing through the lumen an excellent growth substrate for microorganisms (Stadnyk 2002, Tlaskalova-Hogenova et al. 2005).

Colonization of Probiotics in the Gut

It is generally agreed that to permanently establish a bacterial strain in the host intestine, the microorganism must be able to attach to the intestinal mucosal cells (O'Sullivan et al. 1992). Moreover, many pathogens cannot exert their deleterious effects on the gut unless they become so attached (Hoepelman and Tuomanen 1992) and the beneficial action of probiotics has been studied for their purported ability to interfere with the adherence of pathogens to intestinal mucosal cells (Fuller 1991). The normal colonization of the sterile newborn intestine is a complex process. Initial colonization is achieved with maternal vaginal and fecal bacterial flora. The first colonizers have a high reductive potential and include species such as enterobacter, streptococcus, and staphylococcus. These metabolize oxygen, thus encouraging the growth of anaerobic bacteria including lactobacilli and bifidobacteria.

Effects of Probiotics on Health

There are lots of studies on the health benefits of fermented foods and probiotics. However, in most of these studies, researchers did not use sufficient test subjects or the used microorganisms were not identified definitely (Mohania et al. 2008). So, while a number of reported effects have been only partially established, some can be regarded as well established and clinically well documented for specific strains. These health-related effects can be considered as indicated below (Schrezenmeir and De Vrese 2001, Dunne et al. 2001, Dugas et al. 1999, Nagpal et al. 2007, 2010, 2011, 2012a,b, Kumar et al. 2009a,b, 2010, 2011, 2012).

- Managing lactose intolerance
- Improving immune system
- Prevention of colon cancer
- Reduction of cholesterol and triacylglycerol plasma concentrations
- Lowering blood pressure
- Reducing inflammation
- Reduction of allergic symptoms
- Beneficial effects on mineral metabolism, particularly bone density and stability

- Reduction of *Helicobacter pylori* infection
- Suppression of pathogenic microorganisms (antimicrobial effect)
- Prevention of osteoporosis
- Prevention of urogenital infections

Lactose intolerance

Most humans, commonly non-Caucasians, become lactose intolerant after weaning. These lactose intolerant people cannot metabolize lactose due to the lack of an essential enzyme, β-galactosidase. If lactose passes through the small intestine, it is converted to gas and acid in the large intestine by the colonic microflora. Also, the presence of breath hydrogen is a signal for lactose maldigestion. The studies provide that the addition of certain starter cultures to milk products allows lactose intolerant people to consume these products without the usual rise of breath hydrogen or associated symptoms (Fooks et al. 1999, Scheinbach 1998, Quewand and Salminen 1998, Lin et al. 1991). The beneficial effects of probiotics on lactose intolerance are due to the lower lactose concentration in the fermented foods due to the high lactase activity of bacterial preparations used in the production. Given that lactose is converted to, lactic acid by β-galactosidase enzyme, which is contained in yogurt, this becomes more suitable than milk for individuals with lactose intolerance. Furthermore, the LAB which is used to produce yogurt, *Lactobacillus bulgaricus* and *Streptococcus thermophilus*, are not resistant to gastric acidity. Hence, products with probiotic bacteria are more efficient for lactose intolerant humans.

It is thought that the major factor that improves the digestibility by the hydrolyses of lactose is the bacterial enzyme β-galactosidase. Another factor is the slower gastric emptying of semi-solid milk products such as yogurt. So the β-galactosidase activity of probiotic strains and other lactic acid bacteria used in dairy products is really important. β-galactosidase activity within probiotics varies in a huge range. It has to be considered both the enzyme activity of probiotic strain and the activity left in the final product for their use in lactose intolerant subjects (Salminen et al. 2004).

Immuno-modulation

The effect of probiotics on the immune system are promising. However, the mechanism is not well understood. Human studies have shown that probiotic bacteria can have positive effects on the immune system of their hosts (Mombelli and Gismondo 2000). Several reserchers have studied the effects of probiotics on immune system stimulation. Some *in vitro* and *in vivo* searchs have been carried out in mice and with humans. The data indicates that oral bacteriotherapy and living bacteria feeding in fermented milks supported the immune system against some pathogens (Scheinbach 1998, Dugas et al. 1999). Probiotics affect the immune system in different ways such as: producing cytokines, stimulating macrophages and increasing secretory IgA concentrations (Scheinbach 1998, Dugas et al. 1999). Some of these effects are related to adhesion while some of them are not (Quwehand et al. 1999). Link-Amster et al. (1994) examined whether eating fermented milk containing *Lactobacillus acidophilus* La1 and bifidobacteria could modulate the immune response in humans. They gave volunteers the test fermented milk over a period of three weeks during which attenuated *Salmonella typhi* Ty21a was administered to mimic an enteropathogenic infection. After three weeks, the specific serum IgA titre rise to *S. typhi* Ty21a in the test group was >4-fold and significantly higher (p=0.04) than in the control group which did not eat fermented foods but received *S. typhi* Ty21a. These results showed that LAB which can survive in the gastrointestinal tract can act as adjuvants to the humoral immune response (Lime-Amster et al. 1994, Quwehand et al. 1999).

Perdigon et al. (1988) fed the mice with lactobacilli or yogurt and it stimulated macrophages and increased secretory IgA concentrations (Scheinbach 1998). Also in a human trial Halpern et al. (1996) fed humans or human subjects with 450 g of yogurt per day for 4 months and at the end a significant increase was observed in the production of γ-interferon (Fooks et al. 1999). Mattila-Sandholm and Kauppila (1998) showed that *Lactobacillus rhamnosus* GG and *Bifidobacterium lactis* Bb-12 derived extracts suppress lymphocyte proliferation *in vitro*. A very fine study recently published, showed that Streptococcus thermophilus MN-ZLW-002 fermented milk may stimulate non-specific cell-mediated immunity at plasma and epigenomic pulmonary levels which are involved in the protection of the mammals from respiratory infections, thus paving the way to a promising clinical application (Kang et al. 2012). Finally, a recent editorial from Rastmanesh et al. (2012) has envisaged the potential use of probiotics in enhancing the efficacy of HIV vaccination.

Diarrhea

Diarrhea has many causes and many types so it is difficult to evaluate the effects of probiotics on diarrhea. But there is a lot of research and evidence that probiotics have beneficial effects on some types of diarrhea. Diarrhea is a major cause/reason of childrens deaths worldwide and rotavirus is its common cause (Scheinbach 1998). In the treatment of rotavirus diarrhea, *Lactobacillus* GG is reportedly really effective. The best documented probiotic effect is a shortened duration of rotavirus diarrhea using *Lactobacillus* GG. This has been proved in several studies around the world by some researchers like Guandalini (2000) and Pant et al. (1996). Also, *Lactobacillus acidophilus* LB1, *Bifidobacterium lactis* and *Lactobacillus reuterii* are reported to have beneficial effects on shortening the duration of diarrhea (Salminen et al. 2004). One type of diarrhea is travellers' diarrhea (TD) which affects healthy travellers not only in developing countries but also in Europe. Probiotics have beneficial effects in preventing some forms of TD. Oksanen et al. (1990) evaluated the efficacy of *Lactobacillus* GG in preventing diarrhea in 820 people travelling from Finland to Turkey. In a double-blind study by Black et al. (1989) lyophylised bacteria (*L. acidophilus, B. bifidum, L. bulgaricus, S.thermophilus*) were given to 56 Danish tourists on a 2-week trip to Egypt. The occurrence of diarrhea in the group receiving the lactic acid bacteria was 43% while it was 71% in the placebo group (Gismondo et al. 1999).

Antibiotic therapy causes outbreaks of diarrhea. The normal microflora is suppresed during the microbial therapy and may result in a predominance of pathogenic strains. The changes in microflora may also encourage resistant strains like *Clostridium difficile* which is the reason for antibiotic associated diarrhea (ADD). Several clinical trails (McFarland 1998, McFarland and Elmer 2005) have used *Saccharomyces boulardii, Lactobacillus* spp. and *Bifidobacterium* spp. in ADD. Probiotics which are able to restore and replace the normal flora should be used. Also, they should be used in high risk patients such as old, hospitalized or immunocompromised. Studies have shown that *Clostridium difficile* decreases in the presence of *Saccharomyces boulardii* (Gismondo et al. 1999).

Colon and other cancer treatment

Probiotic bacteria play an important role in retarding colon carcinogenesis by possibly influencing metabolic, immunologic, and protective functions in the colon. The use of probiotics in prevention and cancer treatment has been undergoing a recent evaluation in a number of studies. Although we should not expect miraculous outcomes in cancer treatment following probiotics administration, their immunomodulatory properties have been tested and need to be brought to the publics' attention. It is important to note that the desired effects are strain and dose specific and therefore more clinical studies are needed to screen each strain and corresponding disorder. In animals, LAB ingestion was shown to prevent carcinogen induced preneoplastic lesions and tumors. In the study by McIntosh et al. (1999), *Lactobacillus acidophilus* (Delvo Pro LA-1), *Lactobacillus rhamnosus* (GG), *Bifidobacterium animalis* (CSCC1941), and *Streptococcus thermophilus* (DD145) strains were examined for their influence on 1, 2-dimethylhydrazine (DMH) induced intestinal tumors in 100 male Sprague-Dawley rats when added as freeze-dried bacteria. This study concluded that the strain of *L. acidophilus* supplied as freeze-dried bacteria in the diet was protective because it significantly inhibited tumors within the rat colon. There is a substantial amount of study done by Perdignon et al. dealing with anti-inflammatory properties of probiotic bacteria. In the study by Galdeano and Perdigon (2006), the probiotic bacterium *Lactobacillus casei* was screened for its influence on the expression of receptors involved in the innate immune response in colorectal cancer BALB/c model mice. Further, a complex nature of kefir was studied in BALB/c mice. Kefir is fermented milk produced by the action of lactic acid bacteria, yeasts and acetic acid bacteria, trapped in a complex matrix of polysaccharides and proteins. In addition, it is an excellent source of proteins and calcium. A study concluded that since LAB contained in kefir along with yeasts and acetic acid bacteria have an *in vivo* role as oral biotherapeutic substances capable of stimulating immune cells of the innate immune system they are able to promote cell-mediated immune responses against tumors and also against intracellular pathogenic infections. In another kefir related study by Vinderola et al. (2000), the immunomodulating capacity of kefir on the intestinal mucosal immune response in mice of viable or heat-inactivated bacteria at different doses was determined. However, in humans, there is no evidence available on whether probiotics can prevent the initiation of colon cancer. Epidemiologic studies are contradictory; some studies could not find an association between the consumption of fermented-milk products and the risk of colon cancer whereas other studies showed a lower incidence of colon cancer in persons consuming fermented-milk products or yogurt.

There is *in vitro* and *in vivo* evidence not only from animal studies but also from human studies that probiotics have beneficial effects on suppression of cancer. Oral administration of lactic acid bacteria has been shown to reduce DNA damage caused by chemical carcinogens, in gastric and colonic mucosa in rats (Marotta et al. 2003). Moreover, improved gastrointestinal motility due to probiotic consumption (Metugriachuck et al. 2006) may be a further protecting factor against endoluminal carcinogens. The consumption of lactobacilli by healthy volunteers has been demonstrated to reduce the mutagenicity of urine and feaces associated with the ingestion of carcinogens in cooked meat. When it comes to epidemiological studies, they show an association between fermented dairy products and colorectal cancer. The consumption of a large quantity of dairy products especially fermented foods like yogurt and fermented milk containing *Lactobacillus* or *Bifidobacterium* may be related to a lower occurrence of colon cancer (Rafter 2003, Hirayama and Rafter 2000). A number of studies have shown that predisposing factors (increase in enzyme activity that activate carcinogens, increase procarcinogenic chemicals within the colon or in population of certain bacterial genera and species) are altered positively by consumption of certain probiotics (Brady et al. 2000, Kumar et al. 2010, 2011, 2012).

Cholesterol reduction

Lots of researchers have proposed that probiotics have cholesterol reduction effects. However, the mechanism of this effect has not been explained definitely. There are two hypotheses trying to explain the mechanism. One of them is that bacteria may bind or incorporate cholesterol directly into the cell membrane. The other one is, bile salt hydrolysis enzymes deconjugate the bile salts which are more likely to be exerted resulting in increased cholesterol breakdown (Çakır 2003, Scheinbach 1998, Prakash and Jones 2005, Nagpal et al. 2010). A study on the reduction of cholesterol showed that *Lactobacillus reuteri* CRL 1098 decreased total cholesterol by 38% when it was given to mice for 7 days in the rate of 10^4 cells/day. This dose of *Lactobacillus reuteri* caused a 40% reduction in triglycerides and a 20% increase in the ratio of high density lipoprotein to low density lipoprotein without bacterial translocation of the native microflora into the spleen and liver (Kaur et al. 2002).

Functional Foods and Probiotics

Although the primary purpose of food is to provide enough nutrients to fulfill body requirements, various functions of the body are modulated by diet. In order to compensate for deficiency of certain nutrients in the diet due to changes in nutritional habits of developed industrial countries, the concept of functional food has been developed. A food can be regarded as functional if it is satisfactorily demonstrated to affect beneficially one or more target functions in the body, beyond adequate nutritional effect, in a way which is relevant to either an improved state of health and well being and/or reduction of disease risk" (ILSI Europe 1999). Functional food is intended for a population generally in normal health and must demonstrate beneficial effects in amounts that are usually consumed in the diet. Functional food is a natural food, to which a component has been added/removed or a food in which the bioavailability of the components has been modified by technological or biotechnological means (Korhonen 2002). Functional food can be classified into different groups according to their effect: fat replacers, probiotics, probiotics and dietary fibres, antioxidants, vitamins, polyphenols, plant sterols, polyunsaturated fatty acids and minerals. The most promising targets for functional food are the functions and particularly control of nutrient bioavailability (Roberfroid 2000). The gastrointestinal (GI) functions include balanced colonic microflora, control of transit time and mucosal motility, bowel habits, modulation of epithelial cell proliferation, balance of redox and antioxidant systems, metabolism of macronutrients, especially amino acids, carbohydrates and fatty acids. The term functional food originates from the 1980s (Sanders 1999). In 1991, a legal status to functional foods was granted in Japan, indicating foods for special health use. The first functional food probiotic fermented milk drink Yakult has been available in Japan since 1935 (Karimi and Peña 2003). Currently various probiotic supplemented functional foods, i.e., dairy products are available in the market, worldwide, but very few of them have been studied for their claimed health beneficial effects; therefore we think systemic studies should be conducted before claiming the health benefits of known functional foods. National Dairy Research Institute (India) developed a probiotic dahi (Indian yogurt) supplemented with two health beneficial probiotic bacterial strains named *Lactobacillus casei* and *Lactobacillus acidophilus* (Yadav et al. 2005) and has been systematically studied for health beneficial effects, i.e., anti-diabetic (Yadav et al. 2007, 2008) and immunomodulatory (Jain et al. 2009a,b) effects, before taking it to the market. Such studies support the health beneficial effects of foods available for consumers and provide more consumer confidence in market.

Safety of Probiotics

The best evidence for the general safety of lactic acid bacteria and bifidobacteria is their long tradition of use without any harmful effects on human health. With the exception of one strain belonging to the *L. rhamnosus* species, lactobacilli and bifidobacteria used for food production are "generally recognized as safe" (GRAS) by the Food and Drug Administration in USA. Moreover, certain strains of probiotic bacteria have been proven to be free of risk factors like: transferable antibiotic resistances, cancer-promoting and/or putrefactive enzymes and metabolites, hemolysis, activation of thrombocyte-aggregation or mucus degradation in the mucus layer of the gastrointestinal tract. Despite the absence of a pathogenic potential, lactic acid bacteria were found in < 0.1% (enterococci 1%) of clinical samples from severe infections (endocarditis, meningitis, or bacteremia (Gasser 1994). Most probably these bacteria originated from the indigenous microflora, whereby in many cases the translocation was facilitated by underlying disease, lesions or inflammations in the oral cavity and in the gastrointestinal tract, or by an impaired immune system. However, there is no evidence for a higher risk due to the ingestion of probiotic products in comparison with conventional products. This conclusion is supported by a study from Finland, where the consumption of *L. rhamnosus* GG has increased considerably during the last two decades without an increase in the incidence of infections by lactobacilli (Rautio 1999). Moreover, studies in immuno-compromised persons (HIV-positive subjects, patients with leukemia) did not show undesired effects, but rather positive effects, e.g., lower incidence of Candida during chemotherapy. Health risks due to overdosage or long term ingestion have also not been observed.

Prebiotics: Foods for Probiotics

A prebiotic is a non-digestible food ingredient that beneficially affects the host by selectively stimulating growth, activity, or both, of one or a limited number of bacterial species already resident in the colon (Gibson and Roberfroid 1995, Nagpal et al. 2007, Nagpal and Kaur 2011). To exhibit such effects, a prebiotic must neither be hydrolysed nor absorbed in the upper part of the gastrointestinal tract, and must be selectively fermented by one or a limited number of potentially beneficial bacteria residing in the colon (Collins and Gibson 1999). The number of probiotics in the human gut tend to decrease with age (Mitsuoka 1996). Two major strategies have been proposed to maintain a high level of probiotics to sustain beneficial health effects; 1) continuous ingestion of probiotics containing foods or 2) supplementation of food with prebiotics (Gomes and Malcata 1999). These prebiotics are fermented by one or a limited number of potentially beneficial bacteria form the resident colonic microflora. A prebiotic is expected to improve the composition of the colonic microbiota and through this serve as beneficial to the host health (Gibson 1999).

Since the 1980s, awareness of the healthier food and drink market has increased all over the world, and these are named as Functional Foods (Roberfroid 2002). The popularity of these foods reflects nutritional guidelines recommending an increase in the dietary fibre intake. The uses of insoluble fibre ingredients (Gibson 2004), such as bran, have been used in products such as breakfast cereals, bread and pasta, but the acceptability of these materials is limited in different systems, which decreases their incorporation into foods. Soluble fibre ingredients such as oligosaccharides are currently of more interest in formulation of healthy foods because they are more acceptable. Moreover, some of them can be used as thickening in food system to add viscosity or form gel (Dreher 1999). The main reason of prebiotics supplementation to human diet is to beneficially enhance the gut microflora (Kolida et al. 2002), which is *Bifidobacterium* spp., the most dominant and important flora in breast-fed and healthy infants. The beneficial effects of the presence of bifidobacteria in the gastrointestinal tract are dependent on their viability and metabolic activity. Their growth is dependent on the presence of complex carbohydrates known as oligosaccharides. Some oligosaccharides, because of their chemical structure, are resistant to digestive enzymes and therefore pass into the large intestine. Therefore, prebiotics are used as bifidogenic factors in diet applications, especially because of their ability not to degrade in the stomach and small intestine (Crociani et al. 1994). Inulin and oligofructose are recognized as safe ingredient supplements to food without limitation but the European Commission confirmed that oligofructose (FOS) and inulin could be used in foods targeted towards infants older than six months of age at a concentration of 0.8 g/day (Rao 2002). Kaplan and Hutkins (2000) have shown the ability of a selection of twenty-eight lactic acid bacteria and bifidobacteria to ferment inulin and oligofructose on MRS agar.

A range of dietary compounds have been suggested as prebiotics, and these have been selected for their health benefits on the host. Gibson et al. (1995) presented the popularity of Inulin, Fructo-oligosaccharides (FOS) and Galcto-oligosaccharides (GOS) as health benefit subtrates. In human studies the addition to bread of 7g of FOS has been shown to beneficially affect the dominant bifidobacteria as compared to common bread. The authors clearly proved that the use of FOS exerted a profound effect upon bifidobacteria (Gibson 2004). The only prebiotics for which sufficient data has been generated to allow an evaluation of their possible classification as functional food ingredients are the inulin type fructans, which include native inulin, enzymatically hydrolyzed inulin or oligofructose, and synthetic fructooligosaccharides (Roberfroid et al. 1998a, 1998b, Nagpal et al. 2007, 2011). The two basic types of fermentations taking place in the gut are saccharolytic fermentation and proteolytic fermentation. The main end products of carbohydrate metabolism are the short chain fatty acids: acetate, propionate and butyrate. These may be further metabolised systematically or locally to generate energy for the host. The end products of the proteolytic fermentation include more or less toxic compounds as amines, ammonia and phenolic compounds. Fermentation in the gut can be modulated towards saccharolytic by prebiotic consumption. Much of the interest is aimed at non-digestible oligosaccharides and indeed more than 36000 plants worldwide contain FOS; some common sources of inulin are onion (2–6%), garlic (9–16%), leek (3–10%), banana (0.3–0.7%), asparagus (10–15%), Jerusalem artichokes (15–20%), chicory (13–20%), and even wheat (1–4%). Yet the levels are too low for a significant tract effect (Crow 2004). Consumption of more than 4 grams of FOS daily is needed to induce changes in LAB levels in the gut, though estimated daily consumption differs in the US and Europe (Roberfroid 2000, Gibson 2001).

Prebiotics are increasingly used in development of new food products, e.g., drinks, yoghurts, biscuits and table spreads (Gibson and Roberfroid 1995, Gibson 1999). Several prebiotics are available in Europe. The positive effects of prebiotic consumption are: improvement of bowel habit; reduction of diarrhoea and constipation; modulation of lipid metabolism by normalizing cholesterol values; reduction of osteoporosis by improved mineral absorption; reduction of allergy risk through immune system modulation; reduction of colon cancer risk (Roberfroid 2000, Conway 2001). Still, many of the above mentioned health claims require further research.

Synbiotics: Blending Probiotics and Prebiotics

Another possibility of gut microflora management is the use of synbiotics, where probiotics and prebiotics are used in combination. The combination of suitable probiotics and prebiotics enhances survival and activity of the organism, for example a FOS in conjunction with a *Bifidobacterium* strain or lactitol in conjunction with *Lactobacillus* (Gibson and Roberfroid 1995). The combination of prebiotic and probiotic has synergistic effects because in addition to promoting growth of existing strains of beneficial bacteria in the colon, synbiotics also act to improve the survival, implantation and growth of newly added probiotic strains (Nagpal et al. 2007, Nagpal and Kaur 2011). The synbiotic concept has been widely used by European dairy drink and yoghurt manufacturers such as Aktifit (Emmi, Switzerland), Proghurt (Ja Naturlich Naturprodukte, Austria), Vifit (Belgium, UK) and Fysiq (Netherlands) (Niness 1999). The combination of *Bifidobacterium* and oligofructose was reported to synergistically improve colon carcinogenesis in rats compared to when both were given individually (Gallaher and Khil 1999). Another study reported that a synbiotic containing *Pediococcus pentoseceus*, *Leuconostoc mesenteroides*, *Lactobacillus paracasei*, and *L. plantarum* with four fermentable fibres namely β-glucan, inulin, pectin, and resistant starch, reduced the occurrence of post-operation infections from 48% to 13% in 66 liver transplant patients (Rayes et al. 2005). Most of the claims on benefits of different synbiotics are on general health (Gibson and Roberfroid 1995). There have yet to be any clinical trials on suitable combinations of synbiotics that specifically target reduction of serum cholesterol levels in animals and humans. Bifidobacteria and lactobacilli are the most frequent target organisms for prebiotics. Probiotics and prebiotics used in synergistic combination are termed synbiotics. Synbiotics are mixtures that improve the survival and implantation of live microbial dietary supplements in the tract, either by stimulating growth or by metabolically activating the health promoting bacteria (Kaur et al. 2002). Although there is growing interest in the development of new functional foods with synbiotics, combination of prebiotics and probiotics into a synbiotic has been studied to a limited extent and needs further investigations, because of the afore mentioned different substrate requirements for individual probiotic LAB species and strains. Only a few human studies have been carried out on the effectiveness of synbiotics (Morelli et al. 2003, Tsuchiya 2004, Lighthouse 2004, Lamiki 2010).

References

Bezkorovainy, A., R. Miller-Catchpole and E. Kot. 1997. Health benefits of bifidobacteria. Tecnologia Lactea Latinoamericana 10: 34–41.

Black, F.T., P.L. Anderson, J. Orskov, F. Orskov, K. Gaarslev and S. Lauland. 1989. Prophylactic efficacy of lactobacilli on travelers' diarrhoea. Travel Med. 23: 333–335.

Brady, L.J., D.D. Gallaher and F.F. Busta. 2000. The role of probiotic cultures in the prevention of colon cancer. The Journal of Nutrition. 130: 410S–414S.

Brandtzaeg, P. 1995. Basic mechanisms of mucosal immunity, a major adaptive defence system. The Immunologist 3: 89–96.

Castagliuoluolo, I., M.F. Riegler, L. Valenick, J.T. Lamont and C. Pothoulakis. 1999. *Saccharomyces boulardii* protease inhibits the effects of *Clostridium difficile* toxins a and b in human colonic mucosa. American Society for Microbiology 67(1): 302–307.

Chuayana, Jr. E.L., C.V. Ponce, M.R.B. Rivera and E.C. Cabrera. 2003. Antimicrobial activity of probiotics from milk products. Phil J. Microbiol. Infect. Dis. 32(2): 71–74.

Chukeatirote, E. 2002. Potenital use of probiotics. Songklanakarin Sci. Technol. 25: 275–282.

Collins, J.K., G. Thornton and G.O. Sullivan. 1998. Selection of probiotic strains for human applications. International Dairy Journal 8: 487–490.

Collins, M.D. and G.R. Gibson. 1999. Probiotics, prebiotics, and synbiotics: approaches for modulating the microbial ecology of the gut. American Journal of Clinical Nutrition 69: 1052S–7S.

Conway, P.L. 2001. Prebiotics and human health: the state-of-the-art and future perspectives. Scand. J. Nutr. 45: 13–21.

Crociani, F., A. Alessandrini, M.M. Mucci and B. Baavati. 1994. Degradation of complex carbohydrates by *Bifidobacterium* spp. *International* Journal of Food Microbiology 24: 199–210.

Crow, D. 2004. Inulin: a comprehensive scientific review. Retrieved April 28 (from http://members.shaw.ca/duncancrow/inulin_review.html).

De Groote, M.A., D.N. Frank, E. Dowell, M.P. Glode and N.R. Pace. 2005. Lactobacillus rhamnosus GG bacteremia associated with probiotic use in a child with short gut syndrome. Pediatr Infect. Dis. J. 24: 278–280.

Dommett, R., M. Zilbauer, J.T. George and M. Bajaj-Elliott. 2005. Innate immune defence in the human gastrointestinal tract. Mol. Immunol. 42: 903–912.

Dreher, M. 1999. Food sources and uses of dietary fibre. In complex carbohydrates in foods, [SS Cho, editor]. New York: Marcel Dekker, pp. 385–394.

Dugas, B., A. Mercenier, I. Lenoir-Wijnkoop, C. Arnaud, N. Dugas and E. Postaire. 1999. Immunity and probiotics. Trends Immunology Today 20(9): 387–390.

Dunne, C., L. O'Mahony, L. Murphy, G. Thornton, D. Morrissey, S. O'Halloran, M. Feeney, S. Flynn, G. Fitzgerald, C. Daly, B. Kiely, G. C O'Sullivan, F. Shanahan and J.K. Collins. 2001. *In vitro* selection criteria for probiotic bacteria of human origin: Correlation with *in vivo* findings. Am. J. Clin. Nutr. 73: 386S–392S.

Fooks, L.J., R. Fuller and G.R. Gibson. 1999. Prebiotics, probiotics and human gut microbiology. International Dairy Journal 9: 53–61.

Fuller, R. 1991. Probiotics in human medicine. Gut 32: 439–442.

Fuller, R. 1989. Probiotics in man and animals. The Journal of Bacteriology 66: 365–378.

Galdeano, C.M. and G. Perdigon. 2006. The probiotic bacterium *Lactobacillus casei* induces activation of the gut mucosal immune system through innate immunity. Clin. Vaccine Immunol. 13: 219–226.

Gallaher, D.D. and J. Khil. 1999. The effect of synbiotics on colon carcinogenesis in rats. J. Nutr. 129: 1483S–1487S.

Gasser, F. 1994. Safety of lactic acid bacteria and their occurence in human clinical infections. Bull Inst. Pasteur 92: 45–67.

Gibson, G. 2001. Prebiotics for improved gut health. Retrieved November 20 (from: http://www.ifis.org/forum/ /March_2001/prebiotics4health_real.html).
Gibson, G.R. 2004. Fibre and effects on probiotics (the prebiotic concept). Clinical Nutrition Supplements 1: 25–31.
Gibson, G.R. and M.B. Roberfroid. 1995. Dietary modulation of the human colonic microflora: introducing the concept of prebiotics. J. Nutr. 125: 1401–1412.
Gibson, G.R. 1999. Dietary modulation of the human gut microflora using the probiotics oligofructose and inulin. J. Nutr. 129: 1438S–1441S.
Gibson, G.R., E.R. Beatty, X. Wang and J.H. Cummings. 1995. Selective stimulation of bifidobacteria in the human colon by oligofructose and inulin. Gastroenterology 108: 975–982.
Gismondo, M.R., L. Drago and A. Lombardi. 1999. Review of probiotics available to modify gastrointestinal flora. International Journal of Antimicrobial Agents. 12: 287–292.
Gomes, A.M.P. and F.X. Malcata. 1999. *Bifidobacterium* spp. and Lactobacillus acidophilus: biochemical, technological and therapeutical properties relevant for use as probiotics. Trends in Food Science and Technology 10: 139–157.
Guandalini, S. 2000. Treatment of acute diarrhea in the new millennium. J. Pediatr. Gastroenterol. Nutr. 30: 486–489.
Guarner, F. and G.J. Schaafsma. 1998. Probiotics. International Journal of Food Microbiology 39: 237–238.
Guarner, F., G. Perdigon, G. Corthier, S. Salminen, B. Koletzko and L. Morelli. 2005. Should yoghurt cultures be considered probiotic?. British Journal of Nutrition 93: 783–786.
Halpern, G.M., T. Prindiville, M. Blankenburg, T. Hsia and M.E. Gershwin. 1996. Treatment of irritable bowel syndrome with Lacteol Fort: a randomized, doubleblind, cross-over trial. Am. J. Gastroenterol. 91: 1579–1585.
Havenaar, R. and J.H.J. Huis in't Veld. 1992. Probiotics: A general review. In The Lactic Acid Bacteria in Health and disease (Edited by Wood B.); 1: 151–170. Elsevier, Barking.
Herbrecht, R. and Y. Nivoix. 2005. *Saccharomyces cerevisiae* fungemia: an adverse effect of *Saccharomyces boulardii* probiotic administration. Clin. Infect. Dis. 40: 1625–1634.
Hirayama, K. and J. Rafter. 2000. The role of probiotic bacteria in cancer prevention. Microbes and Infection 2: 681–686.
Hoepelman, A.I. and E.I. Tuomanen. 1992. Consequences of microbial attachment: directing host cell functions with adhesions. Infect. Immun. 60: 1729–1733.
Holzapfel, W.H., P. Haberer, J. Snell, U. Schillinger and J. Huis in't Veld. 1998. Overview of gut flora and probiotics. International Journal of Food Microbiology 41: 85–101.
Holzapfel, W.H. and U. Schillinger. 2002. Introduction to pre- and probiotics. Food Research International 35: 109–116.
ILSI Europe. 1999. Scientific concepts of functional foods in Europe: consensus document. Br. J. Nutr. 81: 1S–27S.
Jan, G., P. Leverrier, I. Proudy and N. Roland. 2002. Survival and beneficial effects of propionibacteria in the human gut: *in vivo* and *in vitro* investigations. Lait. 82: 131–144.
Jain, S., H. Yadav and P.R. Sinha. 2009a. Probiotic dahi containing *Lactobacillus casei* protects against *Salmonella enteritidis* infection and modulates immune response in mice. Journal of Medicinal Food 12: 576–583.
Jain, S., H. Yadav, P.R. Sinha and F. Marotta. 2009b. Modulation of cytokine gene expression in spleen and Peyer's patches by feeding dahi containing probiotic *Lactobacillus casei* in mice. Journal of Digestive Diseases 10: 49–54.
Kang, X., G. Sun, J. Sun, N. Ling, Qi Zhoua, L. Zhang and Q. Sheng. 2012. Effects of orally administered Streptococcus thermophilus MN-ZLW-002 fermented milk on murine cell-mediated immune responses. International J Probiotics & Prebiotics (in press).
Kaplan, H. and R.W. Hutkins. 2000. Fermentation of fructooligosaccharides by lactic acid bacteria and bifidobacteria. Applied and Environmental Microbiology 66: 2682–2684.
Karimi, O. and A.S. Peña. 2003. Probiotics: isolated bacteria strain or mixtures of different strains? Drugs Today. 39: 565–597.

Kaur, I.P., K. Chopra and A. Saini. 2002. Probiotics: Potential pharmaceutical applications. European Journal of Pharmaceutical Sciences 15: 1–9.

Kolida, S., K. Tuohy and G.R. Gibson. 2002. Prebiotic effects of inulin and oligofructose. British Journal of Nutrition 87; Suppl. 2: S193–S197.

Kollath, W. 1953. Ernahrung und Zahnsystem (Nutrition and the tooth system). Dtsch Zahnarztl Z. 8(11): 7–16.

Korhonen, H. 2002. Technology options for new nutritional concepts. Int. J. Dairy Technol. 55: 79–88.

Kumar, M., P.V. Behare, D. Mohania, S. Arora, A. Kaur and R. Nagpal. 2009a. Health-Promoting Probiotic Functional Foods: Potential and Prospects. AgroFOOD Industry Hi-tech 20: 29–33.

Kumar, M., A. Kumar, R. Nagpal, D. Mohania, P. Behare, V. Verma, P. Kumar, D. Poddar, P.K. Aggarwal, C.J. Henry, S. Jain and H. Yadav. 2010. Cancer-preventing attributes of probiotics: an update. International Journal of Food Science & Nutrition 61(5): 473–496.

Kumar, M., D. Mohania, D. Poddar, P.V. Behare, R. Nagpal, A. Kumar and P.K. Aggarwal. 2009b. A probiotic fermented milk prepared by mixed culture combination reduces pathogen shedding and alleviates disease signs in rats challenged with pathogens. International Journal of Probiotics and Prebiotics 4(3): 211–218.

Kumar, M., V. Verma, R. Nagpal, A. Kumar, P.V. Behare, B. Singh and P.K. Aggarwal. 2012. Anticarcinogenic effect of probiotic fermented milk and Chlorophyllin on aflatoxin-B1 induced liver carcinogenesis in rats. British Journal of Nutrition (in-press; doi: 10.1017/S0007114511003953) [Epub ahead of print].

Kumar, M., V. Verma, R. Nagpal, A. Kumar, S.K. Gautam, P.V. Behare, C.R. Grover and P.K. Aggarwal. 2011. Effect of Probiotic Fermented Milk and Chlorophyllin on Gene Expressions and Genotoxicity during AFB1-induced Hepatocellular Carcinoma. Gene 490: 54–59.

Lamiki, P., J. Tsuchiya, S. Pathak, R. Okura, U. Solimene, S. Jain, S. Kawakita and F. Marotta. 2010. Probiotics in diverticular disease of the colon: an open label study. J Gastrointestin Liver Dis. 19: 31–36.

Lilly, D.M. and R.H. Stillwell. 1965. Probiotics: Growth-Promoting factors produced by microorganisms. Science 147: 747–748.

Lighthouse, J., Y. Naito, A. Helmy, P. Hotten, H. Fuji, C.H. Min, M. Yoshioka and F. Marotta. 2004. Endotoxinemia and benzodiazepine-like substances in compensated cirrhotic patients: a randomized study comparing the effect of rifaximine alone and in association with a symbiotic preparation. Hepatol Res. 28: 155–160.

Lin, M., D. Savaiano and S. Harlender. 1991. Influence of nonfermented dairy products containing bacterial starter cultures on lactose maldigestion in humans. Journal of Dairy Science 74: 87–95.

Link-Amster, H., F. Rochat, K.Y. Saudan, O. Mignot and J.M. Aeschlimann. 1994. Modulation of a specific humoral immune response and changes in intestinal flora mediated through fermented milk intake. FEMS Immunol. Med. Microbiol. 10: 55–63.

Marotta, F., Y. Naito, E. Minelli, H. Tajiri, J. Bertuccelli, C.C. Wu, C.H. Min, P. Hotten and E. Fesce. 2003. Chemopreventive effect of a probiotic preparation on the development of preneoplastic and neoplastic colonic lesions: an experimental study. Hepatogastroenterology. 50: 1914–1918.

Mattila-Sandholm, T. and T. Kauppila (Eds.). 1998. Functional Food Research in Europe 3rd Workshop, FAIR CT96-1028, PROBDEMO, VTT Symposium 187. Haikko, Finland, p. 125.

McFarland, L.V. and G.M. Elmer. 2005. Properties of evidence-based probiotics for human health. *In:* I. Goklepe and V. Juneja (eds.). Probiotics in food safety and human health: New York Marcel Dekker, Inc. 109–37.

McFarland, L.V. 1998. Risk factor for antibiotic-associated diarrhea. A review of the literature. Ann Med Interne (Paris). 149: 261–266.

McIntosh, G.H., P.J. Royle and M.J. Playne. 1999. A probiotic strain of *L. acidophilus* reduces DMH-induced large intestinal tumors in male Sprague-Dawley rats. Nutr Cancer 35: 153–159.

Mercenier, A., P. Hols, Y. Roussel, G. Perez-Martinez, J. Buesa, M. Wilks, G. Pozzi, E. Remaut, L. Morelli, C. Grangette, V. Monedero, E. Palumbo, B. Foligne, L. Steidler and S. Nutten. 2004. Screening and construction of probiotic strains with enhanced protective properties against intestinal disorders. Microb. Ecol Health Dis. 16: 86–95.

Mercenier, A., S. Pavan and B. Pot. 2002. Probiotics as biotherapeutic agents: present knowledge and future prospects. Curr. Pharm. Des. 8: 99–110.

Metchnikoff, E. 1907. The Prolongation of Life (London, UK: Heinemann).

Metugriachuk, Y., F. Marotta, K. Pavasuthipaisit, O. Kuroi, J. Tsuchiya, A. Lorenzetti, E. Fesce and E. Minelli. 2006. The aging gut motility decay: may symbiotics be acting as "implantable" biologic pace-makers? Rejuvenation Res. 9:342–345.

Mitsuoka, T. 1996. Intestinal flora and human health. Asia pacific Journal of Clinical Nutrition 5: 2–9.

Mohania, D., R. Nagpal, M. Kumar, A. Bhardwaj, M. Yadav, S. Jain, F. Marrota, V. Singh, O. Parkash and H. Yadav. 2008. Molecular methods for identification and characterization of lactic acid bacteria: a review. Journal of Digestive Diseases 9: 190–198.

Mombelli, B. and M.R. Gismondo. 2000. The use of probiotics in medical practice. International Journal of Antimicrobial Agents 16: 531–536.

Morelli, L. 2000. *In vitro* Selection of Probiotic Lactobacilli: A Critical Appraisal. Current Issues in Intestinal Microbiology 1: 59–67.

Morelli, L. and E. Campominosi. 2002. Genetic stability of Lactobacillus paracasei subsp. paracasei F19. Microb Ecology Health Dis. 14:14–16.

Morelli, L., D. Zonenschain, M.L. Callegari, E. Grossi, F. Maisano and M. Fusillo. 2003. Assessment of a new synbiotic preparation in healthy volunteers: survival, persistence of probiotic strains and its effect on the indigenous flora. Nutr. J. 2, 11. Retrieved from: http://www.nutritionj.com/content/2/1/11.

Nagpal, R., P. Behare, R. Rana, A. Kumar, M. Kumar, S. Arora, F. Morotta, S. Jain and H. Yadav. 2011. Bioactive peptides derived from milk proteins. Food and Function 2: 18–27.

Nagpal, R., P.V. Behare, M. Kumar, D. Mohania, M. Yadav, S. Jain, S. Menon, O. Parkash, F. Marotta, E. Minelli, C.J.K. Henry and H. Yadav. 2012a. Milk, milk products and disease free health: an updated overview. Critical Reviews in Food Science & Nutrition 52: 321–333.

Nagpal, R. and A. Kaur. 2011. Synbiotic effect of various prebiotics on *in vitro* activities of probiotic lactobacilli. Ecology of Food & Nutrition 50(1): 63–68.

Nagpal, R., A. Kumar and S. Arora. 2010. *In vitro* probiotic potential of lactobacilli from indigenous milk products. International Journal of Probiotics and Prebiotics 5(2): 103–110.

Nagpal, R., A. Kumar and M. Kumar. 2012b. Fortification and fermentation of fruit juices by probiotic lactobacilli. Annals of Microbiology (doi: 10.1007/s13213-011-0412-5) [Epub ahead of print].

Nagpal, R., H. Yadav, A.K. Puniya, K. Singh, S. Jain and F. Marotta. 2007. Potential of probiotics and prebiotics for synbiotic functional dairy foods. International Journal of Probiotics and Prebiotics 2: 75–84.

Naidu, A.S., W.R. Bidlack and R.A. Clemens. 1999. Probiotic spectra of lactic acid bacteria (LAB). Critical Reviews in Food Science and Nutrition 39(1): 13–126.

Niness, K.R. 1999. Inulin and oligofructose: what are they? Journal of Nutrition 129: 1402S–1406S.

Nobuta, Y., T. Inoue, S. Suzuki, C. Arakawa, T. Yakabe M. Ogawa and N. Yajima. 2009. The efficacy and the safety of Lactobacillus brevis KB290 as a human probiotics. International J. Probiotics & Prebiotics 4: 263–270.

Oksanen, P.J., S. Salminen, M. Saxelin, P. Hamalainen, A. Ihantola-Vormisto, L. Muurasniemi-Isoviita, S. Nikari, T. Oksanen, I. Porsti, E. Salminen, S. Siitonen, H. Stuckey, A. Topilla

and H. Vapaatalo. 1990. Prevention of travellers' diarrhoea by Lactobacillus GG. Annals of Medicine 22: 53–56.
O'Sullivan, M.G., G. Thornton, G.C. O'Sullivan and J.K. Collins. 1992. Probiotic bacteria: myth or reality? Trends in Food Science and Technology 3: 309–314.
Ouwehand, A. and S. Salminen. 1998. The health effects of cultured milk products with viable and non-viable bacteria. Int. Dairy J. 8: 749–758.
Pant, A.R., S.M. Graham and S.J.Allen. 1996. Lactobacillus GG and acute diarrhea in young children in the tropics. J. Trop Pediatr. 42: 162–5.
Paone, F.M. 2012. Bacteria and intestinal health: from babies to adults. Update on probiotics and inflammatory bowel disease in children. International J. Probiotics & Prebiotics (submitted).
Parker, R.B. 1974. Probiotics, the other half of the antibiotic story. Animal Nutrition and Health 29: 4–8.
Perdigon, G., M.E. de Macias, S. Alvarez, G. Oliver and A.P. de Ruiz Holgado. 1988. Systemic augmentation of the immune response in mice by feeding fermented milks with Lactobacillus casei and Lactobacillus acidophilus. Immunology 63: 17–23.
Prakash, S. and M.L. Jones. 2005. Artificial cell therapy: New strategies for the therapeutic delivery of live bacteri. Journal of Biomedicine and Biotechnology 1: 44–56.
Quewand, A.C. and S.J. Salminen. 1998. The health effects of cultured milk products with viable and non-viable bacteria. International Dairy Journal 8: 749–758.
Quwehand, A.C., P.V. Kirjavainen, C. Shortt and S. Salminen. 1999. Probiotics: Mechanisms and established effects. Int. Dairy J. 9: 43–52.
Rafter, J. 2003. Probiotics and colon cancer. Best Practice and Research Clinical Gastroenterology 17(5): 849–859.
Rao, A.V. 2002. Prebiotics and Probiotics: New Concepts in Nutrition and Health. Heinz Infant Nutrition Institute 19: 1–4.
Rastmanesh, R., R. Catanzaro, A. Bomba, F. Allegri and F. Marotta. 2012. Potential of Prebiotics and Probiotics to Enhance the Efficacy of HIV Vaccination: A Working Hypothesis. Clinic Pharmacol Biopharmaceut, 1:1 http://dx.doi.org/10.4172/cpb.1000e102
Rautio, M., H. Jousimies-Somer, H. Kauma, I., Pietarinen M. Saxelin, S. Tynkkynen and M. Koskela. 1999. Liver abscess due to a *Lactobacillus rhamnosus* strain indistinguishable from *L. rhamnosus* strain GG. Clin. Infect. Dis. 28: 1159–1160.
Rayes, N., D. Seehofer, T. Theruvath, R.A. Schiller, J.M. Langrehr, S. Jona, S.Bengmark and P. Neuhaus. 2005. Supply of pre- and probiotics reduces bacterial infection rates after liver transplantation-a randomized, double-blind trial. Am. J. Transplant 5: 125–130.
Roberfroid, M.B. 2002. Functional foods: concepts and application to inulin and oligofructose. British Journal of Nutrition 87: S139–S143.
Roberfroid, M.B. 2000. Prebiotics and probiotics: are they functional foods? Am. J. Clin. Nutr. 71: 1682S–1687S.
Roberfroid, M.B. and N. Delzenne. 1998a. Dietary fructans. Annual Review of Nutrition 18: 117–43.
Roberfroid, M.B., J. Van Loo, and G.R. Gibson. 1998b. The bifidogenic nature of chicory inulin and its hydrolysis products. Journal of Nutrition 128: 11–9.
Rolfe, R.D. 2000. The Role of probiotic cultures in the control of gastrointestinal health. J. Nutr. 130: 396S–402S.
Romond, M.-B., A. Ais, F. Guillemot, R. Bounouader, A. Cortot and C. Romond. 1998. Cell-free whey from milk fermented with Bifidobacterium breve C50 used to modify the colnic microflora of healthy subjects. J. Dairy Sci. 81: 1229–1235.
Salminen, S. and M. Gueimonde. 2004. Human Studies on Probiotics: What is Scientifically Proven. Journal of Food Science 69: M137–M140.
Salminen, M.K., H. Rautelin, S. Tynkkynen, T. Poussa, M. Saxelin, V. Valtonen and A. Järvinen. 2004. Lactobacillus bacteremia, clinical significance, and patient outcome, with special focus on probiotic L. rhamnosus GG. Clin Infect Dis. 38: 62–69.

Salminen, S. and A. von Wright. 1998. Current probiotics-safety assured? Microbial Ecology in Health and Disease 10: 68–77.
Salminen, S., S. Gorbach, Y.K. Lee, and Y. Benno. 2004. Human studies on probiotics: What is scientifically proven today? Lactic Acid Bacteria Microbiological and Functional Aspects. New York: Marcel Dekker Inc.
Salminen, S., E. Isolauri and E. Salminen. 1996. Clinical uses of probiotics for stabilizing the gut mucosal barrier: Successful strains and future challenges. Antonie von Leeuwenhoek 70: 347–358.
Salminen, S., A. Ouwerhand, Y. Benno and Y.K. Lee. 1999. Probiotics: how should they be defined? Trends Food Sci. Technol. 10: 107–110.
Sanders, M.E. 1999. Probiotics. Food Technol. 53: 67–77.
Sanders, M.E. 2003. Probiotics: Considerations for human health. Nutrition Reviews 61: 91–99.
Schaafsma, G. 1996. State of art concerning probiotic strains in milk products. Int. Dairy Fed. Nutr. Newsl., 5: 23–24.
Scheinbach, S. 1998. Probiotics: Functionality and commercial status. Biotechnology Advances 16(3): 581–608.
Schrezenmeir, J. and M. de Vrese. 2001. Probiotics, prebiotics, and synbiotics: approaching a definition. Am. J. Clin. Nutr. 73: 361S–364S.
Shah, N.P. 2000. Probiotic bacteria: selective enumeration and survival in Dairy food. Journal of Dairy Science 83: 894–907.
Simakachorn, N., V. Pichaipat,P. Rithiopornpaisarn, C. Kongkaew, P. Tongpradit and W. Varavithya 2000. Clinical evaluation of the addition of lyophilised, heat-killed *Lactobacillus acidophilus* LB to oral rehydration therapy in the treatment of acute diarrhea in children. J. Pediatr. Gastroenterol. Nutr. 30: 68–72.
Stadnyk, A.W. 1994. Cytokine production by epithelial cells. FASEB J. 8: 1041–1047.
Stadnyk, A.W. 2002. Intestinal epithelial cells as a source of inflammatory cytokines and chemokines. Can J. Gastroenterol 16: 241–246.
Stanton, C., G. Gardiner, H. Meehan, K. Collins, G. Fitzgerald, P.B. Lynch and R.P. Ross 2001. Market potential for probiotics. American Journal of Clinical Nutrition 73: 476s–483s.
Sullivan, A., A. Johansson, B. Svenungsson and C.E. Nord. 2004. Effect of Lactobacillus F19 on the emergence of antibiotic-resistant microorganisms in the intestinal microflora. J. Antimicrob. Chemother. 54: 791–797.
Sullivan, A., L. Barkholt and C.E. Nord. 2003. Lactobacillus acidophilus, Bifidobacterium lactis and Lactobacillus F19 prevent antibiotic-associated ecological disturbances of Bacteroides fragilis in the intestine. J. Antimicrob. Chemother. 52: 308–311.
Tlaskalova-Hogenova, H., L. Tuckova, J. Mestecky, J. Kolinska, P. Rossmann, R. Stepankova, H. Kozakova, T. Hudcovic, T. Hrncir, L. Frolova and M. Kverka. 2005. Interaction of mucosal microbiota with the innate immune system. Scand J Immunol 62 Suppl. 1: 106–113.
Tsuchiya, J., R. Barreto, R. Okura, S. Kawakita, E. Fesce and F. Marotta. 2004. Single-blind follow- up study on the effectiveness of a symbiotic preparation in irritable bowel syndrome. Chin. J. Dig. Dis. 5: 169–74.
Turner, J.R. 2003. Functional morphology of the intestinal mucosae: from crypts to tips. *In*: G.A. Hecht (Ed.). Microbial pathogenesis and the intestinal epithelial cell. ASM Press, Washington, DC. 1–22.
Vergio, F. 1954. Anti-und Probiotika. Hipocrates 25: 116–119.
Vinderola, C.G., W. Prosello, D. Ghiberto and J.A. Reinheimer. 2000. Viability of probiotic (*Bifidobacterium, Lactobacillus acidophilus* and *Lactobacillus casei*) and nonprobiotic microflora in Argentinian Fresco cheese. J. Dairy Sci. 83: 1905–1911.
Xiao, S.D., D.Z. Zhang, H. Lu, S.H. Jiang, H.Y. Liu, G.S. Wang, G.M. Xu, Z.B. Zhang, G.J. Lin and G.J. Wang. 2002. Multicenter randomised controlled trial of heat-killed Lactobacillus acidophilus LB in patients with chronic diarrhea. Chin. J. Dig. Dis. 3: 167–171.
Yadav, H., S. Jain and P.R. Sinha. 2005. Preparation of low fat probiotic dahi. Journal of Dairying, Foods and Home Sciences 24: 172–177.

Yadav, H., S. Jain and P.R. Sinha. 2007. Antidiabetic effect of probiotic dahi containing *Lactobacillus acidophilus* and *Lactobacillus casei* in high fructose fed rats. Nutrition 23: 62–68.

Yadav, H., S. Jain and P.R. Sinha. 2008. Oral administration of dahi containing probiotic *Lactobacillus acidophilus* and *Lactobacillus casei* delayed the progression of streptozotocin-induced diabetes in rats. Journal of Dairy Research 75: 189–195.

2

Sources, Production and Microencapsulation of Probiotics

Dima Stefan, Bahrim Gabriela Elena and *Iordăchescu Gabriela**

Introduction

The interest for probiotics has been growing exponentially over the past fifteen years, based on their valuable contribution for food functionality, safety and health improvement (O'Sullivan 2005).

Derived from Greek, probiotics is defined as, living microorganisms which exert positive effects on reaching the intestines in sufficient numbers (administered via food). With the discovery of more and more microorganisms and their applications, new definitions have been framed: 1) mono or mixed cultures of living micro-organisms which, when applied to animals or humans, beneficially affect the host by improving the properties of the indigenous microflora; 2) living microorganisms, which upon ingestion in certain numbers, exert health benefits beyond inherent basic nutrition; 3) supporting or favoring life, are commonly defined as living microorganisms (bacteria or yeasts) that exert a beneficial effect on

111 Domnească Str. 800201, Galati, Romania.
Emails: dimastefan@yahoo.com; gabriela.bahrim@ugal.ro; gabriela.iordachescu@ugal.ro
* Corresponding author

the health of the host, when ingested; 4) a microbial preparation which contains living and/or dead cells and their metabolites, which is intended to improve the microbial or enzymatic balance at mucosal surfaces or to stimulate immune mechanisms (Farnworth 2006, Holzapfel 2006, Seth and Maulik 2011).

Probiotic Microorganisms

For a microorganism to be considered a probiotic, the validation of health benefits, strain identification, and other characteristics are required (Kailasapathy 2010). For a long time, only a very limited number of microbial strains that are currently used in food products or as supplements were considered as probiotics based on their relevant properties (Grattepanche and Lacroix 2010). The large variety of functional fermented products and the modernisation of the biochemical and genetic investigations of microorganisms has lead to an increase in the number of microorganisms with probiotic potential (Table 1).

Lactic acid bacteria and bifidobacteria

Lactic acid bacteria (LAB), in particular the genus *Lactobacillus* spp., and bifidobacteria (genus *Bifidobacterium* spp.) are considered to be classic probiotic microorganisms, which are most frequently used in food and feed products or as nutraceuticals or pharmaceuticals (Salminen and Ouwehand 2004, O'Sullivan 2005, Hori 2010, Grattepanche and Lacroix 2010).

Numerous species belonging to these two genera are generally recognized as human and animal probiotics and the reason for this is that it appears likely that all are primarily of intestinal origin. Many probiotic strains belonging to both these genera are dominant inhabitants of the intestinal tracts or vaginas of animals or humans, infants or adults (Marks 2004).

The strains with the most published clinical data are *L. rhamnosus* GG (ATCC 53103), *L. casei* Shirota, *L. johnsonii* La1, *L. acidophilus* NCFB 1478, *Bifidobacterium animalis* Bb-12 and *L. reuteri* (Donohue 2004). Other genera used are *Lactococcus, Streptococcus, Leuconostocs, Pediococcus, Carnobacterium, Enterococcus, Vagococcus* and *Weisella* but for these the probiotic properties and applications must be demonstrated by *in vitro* and *in vivo* studies (Antoine 2011).

Research conducted in recent years has been dominated by advances in molecular biology and the application of the latest genetic tools, allowing the identification of new species or the reclassification of known species.

Table1. Microorganisms used as probiotics (adapted from: Farnworth 2006, Holzapfel 2006, Shah 2010, Seth and Maulik 2011, Hui 2012, Martínez et al. 2012, Pérez Martínez et al. 2012).

Micoorganisms	Genus	Species
Lactic acid bacteria	*Lactobacillus* spp.	*L. acidophilus* *L. amylovorus* *L. brevis* *L. bulgaricus* *L. casei* *L. crispatus* *L. delbrueckii* spp. *bulgaricus*[b] *L. fermentum* *L. gallinarum*[a] *L. gasseri* *L. helveticus* *L. johnsonii* *L. lactis* *L. paracasei* *L. plantarum* *L. reuteri* *L. rhamnosus* *L. salivarius*
	Streptococcus spp.	*S. salivarius* subsp. *thermophilus*[b]
	Lactococcus spp.	*L. lactis*
	Leuconostoc spp.	*Lc. mesenteroides*
	Pediococcus spp.	*P. pentosaceus* *P. acidilactici*
Bifidobacteria	*Bifidobacterium* spp.	*B. adolescentis* *B. animalis*[c] *B. bifidum* *B. breve* *B. essensis* *B. infantis* *B. laterosporus* *B. longum*
Propionibacteria	*Propionibacterium* spp.	*P. acidipropionici* *P. freudenreichii* *P. jensenii* *P. thoenii*
Enterobacteria	*Enterococcus* spp.	*E. faecalis*[a] *E. faecium*
Sporulated bacteria	*Bacillus* spp.	*B. alcalophilus* *B. cereus* *B. clausii* *B. coagulans* *B. subtilis*
Other bacteria	*Escherichia* spp. *Sporolactobacillus* spp.	*E. coli* *S. inulinus*[a]
Yeasts	*Saccharomyces* spp.	*S. cerevisiae (boulardii)*[d]

[a] Mainly for animals; [b] *L. delbrueckii* subsp. *bulgaricus* is typically used as a starter culture for yogurt; Mainly in pharmaceutical preparations; [c] Some *Bifidobacterium animalis* strains are commonly referred to in commercial labels as *Bifidobacterium lactis;* [d] Probiotic *Saccharomyces cerevisiae* strain is marketed as *Saccharomyces boulardii.*

This was based on the understanding of their taxonomy, metabolism, and interactions with other microbes and, in probiotic applications, ultimately with the host (Table 2) (Crittenden 2004). Several probiotic strains have already been completely sequenced. The genome of *Bifidobacterium longum* revealed a large number of genes potentially coding for enzymes in the metabolism of prebiotic carbohydrates. This opens the door to opportunities to develop new combinations of symbiotic products by combining probiotics with prebiotics (Schmid et al. 2006).

Table 2. Recent changes in the taxonomy of the genus *Bifidobacterium* (adapted from Crittenden 2004).

New designation	Old designation
Bifidobacterium infantis	*Bifidobacterium longum*
B. lactis	*B. animalis*
B. suis	*B. longum*
B. globosum	*B. pseudolongum* subsp. *globosum*
B. denticolans	*Parascardovia denticolens*
B. inopinatum	*Scardovia inoponata*
Recent additions to the genus	
B. coryneforme	
B. gallicum	
B. gallinarum	
B. merycicum	
B. ruminantium	
B. saeculare *B. scardovii*	
B. thermacidophilum	

Propionic bacteria as probiotics

The literature offers very limited information about the potential probiotic properties of propionibacteria (PAB) (*Propionibacterium* spp.) in comparison with lactic acid bacteria and bifidobacteria. The most cited probiotic species which are also available on the market mainly as pharmaceutical preparations and as animal feed supplements are: *P. acidipropionici*, *P. freudenreichii*, *P. jensenii* and *P. thoenii* (Ouwehand 2004, Holzapfel 2006). PAB have a number of properties that make them good probiotic candidates but it is still uncertain if these bacteria represent an important fraction of the intestinal microbiota. Some PAB strains are able to survive at low pH levels. This property is an important selection criterion for probiotics. The ability of PAB to survive at a low pH can be significantly improved in a food matrix and by a short exposure to a nonlethal pH (e.g., pH 5,0). The mechanism by which the cells preadaptation can be maintained in a product to facilitate better survival of the gastric transit remains to be determined. Strains of PAB, which have also been observed to resist Enterococci (genus *Enterococcus* spp.), are not typical LAB indigenous in the microbiota of the gastrointestinal tract. They occur naturally in some foods, and are common in veterinary probiotics. Some enterococci demonstrate antibiotic resistance, with the exception of strains such as *E. faecium* and *E. faecalis*, which are not considered safe due to their use history. Strains of these species are being marketed under the name of "Symbioflor" (Donohue 2004, Holzapfel 2006). It has been proven that *E. faecium* is more pH stable than *L. acidophilus* and produces bacteriocins against some enteropathogens (Shah 2010). In spite of that, the tendency of some strains to exchange genetic material, the pathogenity and the acquiring of antibiotic resistance renders their use as probiotics questionable resistant to bile salts.

Enterococcoci as probiotics

Spore-forming probiotic bacteria

Although considered nonconventional probiotics, species of the genera *Bacillus* (*B. cereus, B. clausii, B. pumilis)* and *Clostridium* (*Clostridium butyricum*) are proposed as potential probiotics (human and animal). They do not colonize the intestine but have a transient presence in ingested foods and based on their abilities to form endospores can survive passage through the stomach and duodenum. These are attractive properties for *in vivo* use. Probiotics containing *B. subtilis* and *B. alcalophilus* strains are sold in Europe and Asia. The strain *Bacillus* IP5832 (identified as *Bacillus cereus*) is used for production of "Bactisubtil" (Synthelabo Belgium) (O'Sullivan 2005). *B. clausii* (previously classified as *B. subtilis* species, a constituent of the probiotic Enterogermina) is a mixture of spore form of strains of *B. subtilis* given orally as a pharmaceutical probiotic (Donohue 2004).

Other probiotic bacteria

A limited number of investigations have also been carried out into the potential properties of genera *Escherichia* (*E. coli* the "Nissle") and *Sporolactobacillus* (*S. inulinus*) in their use as probiotics (Holzapfel 2006, Hori 2010, Shah 2010, Hui 2012).

Fungi as probiotics

Some yeasts strains are frequently used as probiotic microorganisms. Although not exclusively so, for example *Saccharomyces cerevisiae* (*boulardii*) is a probiotic yeast first isolated from litchi fruit (Holzapfel 2006, Grattepanche and Lacroix 2010, Hori 2010). *Saccharomyces kefir* is part of the core of Kefir, a very common fermented milk consumed in Eastern Europe. *Kluyveromyces* strains are used in a Sudanese traditional fermented dairy product named *Rob.* Also moulds such as *Aspergillus* strains are found in some traditional fermented milks (Antoine 2011).

Isolation and Characterization of Probiotics

The most important sources for the isolation of probiotics are gut microbiota, body fluids and faeces of healthy humans and animals (Martínez 2012). Other sources can be fermented food, usually obtained from traditional processing.

Typically, probiotics are isolated in pure cultures by cultivation on selective media. Most probiotic microorganisms are nutritionally fastidious, require expensive culture mediums and addition of growth-promoting factors for propagation. The specific culture mediums are efficient for the maintenance of pure strains but are less effective for isolating them from complex flora since they often permit the growth of other genera. This happens because of their difficulty to propagate outside their natural environment (Ballongue 2004).

In order to accurately identify probiotic strains, there are many steps to be followed in the characterization of a species (Gueimonde 2011, Martínez et al. 2012). Methodologies for fast and efficient identification and enumeration of probiotics in food have been implemented (Martínez et al. 2012). Correct taxonomic and functional characterization of isolated strains is based on biochemical properties, physiological and functional characteristics evaluation (Table 3).

For proper species identification, it is extremely important to consider the fact that probiotic effects are strain specific and thus it is necessary to identify the microorganisms at the strain level. A reliable identification by faster and modern methods should confirm the identity of each strain in commercial use. Such identification is also necessary for the proper labeling of probiotic products. An accurate identification allows linking the microorganism to what is already known about the corresponding microbial group, permitting the prediction of some of its properties as described in the safety and efficacy assessment parts (Salminen and Atte von Wright 2011).

Table 3. Methods used to characterize probiotics (adaptation from Gueimonde 2011, Martínez et al. 2012).

No	General characteristics	Specific properties	Techniques used
1.	Safety assessment	Strain and species identification	- Pure culture isolation by cultivation on selective media. - Strain identification by phenotypic and genotypicmethods: DNA–DNA hybridization, Randomly Amplified Polymorphic DNA (RAPD-PCR) or Pulsed Field Gel Electrophoresis (PFGE). - Tests for survival properties evaluation in extreme environmental conditions (acidic pH, presence of bile, competitive physico-chemical and biological conditions).
		Biogen amine formation	- Qualitative evaluation based on biochemical properties, by the change of the bromocresol purple indicator to purple in the presence of histidine, lysine, ornithine, and tyrosine added to the specific medium. - Quantivative evaluation based on high-performance liquid chromatography (HPLC) with *o*-phtalaldehyde postcolumn derivatization.
		Antibiotic resistance testing	- Agar disk diffusion test. - Microdilution antimicrobial susceptibility testing by determination of minimal inhibitory concentration (MIC) by use of microtiter ELISA plates.
2.	Determination of strain survival and colonization potential	Tolerance to gastrointestinal conditions	- Test of the sensitivity to bile salts and to gastric acidity based on bacterial plate systems assay. - Microtiter plates assay, similary to the MAST.
		Adhesion to mucus and extracellular matrix proteins	- *In vitro* test to assess the adhesion properties by using human epithelial cell lines (mostly Caco-2 and HT-29), or mucus-secreting HT-29-MTX cells as well as intestinal mucus isolated from faeces, ileostomy, or resected human intestinal tissue.
		Antimicrobial activity against potentially pathogenic bacteria	- Diffusion methods—qualitative techniques to show the presence or absence of substances with antagonistic activity. - Dilution methods—quantitative assays which provide information on the minimal inhibitory concentrations. - Microdilution antimicrobial susceptibility test (MAST) by using microplates.

Table 3. contd....

Table 3. contd....

No	General characteristics	Specific properties	Techniques used
3.	Production of metabolites, exopolysaccharides, and enzymes	Exopolysaccharides (EPS)	- Growing probiotic bacteria on a medium with abundant carbohydrates; the ruthenium (0.08 g/L) red milk plates method is highly recommended. - Quantitative procedures to distinguish high producers, or even to purify the polysaccharide for fine compositional analysis; deproteinization steps and peptide precipitation with trichloroacetic acid followed by precipitation of the EPS.
		Folate (vitamin K)	- Microbiological bioassay for folate production using Bacto folic acid assay medium, cell extracts and culture supernatants.
		Bile salt hydrolases (BSH)	- Qualitative analysis by cultivation on agar dishes, on MRS agar supplemented with taurodeoxycholate or glycodeoxycholate and $CaCl_2$. BSH activity will be revealed by a white halo surrounding the producing colony, as a consequence of the precipitation of the calcium salt of the deconjugated bile salt. - Quantitative determination of BSH activity, by use of HPLC methods.
4.	Functional characterization	Biomedical and immunological tests[a]	
		In vitro assays	- *Functional properties of probiotics tested with intestinal epithelial cell cultures* (IECs). The most used cell lines are HT-29, Caco-2 T83, in continuous cultures using special culture rooms, cabins, and 5% CO_2 incubators. - *Probiotic adhesion to enterocytes and inhibition of pathogen adhesion.* The exclusion assays and displacement assays are used. Adhesion is calculated as the ratio between the radioactivity bound to the cells and the total radioactivity of the initial amount of bacteria added at the start. The quantification of attached bacteria by qPCR. - *Anti-inflammatory effect.* The enzyme-linked immunosorbent assays (ELISA), electrophoretic mobility shift assay and Western blotting method are used. - *Stimulation/Inhibition of innate immunity.* The qPCR method and immunolabeling - flow cytometry techniques are current used.

Table 3. contd....

Table 3. contd.

No	General characteristics	Specific properties	Techniques used
			- *Effect of DNA and secreted metabolites.* *In vitro* experimental systems and particularly epithelial cells culture methods are used. - *Anticancer and other effects. In vitro* systems for cytotoxic induction of apoptosis of cultured colon cancer cell lines with qPCR evaluation. - *Use of cultured lymphocytes.* Flow cytometry technique is recommended.
		Intestine explants: mice or human intestine	The use of cultured intestinal explants in an organ culture system allows the study of probiotic effects on a whole tissue level.
		In vivo tests	They are very similar to those used in the study of drug effects. They are limited based on constrictions regarding the biodiversity of organisms and bioethics.

[a]*Current Protocols in Immunology*//onlinelibrary.wiley.com/book/10.1002/0471142735/toc

The characterization of microorganisms by molecular spectroscopy involves analysis of a culture, usually during the growth phase, by Fourier transform mid-infrared spectroscopy.

Chemometric methods for pattern recognition incorporating biological and morphological measurements are being developed for automated classification of microorganisms. Fermentation tests, particularly those involving the differential fermentation of polysaccharides and mucin, have been utilized as a method of identification of the various species of probiotics (Donohue 2004).

Recently, molecular techniques have replaced or complemented most traditional phenotypic methods. DNA–DNA hybridization, Randomly Amplified Polymorphic DNA (RAPD-PCR), or Pulsed Field Gel Electrophoresis (PFGE) are recommended for typing probiotic strains. The main technique used recently involves a 16S rRNA gene-targeted, species-specific polymerase chain reaction (PCR) method (Marks 2004, Salminen and Atte von Wright 2011, Martínez et al. 2012).

The diversity of the phisiological properties that need to be detected in order to define a microorganism as probiotic, in the context of the expanding range of probiotic foods, is hindering the identification and selection processes. Many questions have to be answered before considering a microorganism as a probiotic candidate. Once a species is chosen, it is then necessary to find a strain of that species with all the traits necessary for an efficient effect (O'Sullivan 2005).

Criteria of Selection of Probiotics

For microorganisms to be considered as probiotics, the following criteria need to be fulfilled: a) *in vitro* properties—competition for organic nutrients; competition for iron; biopreservative compounds production; enzyme production potential, survival during processing and storage; b) *in vivo* properties—adherence to intestinal cells; tolerance to acid and bile, survival and efficacy in the human intestine (Seth and Maulik 2011).

In the past, probiotics have been selected on the basis of their suitability to the product's specific environment and to technological procedures, as well as the survival rate during the gastrointestinal tract passage and their colonization potential. These criteria are still used. The safety issues are also a general concern when microorganisms are isolated from body fluids and faeces (Pérez-Martínez 2012). Thus, selection criteria for probiotic strains have concentrated on performance during manufacture of the strain, its incorporation into the food matrix, and viability over the shelf life of the product. Because evidence shows that the health properties imparted by a probiotic are dependent on the particular strain and are not a property necessarily common to all strains of a particular species, manufacturers are now promoting their products as containing a particular strain. Definitive fingerprinting of strains is becoming part of the selection criteria for probiotics (Salminen et al. 2004).

The proper procedure of selection used must be in each case correlated with the functionality of the probiotic *in vitro* and *in vivo* (Martínez et al. 2012). Classically, probiotics are selected from the isolation from faeces directly on selective media. Research during the past two decades focused mainly on functional features of strains selected. At present, the characterization of probiotic properties implies many qualitative or quantitative tests (Table 4) by which such strains can be characterized and particular claims be sustained—either by *in vivo* or validated *in vitro* tests—even when all the mechanisms involved have not yet been fully elucidated (Salminen et al. 2004, Marks 2004, Farnworth 2006, Holzapfel 2006, Seth and Maulik 2011).

Five major aspects are considered most important for the selection of a probiotic strain (Holzapfel 2006, Diez-Gonzalez and Schamberger 2006, Kailasapathy 2010):

1. *Ecological, genetic and biochemical aspects*, e.g., origin, identity, biochemical properties, metabolic and genetic stability.
2. *Technological aspects*, including growth properties *in vitro*, adaptation persistence and survival during storage. For the producer, the obtaining of some probiotic strains, and also the sensory properties of the resulting products, are still major obstacles toward the large-scale production of functional foods containing probiotic strains.

Table 4. New challenges in obtaining probiotics (adaptation from Salminen et al. 2004, Marks 2004, Farnworth 2006, Seth and Maulik 2011).

Steps	Actions
Strains isolation, identification and characterisation	Selection of sources for strains isolation Pure cultures obtaining and preservation **Physiological properties evaluation:** - phenotypic carbohydrate fermentation profiles tests - phenotypic enzyme profile tests - catalase test - β-galactosidase activity - relative lactic acid production - ability to produce D-lactic acid - ability to produce hydrogen peroxide - ability to utilize prebiotics (e.g., oligosaccharides, inulin, resistant starch) for growth - ability to synthesize or utilize vitamins (B-group, folate, vitamin K) - bile acid deconjugation properties **Genetic characterization:** - DNA-based tests: PCR based; RAPD, PFGE for strains, and 16sRNA sequences and ITS region tests
Strains selection	**Criteria to be fulfilled:** - it should be isolated from the same species as its intended host - it should have scientifically proven efficacy - it must be safe for human consumption (GRAS category microorganism) - a large number of viable bacteria must be able to survive prolonged periods *in vitro* - it must resist destruction by gastric and intestinal juices - it must show reasonable persistence in the intestinal tract - it should demonstrate antagonism to harmful intestinal bacteria - it should demonstrate a desirable enzyme pattern - it should demonstrate immunomodulation activity - it must show desirable characteristics in animal trials
Manufacturing and practical approaches	**Testing the abilities:** - to grow quickly to high numbers in a simple and cheap fermentation medium - to grow and survive in microaerophilic or aerobic conditions - to withstand centrifugation, filtration and freezing/lyophilization without significant loss of numbers - to become "active" quickly following application - to survive incorporation into a wide variety of food matrices, including being subject to processing temperatures above 45°C and raised concentrations of ethanol and sodium chloride - to be done on the organism in the actual food matrix

Table 4. contd....

Table 4. contd.

Steps	Actions
Quality control	**Physiological properties evaluation:** - assessment of viability of probiotic at intervals over shelf life of products, under different storage environments (e.g., temperature) - tolerance to acid–pepsin solution at pH 2 for 2 hours (measurement of survival after exposure) - tolerance to bile salts at physiological concentrations (measurement of growth), and at higher concentrations (measurement of survival after exposure) **Health properties analysis:** - ability to inhibit pathogens (e.g., *Salmonella typhimurium, Clostridium perfringens, Clostridium difficile, Escherichia coli, Candida albicans*) *in vitro* and *in vivo* - ability to adhere to Caco2 cells, HT29 cells and fecal and ileostomy mucus - autoaggregation of the productive cells **Safety properties:** - safe history of origin and/or use - dose–response curves in animal models - platelet aggregation tests
	Other properties: - aflatoxin removal - biofilm surface protection

3. *Physiological aspects*, resistance against environmental stress and to the antimicrobial factors prevailing in the upper gastrointestinal tract as encountered during the stomach-duodenum passage (pH 2,5 gastric juice, bile acid, pancreatic juice), adhesion potential to the intestinal epithelium.
4. *Functional aspects and health beneficial features*, adhesion, colonization potential of the mucosa, competitiveness, specific antimicrobial antagonism against pathogens, stimulation of immune response, selective stimulation of beneficial autochthonous bacteria, restoration of the "normal" population.
5. *Safety aspects*, generally regarded as safe (GRAS) characteristics, no invasive potential, no transferable resistance against therapeutic antibiotics, no virulence factors.

An ideal candidate used as probiotic must have good, stable properties so that it can be cultured and incorporated into food products without loosing viability and functionality or creating unpleasant flavors or textures in the product. The adhesion abilities to the intestinal epithelial cells are also considered a classification requirement for probiotic bacteria. In addition, these abilities of probiotic bacterial cells to adhere to intestinal epithelial cells potentially stimulates the immune system.

New probiotic strains with improved technological and functional properties and/or development of fermentation are therefore clearly needed, along with stabilization technologies, to control cell physiology and to protect cells during downstream processing, storage, and in gastrointestinal tract following ingestion (Naidu and Clemens 2000).

Probiotics Production and Quality Evaluation

Probiotics are produced and used as single, mixed or multiple starter cultures. For practical reasons, using a combination of strains might be ideal regarding technological aproaches to *in vitro* and *in vivo* functionality and viability preserving. The constitutive strains and concentration of each individual probiotic strain should be optimized and well defined before such combinations are marketed.

Probiotic preparations may be administered in different formulations depending on the condition to be treated. The most common forms in which probiotics are produced and marketed are as concentrated frozen cultures in milk or whey or liphophylised powder as capsules or tablets (Naidu and Clemens 2000, Seth and Maulik 2011).

The principal technological steps for probiotic production are presented in Table 5. Recently, developments in probiotic production have mainly focused on achieving high viable cell yields while keeping costs low (Grattepanche and Lacroix 2010).

Table 5. Principal steps in probiotics production and practical use (adaptation from Grattepanche and Lacroix 2010).

No.	Step	Efficiency criteria
1.	Starters cultivation and multiplication	- Good ability to grow and multiply in milk. - Biotechnological reproducibility of the fermentation process. - Easy and inexpensive cultivation to obtain a high yield of biomass. - Good resistance of the cells to shear forces encountered in the bioreactor. - Bacteriophage resistant.
2.	Downstream process	- Stability during freezing and spray-drying. - Cells stability assurance by conservation and encapsulation.
3.	Preservation	- Cells tolerance to oxygen and low pH. - Safety, functionality and technological properties preservation.
4.	Probiotics food production	- Symbiotic activity. - Stability over shelf life of the food products. - Absence of antagonistic activity when used in mixed culture. - Preservation of the particular organoleptic properties of the food products.

Novel cultivation approaches should be developed with the aim to maximize viability and health functionality, and most importantly to broaden the range of probiotic strains to sensitive probiotics exhibiting high functionality that cannot be produced with traditional technologies. Research for new technologies such as continuous cultures to produce cells with controlled physiology, and cell immobilization to increase cell density and productivity, enhance process stability, protect cells from environmental stresses, and the inductance of stress tolerance should be pursued, also considering the challenges to transfer these technologies to an industrial setting.

Encapsulation technologies promise to sustain both functional properties and viability of probiotic cells when added to products and during the transit of the upper digestive tract.

Additionally, encapsulation matrices could be developed to target the delivery of probiotic cells to specific regions of the gut (e.g., small/large bowel), enhancing their efficacy while protecting the cells from environmental stresses in earlier stages. Targeted delivery and the activity of probiotics in specific locations of the gastrointestinal tract requires knowledge of the interaction between receptors and molecules on cell surfaces (Kailasapathy 2010).

Synbiotics preparations combine probiotics and prebiotics in a single product on the basis that their simultaneous administration should encourage the growth and persistence of probiotics within the intestinal tract and hence improve the therapeutic value (Marks 2004, Mäyrä-Mäkinen and Bigret 2004).

The quality of such a probiotic starter culture depends on many factors: food system composition, added additives and ingredients, additional starter cultures used, processing conditions, and other parameters such as storage, packaging, etc. (Hui 2012).

Encapsulation of probiotic cells

Nowadays more and more consumers prefer foods that provide various benefits for their personal health. The most common bioactive components in functional foods are probiotics.

It is estimated that the market for foods with added probiotics in Europe will increase to 130 millions euros in 2013 (Sanchez et al. 2012). In the last years the types of foods that contains probiotics are more and more diversified. Besides dairy products (yogurt, cheese, ice cream, dairy dessert), probiotics have also been introduced in other food systems like chocolates, juices, dry beverages, sausages, biscuits, etc. (Burgain et al. 2011).

According to the definition given by FAO/WHO (2001), probiotics are microorganisms that accomplish the following conditions: are living cells, are administered in adequate amounts, and confer a health benefit to the host.

Factors that ensure the effectiveness of probiotics are: viability, better growth and survivability during food manufacturing and storage as well as in the gastrointestinal tract, resistance to acid, bile and gastrointestinal enzymes, adhesion to intestinal epithelium, antimicrobial properties and antibiotic resistance (Randheera et al. 2010).

One way of introducing probiotic bacteria in foods that provide the above conditions is by using microcapsules.

Encapsulation is defined as a physico-chemical process in which solids, liquids and gasous substances are surrounded by a coating or embedded in homogeneous or heterogeneous particles, to give small capsules with diameters ranging from a few nm to a few mm.

By encapsulation the isolation, protection, transport and release of the active components is achieved.

The encapsulation process involves two types of materials: the material that is encapsulated, called core material or internal phase, and the material that facilitates the encapsulation called carrier material, wall material, suport material, shell, membrane, external phase or matrix (Zuidam and Shimoni 2009).

In terms of morphology, microcapsules can be of two types (Fig. 1.):

- microcapsules type reservoir (or capsules) in which the active substance is included in a homogeneous space, called the core, surrounded by a protective membrane. In a microcapsule there may be one or more cores (microreservoirs) which contains the active substance;
- microcapsules type matrix, called microspheres (micro beads), in which the active substance is dispersed in polymeric network spaces. Sometimes microspheres (micro beads) can be recovered by the coating material.

Microencapsulation ensures the maintaining of micro-organisms viability that both factors are subject to aggression during processing and storage (temperature, pressure, humidity, high concentration of ions, the presence of oxygen), and especially during consumption and their passing throught gastro-intestinal tract (extreme pH variations, presence of bile salts).

Co-microencapsulation of probiotics with prebiotics provides cell multiplication and achievement of the minimum limit of 10^7 viable probiotic bacteria per gram of probiotic product for better efficacy in regulating beneficial effects (Ranadheera et al. 2010, Homayoni et al. 2012). Moreover,

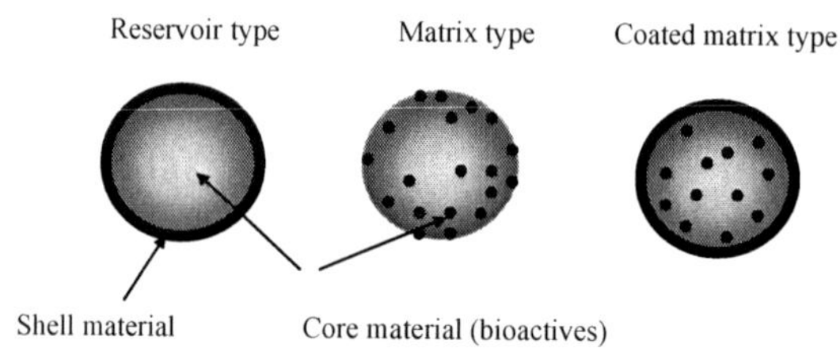

Fig. 1. Schematic representation of microcapsules morphology.

the microencapsulated bacteria consumption also ensures their release in the gastro intestinal tract (GIT) active area where there are Peyer's patches and other mucosa-associated lymphatic tissues which play a critical role in immunostimulation (Manojlović 2010).

Carrier materials and technologies used to encapsulate probiotic cells

Frequent materials used to encapsulate probiotic bacteria are: carbohydrates (starch and derivatives, cellulose and derivatives, gum arabic, gum tragacanth, alginate, carrageenan, gellan, xanthan, chitosan) and proteins (gelatin, milk proteins, whey proteins, gluten, soy protein, etc.). At encapsulation a single material can be used or a mixture of materials (Livney 2010, Burgain et al. 2011).

During the selection process of carrier materials to be used for encapsulation of food ingredients, some general properties should be considered (Ubbink and Krüger 2006):

- to be food grade;
- to provide maximal protection of the bioactive component against environmental conditions (oxygen, water vapour, temperature, pH, moisture, enzymes, UV light);
- to have good rheological characteristics at high concentration;
- to have a good emulsification activity;
- to have a good sensory quality,
- to require low cost production and to be available in large quantities.

From the large number of microencapsulation techniques for bioactive components, a smaller number of them are used for probiotics encapsulation (Dima 2009, Zuidam and Shimoni 2009).

The constraints of microencapsulation techniques of probiotic bacteria is on the one hand due to the size of the bacteria (1–5 μm), that does not allow the use of nanotechnologies, and on the other hand due to special conditions required for their viability.

Several methods of microencapsulation of probiotic bacteria have been reported (Champagne and Fustier 2007, De Vos 2007, Manojlović et al. 2009, Bourgain et al. 2011). These methods include: spray-drying, spray freeze drying, fluid bed coating, extrusion and emulsification /ionotropic internal gelation.

Spray-drying method represents one of the oldest and the most widely used encapsulation techniques used in the food industry area (Gharasallaoui et al. 2007, Drusch 2007). The method presents the following advantages: it can be applied on industrial scale, it is a fast high-yield, it can be applied to various biocomponents encapsulation, and microcapsules are readily dispersible in water.

The disadvantages of this method are: the complexity of the equipment, the non-uniform conditions in the drying chamber, the method is limited to shell materials soluble or dispersible in water, and spray-dried capsules carry a lower loading (20–30%).

Probiotic microencapsulation spray-drying method is done with great care, following carefully: drying temperature, drying time, type of atomization, shell materials, storage conditions (Manojlović et al. 2009).

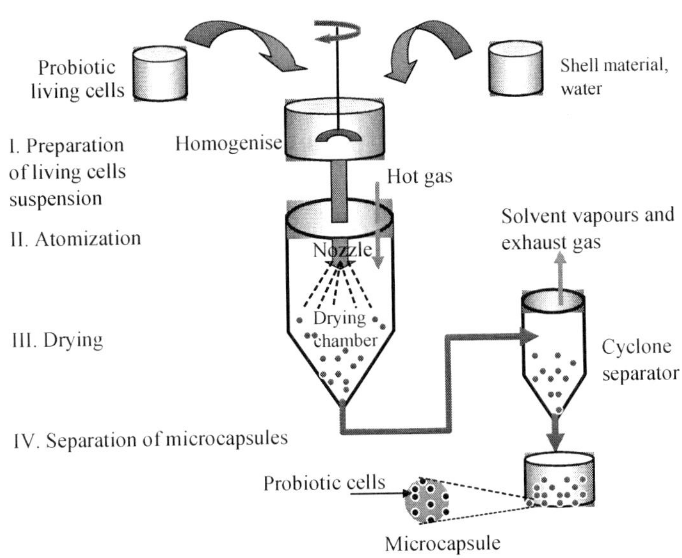

Fig. 2. Schematic presentation of the spray-drying probiotic encapsulation process.

Probiotic microencapsulation spray-drying stages are schematically represented in Fig. 2.

For spray-drying probiotics microencapsulation have been used different species of bacteria encapsulated in different shell materials such as: *Lactobacillus paracasei* and skim milk (Gardiner et al. 2002), *Lactobacillus acidophilus* and maltodextrin/arabic gum (Su et al. 2007), *Bifiobacterium ruminantium* and starch (O'Riordan et al. 2001).

Spray freeze drying is a similar method to spray drying. The difference is represented by how the droplets produced by atomization are strengthened.

In freeze drying method the droplets with probiotic cells in shell material are frozen into the vapours of a cryogenic liquid such as liquid nitrogen. Frozen droplets are then dried in a freeze dryer (De Vos et al. 2010, Burgain et al. 2011). In order to improve probiotic stability, different cryoprotectants are used in freeze drying method: starch, sucrose, fructose, lactose, mannose, monosodium glutamate, sorbitol, trehalose, soy protein isolate and 20% maltodextrin (Capela et al. 2006, Chavez and Ledeboer 2007). The disadvantage of this method is the high energy consumption which implies a much higher cost than spray drying method. However, freeze drying method is widely used in micro probiotics, both in food and pharmaceuticals industries.

Fluid bed coating or spray coating method represents an encapsulation technology that utilises a spray process to deliver film material to a core particle, and fluidize air to circulate materials. Fluid-bed technology is restricted to solid core materials ranging from 30 μm to several centimetres in diameter (Frey and Hall 2004). Solid forms of probiotics obtained by spray-drying or freeze-drying are moved by the fluidizing air and a liquid coating material is sprayed through a nozzle over the core material in a hot environment. These processes are different by the position of the nozzle inside the fluid-bed chamber: bottom spray (Wurster) process, top-spray process and tangential-spray process (Fig. 3).

This method is especially used at the microcapsule coating with several layers of shell material: lipids, fatty acids, casein, cellulose derivatives, carrageenan, alginate (Champagne and Fustier 2007). The multilayer microparticles allow an increase in probiotic viability during shelf life processing and during its passage through the gastro intestinal tract (GIT).

Extrusion technique is the simplest method used to produce the probiotic microspheres. The principle of this technique is the external gelation of hydrocolloids using different gelification agents (calcium chloride solution for alginate, potasium chloride for carrageenan and tripolyphosphat for chitosan, transglutaminase-enzyme for caseinate).

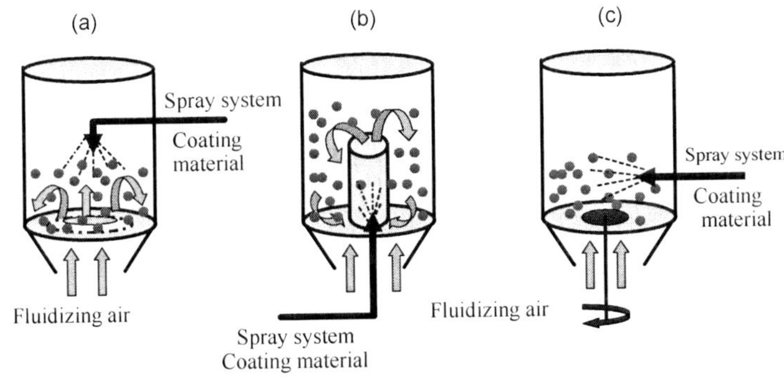

Fig. 3. Schematic presentation of the fluid bed coating technology: (a) fluid bed top spray-coating; (b) fluid bed bottom spray-coating with the Würster device; (c) fluid bed tangential spray-coating. Adapted from Champagne and Fustier (2007) and Burgain et al. (2011).

A suspension of living cells and hydrocolloid solution is extruded through a needle to produce droplets that are collected in a bath where gelation occurs (ionotropic or thermal). The extrusion technique produces spherical polymer beads, ranging from 2 to 3 mm diameter. Using additional drag forces (coaxial flow, electrostatic field) smaller polymer beads are obtained (down to 100 μm). The extrusion technology represents an adequate method for encapsulation of living cells, because it does not involve deleterious solvents and can be accomplished under both aerobic and anaerobic conditions (Krasaekoopt et al. 2003, Chen and Chen 2007, Kailasapathy 2006, De Vos et al. 2010).

The most common hydrocolloid used to produce the probiotic microencapsules is represented by alginate. Alginate is a linear polymer of two uronic acids: β-D-manuronic acid and α-L-guluronic, obtained by extraction from brown algae. During gelation, the calcium ions occupy the space between two alginate polymer chains, and the strong interchain binding results in a conformation called the "egg-box" model (Poncelet and Markvicheva 2004).

The size of alginate beads depends on differents parameters, such as: the alginate structure, the alginate and calcium concentration, the gelification time, the needle diameter and finally the distance between the outlet and the coagulation solution.

Several researchers have reported that the survivability of probiotic bacteria depends on the size of alginate beads and the cells concentration (Lee and Heo 2000, Truelstrup et al. 2002). Mandal et al. (2006) reported the increase of viability of *Lactobacillus casei* with alginate concentration increasing from 2% to 4%.

Milk proteins can be used as carrier materials for microencapsulation of probiotic bacteria by extrusion technology.

Nag et al. (2011) have encapsulated *Lactobacillus casei* into a mixture of sodium caseinat (10 % w/w), gellan gum (0.25%) and cells (2.5%) using a combination of gelation and water-in-oil emulsion. The average diameters of the microspheres were found to be about 287 and 399 µm respectivly and the viability of encapsulated cells in simulated gastric fluid (SGF) was reduced by only about 3,1 log CFU after 120 min of incubation. Using sodium caseinate gelled with transglutaminase enzyme for *Lactobacillus paracasei* encapsulation by extrusion method, Heidebach et al. (2009) found about 3,0 log CFU reduction after 90 min of incubation at pH 2,5.

Doherty et al. (2011) have encapsulated *Lactobacillus rhamnosus GG* in whey protein isolate as carrier material using laminar jet break-up extrusion technology.

Chitosan was used as a coating material in order to improve the encapsulation of probiotic cells in calcium alginate beads. For example, Chávarri et al. (2010) have investigated the co-microencapsulation of *Lactobacillus gasseri* and *Bifidobacterium bifidum* as probiotics and quercetin as prebiotic, using alginate as the supporting matrix. In order to form beads, the sodium alginate solution (20g/L) was extruded into chitosan/$CaCl_2$ 0.1M solution. The stability test in SGF showed that after exposure to SGF for 5 mins, the survivavility was 95%, 94%, 78% and 66% from the initial population in case of chitosan-coated alginate microspheres with *B. bifidum*, with *L. gasseri*, free *L. gasseri* and free *B. bifidum*, respectively.

For enhancing survival during exposure to the conditions of GIT, *Lactobacillus acidophilus PTCC1643* and *Lactobacillus rhamnosus PTCC1637* probiotic cells were encapsulated into uncoated calcium alginate beads. The same beads were coated with one or two layers of sodium alginate. After incubation in SGF for 60 min and intestinal juices (pH 7,25/2h), the number of surviving cells were 6,5 log CFU mL^{-1} for *L. Acidiphilus* and 7,6 log CFU mL^{-1} for *L. Rhamnosus* by double layer coated alginate microsphere, respectively, while 2,3 and 2,0 log CFU mL^{-1} were obtained for free cells, respectively (Mokarram et al. 2009).

Emulsification and internal ionic gelification represents a chemical technique used to encapsulate the probiotic bacteria in microspheres with less than 100 µm (Poncelet et al. 1992, 1995).

The principle of this method is based on emulsification of hydrocolloid (alginate), calcium carbonate and probiotic cells aqueous suspension into mineral or vegetable oil, in the presence of lipophilic surfactant (Span 80). In this way will be obtained a water-in-oil emulsion in which the internal phase contains droplets of both alginate and living cells. In order to achieve the internal gelation, the acetic acid (100 µL/100 mL, diluted in a small amount of oil) is added into the formed emulsion. The water-in-oil emulsion is destroyed by a $CaCl_2$ solution (0.05M) and probiotic alginate beads remain inside the aqueous phase. The alginate beads are then separated from oil and washed (Dima and Bahrim 2007). A large number of parameters may influence the characteristics of the microspheres during emulsification/internal gelation process, such as: internal phase ratio, emulsifier concentration, calcium/alginate molar ratio and alginate concentration. The size of alginate beads depends especially on the emulsifier and alginate concentrations. An increase in the emulsifier concentration from 0 to 1% resulted in a decrease in the microspheres mean size from 288 to 53 µm, and when alginate concentration was increased from 2 to 3% (w/v) the particle diameter increased from 53 to 112 µm (Silva et al. 2006).

Future Prospects

Synbiotic foods play an important role in the health of consumers. Development of innovative health-based fermented products incorporating highly efficient strains of probiotic bacteria and highly bioactive prebiotic substances will increase in the near future. Simultaneously, there is a need for improved and cost-effective fermentation technology for maximizing the synbiotic health properties of new and innovative fermented foods.

Studies must focus on the relationship between the constitutive strains of a mixture. Understanding the synbiotic and inhibitory phenomena between strains is of prime importance for the control of the cultures used in fermented products.

The second main direction followed by researchers is in genetics. These programs aim at implementing new characteristics in technologically interesting strains. The concept of genetic manipulation of bacteria for a specific probiotic function is appealing. Consumer resistance to genetically modified organisms (GMO) in foods is so great that GMO probiotics are unlikely to be used in the near future, with the possible exception of clinical applications.

Microencapsulation offers the potential for developing the innovative functional foods area.

Recently, probiotic cells have been co-encapsulated with nondigestible food ingredients that induce the growth of probiotic cells in GIT (prebiotics), and also using other different biocomponents, such as: antioxidants, bacteriocins, etc. in order to enhance the antimicrobial behaviour of the probiotic bacteria.

One of the technological challenges in the manufacturing process of probiotic microcapsules is represented by particle size reduction. This will help the producer to offer a healthy food with good sensorial properties.

More and more recent studies are highlighting the microencapsulation of probiotic cells process complexity and permanent interest of researchers to diversify the wall material of food systems used for the management of carrier food matrices.

References

Antoine, J.M. 2011. Current Challenges for Probiotics in Food. *In*: S. Lahtinen, A.C. Ouwehand, S. Salminen and A. von Wright [eds.]. Lactic Acid Bacteria Microbiological and Functional Aspects. Fourth Edition. CRC Press, Taylor & Francis Group, LLC, Boca Raton, London, New York, pp. 213–226.

Ballongue, J. 2004. Bifidobacteria and Probiotic Action. *In*: S. Salminen, A. von Wright and A. Ouwehand [eds.]. Lactic Acid Bacteria Microbiological and Functional Aspects. Third Edition, Revised and Expanded. Marcel Dekker, Inc. New York, USA, pp. 67–124.

Burgain, J., C. Gaiani, M. Linder and J. Scher. 2011. Encapsulation of probiotic living cells: From laboratory scale to industrial applications. J. of Food Eng. 104: 467–483.

Capela, P., T.K.C. Hay and N.P. Shah. 2006. Effect of cryoprotectants, prebiotics and microencapsulation on survival of probiotic organisms in yoghurt and freeze dried yoghurt. Food Res. Int. 39(2): 203–211.

Champagne, P.C. and P. Fustier. 2007. Microencapsulation for the improved delivery of bioactive compounds into foods. Curr. Opin. in Biotechnol. 18: 184–190.

Chávarri, M., I. Marañón, R. Ares, F.C. Ibáñez, F. Marzo and C. Villarán Mdel. 2010. Microencapsulation of a probiotic and prebiotic in alginate-chitosan capsules improves survival in simulated gastro-intestinal conditions. Int. J. Food Microbiol. 142(1-2): 185–189.

Chavez, B.E. and A.M. Ledeboer. 2007. Drying of probiotics: optimization of formulation and process to enhance storage survival. Dry. Technol. 25: 1193–1201.

Chen, M.J. and K.N. Chen. 2007. Application of Probiotic Encapsulation in Dairy Product. *In*: J.M. Lakkis [ed.]. 2007. Encapsulation and Controlled Release Technologies in Food Systems. Blackwell Publishing, pp. 83–112.

Crittenden, R. 2004. An Update on Probiotic Bifidobacteria. *In*: S. Salminen, A. von Wright and A. Ouwehand [eds.]. Lactic Acid Bacteria Microbiological and Functional Aspects. Third Edition. Revised and Expanded. Marcel Dekker, Inc. New York, USA, pp. 125–158.

De Voss, P., M.M. Faas, M. Spasojevic and J. Sikkema. 2010. Encapsulation for preservation of functionality and targeted delivery of bioactive food components. Int. Dairy J. 20: 292–302.

Diez-Gonzalez, F. and G.P. Schamberger. 2006. Use of Probiotics in Preharvest Food Safety Applications. *In*: M. Ahmedna, I. Goktepe and V.K. Juneja [eds.]. Probiotics in Food Safety and Human Health. CRC Press, Taylor & Francis Group, LLC, Boca Raton, London, New York, pp. 251–272.

Dima, S. and G. Bahrim. 2007. Intelligent invertase micro and nanocapsules make and use. Romanian Biotechnol. Letters 12(3): 3287–3293.

Dima S. 2009. Bazele microîncapsulării. *In*: T. Florea, S. Dima and G.M. Costin [eds.]. Microîncapsularea pentru sisteme alimentare. Academica, Galati, Romania, pp. 2–105.

Doherty, S.B., V.L. Gee, R.P. Ross, C. Stanton, G.F. Fitzgerald and A. Brodkorb. 2011. Development and characterisation of whey protein micro-beads as potential matrices for probiotic protection. Food Hydrocolloids 25: 1604–1617.

Donohue, D.C. 2004. Safety of Novel Probiotic Bacteria. *In*: S. Salminen, A. von Wright and A. Ouwehand [eds.]. Lactic Acid Bacteria Microbiological and Functional Aspects. Third Edition. Revised and Expanded. Marcel Dekker, Inc. New York, USA, pp. 531–547.

Drusch, S. 2007. Sugar beet pectin: A novel emulsifying wall component for microencapsulation of lipophilic food ingredients by spray-drying. Food Hydrocolloids 21: 1223–1228.

FAO/WHO Experts' Report. 2001. Health and nutritional properties of probiotics in food including powder milk with live lactic acid bacteria.
Farnworth, E.R. 2006. Probiotics and Prebiotics. *In*: R.E.C. Wildman [ed.]. Handbook of Nutraceuticals and Functional Foods. Second Edition. CRC Press, Taylor & Francis Group, LLC, Boca Raton, London, New York, pp. 335–352.
Frey, R.C. and H.S. Hall. 2004. Single-core encapsulation. Filmcoating. *In*: P. Vilstrup [ed.]. Microencapsulation of Food Ingredients. Leatherhead International Limited, UK, pp. 83–100.
Gardiner, G.E., P. Bouchier, E. O'Sullivan, J. Kelly, K.J. Collins, G. Fitzgerald, R.P. Ross and C. Stanton. 2002. A spray-dried culture for probiotic Cheddar cheese manufacture. Int. Dairy J. 12(9): 749–756.
Gharasalloui, A., G. Roudaut, O. Chambin, A. Voilley and R. Saurel. 2007. Application of spray-drying in microencapsulation of food ingredients: An overview. Food Res. Int. 40: 1107–1121.
Grattepanche, F. and C. Lacroix. 2010. Production of High-Quality Probiotics Using Novel Fermentation and Stabilization Technologies, *In*: D. Bagchi , C.L. Francis and D.K. Ghosh [eds.]. Biotechnology in Functional Foods and Nutraceuticals. CRC Press, Taylor & Francis Group, LLC, Boca Raton, USA, pp. 361–388.
Gueimonde, M.C.G., L.R. Gavilán and B. Sánchez. 2011. Stability of Lactic Acid Bacteria in Foods and Supplements. *In*: S. Lahtinen, A.C. Ouwehand, S. Salminen and A. von Wright [eds.]. Lactic Acid Bacteria Microbiological and Functional Aspects. Fourth Edition. CRC Press, Taylor & Francis Group, LLC, Boca Raton, London, New York, pp. 361–383.
Heidebach, T., P. Forst and U. Kulozik. 2009. Transglutaminase-induced caseinate gelation for the microencapsulation of probiotic cell. Int. Dairy J. 19: 77–84.
Holzapfel, W.H. 2006. Introduction to Prebiotics and Probiotics. *In*: M. Ahmedna, I. Goktepe and V.K. Juneja [eds.]. Probiotics in Food Safety and Human Health. CRC Press, Taylor & Francis Group, LLC, Boca Raton, London, New York, pp. 1–34.
Homayoni Rad, A., E.V. Mehrabany, B. Alipoor, L.V. Mehrabany and M. Javadi. 2012. Do probiotics act more efficiently in foods than in supplements? Nutr. 28: 733–736.
Hori, T. 2010. Probiotics. Recent Human Studies Using *Lactobacillus casei* strain Shirota. *In*: S. Sungsoo Cho and E.T. Finocchiaro [eds.]. Handbook of Prebiotics and Probiotics Ingredients Health Benefits and Food applications 2010. CRC Press, Taylor & Francis Group, LLC, Boca Raton, USA, pp. 193–205.
Hui, Y.H. 2012. Probiotics: An Overview. *In*: Y.H. Hui and E. Özgül Evranuz [eds.]. Handbook of Animal-Based Fermented Food and Beverage Technology. Second Edition. CRC Press, Taylor & Francis Group, LLC, Boca Raton, USA, pp. 741–748.
Kailasapathy, K. 2006. Survival of free and encapsulated probiotic bacteria and their effect on the sensory properties of yoghurt. LWT. Food Sci. and Tech. 39(10): 1221–1227.
Kailasapathy, K. 2010. Probiotic and Prebiotic Fermented Food. *In*: J.P. Tamang and K. Kailasapathy [eds.]. Fermented Foods and Beverages of the World. CRC Press, Taylor & Francis Group, LLC, Boca Raton, London, New York, pp. 377–390.
Krasaekoopt, W., B. Bhandari and H. Deeth. 2003. Evaluation of encapsulation techniques of probiotics for yoghurt. Int. Dairy J. 13(1): 3–13.
Lee, K.Y. and T.R. Heo. 2000. Survival of *Bifidobacterium longum* immobilised in calcium alginate beads in simulated gastric juices and bile salt solutions. Appl. Environ. Microbiol. 66(2): 869–873.
Livney, D.Y. 2010. Milk proteins as vehicles for bioactives. Current Opinion in Colloid and Interface Science 15: 73–83.
Mandal, S., A.K. Puniya and K. Singh. 2006. Effect of alginate concentrations on survival of microencapsulated *Lactobacillus casei* NCDC-298. Int. Dairy J. 16: 1190–1195.
Manojlović, V.N., V.A. Nedović, K. Kailasapathy and N.J. Zuidam. 2009. Encapsulation of Probiotics for use in Food Products. *In*: N.J. Zuidam and V.A. Nedović [eds.]. Encapsulation Technologies for Active Food Ingredients and Food Processing. Springer: Dordrecht, The Netherlands, pp. 269–302.

Marks, J. 2004. Successful Probiotic Bifidobacteria. *In*: C. Shortt and J. O'Brien [eds.]. Handbook of Functional Dairy Products. CRC Press, Taylor & Francis Group, LLC, Boca Raton, London, New York, Washington, D.C., pp. 33–69.

Martínez, G.P., C. Bäuerl and M. Collado. 2012. Selection and Evaluation of Probiotics. *In*: M.L. Nollet and F. Toldrá [eds.]. Handbook of Analysis of Active Compounds in Functional Foods. CRC Press, Taylor & Francis Group, LLC, Boca Raton, USA, pp. 607–638.

Mäyrä-Mäkinen, A. and M. Bigret. 2004. Industrial Use and Production of Lactic Acid Bacteria. *In*: S. Salminen, A. von Wright and A. Ouwehand [eds.]. Lactic Acid Bacteria Microbiological and Functional Aspects. Third Edition. Revised and Expanded. Marcel Dekker, Inc. New York, USA, pp. 175–198.

Mokarram, R.R., S.A. Mortazavi, M.B. Habibi Najafi and F. Shahidi. 2009. The influence of multi stage alginate coating on survivability of potential probiotic bacteria in simulated gastric and intestinal juice. Food Res. Intern. 42: 1040–1045.

Nag, A., K.S. Han and H. Singh. 2011. Microencapsulation of probiotic bacteria using pH-induced gelation of sodium caseinate and gellan gum. Int. Dairy J. 21: 247–253.

Naidu, A.S. and R.A. Clemens. 2000. Probiotics. *In*: A.S. Naidu [ed.]. Natural Food Antimicrobial Systems. CRC Press, Taylor & Francis Group, LLC, Boca Raton, London, New York, Washington, D.C. USA, pp. 431–462.

O'Sullivan, D.J. 2005. Primary Sources of Probiotic Cultures. *In*: M. Ahmedna, I. Goktepe and V.K. Juneja [eds.]. Probiotics in Food Safety and Human Health. CRC Press, Taylor & Francis Group, LLC, Boca Raton, USA, pp. 91–108.

Ouwehand, A.C. 2004. The Probiotic Potential of Propionibacteria. *In*: S. Salminen, A. von Wright and A. Ouwehand [eds.]. Lactic Acid Bacteria Microbiological and Functional Aspects. Third Edition. Revised and Expanded. Marcel Dekker, Inc. New York, USA, pp. 159–174.

Pérez-Martínez, G., C. Bäuerl and M.C. Collado Amores. 2012. Selection and Evaluation of Probiotics. *In*: L.M.L. Nollet and F. Toldrá [eds.]. Handbook of Analysis of Active Compounds in Functional Foods. CRC Press, Taylor & Francis Group, LLC, Boca Raton, USA, pp. 607–638.

Poncelet, D., R. Lencki, C. Beaulieu, J.P. Halle, R.J. Neufeld and A. Fournier. 1992. Production of alginate beads by emulsification/internal gelation. Methodology Appl Microbiol Biotechnol. 38: 39–45.

Poncelet, D., D.S. Poncelet, B. Beaulieu, M.L. Huguet, A. Fournier and R.J. Neufeld. 1995. Production of alginate beads by emulsification/internal gelation. II. Physicochemistry. Appl. Microbiol. Biotechnol. 43: 644–650.

Poncelet, D. and E. Markvicheva. 2004. Multiple-core encapsulation. Microencapsulation and alginate. *In*: P. Vilstrup [ed.]. Microencapsulation of Food Ingredients. Leatherhead International Limited. UK, pp. 215–223.

Randheera, R.D.C.S., S.K. Baines and M.C. Adams. 2010. Importance of food in probiotic efficacy. Food Res. Int. 43: 1–7.

Salminen, S., M. Playne and Y.K. Lee. 2004. Successful Probiotic Lactobacilli: Human Studies on Probiotic Efficacy. *In*: C. Shortt and J. O'Brien [eds.]. Handbook of Functional Dairy Products. CRC Press, Taylor & Francis Group, LLC, Boca Raton, London, New York, Washington, D.C., pp. 13–32.

Salminen, S. and A.C. Ouwehand. 2004. Future Aspects of Research and Product Development of Lactic Acid Bacteria. *In*: S. Salminen, A. von Wright and A. Ouwehand [eds.]. Lactic Acid Bacteria Microbiological and Functional Aspects. Third Edition. Revised and Expanded. Marcel Dekker, Inc. New York, USA, pp. 439–508.

Salminen, S. and A. von Wright. 2011. Probiotics: Safety and Efficacy. *In*: S. Lahtinen, A.C. Ouwehand, S. Salminen and A. von Wright [eds.]. Lactic Acid Bacteria Microbiological and Functional Aspects. Fourth Edition. CRC Press, Taylor & Francis Group, LLC, Boca Raton, USA, pp. 689–703.

Sanchez, B., L. Ruiz, M. Gueimonde, P. Ruas-Madiedo and A. Margolles. 2012. Toward improving technological and functional properties of probiotics in foods. Trends in Food Sci. & Tech. 26: 56–63.

Schmid, K., R.C. Schlothauer, U. Friedrich, C. Staudt, J. Apajalahti and B.E. Hansen. 2006. Development of Probiotic Food Ingredients. *In:* I. Goktepe, V.K. Juneja and M. Ahmedna [eds.]. Probiotics in Food Safety and Human Health. CRC Press, Taylor & Francis Group, LLC, Boca Raton, USA, pp. 35–66.

Seth, S.D. and M. Maulik. 2011. Probiotics a Pharmacologist's Perspective. *In*: G. Balakrish Nair and T. Yoshifumi [eds.]. Probiotic Foods in Health and Disease. CRC Press, Boca Raton, New York, Oxon, Science Publishers, Enfield, New Hampshire, USA, pp. 41–47.

Shah, N.P. 2010. Probiotics: Health Benefits, Efficacy and Safety. *In*: D. Bagchi, F.C. Lau and D.K. Ghosh [eds.]. Biotechnology in Functional Foods and Nutraceuticals. CRC Press, Taylor & Francis Group, LLC, Boca Raton, USA, pp. 485–496.

Silva, C.M., A.J. Ribeiro, I.V. Figueiredo, A.R. Gonçalves and F. Veiga. 2006. Alginate microspheres prepared by internal gelation: Development and effect on insulin stability. Int. J. Pharm. 311: 1–10.

Truelstrup Hansen, L.T., P.M. Allan-Wojtas, Y.L. Jin and A.T. Paulson. 2002. Survival of Ca–alginate microencapsulated *Bifidobacterium* spp. in milk and simulated gastrointestinal condition. Food Microbiol. 19: 35–45.

Ubbink, J. and J. Krüger. 2006. Physical approaches for the delivery of active ingredients in foods. Trends Food Sci. Technol. 17: 244–254.

Zuidam, J.N. and E. Shimoni. 2009. Overview of Microencapsulation for Use in Food Product or Processes and Methods to Make Them. *In*: N.J. Zuidam and V.A. Nedović [eds.]. Encapsulation Technologies for Active Food Ingredients and Food Processing. Springer: Dordrecht, The Netherlands, pp. 3–30.

3

Sources and Production of Prebiotics

Linglin Fu and *Yanbo Wang**

Introduction

The concept of prebiotics was introduced in 1995 by Gibson and Roberfoid as an alternative approach to the modulation of the gut microbiota (Gibson and Roberfroid 1995). A more recent definition of the term is "a selectively fermented ingredient that allows specific changes, both in the composition and/or activity in the gastrointestinal microbiota that confers benefits upon host wellbeing and health" (Gibson et al. 2004). An ingredient must fulfil three fundamental conditions in order to be considered as a prebiotic: (i) resistance to the digestion process, which involves gastric acids, intestinal brush border, pancreatic enzymes, etc.; (ii) fermentation by the large intestinal microbiota; (iii) a selective effect on the microbiota that has associated health-promoting effects (Charalampopoulos and Rastall 2012). Although all of these criteria are important, the third one is the most difficult to fulfill (Roberfroid 2007). Currently, a number of 'prebiotic candidates' that meet the first two criteria still need to be validated as selective stimulators of intestinal bacteria associated with health and wellbeing.

Food Quality & Safety Department, Zhejiang Gongshang University, Hangzhou 310035, China.
* Corresponding author

Most prebiotics are short-chain carbohydrates with a degree of polymerisation of 2 or more, which are not susceptible to digestion by pancreatic and brush border enzymes (Steed and Macfarlane 2009). The substances classified as prebiotics are fructans, lactulose, xylooligosaccharides (XOS) and mannanoligosaccharides (MOS). Fructans are composed of one or as many as 70 units of fructose linked or not linked to a terminal sucrose molecule, such as inulin, fructooligosaccharides (FOS) and galactooligosaccharides (GOS) (Carabin and Flamm 1999). Lactulose (4-0-β-D-galactopyranosyl-D-fructofuranose) are disaccharides composed of galactose and fructose, while XOS are sugar oligomers formed by units of xylose (Gibson et al. 2004). In terms of their production, most prebiotics are obtained by (i) extraction from plants, (ii) enzymatic hydrolysis of plant polysaccharides, and (iii) transgalactosylation reactions catalysed by an enzyme, using either a mono-saccharide or a di-saccharide as the substrate. For example, inulin, XOS and GOS are produced by extraction from chicory, enzymatic hydrolysis of xylans from cereal grains and lactose using β-galactosidase as the biocatalyst, respectively (Sangwan et al. 2011, Torres et al. 2010, Voragen 1998). In addition, MOS are obtained from the cell walls of yeast (*Saccharomyces cerevisiae*) (Bychkov et al. 2010).

Prebiotics can be formulated either as a powder or syrup and marketed as supplements or incorporated into food products, most commonly yogurts and breads (Heller 2001). Most studies have been focused on fructans since they are manufactured at relatively low cost, and are also valuable functional ingredients for the food industry with the potential to improve the sensory properties of food (Macfarlane et al. 2006). Lactulose are usually found in milk and milk products which undergo thermal treatments and are used in the baby food and pharmaceutical industries. XOS, found naturally in fruit, vegetables, milk and honey, can be used for various purposes, among which are applications in the food and pharmaceutical industries. Moreover, XOS are moderately sweet, stable over a wide range of pH and temperatures and have organoleptic characteristics suitable for incorporation into foods (Otieno and Ahring 2012). Due to the specific effects on human health, prebiotics have become a growing segment in the world market in recent years (Granato et al. 2010). This growth has been enhanced by technological innovations, development of new products, and the increasing number of health-conscious consumers interested in products that improve life quality.

The Rationale for use of Prebiotics

Prebiotics have been associated with a variety of health benefits including an increase in the bioavailability of minerals, modulation of the immune system, prevention of gastrointestinal infections, modification of inflammatory conditions, regulation of metabolic disorders and reduction of risk of cancer (Roberfroid et al. 2010). The mechanisms through which prebiotics affect the host largely attribute to i) promotion of beneficial microbiota and inhibition of the growth of potential pathogens/harmful microorganisms, and ii) strengthening of the barrier function of the epithelia and immune stimulation (Lomax and Calder 2009). Here we focus on the available knowledge of interactions among prebiotics, gut microbiota and mucosal immune system to exhibit the rationale for use of prebiotics.

Promotion of beneficial microbiota and host-microbe crosstalk

The gut microbiota is now perceived as a key player in health and well-being with a composition in which potentially health-promoting dominant microorganisms (especially the saccharolytic genera/species, e.g., bifidobacteria) are elevated and/or more active than the potentially harmful ones (especially the proteolytic/putrefactive genera/species) (Roberfroid et al. 2010). The composition and activity of the intestinal microbiota can influence health and disease through its involvement in nutrition, host physiological functions, and pathogenesis of certain disease conditions (Ringel and Carroll 2009, Roberfroid 2008). A large number of human intervention studies that have been performed show that prebiotics can result in statistically significant changes in the composition of the gut microbiota, especially an increase of faecal concentrations of beneficial microbiota (e.g., bifidobacteria), in line with the prebiotic concept (Andersen et al. 2011, Cervera-Tison et al. 2012, Maccaferri et al. 2012, Shimizu et al. 2012, Toward et al. 2012). Intestinal bacteria may also contribute to the colonization resistance to bacterial pathogens (Fukuda et al. 2011, Gibson et al. 2005). If prebiotics are used to increase bifidobacteria or lactobacilli towards being the numerically predominant genus in the colon, an improved colonisation resistance will definitely result.

Moreover, a host-microbe cross-talk also exists in the gut and has developed by the coevolvement between human beings and this abundant gut microbiota (Zaneveld et al. 2008). Key in this cross-talk is that the host continuously detects microbial signals through strategically localized host receptors (Medzhitov and Janeway 2002). As a result of the continuous detection of microbes, host defence molecules are continuously secreted and trapped in the overlaying mucus layer, which allows the host to particularly control the composition and abundance of the mucosa-associated microbiota. Given that humans closely interact with their co-evolved luminal and mucosal intestinal microbiota, there is great interest in dietary interventions, e.g., prebiotic compounds that are able to modulate both the luminal and mucosal microbial composition and activity (Langlands et al. 2004, Van den Abbeele et al. 2011). In this case, prebiotics may beneficially regulate the host-microbe interactions.

Besides, recent data, both from experimental models and from human studies, support the beneficial effects of prebiotics on changes of gut microbiota composition (especially the number of bifidobacteria) which may contribute to modulate metabolic processes associated with obesity and diabetes type 2 (Cani et al. 2007, Cani and Delzenne 2009, Delzenne et al. 2011, Roberfroid et al. 2010).

Influence on the mucosal barrier and immune stimulation

Prebiotics act like growth factors to particular commensal bacteria (e.g., *Lactobacillus* spp.), which can both improve gut barrier function through protection of the epithelial tight junctions during external stress (Montalto et al. 2004, Seth et al. 2008), and activate the immune responses in the gut-associated lymphoid tissue (GALT) (Vulevic et al. 2008).

The mucus layer normally consists of a double protective layer: a very dense, firmly attached and quite sterile inner mucus layer and a less dense, loosely attached, more strongly colonized outer mucus layer (Johansson et al. 2010). Prebiotics are typically shown to increase mucin-levels which improves intestinal barrier function (Stoidis et al. 2010, Zhong et al. 2009). Besides, modulation of the microbiota at the gastrointestinal tract by prebiotics also has a broad influence on the immune response of the host (Fig. 1).

The general observations are that inulin consumption increases the phagocytic capacity of macrophages and the production of secretory immunoglobulin A (IgA-s), which plays an important role in the defense

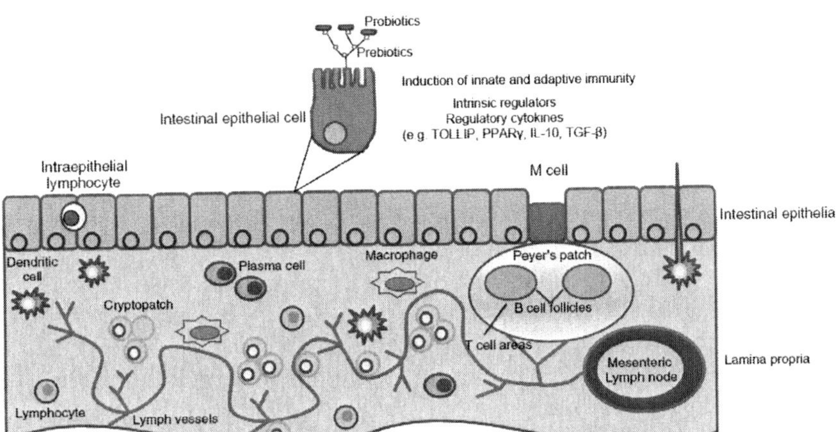

Fig. 1. Interactions of prebiotics and immune system in the intestinal mucosa, which display immunomodulatory functions (Modified from Choque Delgado et al. 2011).

Color image of this figure appears in the color plate section at the end of the book.

of the gastrointestinal tract (Van Loo 2004). A recent study showed that the addition of a new developed β-galactomannan (βGM) prebiotic inhibited Salmonella-induced proinflammatory mRNA (cytokines tumor necrosis factor alpha [TNF-α], interleukin-1α [IL-1α], IL-6, and granulocyte-macrophage colony-stimulating factor [GM-CSF] and chemokines CCL2, CCL20, and CXCL8) and at protein levels (IL-6 and CXCL8) in porcine ileum intestinal epithelial cells (IECs), but may promote dendritic cells (DCs) activation (Badia et al. 2012).

Sources and Production of Prebiotics

Sources and physicochemical properties of prebiotics

In order to be effective, prebiotics need to reach the large bowel with their chemical and structural properties essentially unchanged to further selectively stimulate the microbiota (Figueroa-González et al. 2011). Prebiotic oligosaccharides may be manufactured by extraction from plant materials, microbial/enzymatic synthesis and enzymatic hydrolysis of polysaccharides. Natural sources of prebiotics exist (e.g., galactooligosaccharides in breast milk, fructans in onion (*Allium cepa*), leeks (*Allium porrum*) and garlic (*Allium sativum*), soya-oligosaccharides in soyabean) but, because of their prebiotic effects, biotechnology (enzymic or thermal processes) has been applied to obtain new types by either enzymic synthesis from simple sugars or enzymic hydrolysis from more complex carbohydrates (Murphy 2001). Short-chain fructo-oligosaccharides, for example inulin, may be thus obtained by synthesis from saccharose, or through controlled and partial hydrolysis from chicory (*Cichorium intybus*) roots (Roberfroid and Slavin 2000). Nowadays, various prebiotics are produced at industrial scale and are widely available in the market (Grajek et al. 2005). A brief summary of sources, chemical structure, manufacture methods and physicochemical properties of main candidates for prebiotic status are provided in Table 1.

General production of prebiotics

Extraction from biological materials

Some prebiotics and candidate prebiotics are naturally present in plant materials. Fructans such as inulin can be readily extracted from sources such as chicory, the main industrial source, and agave. Soy oligosaccharides are extracted from soybeans (Crittenden and Playne 1996). Extraction from an easily grown crop such as chicory provides an economic advantage for inulins as prebiotic products (Roberfroid 2005).

Table 1. Sources, manufacture methods and properties of main candidates for prebiotic status.

Carbohydrate	Natural source	Chemical structure	Method of manufacture	Physicochemical property
Inulin	Fruits and vegetables (onions, banana, garlic, etc.)	β(2-1)-Fructans	Extraction from chicory root and *Agave tequilana*	Water solubility of 28% (w/v) at 80°C; Approx. 10% sweetness of sugar/sucrose; Degree of polymerization (DP) of <= 10; Temperature (T_p) of melting, 184.5°C with 5.3% (w/w) H_2O
Fructo-oligosaccharides	Fruits and vegetables (onions, banana, garlic, etc.)	β(2-1)-Fructans	Transfructosylation from sucrose or hydrolysis of chicory inulin	Highly hygroscopic; Viscosity and thermal stability, higher than that of sucrose; Highly stable in pH range of 4.0–7.0; Solubility, freezing and boiling points, similar to sucrose
Galacto-oligosaccharides	Human milk	Galactose oligomers and some glucose/lactose/galactose units	Produced from lactose by β-galactosidase	Water-soluble, about 80% (w/w); Viscosity, similar to high-fructose corn syrup; Stable to 37°C at pH 2 for several months; Sweetness, typically 0.3 to 0.6 times that of sucrose
Soya-oligosaccharides	Soyabean	Mixture of raffinose and stachyose	Extracted from soyabean whey	Sweetness of sucrose 70%; Stable below 15°C; Water activity close to sucrose; Lower viscosity than maltose; Heat value of 50% sucrose
Xylo-oligosaccharides	Bamboo shoots, fruits, vegetables, milk and honey	β(1–4)-Linked xylose	Enzymatic hydrolysis of xylan. Enzyme treatments of native lignocellulosic materials. Hydrolytic degradation of xylan by steam, water or dilute solutions of mineral acids	Sweetness, equivalent to 30% sucrose; Stable over a wide range of pH (2.5–8.0) and temperatures (up to 100°C); Water activity, similar to glucose; Antifreezing activity of on water at temperatures higher than −10°C
Isomalto-oligosaccharides	Starch (wheat, barley, corn, pulses oats, tapioca, rice, potato, etc.)	α(1–4)-Glucose and branched α(1–6)-glucose	Microbial or enzymatic transgalactosylation of maltose. Enzymatic synthesis from sucrose	Stable in pH 2–9 and normal baking temperatures; ~60% as sweet as sucrose; Water soluble and mildly sweet; High moisture-retaining capacity
Pyrodextrins	Starch (lentil, seeds of sorghum, sagu roots, etc.)	Mixture of glucose-containing oligosaccharides	Pyrolysis of potato or maize starch	Typically amorphous; Water-soluble; Low intrinsic viscosity

Alvani et al. 2011, Crittenden and Playne 1996, De Gennaro et al. 2000, Figueroa-González et al. 2011, Laurentín et al. 2003, Mussatto and Mancilha 2007, Torres et al. 2010, Vázquez et al. 2000, Yun 1996.

Production by polysaccharide hydrolysis

Fructo-oligosaccharides can be manufactured by the hydrolysis of inulin. Chicory inulin is partially hydrolyzed by endo-inulinase (EC3.2.1.7) to produce a mixture of fructo-oligosaccharides with an average degree of polymerization (DP) of 4 (Yun 1996).

The xylooligosaccharides production at an industrial scale is carried out from lignocellulosic materials (LCMs). Starting from a xylan-rich feedstock, some of the heterocyclic ether bonds of the xylan backbone have to be hydrolyzed to give compounds of lower polymerization degree (Vázquez et al. 2000). The key procedure for XO production is as follows (Fig. 2): i) Enzyme treatments of native, xylan-containing LCM, ii) Chemical fractionation of a suitable LCM to solubilize xylan, with further enzymatic hydrolysis of this polymer to XOS, and iii) Hydrolytic degradation of xylan to XOs by steam, water or dilute solutions of mineral acids.

The separation of XOS within a given DP range has been carried out by membrane techniques, in order to remove both XOS within the undesired DP range and non-saccharide compounds (Crittenden and Playne 1996). Adsorption has also been used for purification of XO-containing liquors (Pellerin et al. 1991). The economical autohydrolysis approach is also potentially developed for the manufacture of XOS from brewery spent grains (Carvalheiro et al. 2004, 2005). Recently, Moura et al. (2007) selectively produced XOS by means of autohydrolysis techniques using corn cobs as the raw substrate. Moreover, the obtained oligosaccharides promoted the growth of both *Bifidobacterium* and *Lactobacillus* species, and exhibited a potential bifidogenic capability similar to commercial XOS.

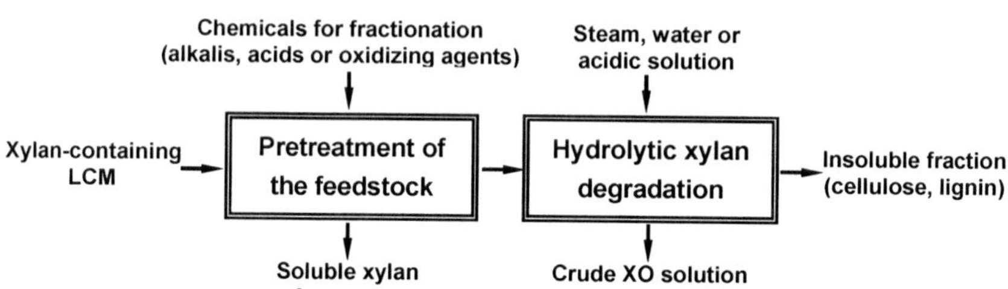

Fig. 2. Procedure for XO production by hydrolysis (Modified from Vázquez et al. 2000).

Production by enzymatic glycosyl transfer

The third general approach to production of prebiotics involves enzyme-catalyzed transfer reactions. Typically, a readily available substrate such as sucrose or lactose is used, and a suitable glycosyltransferase or glycosidase enzyme is used to produce novel oligosaccharides.

The preferred mode for galacto-oligosaccharides (GOS) synthesis is by enzymatic catalysis from lactose using glycosyltransferases (EC 2.4) or glycoside hydrolases (EC 3.2.1) (De Roode et al. 2003). Glycosyltransferases and glycoside hydrolases are enzymes that are responsible for the transfer of glycosyl moieties from a donor sugar to an acceptor (Ly and Withers 1999). However, GOS are industrially produced using the catalytic activity of glycoside hydrolases rather than glycosyltransferases due to the unavailability of the latter (Tzortzis and Vulevic 2009). The reaction is kinetically controlled, and so the optimum yield of products is dependent on lactose concentration and reaction time. Commercial GOS products are usually approximately 55% oligosaccharides with the balance being made up of lactose, glucose, and galactose (Crittenden and Playne 1996).

Chemical synthesis

Lactulose is unusual among prebiotics as it is the only one manufactured by chemical synthesis (Timmermans 2007). It is manufactured by a Lobry de Bruyn-Alberda van Ekenstein isomerization of lactose catalyzed by sodium hydroxide or borate, converting the glucosyl moiety into fructose.

Novel techniques and economical sources for prebiotic production

The current processes to obtain oligosaccharides with prebiotic status have very low yields affected by several factors such as the enzyme source, the concentration and nature of the substrate and the reaction conditions, thus increasing the production cost. Thus, the yield of GOS synthesis from lactose using glycoside hydrolases can be increased by: i) using highly concentrated starting lactose solution ii) decreasing water thermodynamic activity iii) removing the final product and/or inhibitors from the reaction medium and iv) modifying the enzyme (Torres et al. 2010). Panesar et al. (2006) noted that the yield of oligosaccharides can be increased by decreasing the water content in the reaction medium. Besides, the glucose/galactose ratio at maximum GOS yield using a crude enzyme fraction from *A. aculeatus* increased from 2.2 to 12.7 when an aqueous 2-phase system is used, reflecting an advantageous environment for galactosyl transfer reactions (Del-Val and Otero 2003). Therefore the enzymatic production of prebiotics in organic media represents an interesting research field to improve the yield of prebiotics over traditional synthesis in aqueous media. In addition, optimization of the enzyme structure also can contribute to increasing the maximum GOS yield from lactose. A protein engineering approach was applied to β-glucosidase from *G. stearothermophilus*. An increase in GOS yield was observed by changing arginine 109 residue to lysine, valine or tryptophan on the active site (Placier et al. 2009).

It is generally accepted that enzymatic application for oligosaccharide production from biomass is a very expensive process at commercial level. There are different options for reducing commercial production costs of using enzyme in hydrolysis of biomass into oligosaccharides (Otieno and Ahring 2012): i) use of crude enzyme mixtures ii) immobilization of enzymes that offers the advantage of reusable enzymes through recycling of enzyme for usability in subsequent batches iii) use of selective xylanases that produce higher yields of xylooligosaccharides and less xylose.

Furthermore, the production of resistant starch is another interesting option for low-cost prebiotic production at industrial scale. Resistant starch, which can be found naturally in cereal grains, is a substantial component of corn, wheat, rice and oat (Snow and O'Dea 1981). Industrial methods for manufacturing resistant starch include partial acid hydrolysis, hydrothermal treatment, heating, retrogradation, extrusion cooking, chemical modification and repolymerisation (Charalampopoulos et al. 2002). However, Topping et al. (2003) recognised that although there is a great deal of promise, further research needs to assess the potential of resistant starch as a prebiotic in humans.

Food Applications of Prebiotics

Prebiotics can be incorporated into many foodstuffs as agents to improve or maintain a balanced intestinal microflora to enhance health and well-being. In food formulations, they can also significantly improve organoleptic characteristics, upgrading both taste and mouthfeel. Food applications of prebiotics are illustrated in Table 2.

Table 2. Food applications of prebiotics (Wang 2009).

Applications	Functional properties
Yoghurts and desserts	Sugar replacement, texture and mouthfeel, fiber, and prebiotics
Beverages and drinks	Sugar replacement, mouthfeel, foam stabilization, and prebiotics
Breads and fillings	Fat or sugar replacement, texture, fiber, and prebiotics
Meat products	Fat replacement, texture, stability, and fiber
Dietetic products	Fat or sugar replacement, fiber, and prebiotics
Cake and biscuits	Sugar replacement, moisture retention, fiber, and prebiotics
Chocolate	Sugar replacement, heat resistance and fiber
Sugar confectionary	Sugar replacement, fiber, and prebiotics
Soups and sauces	Sugar replacement, and prebiotics
Baby food	Texture, body and mouthfeel, fiber, stability, and prebiotics

Served as functional food ingredients, firstly, prebiotics must be definitely safe. The safety of inulin-type fructans and galactose and their sources as food or food ingredients is not debated based on their long-term use. However, the novel emerging prebiotics still need the further consideration of their safety in humans. The ecological consideration has significance for the safety of prebiotics (Hammes and Hertel 2002). Generally, prebiotics are assumed safe. However, alterations in the intestinal microflora could result in adverse effects, depending on what bacterial populations are stimulated. Moreover, the shift of the metabolism towards enhanced butyrate formation may have some disadvantageous consequences, e.g., by supporting the growth of undesired clostridia (Wang 2009). A recent study was carried out to assess the tolerance and safety of a formula containing an innovative mixture of oligosaccharides in early infancy (Piemontese et al. 2011). Formula-fed infants were randomly fed with a regular formula containing a mixture of neutral oligosaccharides and pectin-derived acidic oligosaccharides (prebiotic formula group). The authors demonstrated the tolerability and the long term safety of such potential prebiotic formula in a large cohort of healthy infants.

Another important aspect for food application of prebiotics is that they must be chemically stable to food processing treatments, such as heat, low pH, and Maillard reaction conditions. Otherwise, the prebiotics are unavailable for bacterial metabolism, and would no longer provide selective stimulation of beneficial microorganisms (Wang 2009). Huebner et al. (2008) determined the effect of processing conditions on the prebiotic activity of commercial prebiotics using a prebiotic activity assay. The results showed that prebiotics were considered functionally stable at the food processing conditions of low pH (pH 3–6) and Maillard reaction (up to 6 h at 85°C with 1% glycine, pH 7), while only heating at low pH (30 min at 85°C, pH 4–7) caused a significant reduction in prebiotic activity.

Generally, the Asian continent has been the leading consumer of fermented functional foods with Japan being the leading market in the region (Nakakuki 2002). The worldwide market of prebiotics was estimated to reach US$ 155.41 billion after 2010, with a yearly growth potential of 10% (Otieno and Ahring 2012). The European and the U.S. market for prebiotics is projected to reach nearly $1.2 billion and $225 million, respectively, by the year 2015, according to a new report from Global Industry Analysts (GIA). Developing countries are also considered as an emerging prebiotics market where cultural factors, low levels of nutritional awareness, and income constraints have previously limited the penetration of such products. Specifically, China, Brazil and Mexico are the most important emerging markets and a growing consumer base with a strong and growing economy (Justfood 2006).

More recently, researchers have begun to unravel at least some of the functional and nutritional interactions between members of the colonic ecosystem (Dethlefsen et al. 2006, Flint et al. 2007, Falony et al. 2009a,b), and this is providing us with new outcomes to target in prebiotic interventions. This coincides with a new focus on the products of bacterial metabolism as important indicators of prebiotic action (Flint et al. 2007). Our understanding of the biological properties of metabolites is increasing, as is our ability to determine the systemic metabolic consequences of prebiotic action through metabonomic approaches (Waldram et al. 2009).

Acknowledgements

This work is supported by the National Natural Science Foundation of China (NNSFC, No. 31072221 and 31101921), the Project of the Scientific Research Foundation for the Returned Overseas Chinese Scholars, State Education Ministry (SRF for ROCS, SEM) and Agricultural Scientific Program of Hangzhou Science and Technology Bureau (No. 20120232B58).

References

Alvani, K., X. Qi and R.F. Tester. 2011. Use of carbohydrates, including dextrins, for oral delivery. Starch 63: 424–431.

Andersen, J.M., R. Barrangou, M. Abou Hachem, S. Lahtinen, Y.J. Goh, B. Svensson and T.R. Klaenhammer. 2011. Transcriptional and functional analysis of galactooligosaccharide uptake by lacS in *Lactobacillus acidophilus*. Proceedings of the National Academy of Sciences of the United States of America 108: 17785–17790.

Badia, R., M.T. Brufau, A.M. Guerrero-Zamora, R. Lizardo, L. Dobrescu, R. Martin-Venegas, R. Ferrer, H. Salmon, P. Martínez and J. Brufau. 2012. β-Galactomannan and *Saccharomyces cerevisiae* var. boulardii modulate the immune response against *Salmonella enterica* serovar Typhimurium in porcine intestinal epithelial and dendritic cells. Clinical and Vaccine Immunology 19: 368–376.

Bychkov, A.L., K.G. Korolev and O.I. Lomovsky. 2010. Obtaining mannanoligosaccharide preparations by means of the mechanoenzymatic hydrolysis of yeast biomass. Applied Biochemistry and Biotechnology 162: 2008–2014.

Cani, P.D. and N.M. Delzenne. 2009. Interplay between obesity and associated metabolic disorders: new insights into the gut microbiota. Current Opinion in Pharmacology 9: 737–743.

Cani, P.D., J.J. Holst, D.J. Drucker, N.M. Delzenne, B. Thorens, R. Burcelin and C. Knauf. 2007. GLUT2 and the incretin receptors are involved in glucose-induced incretin secretion. Molecular and Cellular Endocrinology 276: 18–23.

Carabin, I.G. and W.G. Flamm. 1999. Evaluation of safety of inulin and oligofructose as dietary fiber. Regulatory Toxicology and Pharmacology 30: 268–282.

Carvalheiro, F., M.P. Esteves, J.C. Parajó, H. Pereira and F.M. Girio. 2004. Production of oligosaccharides by autohydrolysis of brewery's spent grain. Bioresource Technology 91: 93–100.

Carvalheiro, F., G. Garrote, J.C. Parajó, H. Pereira and F.M. Girio. 2005. Kinetic modeling of brewery's spent grain autohydrolysis. Biotechnology Progress 21: 233–243.

Cervera-Tison, M., L. Tailford, C. Fuell, L. Bruel, G. Sulzenbacher, B. Henrissat, J.G. Berrin, M. Fons, T. Giardina and N. Juge. 2012. Functional analysis of family GH36 α-galactosidases from *Ruminococcus gnavus* E1: insights into plant oligosaccharides metabolism of a human gut symbiont. Applied and Environmental Microbiology (in press).

Charalampopoulos, D. and R.A. Rastall. 2012. Prebiotics in foods. Current Opinion in Biotechnology 23: 187–191.

Charalampopoulos, D., R. Wang, S.S. Padiella and C. Webb. 2002. Application of cereals and cereal components in functional food: a review. International Journal of Food Microbiology 79: 131–141.

Choque Delgado, G.T., W.M.S.C. Tamashiro, M.R.M. Junior, Y.M.F. Moreno and G.M. Pastore. 2011. The putative effects of prebiotics as immunomodulatory agents. Food Research International 44: 3167–3173.

Crittenden, R.G. and M.J. Playne. 1996. Production, properties and applications of food-grade oligosaccharides. Trends in Food Science and Technology 7: 353–361.

De Gennaro, S., G.G. Brich, S.A. Parke and B. Stancher. 2000. Studies on the physicochemical properties of inulin and inulin oligomers. Food Chemistry 68: 179–183.

De Roode, B.M., A.C.R. Franssen, A. van der Padt and R.M. Boom. 2003. Perspectives for the industrial enzymatic production of glycosides. Biotechnology Progress 19: 1391–402.

Del-Val, M.I. and C. Otero. 2003. Biphasic aqueous media containing polyethylene glycol for the enzymatic synthesis of oligosaccharides from lactose. Enzyme and Microbial Technology 33: 118–126.

Delzenne, N.M., A.M. Neyrinck, F. Bäckhed and P.D. Cani. 2011. Targeting gut microbiota in obesity: effects of prebiotics and probiotics. Nature Reviews Endocrinology 7: 639–646.

Dethlefsen, L., P.B. Eckburg, E.M. Bik and D.A. Relman. 2006. Assembly of the human intestinal microbiota. Trends in Ecology and Evolution 21: 517–523.

Falony, G., T. Calmeyn, F. Leroy and L. De Vuyst. 2009a. Coculture fermentations of Bifidobacterium species and Bacteroides thetaiotaomicron reveal a mechanistic insight into the prebiotic effect of inulin-type fructans. Applied and Environmental Microbiology 75: 2312–2319.

Falony, G., K. Lazidou, A. Verschaeren, S. Weckx, D. Maes and L. De Vuyst. 2009b. *In vitro* kinetic analysis of fermentation of prebiotic inulin-type fructans by Bifidobacterium species reveals four different phenotypes. Applied and Environmental Microbiology 75: 454–461.

Figueroa-González, I., G. Quijano, G. Ramírez and A. Cruz-Guerrero. 2011. Probiotics and prebiotics-perspectives and challenges. Journal of the Science of Food and Agriculture 91: 1341–1348.

Flint, H.J., S.H. Duncan, K.P. Scott and P. Louis. 2007. Interactions and competition within the microbial community of the human colon: links between diet and health. Environmental Microbiology 9: 1101–1111.

Fukuda, S., H. Toh, K. Hase, K. Oshima, Y. Nakanishi, K. Yoshimura, T. Tobe, J.M. Clarke, D.L. Topping, T. Suzuki et al. 2011. Bifidobacteria can protect from enteropathogenic infection through production of acetate. Nature 469: 543–547.

Gibson, G.R. and M.B. Roberfroid. 1995. Dietary modulation of the human colonic microbiota–introducing the concept of prebiotics. Journal of Nutrition 125: 1401–1412.

Gibson, G.R., A.L. McCartney and R.A. Rastall. 2005. Prebiotics and resistance to gastrointestinal infections. British Journal of Nutrition 93: S31–S34.

Gibson, G.R., H.M. Probert, J. Van Loo, R.A. Rastall and M.B. Roberfroid. 2004. Dietary modulation of the human colonic microbiota: updating the concept of prebiotics. Nutrition Research Reviews 17: 259–275.

Grajek, W., A. Olejnik and A. Sip. 2005. Probiotics, prebiotics and antioxidants as functional foods. Acta biochimica Polonica 52: 665–671.

Granato, D., G.F. Branco, A.G. Cruz, J.A.F. Faria and N.P. Shah. 2010. Probiotic dairy products as functional foods. Comprehensive Reviews in Food Science and Food Safety 9: 455–470.

Hammes, W.P. and C. Hertel. 2002. Research approaches for pre- and probiotics: Challenges and outlook. Food Research International 35: 165–170.
Heller, K.J. 2001. Probiotic bacteria in fermented foods: product characteristics and starter organisms. American Society for Clinical Nutrition 72: 374S–379S.
Huebner, J., R.L. Wehling, A. Parkhurst and R.W. Hutkins. 2008. Effect of processing conditions on the prebiotics activity of commercial prebiotics. International Dairy Journal 18: 287–293.
Johansson, M.E., J.K. Gustafsson, K.E. Sjoberg, J. Petersson, L. Holm, H. Sjovall and G.C. Hansson. 2010. Bacteria penetrate the inner mucus layer before inflammation in the dextran sulfate colitis model. PLoS One 5:e12238.
Justfood. 2006. Global market review of functional foods-forecasts to 2012. Aroq Publication, Worcs, UK.
Langlands, S.J., M.J. Hopkins, N. Coleman and J.H. Cummings. 2004. Prebiotic carbohydrates modify the mucosa associated microflora of the human large bowel. Gut 53: 1610–1616.
Laurentín A., M. Cárdenas, J. Ruales, E. Pérez and J. Tovar. 2003. Preparation of indigestible pyrodextrins from different starch sources. Journal of Agricultural and Food Chemistry 51: 5510–5515.
Lomax, A.R. and P.C. Calder. 2009. Prebiotics, immune function, infection and inflammation: a review of the evidence. British Journal of Nutrition 101: 633–658.
Ly, H.D. and S.G. Withers. 1999. Mutagenesis of glycosidases. Annual Review of Biochemistry 68: 487–522.
Maccaferri, S., A. Klinder, S. Cacciatore, R. Chitarrari, H. Honda, C. Luchinat, I. Bertini, P. Carnevali, G.R. Gibson, P. Brigidi and A. Costabile. 2012. *In vitro* fermentation of potential prebiotic flours from natural sources: Impact on the human colonic microbiota and metabolome. Molecular Nutrition and Food Research 56: 1342–1352.
Macfarlane, S., G. Macfarlane and J.H. Cummings. 2006. Review article: Prebiotics in the gastrointestinal tract. Alimenary Pharmacology Theraphy 24: 701–714.
Medzhitov, R. and C.A. Janeway Jr. 2002. Decoding the patterns of self and nonself by the innate immune system. Science 296: 298–300.
Montalto, M., N. Maggiano, R. Ricci, V. Curigliano, L. Santoro, F. Di Nicuolo, F.M. Vecchio, A. Gasbarrini and G. Gasbarrini. 2004. *Lactobacillus acidophilus* protects tight junctions from aspirin damage in HT-29 cells. Digestion 69: 225–228.
Moura, P., F. Carvalheiro, F. Gírio, M.C. Loureiro and M.P. Esteves. 2007. *In vitro* fermentation of xylo-oligosaccharides from corn cobs autohydrolysis by *Bifidobacterium* and *Lactobacilllus* strain. LWT-Food Science and Technology 40: 963–972.
Murphy, O. 2001. Non polyol low-digestible carbohydrates: food applications and functional benefits. British Journal of Nutrition 85: S47–S53.
Mussatto, S.T. and I.M. Mancilha. 2007. Non-digestible oligosaccharides: A review. Carbohydrate Polymers 68: 587–597.
Nakakuki, T. 2002. Present status and future of functional oligosaccharide development in Japan. Pure and Applied Chemistry 74: 1245–1251.
Otieno, D.O. and B.K. Ahring. 2012. The potential for oligosaccharide production from the hemicellulose fraction of biomasses through pretreatment processes: xylooligosaccharides (XOS), arabinooligosaccharides (AOS), and mannooligosaccharides (MOS). Carbohydrate Research 360: 84–92.
Panesar, P.S., R. Panesar, R.S. Singh, J.F. Kennedy and H. Kumar. 2006. Microbial production, immobilization and applications of β-D-galactosidase. Journal of Chemical Technology and Biotechnology 81: 530–543.
Pellerin, P., M. Gosselin, J.P. Lepoutre, E. Samain and P. Debeire. 1991. Enzymatic production of oligosaccharides from corncob xylan. Enzyme and Microbial Technology 13: 617–621.
Piemontese, P., M.L. Giannì, C.P. Braegger, G. Chirico, C. Grüber, J. Riedler, S. Arslanoglu, van M. Stuijvenberg, G. Boehm, J. Jelinek et al. 2011. Tolerance and safety evaluation

in a large cohort of healthy infants fed an innovative prebiotic formula: a randomized controlled trial. PLoS One 6:e28010.
Placier, G., H. Watzlawick, C. Rabiller and R. Mattes. 2009. Evolved beta-galactosidases from *Geobacillus stearothermophilus* with improved transgalactosylation yield for galacto-oligosaccharide production. Applied and Environmental Microbiology 75: 6312–6321.
Ringel, Y. and I.M. Carroll. 2009. Alterations in the intestinal microbiota and functional bowel symptoms. Gastrointestinal Endoscopy Clinics of North America 19: 141–150.
Roberfroid, M. and J. Slavin. 2000. Non digestible oligosaccharides. Critical Reviews in Food Sciences and Nutrition 46: 461–480.
Roberfroid, M. 2007. Prebiotics: the concept revisited. Journal of Nutrition 137: 830S–837S.
Roberfroid, M., G.R. Gibson, L. Hoyles, A.L. McCartney, R. Rastall, I. Rowland, D. Wolvers, B. Watzl, H. Szajewska, B. Stahl et al. 2010. Prebiotic effects: metabolic and health benefits. British Journal of Nutrition 104: S1–S63.
Roberfroid, M.B. 2005. Introducing inulin-type fructans. British Journal of Nutrition 93: S13–S25.
Roberfroid, M.B. 2008. General introduction: prebiotics in nutrition. *In*: G.R. Gibson and M.B. Roberfroid [eds.]. Handbook of prebiotics. CRC Press, Boca Raton, Florida, USA, pp. 1–11.
Sangwan, V., S.K. Tomar, R.R.B. Singh, A.K. Singh and B. Ali. 2011. Galactooligosaccharides: novel components of designer foods. Journal of Food Science 76: R103–R111.
Seth, A., F. Yan, D.B. Polk and R.K. Rao. 2008. Probiotics ameliorate the hydrogen peroxide-induced epithelial barrier disruption by a PKC- and MAP kinase-dependent mechanism. American Journal of Physiology-Gastrointestinal and Liver Physiology 294: G1060–G1069.
Shimizu, K., H. Ogura, T. Asahara, K. Nomoto, M. Morotomi, O. Tasaki, A. Matsushima, Y. Kuwagata, T. Shimazu and H. Sugimoto. 2012. Probiotic/Synbiotic therapy for treating critically ill patients from a gut microbiota perspective. Digestive diseases and sciences (in press).
Snow, P. and K. O'Dea. 1981. Factors affecting the rate of hydrolysis of starch in food. The American Journal of Clinical Nutrition 34: 2721–2727.
Steed, H. and S. Macfarlane. 2009. Mechanisms of prebiotic impact on health. *In*: D. Charalampopoulos and R.A. Rastall [eds.]. Prebiotics and Probiotics Science and Technology. Springer, New York, USA, pp. 135–161.
Stoidis, C.N., E.P. Misiakos, P.Patapis, C.I. Fotiadis and B.G. Spyropoulos. 2010. Potential benefits of pro- and prebiotics on intestinal mucosal immunity and intestinal barrier in short bowel syndrome. Nutriton Research Reviews 21: 1–9.
Timmermans, E. 2007. Lactose: its manufacture and physico-chemical properties. *In*: H. van Bekkum, A.G. Voragen, H. Röper and F. Voragen [eds.]. Carbohydrates as organic raw materials III. Wiley, New York, USA, pp. 93–113.
Topping, D.L., M. Fukushima and A.R. Bird. 2003. Resistant starch as a prebiotic and synbiotic: state of the art. The Proceedings of the Nutrition Society 62: 171–176.
Torres, D.P.M., M.D.F. Goncalves, J.A. Teixeira and L.R. Rodrigues. 2010. Galacto-oligosaccharides: production, properties, applications, and significance as prebiotics. Comprehensive Reviews in Food Science and Food Safety 9: 438–454.
Toward, R., S. Montandon, G. Walton and G.R. Gibson. 2012. Effect of prebiotics on the human gut microbiota of elderly persons. Gut Microbes 3: 57–60.
Tzortzis, G. and J. Vulevic. 2009. Galacto-oligosaccharide prebiotics. *In*: D. Charalampopoulos, and Rastall, R.A. [eds.]. Prebiotics and probiotics science and technology. Springer, New York, USA, pp. 207–244.
Van den Abbeele, P., T. Van de Wiele, W. Verstraete and S. Possemiers. 2011. The host selects mucosal and luminal associations of coevolved gut microorganisms: a novel concept. FEMS Microbiology Reviews 35: 681–704.
Van Loo, J.A.E. 2004. Prebiotics promote good health: The basis, the potential, and the emerging evidence. Journal of Clinical Gastroenterology 38: 70–75.

Vázquez, M.J., J.L. Alonso, H. Dominguez and J.C. Parajó. 2000. Xylooligosaccharides: Manufacture and applications. Trends in Food Science and Technology 11: 387–393.
Voragen, A.G.J. 1998. Technological aspects of functional food-related carbohydrates. Trends in Food Science and Technology 9: 328–335.
Vulevic, J., A. Drakoularakou, P. Yaqoob, G. Tzortzis and G.R. Gibson. 2008. Modulation of the fecael microflora profile and immune function by a novel trans-galactooligosaccharide mixture (B-GOS) in healthy elderly volunteers. American Journal of Clinical Nutrition 88: 1438–1446.
Waldram, A., E. Holmes, Y. Wang, M. Rantalainen, I.D. Wilson, K.M. Tuohy, A.L. McCartney, G.R. Gibson and J.K. Nicholson. 2009. Top-down systems biology modeling of host metabotype-microbiome associations in obese rodents. Journal of Proteome Research 8: 2361–2375.
Wang, Y. 2009. Prebiotics: Present and future in food science and technology. Food Research International 42: 8–12.
Yun, J.W. 1996. Fructooligosaccharides-occurrence, preparation and application. Enzyme and Microbial Technology 19: 107–117.
Zaneveld, J., P.J. Turnbaugh, C. Lozupone, R.E. Ley, M. Hamady, J.I. Gordon and R. Knight. 2008. Host-bacterial coevolution and the search for new drug targets. Current Opinion in Chemical Biology 12: 109–114.
Zhong, Y., D. Cai, W. Cai, S. Geng, L. Chen and T. Han. 2009. Protective effect of galactooligosaccharide-supplemented enteral nutrition on intestinal barrier function in rats with severe acute pancreatitis. Clinical Nutrition 28: 575–580.

4

Analysis of Probiotics and Prebiotics

*Semih Otles** and *V. Hazal Ozyurt*

Probiotics

Probiotics are live microorganisms that confer health benefits on the host when they are administered in adequate numbers (Coman et al. 2012, da Cruz et al. 2007, FAO/WHO 2002, Saad et al. 2012). Probiotics are supplements or foods that contain viable microorganisms that cause alterations of the microflora of the host. Probiotic microorganisms are typically members of the genera *Lactobacillus*, *Bifidobacterium*, and *Streptococcus* (Farnworth et al. 2007). These bacteria are fermentive, obligatory, or facultative anaerobic organisms. They typically produce lactic acid. Their inherent biological properties enable them to predominate and surpass potential pathogenic microorganisms in the human digestive tract. These microbes produce small molecular metabolic byproducts that set forth beneficial regulatory influence on host biological functions, including short-chain fatty acids such as butyrate. The most studied probiotic bacteria to date include *Lactobacillus rhamnosus* GG (LGG), *Bifidobacterium lactis*, and *Streptococcus thermophilus*. These probiotic bacteria are biologically different from the Gramnegative, motile, non–lactic-acid–producing bacteria such as *Klebsiella*, *Pseudomonas*, *Serratia*, and *Proteus* species, which also may be prominent flora in the human digestive system. These potentially harmful bacteria may

Ege University, Faculty of Engineering, Department of Food Engineering, 35100, Bornova, Izmir, Turkey.
* Corresponding author

translocate across the intestinal epithelium and could result in disease in humans. Some yeasts and yeast byproducts have also been studied and have been frequently used as probiotic agents, such as the yeast *Saccharomyces boulardii* (Thomas and Greer 2010). A brief summary of key probiotics in the literature is provided in Table 1.

Table 1. Summary of key probiotics in the literature (Figueroa-Gonzalez et al. 2011).

Micro-organism	
Lactic acid bacteria	*Lactobacillus rhamnosus* GG *Lactobacillus casei* *Lactobacillus casei Shirota* *Lactobacillus acidophilus* *Lactobacillus johnsonii* *Lactobacillus plantarum*
Bifidobacteria	*Bifidobacterium breve* *Bifidobacterium bifidum* *Bifidobacterium infantis* *Bifidobacterium animalis*
Yeasts	*Saccharomyces cerevisiae Boulardii*

The use of probiotics have numerous health effects. These are well classified:

1. lower frequency and duration of diarrhea associated with antibiotics (Clostridium difficile), rotavirus infection, chemotherapy, and, to a lesser extent, travelers' diarrhea
2. stimulation of humoral and cellular immunity
3. decrease in adverse metabolites, e.g., amonium and procancerogenic enzymes in the colon
4. reduction of Helicobacter pylori infection
5. reduction of allergic symptoms
6. relief from costive and irritable bowel syndrome
7. beneficial effects on mineral metabolism, particularly bone density and stability
8. blocking of cancer
9. reduction of cholesterol and triacylglycerol plasma concentrations (Ejtahed et al. 2012, Schrezenmeir and de Vrese 2001).

Classical approaches to identify the intestinal microflora

Analysis of complex microbial communities such as those found in the GIT of all life forms have been limited because of cultural bias when selective culture medium is used for bacteria isolations (Lee and Salminen 2009).

Fast and reliable quality control of these products is crucial in order to obtain functional and safe probiotic products for human consumption (Temmerman et al. 2003).

Culture-dependent approaches

Classic techniques used to characterize the intestinal microflora (GIT) are independent of classical culture-bound techniques and classical culture techniques. Previous analyses of probiotic products have claimed that the identity and number of recovered microbial species do not always correlate with the information stated on the product labels. However, the cultivation-dependent approaches have proven limitations in terms of recovery rate and reproducibility. Besides, more extensive insight into the production process and the survival capacity of the introduced strains requires analysis of both viable and nonviable bacteria. Currently, analysis of most probiotics is still based on conventional culture-dependent methods involving the use of specific isolation media and identification of a limited number of isolates making this approach relatively insensitive, laborious, and time-consuming (Jany and Barbier 2008, Temmerman et al. 2003). In the culture-dependent approach, a rather high percentage of probiotic products suffered from incorrect labeling and yielded low bacterial counts, which may decrease their probiotic potential (Lee et al. 2007, Temmerman et al. 2003). In conclusion culture-dependent methods consist of isolating and culturing microorganisms prior to their identification according to either morphological, biochemical or genetic characteristics (Jany and Barbier 2008).

Culture-dependent approach's analysis found in Bergey's Manual of Determinative Bacteriology:

- Azoreductase activity
- β-glucuronidase
- Conjugated/unconjugated bile salt ratio and Hydrolisis of bile salt
- Proteolytic activity
- β-glucosidase activity
- Short chain fatty acids
- Urobilinogen

Azoreductase activity

Azo dye compounds stand for a large group of chemicals comprehensively used in commercial industries (Prival et al. 1988). Several human intestinal microbiota possess azoreductase activity which plays an important role in the toxicity and mutagenicity of these azo dye compounds (Macwana et al. 2010). Nevertheless, they are reduced by azoreductases from intestinal bacteria.

The first catabolic step in the reduction of azo dyes is accompanied by a decrease in the visible light absorbance of the dye and then decolorization of the dye, is the reduction of the azo bridge to produce aromatic amines. Aromatic amines are known human carcinogens. A number of azo dyes have been classified as carcinogenic. The ability of the intestinal microflora of human and other animal species to reduce the azo groups of xenobiotic compounds has been known for many years. However, the specific organisms of the intestinal microflora participating in azo dye reduction are poorly understood (Rafii et al. 1990).

Rafii et al. (1990) developed that azo dye reduction for the detection of anaerobic bacteria producing azoreductases. 10 strains of anaerobic bacteria were identified with azo dye as *Eubacterium hadrum* (2 strains), *Eubacterium* spp. (2 species), *Clostridium clostridiiforme*, a *Butyrivibrio* sp., a *Bacteroides* sp., *Clostridium paraputrificum*, *Clostridium nexile*, and a *Clostridium* sp. They, isolated from the feces of a healthy individual, represent several species of bacteria of the organisms capable of producing azoreductase.

Nakanishi et al. (2001) compared the azoreductase from *Staphylococcus aureus* and *E. coli*. While the azoreductase from *E. coli* is a 46 kDa homodimer, requires FMN as a cofactor, and uses NADH as an electron donor, the azoreductase from *Staphylococcus aureus*, which is a tetramer, utilizes NADPH for azo dye reduction.

β-glucuronidase activity

The carcinogenic effect of endogenous toxic and genotoxic compounds is probably influenced by the activity of the bacterial enzymes NAD(P)H dehydrogenase, nitroreductase, β- glucuronidase, β-glucosidase, and 7-α-dehydroxylase. Bifidobacteria and lactobacilli have lower activities of these xenobiotic-metabolizing enzymes than do bacteroides, clostridia, and enterobacteriaceae. For example, β-glucuronidase is most highly present in enterobacteria and clostridia. As a consequence of these enzymes, toxic compounds detoxified in the liver by conjugation are regenerated by the release of toxic aglycones. Moreover, products of hydrolysis of glucuronides may re-enter enterohepatic circulation and therefore delay the excretion of compounds (Wollowski et al. 2001).

LAB strains were shown to influence the activity of nitroreductase and β-glucuronidase. Furthermore, the change of a mixed diet to a lactovegetarian diet resulted in a decrease of β- glucuronidase (Johansson et al. 1990). Goldin et al. (1992), Benno and Mitsuoka (1992) and Bouhnik et al. (1996) used *Bifidobacterium* fermented milk; milk was fermented by procarcinogenic enzyme activity with *Lactobacillus acidophilus*, *Bifidobacterium bifidum*, *Streptococcus lactis*, *Streptococcus cremoris* for determining β-glucuronidase activity. Thus, β-glucuronidase increased.

Conjugated/unconjugated bile salt ratio and Hydrolysis of bile salt

The primary bile acids, cholic and chenodeoxycholic acid, are synthesized de novo in the liver from cholesterol (Arias et al. 1994). Bile salt hydrolase (BSH) activity has been detected in *Lactobacillus*, *Bifidobacterium*, *Enterococcus*, *Clostridium*, and *Bacteroides* spp. *Lactobacilli* and *bifidobacteria*, whereas *Bacteroides*, *Clostridium*, and *Enterococcus* spp. have also been in the gastrointestinal tract (Ahn et al. 2003, Elkins et al. 2001).

The ability of probiotic strains to hydrolyze bile salts has often been included among the criteria for probiotic strain selection. Microbial BSH activity has also been presented to be potentially harmful to the human host (Begley et al. 2006).

A process termed enterohepatic recirculation keeps bile acids effeciently under normal conditions. Conjugated and unconjugated bile acids are absorbed by passive diffusion along the entire gut and by active transport in the terminal ileum. Reabsorbed bile acids enter the portal bloodstream and are taken up by hepatocytes, reconjugated, and resecreted into bile. Approximately 5% of the total bile acid pool (0.3 to 0.6 g) per day eludes epithelial absorption and can be extensively modified by the indigenous intestinal bacteria. One important transformation is deconjugation, a reaction that must occur before further modifications are possible. BSH enzymes catalyze deconjugation and hydrolyze the amide bond and liberate the glycine/taurine moiety from the steroid core. The resulting acids are called unconjugated or deconjugated bile acids (Begley et al. 2006).

Proteolytic activity

Proteolysis is the most important biochemical process occurring in sour milk products during fermentation and storage. Extracellular proteinases are involved in the initial degradation of caseins, producing a large number of oligopeptides. As a result, further split by intracellular peptidases is important to realise the needs for essential and growth stimulating amino acids and peptides (Donkor et al. 2007). Lactobacilli and bifidobacteria have been shown to possess several proteolytic and peptidolytic enzymes, and therefore have the potential to influence proteolysis (Bergamini et al. 2009, Shihata and Shah 2000). Aminopeptidases are thought to be of ultimate importance for the development of flavour in fermented milk products, since they are capable of releasing single amino acid residues from oligopeptides formed by extracellular proteinase activity. However, milk does not contain sufficient free amino acids and peptides to allow growth of lactic acid bacteria (LAB). Thus, these LAB have a complex system of proteinases and peptidases enabling them to use milk casein as a source of amino acids and nitrogen. The first step in casein degradation is intervened by cell wall located proteases, which seperate casein to oligopeptides. In addition degradation to smaller peptides and amino acids passing through the cell membrane is verified by peptidases (Shihata and Shah 2000).

β-glucoside activity

β-glucosidase is an important enzyme that could be used in the bioconversion of the predominant soy isoflavone glucosides to their bioactive aglycone forms (Otieno et al. 2005). The β-glucoside forms are not absorbed and require hydrolysis for bioavailability and subsequent metabolism. Hydrolysis occurs along the entire length of the intestinal tract by the action of both the brush border membrane- and the bacterial β-glucosidases. The aglycones are released and further metabolism of daidzein and genistein occur. Intestinal biotransformations include dehydroxylation, reduction, C-ring cleavage, and demethylation. However, the effectiveness of the microbial biotransformation are not well known (Donkor and Shah 2008).

Marteau et al. (1990) studied 9 subjects who consumed *L. acidophilus* and *Bifidobacterium bifidum* for 3 wk, and found there was a decrease only in the fecal activity of nitroreductase, whereas β-glucosidase activity increased. An increase in β-glucosidase could potentially be regarded as an advantage of health by releasing flavonoids with antimutagenic, antioxidative, anticancerogenic, and immune stimulatory effects (Cai et al. 1998).

Short chain fatty acids

Dairy propionibacteria can produce short chain fatty acids (SCFAs), propionic and acetic acids. So, probiotics are used in cancer treatments (Jan et al. 2002). Propionibacterial metabolites might exert anti-inflammatory effects *in situ*. However, such effects require high populations of live and metabolically active propionibacteria in the colon (Sanders and Marco 2010).

Dietary carbohdrates, specificially resistant starches and dietary fiber, are substrates for fermentation that produce SCFAs, primarily acetate, propionate and butyrate, as end products. The rate and amount of SCFA production depends on the species and amounts of microflora present in the colon, the substrate source and gut transit time. SCFAs are readily absorbed. Butyrate is the major energy source for colonocytes. Propionate is largely taken up by the liver. Acetate enters the peripheral circulation to be metabolized by peripheral tissues (Wong et al. 2006).

Cousin et al. (2012) could develop a dairy product fermented by dairy propionibacteria. Viability and SCFA content in the colon indicated survival and metabolic activity of *P. freudenreichii.* Fermented milk allowed *P. freudenreichii* survival and activity *in vivo*. Piglets were fed daily with sterile milk, a lyophilizate or with fermented milk. The SCFA content measured in the colon differed between groups. Significant differences were observed concerning the propionibacterial metabolites: acetic and propionic acids. These two SCFAs were higher in the fermented milk group compared to the two other groups ($P < 0.05$). In contrast, no significant difference was observed between the sterile milk and the lyophilizate group (Cousin et al. 2012).

Urobilinogen test

Urobilinogen is a water soluble and transparent product, the by-product of bilirubin reduction performed by the interstinal bacteria. It is formed by the split of hemoglobin. Whereas half of urobilinogen circulates back to the liver, the other half is excreted through feces as urobilin. Whenever there is hepatic damage, excess of it gets excreted out through the kidneys. This cycle is known as the enterohepatic urobilinogen cycle. To detect the type of damage in the liver, urobilinogen tests are performed by measuring uribilinogen levels in the urine (Cardona et al. 2002).

Culture-independent approaches

Culture-independent methods based on the direct analysis of DNA (or RNA) without any culturing step. These methods are based on total DNA (or RNA) which are directly extracted from the substrate. Culture-independent methods of enumeration with minimal cultural confusion would be desirable, particularly if these methods also enabled identification of OTU (operational taxonomic unit) or phylotypes, since many studies are generally aimed at understanding the variety and prosperity of bacteria species colonized in various locations in the GIT (Lee and Salminen 2009). For this reason, culture-independent analysis has recently been promoted as an alternative and/or complementary approach for quality control measurements of probiotic products (Masco et al. 2005). Moreover, as this method is fast and potentially more exhaustive than the culture-dependent approach, it is well suited for analysing microbial communities over time. The method uses polymerase chain reaction (PCR) amplification of total

DNA (Jany and Barbier 2008). The molecular methods involve extraction of total bacterial DNA directly from the product, PCR amplification of the V3 region of the 16S ribosomal DNA (Hoefel et al. 2005, Lee et al. 2007, Masco et al. 2005, Temmerman et al. 2003), and to identify use either gel or capillary separation or hybridization to specific probes.

Techniques used for monitoring microbial communities include PCR-denaturing gradient gel electrophoresis (PCR-DGGE), PCR-temporal temperature gradient gel electrophoresis (PCR-TTGE), single-strand conformation polymorphism-PCR (SSCP-PCR), terminal restriction fragment length polymorphism (T-RFLP), denaturing high-performance liquid chromatography (DHPLC) and DNA microarrays.

PCR-codenaturing gradient gel electrophoresis (PCR-DGGE)

PCR-DGGE has demostrated to be a useful analytical method for the investigation of complex microbial populations without previous separation of the individual inhabitants (Fasoli et al. 2003, Possemiers et al. 2004). For PCR-DGGE, the denaturing conditions rely on the use of chemical denaturants (formamide and urea) including an acrylamide gel as a linear denaturing gradient. PCR-DGGE electrophoresis is implemented at constant temperature, typically between 55°C and 65°C (Jany and Barbier 2008). DGGE band patterns allowed direct identification of the amplicons at the species level (Fasoli et al. 2003, Hoefel et al. 2005, Temmerman et al. 2003, Villarreal et al. 2010). This whole culture-independent approach can be performed in less than 30 h (Temmerman et al. 2003, Villarreal et al. 2010).

Fontana et al. (2005) appreciated to control fermentation process and to investigate bacterial communities developed in two artisanal Argentinean fermented sausages with different PCR-DGGE protocols. An intense band corresponding to *Lactobacillus sakei* was observed to be present in both samples. *Staphylococcus saprophyticus* was only observed in Tucumán sausage while a band identified as *Brochothrix thermophacta* was detected in Córdoba sausage (Fontana et al. 2005).

DGGE analyses are employed for the separation of double-stranded DNA fragments that are identical in length, but differ in sequence. The technique profits (among other factors) the difference in stability of G-C pairing (3 hydrogen bonds per pairing) as opposed to A-T pairing (2 hydrogen bonds). In general, DNA fragments richer in GC will be more stable and remain double-stranded until reaching higher denaturant concentrations. DNA fragments of differing sequences can be separated in an acrylamide gel (Possemiers et al. 2004). PCR-DGGE are based on the separation of PCR amplicons of the same size but several sequences (Jany and Barbier 2008).

PCR-temporal temperature gradient gel electrophoresis (PCR-TTGE)

PCR-TTGE is similar to PCR-DGGE. Both are based on similar separation techniques. For PCR-TTGE, the denaturing gradient is obtained by varying the temperature over time without chemicals, thus generating more reproducible data (Jany and Barbier 2008). The temperature of a gel plate in TTGE increases gradually and uniformly with time, which makes it easier to modulate the temperature over time. This provides an increased sensitivity as the separation range expands. TTGE have been employed primarily to screen for mutations in a variety of genes or to determine the genetic diversity of complex microbial populations Moreover, TGGE was applied as a tool in bacterial taxonomy (Vasquez et al. 2001).

Ubeda et al. (2009) determined in their study that *Saccharomyces* species and strains could be distinguished using different TTGE melting points. Some degree of discrimination was achieved under different conditions (Ubeda et al. 2009). Halos et al. (2006) evaluated a method allowing the one-step detection of bacterial pathogen DNA in ticks in their study. Firstly, DNA extracts from bacteria known to be tick-borne pathogens were used to establish a TTGE pathogen DNA reference marker. Secondly, they used broad-range PCR-TTGE to detect the presence of DNA from these three pathogens in 55 DNA extracts from pools of 10 nymphal Ixodes ricinus ticks, which had been previously shown to carry DNA from at least one of those bacteria by specific PCR. Thus, broad-range PCR-TTGE allowed the single step detection of DNA (Halos et al. 2006).

Single-strand conformation polymorphism-PCR (SSCP-PCR)

SSCP is an effective method for identifying sequence variation in amplified DNA (Bonifácio et al. 2001). SSCP is also an extremely useful method for both identifying and characterizing genetic polymorphisms and mutations (Han and Robinson 2003).

SSCP-PCR is a technique using either acrylamide gel- or capillary-based automated sequencer and this technique is based on the separation of denatured PCR products. Under non- denaturing conditions, single-stranded DNA folds into tertiary structures according to their nucleotide sequences and their physicochemical environment. This causes differences in electrophoretic mobility in non-denaturing gels. SSCP-PCR is potentially easier to put through than PCR-DG/TTGE since there is no need for gradient gels and so it can be performed using an automated sequencer. However, when using an automated sequencer, one of the disadvantages of this technique lies in the difficulty of appending new data to an existing database: Probiotics cannot be directly sequenced because they are labelled

(Hayashi 1991, Jany and Barbier 2008). The SSCP analysis is sensitive and efficient for discriminating different clones (Xie et al. 2002).

Although SSCP-PCR is an effective method and sensitive, it necessitates the use of radioisotopes. To avoid radioisotopes, silver staining was introduced for band detection (Iwahana and Itakura 1998).

Terminal restriction fragment length polymorphism (T-RFLP)

T-RFLP analysis is a method of comparative community analysis. Besides, it is a quantitative molecular technique and this technique is highly sensitive (Marsh 1999). The T-RFLP method is a recently described fingerprinting technique (Lukow et al. 2000, Smalla et al. 2007).

T-RFLP analysis is based on the restriction endonuclease digestion of fluorescently end-labelled PCR products. The digested products are separated by gel electrophoresis using either acrylamide gel- or capillary-based automated sequencer, with laser detection of the labelled fragments. The method provides distinct profiles (fingerprints) dependent on the species composition of the communities of the samples (Jany and Barbier 2008, Lee and Salminen 2009). This method's principles:

1. Near complete 16S rRNA genes in a sample are amplified using a fluorescently-labelled primer to yield a mixture of labelled 16S rRNA genes.
2. These amplification products are digested with restriction enzymes to produce labelled terminal restriction enzyme fragments (TRFs).
3. These T-RFs are then denatured, and the single stranded DNA thus obtained is separated by electrophoresis under denaturing conditions (e.g., at high temperature) (Lee and Salminen 2009).

T-RFLP is a rapid method and can be automated to process multiple samples in a short time-span. Nevertheless, the variation in 16S rRNA gene copy number in different microbes makes this technique only "semi-quantitative" and reliable lower limit of detection of PCR products in a mixture is low—about 1% (Lee and Salminen 2009). T-RFLP analysis is a highly reproducible and robust technique (Osborn et al. 2000).

Denaturing high-performance liquid chromatography (DHPLC)

DHPLC is demonstrated to be the superior technique for single nucleotide polymorphism (SNP) detection in terms of sensitivity, efficiency, and economy. This method is successfully used to detect mutations involved in a number of diverse diseases, such as breast and ovarian cancers, multiple sclerosis, Marfan syndrome, schizophrenia and hereditary multiple exostoses (Wolford et al. 2000).

DHPLC has been described recently as a method for screening DNA samples for single nucleotide polymorphisms and inherited mutations (Liu et al. 1998). DHPLC allows separation of amplicons which use an ion-pair reversed-phase high-performance liquid chromatography (IP RP HPLC) automated detection system. It was used to detect SNPs in the beginning in clinical applications. DHPLC is a promising approach for microbial community analysis (Wolford et al. 2000).

PCR amplicons are injected into a chromatography column containing alkylated non-porous polystyrene/polydivinylbenzene particles. Separation of the different amplicons relies on the elution of partially denatured PCR products.

DHPLC permits high-throughput automated analyses and, unlike SSCP-PCR or T-RFLP, it allows the collection of elution fractions corresponding to different amplicons that can be directly sequenced even more easily than with PCR-DG/TTGE methods. As to SSCP-PCR verified using an automated sequencer, samples that present unknown profiles cannot be directly sequenced because they are cut and labelled. T-RFLP has been used to study diverse microbial communities and has been extensively used by mycologists since this method is reportedly more sensitive than PCR-DG/TGGE for fungi (Jany and Barbier 2008, Lee and Salminen 2009).

Büchl et al. (2010) established differentiation of probiotic and environmental Saccharomyces cerevisiae strains in animal feed with DHPLC and the result of this study demonstrated that probiotic S. cerevisiae strains in feed could be differentiated successfully from environmental isolates using DHPLC.

DNA (cDNA) microarray

DNA microarrays enable researchers to monitor the expression of thousands of genes simultaneously (Draghici et al. 2006). The complementary DNA (cDNA) microarray (or microchips) technology has significantly changed the way gene expression can be assessed. By using DNA microarrays, the identification of labelled PCR products or directly reacquired RNA relies on their hybridization to oligonucleotide probes attached to a substrate. In contrast to the previous methods, DNA microarray technology potentially allows the contemporaneous application of almost unlimited number of probes in a single hybridization experiment. However, for this approach to work, each probe must specifically hybridize, under given limited conditions. Moreover, the design and sensibility of efficient probes depend on the extensiveness and quality of probe target database. The low quality of some annotated sequences in the available databases complicates probe design (Jany and Barbier 2008, Lee and Salminen 2009).

Compared with culture-dependent analysis, the culture-independent analysis was found to have a much higher sensitivity for detection of microbial strains in probiotic products in a fast, reliable, and reproducible manner (Temmerman et al. 2003). Culture-dependent analysis involved the evaluation and use of Bifidobacterium-selective media, followed by repetitive DNA element (rep)-PCR fingerprinting and Pulsed-Field Gel Electrophoresis (PFGE) of a selection of isolates. In parallel, all products were also subjected to a culture-independent analysis based on DGGE analysis of 16S rDNA nested-PCR products (Masco et al. 2005). The observation that culture-dependent and independent approaches target different organisms has implications for the use of the latter for studies in which taxonomic identification has a predictive value.

Prebiotics

Prebiotics are cost-effective and efficient tools to promote the growth and/or activity of certain bacteria in the indigenous flora of human gastrointestinal tract to beneficially affect host health and well-being (Makras et al. 2005, Oliveira et al. 2009, Rosenberg and Gophna 2011). Prebiotics are specialized plant fibers which beneficially nourish the good bacteria already in the large bowel or colon. Prebiotics are often non-digestible oligosaccharides such as fructo-oligosaccharides, inulin and oligofructose, together with emerging prebiotics such as xylo-oligosaccharides and arabinoxylan oligosaccharides (Buttriss and Stokes 2008, Carvalho-Wells et al. 2010, Kochar et al. 2007, Mandalari et al. 2007, Wichienchot et al. 2011), beneficially affecting the host by selectively stimulating the growth and/or activity of one, or a limited number of, bacteria in the colon (Commane et al. 2005, Palframan et al. 2003).

Prebiotics are different from probiotics. Prebiotics are a special form of dietary fiber, are not affected by heat, cold, acid or time, provide a wide range of health benefits and nourish the good bacteria which everyone already has in their gut. Probiotics are live bacteria in yogurt, dairy products and pills, and the bacteria must be kept alive. They may be killed by heat, stomach acid or simply die with time, must compete with the over 1000 bacteria species already in the gut and have been shown to be helpful for irritable bowel disease and for recurrence of certain bowel infections such as C. difficile (Oliveira et al. 2009, Schrezenmeir and de Vrese 2001, Toward et al. 2012).

There are two main groups of analytical techniques used for the analysis of prebiotics: separation and spectroscopic techniques. Separation techniques (chromatographic and electrophoretic) cause the disintegration of the constituents of a sample permitting the achievement of quantitative information. Spectroscopic techniques are also frequently necessary to provide detailed structural data of an isolated compound or a simple mixture. Combination of several techniques is often necessary to achieve all the required information about composition of complex mixtures. Although colorimetric methods such as determination of total carbohydrate or reducing sugar contents are still in use for oligosaccharide characterization, the separation techniques such as planar chromatography, gas chromatography (GC), high performance liquid chromatography (HPLC) and capillary electrophoresis (CE), which provide qualitative and quantitative information of independent oligosaccharides, are the most widely used and therefore the main aim of this section. These techniques can be coupled to spectroscopic instruments in order to obtain structural information. Moreover, nuclear magnetic resonance (NMR) and mass spectrometry (MS) are directly used for prebiotic structural analysis (Mandalari et al. 2007).

The prebiotic effect of a pectic oligosaccharide rich extract enzymatically derived from bergamot peel was studied using pure and mixed cultures of human faecal bacteria. This was compared to the prebiotic effect of fructo-oligosaccharides (FOS). Addition of the bergamot oligosaccharides (BOS) resulted in a high increase in the number of bifidobacteria and lactobacilli, whereas the clostridial population decreased. A prebiotic index (PI) was calculated for both FOS and BOS after 10 and 24 h incubation. Generally, higher PI scores were obtained after 10 h incubation, with BOS showing a greater value (6.90) than FOS (6.12) (Mandalari et al. 2007).

Planar Chromatography

Planar Chromatography was one of the earliest chromatographic techniques used for carbohydrate analysis, but at present it is hardly utilised and mainly combined with other techniques. This method includes both paper (PC) and thin-layer chromatography (TLC). Planar chromatography includes modern techniques derived from TLC such as HPTLC (High Performance TLC), OPTLC (Over Pressured TLC) and UTLC (Ultra TLC). A combination of PC, HPLC and high performance anion exchange chromatography (HPAEC) is used for the isolation of two octasaccharides, two dodecasaccharides and a tridecasaccharide from samples (Charalampopoulos and Rastall 2009).

HPLC

HPLC is one of the most widespread techniques for oligosaccharide analysis. Qualitative and quantitative characterization of prebiotic carbohydrates have been developed using different operation modes and detectors. Most methods are based on the condensation of a carbonyl group in carbohydrates with primary amines to give a Schiff base which is then reduced to a N-substituted glycosil amine. The primary amine has to posses the desired chromophore or fluorophore substituent, usually an aromatic ring. Reductive amination has been carried out with 2-aminopyridine, different trisulphonates, esters of p-aminobenzoic acid, 2-aminoacridone. Acetylation reactions of oligosaccharides overcome problems of solubility in organic solvents, whereas perbenzoylated derivatives improve the chromatographic properties on reverse phase columns. The alkylated silica-based stationary phases of octadecyl-coated (C18) sorbents are the most commonly utilized. Moreover, columns can present different percentages of bonded alkyl chains which could show a wide effect on carbohydrate resolution. The reverse phase (RP)-HPLC is used as mode. Aminoalkyl-modified silica gel columns provide good resolution; however, their stability is low and can be easily degraded. Cyclodextrin-based columns for the separation of neutral prebiotic carbohydrates has been also proposed. Several stationary phases with highly polar sorbents such as cyano, hydroxyl, diol, derivatives of poly(succinimide), sulfoalkylbetaine, etc. have been also used for carbohydrate analysis. Moreover, the use of size exclusion for HPLC (HPSEC) is also commonly applied. Oligosaccharides are eluting in order of decreasing molecular size from a stationary phase constituted by cross-linked polysaccharide or polyacrylamide. Cation exchange resins are composed by cross-linked polystyrene and silica-based ion exchangers such as calcium or silver. Carbohydrates also elute in order of decreasing molecular size and the chromatographic mechanism is based on both the size exclusion and ligand-exchange. Anion-Exchange Chromatography is used to separate anionic analytes which are either anions in their common form (e.g., amino acids) or analytes that can be ionized at high pH values (e.g., carbohydrates at >pH 12). Therefore, HPAE uses hydroxide-based eluents at high pH to produce anions from analytes that would not be anionic at neutral pH. A range of different columns for carbohydrate separations from mono, di-, tri- to oligo- and polysaccharides is available. Conventional HPLC phases of between 3 and 10 μm diameter of particles are commonly used for oligosaccharide analysis; 3 μm silica columns. Not only is the separation of oligosaccharides a problem in HPLC because of their similar structures, but also to achieve a sensitive detection can be a difficult task (Charalampopoulos and Rastall 2009).

Refractive index (RI) detectors are the most common detectors used for carbohydrate analysis. However, their main disadvantage originates from their dependence on temperature and mobile phase composition changes. Furthermore, UV detectors at low wavelengths (below 210 nm) show similar sensitivity to RI detectors. However, they approve changes in temperature and gradient elution. Flourometric dedectors have been applied to monosaccharide analysis, while only few works have been reported about oligosaccharide analysis. Evaporative light scattering detectors are universal, more sensitive than RI and are compatible with elution gradients. These dedectors utilize a spray which atomizes the column effluent into small droplets. These droplets are evaporated and the solutes as fine particulate matter are suspended in the atomizing gas. These particles diffuse the light originated from a monochromatic or polychromatic source (Charalampopoulos and Rastall 2009).

Šimonová et al. (2010) determined with HPLC with RI detection the amount of released fructose in cabbage juices. Lactobacillus amylovorus CCM 4380, Lactobacillus amylophilus CCM 7001, Lactobacillus plantarum CCM 7039, Bifidobacterium longum CCM 4990, and a mixture of Lactobacillus plantarum CCM 7039 and Bifidobacterium longum CCM 4990 (ratio 1:1, v/v) were used for lactic acid fermentation of cabbage juices with the addition of 2% inulin preparation FRUTAFIT. It was concluded that the bacteria are unable to degrade inulin (Šimonová et al. 2010).

PADs are commonly coupled to HPAEC and HPAEC-PADs allow the detection of non-derivatised carbohydrates at very low picomole levels and enables complete, single step separation of neutral and charged oligo and polysaccharides differing by branch, linkage, and positional isomerism. Because anion exchange chromatography is not a technique commonly associated with the analysis of neutral carbohydrates. This detection provides a high selectivity; only compounds oxidizable at the selected voltages being detected (Charalampopoulos and Rastall 2009, Corradini et al. 2004, Corradini et al. 2012). HPAEC-PAD has been accepted as the most powerful method for direct determination of prebiotics. It provides both the content and the degree of polymerization profiles. However, the analytical anion exchange columns are of relatively high cost. A simple reaction using iodine has been used for the determination of sugars and provides definition of the amount of glucose and fructose in syrup samples (Cho et al. 1999, Saengkanuk et al. 2011). PAD detection may also be defined with the more generic name of "pulsed electrochemical detection" (PED) (Corradini et al. 2012).

Despite the advantages of HPAEC-PED, in this technique the peak identification is obviously difficult to be implemented and peak assignment is often based on a generally accepted assumption that the retention time of a homologous series of carbohydrates increases as the degree of polymerization increases. It means each peak eluted represents a chain with one more unit than the previous peak (Corradini et al. 2012).

The analysis of inulin in Jerusalem artichoke tubers was carried out by Saengkanuk et al. (2011). They extracted the inulin from the artichoke tuber samples using accelerated solvent extraction method, before subsequent hydrolysis in acid condition. The hydrolysates were then analyzed for fructose using spectrophotometry. The inulin content in the samples was found approximate 63–75.5% dry weight, and the degree of polymerization was in the range of 14–20. The inulin contents obtained from this method were not significantly different ($p = 0.05$) from those obtained from HPAEC-PAD. Therefore, Spectrophotometric method was used as an alternative to proved chromatographic analysis (Saengkanuk et al. 2011).

Fructans are first extracted from the product with boiling water. An aliquot of this extract is treated with amyloglucosidase. A part of the hydrolyzate is treated with inulinase; glucose, fructose and sucrose are assayed in the first and second hydrolyzates and in the initial sample by high performance anion-exchange chromatography with pulsed amperometric detection (HPAEC-PAD). The concentration of fructans is calculated by the difference in these determinations (Hoebregs 1997).

Borromei et al. (2009) developed HPAEC-PED methods which could analyze FOS and inulins with a good resolution and relatively short retention times to evaluate structural differences between fructooligosaccharide and inulins and the possible presence of inulooligosaccharides as well as of branching. From the chromatograms generated by HPAEC it was not possible to identify each observed component but it was reported a qualitative comparison of reported chromatographic profiles. In conclusion, in this study, the results compared fructans with different degrees of polymerization (Borromei et al. 2009). Corradini et al. (1994) showed that separation by HPAEC of twelve disaccharides was strongly affected, beside their acidity, by the accessibility of oxyanions to the functional groups of the anion exchanger column. This effect has been observed in particular for the glucobioses trehalose, isomaltose, gentiobiose, nigerose, and maltose (Corradini et al. 1994).

Borromei et al. (2010) presented a method. They used HPAEC-PAD to determine prebiotics in storage of fermented milk. This method permits determination of prebiotics in storage of fermented milk. Furthermore, it provided synergic effect between prebiotics and probiotics. The information obtainable by this method could also be useful for defining the right health claims related to the quantified amount of fructooligosaccharide (FOS) and inulooligosaccharide (IOS) in the final product and not to the amount added. The advantages of this method is the possibility of evaluating the amount of the single component of FOS and IOS (Borromei et al. 2010).

The use of MS detectors coupled to HPLC systems has considerably enriched the field of carbohydrate analysis. MS detectors have been commonly utilised with alkyl- and aminoalkyl-bonded phases (Charalampopoulos and Rastall 2009). Mass spectrometry (MS) analysis were conducted with the aim to establish the correct assignment of the degree of polymerization to fructans and to foods which have limited information about carbohydrates composition are reported in literature (Borromei 2009).

GC

GC has seen widespread use for sugar determination as it is a relatively cheap, simple and powerful analytical technique. Soluble carbohydrates in foods are usually extracted with ethanolic or methanolic solutions. On the other hand, this procedure is bring about to discard insoluble material, lipids and proteins, desalt the sample or remove impurities. So, before their chromatographic analysis, an enrichment of the samples with carbohydrates is provided and they are purified. Due to the polar nature of carbohydrates, a derivatization step previous to GC analysis is required. Acetates, methyl ethers, trifluoroacetates and trimethylsilyl ethers have been the most important derivatives used for carbohydrate determination. The elucidation of structural chemistry of complex carbohydrates requires sophisticated instrumentation such as mass spectrometry (MS) or nuclear magnetic resonance (NMR). The coupling of a MS detector to a gas chromatograph contributes to the identification and quantification of carbohydrates (Charalampopoulos and Rastall 2009).

GC-MS has been applied for the determination of composition and sequence of oligosaccharides after complete hydrolysis and derivatization. It consists of the following steps: Initially, the free hydroxyl groups of polymerized sugars are methylated, forming their correspondent methyl ethers. Then, hydrolysis of the polymer is performed releasing the free hydroxyl groups in places where previously were glycosidic linkages. After all, these hydroxyl groups are converted into more volatile compounds, the most common derivatives being alditol or, aldonononitrile acetates. These samples are analyzed so as to determine the original linkages and to obtain quantitative linkage information on complex polysaccharides by GC-MS. The most common liquid stationary phases used for carbohydrate analysis by GC are those based on polysiloxanes (called "silicones"), because they are stabile and permeable towards solute (Charalampopoulos and Rastall 2009).

Programed temperature is convenient during the chromatographic run for carbohydrates which are complex, so that each compound can be analyzed in a good light. The temperature commonly used for carbohydrate analysis range from 60 to 330°C. Flame ionization detection (FID) is the most frequently used for GC analysis of carbohydrates. However, the identification by GC always requires the use of standard compounds (Charalampopoulos and Rastall 2009).

Lopez-Molina et al. (2005) carried out physico-chemical analysis of the properties of artichoke inulin (*Cynara scolymus* L.) with GC–MS. The main constituent monosaccharide in artichoke inulin was determined as fructose. A comparison of gas chromatograms for different inulins-inulin, chicory, dahlia, and Jerusalem artichoke inulins—is found (Lopez-Molina et al. 2005).

Makras et al. (2005) investigated the ability of lactobacilli to ferment inulin-type fructans. Then, they analyzed fructans. GC-MS is used for analysis of fructans and finally fructans are determined as quantitative. The first gas chromatograph was used to analyze oligofructose and prehydrolyzates of oligofructose-enriched inulin. The second gas chromatograph was used to analyze the hydrolysates of long-chain inulin and oligofructose-enriched inulin. For the analysis of the samples containing oligofructose or oligofructose-enriched inulin, a derivatization procedure involving oxymation and silylation of the sugars was carried out. For the analysis of the samples containing long-chain inulin or oligofructose enriched inulin, a procedure involving the preparation of prehydrolyzates and enzymatic hydrolysis with inulinase was performed. The derivatization of the samples before and after hydrolysis was carried out as described above. Finally, the fructan analyses proved a rapid degradation and metabolism of oligofructose and long-chain inulin (Makras et al. 2005).

Capillary Electrophoresis (CE)

In order to characterize prebiotics, an innovative method to analyze SCFAs in faecal cultures, based on capillary electrophoresis, was developed. Capillary electrophoresis (CE) with indirect UV detection is a valuable detection method for non-UV-absorbing low-molecular-mass ions and reversing the electroosmotic flow (EOF) is essential to achieve rapid CE separations of anionic analytes. The direction of the EOF can be reversed by a chemical modification of the capillary wall or by dynamic coatings adding suitable electrolyte additives (Corradini et al. 2004). CE is an attractive and powerful microanalytical technique to separate a wide range of charged and uncharged compounds. The advantages of CE include the extremely simple operation and the low consumption of sample. But this technique has the lack of sensitivity when low concentration levels are present. There are different operation modes of CE: capillary zone electrophoresis (CZE), capillary gel electrophoresis (CGE), micellar electrokinetic chromatography (MEKC), capillary isolectric focusing (CIEF) and capillary isotachophoresis (CITP). In all cases, the separation is achieved because of differences in migration of different solutes. The most commonly used modes in the analysis of carbohydrates are CZE and MEKC (Charalampopoulos and Rastall 2009). Also, the use of a polycationic electroosmotic modifier such as hexadimethrine bromide (HDB) is an alternative method (Corradini et al. 2004).

Shen et al. (2001) employed a CE method and UV detection (205 nm) to separate three sets of structural isomers of sialylated oligosaccharides in human milk and bovine colostrum. They developed conditions for baseline resolution of specific sets of isomers within a 35 min run (Shen et al. 2001).

Petzelbauer et al. (2006) separated and quantified the major GOS obtained during lactose conversion at 70°C, catalyzed by β-galactosidases from the archea *Sulfolobus solfataricus* and *Pyrocccus furiosus*. Carbohydrates were analyzed using as running buffer phosphate pH 2.5, derivatized using an aminopyridine solution and detected by UV (240 nm). Finally two disaccharides and two trisaccharides were identified (Petzelbauer, et al. 2006).

Electrospray ionization mass spectrometry is a desorption ionization method. Desorption ionization methods can be performed on solid or liquid samples, and allows for the sample to be nonvolatile or thermally unstable. The instrument has a small mass range that it is able to detect, therefore the mass of the unknown injected sample can easily be determined, since it must be in the range of the instrument. This quantitative analysis is done by considering the mass to charge ratios of the various peaks in the spectrum. The purity in a sample is important because this technique does not work well when mixtures are used as the analyte. This method has advantages: One advantage is that handle samples have large masses. Another is also the most available method in ionization methods. However, this method has disadvantages as well. A major disadvantage is that this technique cannot analyze mixtures very well. Moreover, the apparatus is very difficult to clean (Ho et al. 2003).

Matrix-assisted laser desorption/ionization

Matrix-assisted laser desorption/ionization (MALDI) is a soft ionization technique used in mass spectrometry allowing the analysis of biomolecules and large organic molecules, when ionized by more conventional ionization methods. It is similar in character to electrospray ionization. The MALDI is a two step process. First, desorption is induced by a UV laser beam, so matrix material heavily absorbs UV laser light, leading to the removal of upper layer of the matrix material. A hot plume produced during the remove contains many species: neutral and ionized matrix molecules, protonated and deprotonated matrix molecules, matrix clusters and nanodroplets. The second step is ionization. Protonation of analyte molecules occur in the hot plume. Sample size depends on molecular weight, the higher the molecular weight the more the sample that is needed. Samples are dissolved in a suitable solvent. The mechanism of MALDI is still debated (Zenobi and Knochenmuss 1998).

Matrix-assisted laser desorption/ionization mass spectrometry (MALDI-TOF MS) gives better results for some foods rather than HPAEC-PAD because fructans profiles are unknown and so MALDI-TOF MS identifications need less time to optimize analysis and assure correct molecular assignment. Moreover, MALDI-TOF MS is far less inclined to contaminant influence. The limit of MALDI-TOF MS is that similar mass branched and linear isomers may not be distinguished. Also, despite the fact that HPAEC-PED does not permit for structure elucidation, MALDI-TOF MS allows identification of unknown carbohydrates relative to standards (Borromei et al. 2009).

Borromei et al. (2009) analyzed with both HPAEC-PED and MALDI-TOF MS to verify the chain length distribution of the analyzed FOS and inulin. The MALDI-TOF mass spectra exhibited the sodium and potassium adducts and ascribed the degree of polymerization of these fructans.

Nuclear Magnetic Resonance (NMR) spectroscopy

Nuclear Magnetic Resonance (NMR) spectroscopy is the most powerful technique available for determining the structure of organic compounds. This technique endures the ability of atomic nuclei to behave as a small magnet as well as organize themselves with an external magnetic field. When irradiated with a radio frequency signal the nuclei in a molecule can change from being organized with the magnetic field to being opposed to it. Thus, it is called "nuclear" for the instrument works on stimulating the "nuclei" of the atoms to absorb radio waves. The energy frequency is indicated as an NMR spectrum. MS and NMR are different from one another: MS is malign, whereas NMR is not. However, a much smaller amount of material is needed for MS techniques. On the other hand, NMR and Mass Spectrometry (MS) are complementary techniques: whereas MS may tell the weight of a molecule, NMR can ensure that difference in structural isomers, and provide information about connectivities between atoms within a molecule (Chatham and Blackband 2001).

He et al. (2012) has investigated the reddish brown organic haze surrounding Titan using methods including remote observation, direct exploration and laboratory simulations. They reported here the structural investigation of the 13C and 15N labeled, simulated Titan haze aerosol (tholin) using solution-state NMR. These spectra proved a material composed of a mixture of moderate polymer and small molecules (He et al. 2012).

Maina et al. (2008) used NMR spectroscopy techniques to analyze the structures of dextrans produced by *Leuconostoc citreum* E497 and *Weissella confusa* E392. The dextrans were compared to that of *L. mesenteroides* B512F. Dextrans are the main exopolysaccharides (EPS) produced by Leuconostoc species. Generally, *W. confusa* E392 showed better growth and produced more EPS than did *L. citreum* E497 and *L. mesenteroides* B512F. Dextran from *W. confusa* E392 was found to be more linear than that of *L. mesenteroides* B512F. Dextran from *L. citreum* E497 may be useful as a source of prebiotic gluco-oligosaccharides, whereas *W. confusa* E392 could be a suitable alternative to widely used *L. mesenteroides* B512F in the production of linear dextran (Maina et al. 2008).

Milk oligosaccharides are complex: approximately 150–200 oligosaccharides have been determined in human milk. As an example of NMR application to oligosaccharides in milk, four neutral trisaccharides were characterized in colostrum: α-L-Fuc*p*-(l→2)-β-D-Gal*p*-(l→4)-Glc, α-D-Gal*p*-(1→3)-β-D-Gal*p*-(1→4)-Glc, β-D-Gal*p*-(l→3)- β -D-Gal*p*-(l→4)-Glc, and β -D-Gal*p*-(1→6)- β –D-Gal*p*-(1→4)-D-Glc (Urashima et al. 1994).

In spite of advances in analytical techniques in recent years there is still a lack of accurate and precise methods to characterize and quantify prebiotic oligosaccharides comprising complex mixtures with similar structural characteristics. Whereas GC has commonly been used to determine the composition of low molecular weight carbohydrates, high molecular weight carbohydrates oligosaccharides are characterised by HPLC, HPLC, CE and GC are used for the separation and isolation of the different constituents. To determine the structure NMR are used and, for studying their molecular weight MS are used.

References

Ahn, Y.T., G.B. Kim, Y.S. Lim, Y.J. Baek and Y.U. Kim. 2003. Deconjugation of bile salts by *Lactobacillus acidophilus* isolates Int. Dairy J. 13: 303–311.

Arias, I.M., N. Boyer, N. Fausto, W.B. Jackoby, D.A. Schachter and D.A. Shafritz. 1994. The liver: biology and pathobiology. Raven Press Ltd., New York, N.Y.

Begley, M., C. Hill and C.G.M. Gahan. 2006. Bile Salt Hydrolase Activity in Probiotics. Applied And Environmental Microbiology 72(3): 1729–1738.

Benno, Y. and T. Mitsuoka. 1992. Impact of *Bifidobacterium longum* on human fecal microflora. Microbiol. Immunol. 36: 683–694.

Bergamini, C.V., E.R. Hynes, S.B. Palma, N.G. Sabbag and C.A. Zalazar. 2009. Proteolytic activity of three probiotic strains in semi-hard cheese as single and mixed cultures: Lactobacillus acidophilus, Lactobacillus paracasei and Bifidobacterium lactis. International Dairy Journal 19: 467–475.

Bergey's Manual of Determinative Bacteriology:

Bonifácio, C., I.C. Santos, C. Belo and A. Cravador. 2001. Single-Strand Conformation Polymorphism (SSCP) Analysis of α-S_1-Casein, β-Casein and κ-Casein Genes in Charnequeira Portuguese Indigenous Goat Breed. Arch. Zootec. 50: 105–111.

Borromei, C. 2009. Characterization of Carbohydrates as Food Quality, Process And Tipicity Markers by HPAEC-PAD Optimized And Validated Methods, Unıversıta' Degli Studi dı Parma, Doctoral Thesis.

Borromei, C., M. Careri, A. Cavazza, C. Corradini, L. Elviri, A. Mangia and C. Merusi. 2009. Evaluation of Fructooligosaccharides and Inulins as Potentially Health Benefiting Food

Ingredients by HPAEC-PED and MALDI-TOF MS, International Journal of Analytical Chemistry, 9.

Borromei, C., A. Cavazza, C. Corradini, C. Vatteroni, A. Bazzini, R. Ferrari and P. Merusi. 2010. Validated HPAEC-PAD method for prebiotics determination in synbiotic fermented milks during shelf life. Anal. Bioanal. Chem. 397: 127–135.

Bouhnik, Y., B. Flourie, C. Andrieux, N. Bisetti, F. Briet and J.C. Rambaud. 1996. Effects of *Bifidobacterium* sp. fermented milk ingested with or without inulin on colonic bifidobacteria and enzymatic activities in healthy humans. Eur. J. Clin. Nutr. 50: 269–273.

Buttriss, J.L. and C.S. Stokes. 2008. Dietary fibre and health: an overview, British Nutrition Foundation Nutrition Bulletin 33: 186–200.

Büchl, N.R., M. Hutzler, H. Mietke-Hofmann, M. Wenning and S. Scherer. 2010. Differentiation of probiotic and environmental Saccharomyces cerevisiae strains in animal feed. Journal of Applied Microbiology 109: 783–791.

Cai, Q., R.O. Rahn and R. Zhang. 1998. Dietary flavonoids, quercetin, luteolin and genistein, reduce oxidative DNA damage and lipid peroxidation and quench free radicals. Cancer Lett. 119: 99–107.

Cardona, M.E., de V. Vanay, T. Midtvedt and E. Norin. 2002. Effect of probiotics on five biochemical microflora-associated characteristics, *in vitro* and *in vivo*. Scandinavian Journal of Nutrition 46: 73–79.

Carvalho-Wells, A.L., K. Helmolz, C. Nodet, C. Molzer, C. Leonard, B. McKevith, F. Thielecke, K.G. Jackson and K.M. Tuohy. 2010. Determination of the *in vivo* prebiotic potential of a maize-based whole grain breakfast cereal: a human feeding study. British Journal of Nutrition 104: 1353–1356.

Charalampopoulos, D. and R.A. Rastall. 2009. Prebiotics and Probiotics Science and Technology, Springer Science+Business Media, NY, USA.

Chatham, J.C. and S.J. Blackband. 2001. Nuclear Magnetic Resonance Spectroscopy and Imaging in Animal Research 42(3): 189–208.

Cho Sungsoo, S., L. Prosky and M. Dreher. 1999. Complex Carbohydrates in Foods. Marcel Dekker Inc., NY.

Coman, M.M., C.C. Cecchini, M.C. Verdenelli, S. Silvi, C. Orpianesi and A. Cresci. 2012. Functional foods as carriers for SYNBIO®, a probiotic bacteria combination, International Journal of Food Microbiology, In Press 1–21.

Commane, D.M., C.T. Shortt, S. Silvi, A. Cresci, R.M. Hughes and I.R. Rowland. 2005. Effects of Fermentation Products of Pro- and Prebiotics on Trans-Epithelial Electrical Resistance in an *in vitro* Model of the Colon. Nutrition and Cancer 51: 102–109.

Corradini, C., F. Bianchi, D. Matteuzzi, A. Amoretti, M. Rossi and S. Zanoni. 2004. High-performance anion-exchange chromatography coupled with pulsed amperometric detection and capillary zone electrophoresis with indirect ultra violet detection as powerful tools to evaluate prebiotic properties of fructooligosaccharides and inulin. Journal of Chromatography A 1054: 165–173.

Corradini, C., A. Cavazza and C. Bignardi. 2012. High-Performance Anion-Exchange Chromatography Coupled with Pulsed Electrochemical Detection as a Powerful Tool to Evaluate Carbohydrates of Food Interest: Principles and Applications. International Journal of Carbohydrate Chemistry 13.

Corradini, C., D. Corradini, C.G. Huber and G.K. Bonn. 1994. Synthesis of a polymeric-based stationary phase for carbohydrate separation by high-pH anion-exchange chromatography with pulsed amperometric detection. Journal of Chromatography A 685(2): 213–220.

Cousin, F.J., S. Louesdon, M.B. Maillard, S. Parayre, H. Falentin, S.M. Deutsch, G. Boudry and G. Jan. 2012. The first dairy product exclusively fermented by Propionibacterium freudenreichii: A new vector to study probiotic potentialities *in vivo*. Food Microbiology, In Press 1–12.

da Cruz, A.G., J.A.F. Faria and A.G.F. Van Dender. 2007. Packaging system and probiotic dairy foods. Food Research International 40: 951–956.

Donkor, O.N., A. Henriksson, T.K. Singh, T. Vasiljevica and N.P. Shah. 2007. ACE-inhibitory activity of probiotic yoghurt. International Dairy Journal 17: 1321–1331.

Donkor, O.N. and N.P. Shah. 2008. Production of β-Glucosidase and Hydrolysis of Isoflavone Phytoestrogens by *Lactobacillus acidophilus*, *Bifidobacterium lactis*, and *Lactobacillus casei* in Soymilk. Journal of Food Science 73: 15–20.

Draghici, S., P. Khatri, A.C. Eklund and Z. Szallasi. 2006. Reliability and reproducibility issues in DNA microarray measurements. Trends Genet 22: 101–109.

Ejtahed, H.S., J. Mohtadi-Nia, A. Homayouni-Rad, M.D. Mitra Niafar, M. Asghari-Jafarabadi and V. Mofid. 2012. Probiotic yogurt improves antioxidant status in type 2 diabetic patients. Nutrition 28: 539–543.

Elkins, C.A., S.A. Moser and D.C. Savage. 2001. Genes encoding bile salt hydrolases and conjugated bile salt transporters in *Lactobacillus johnsonii* 100-100 and other *Lactobacillus* species. Microbiology 147: 3403–3412.

FAO/WHO. 2002. Guidelines for the evaluation of probiotics in food. London, Ontario: Food and Agriculture Organization of the United Nations and World Health Organization Working Group Report 1–11.

Farnworth, E.R., I. Mainville, M.P. Desjardins, N. Gardner, I. Fliss and C. Champagne. 2007. Growth of probiotic bacteria and bifidobacteria in a soy yogurt formulation. International Journal of Food Microbiology 116: 174–181.

Fasoli, S., M. Marzotto, L. Rizzotti, F. Rossi, F. Dellaglio and S. Torriani. 2003. Bacterial composition of commercial probiotic products as evaluated by PCR-DGGE analysis. International Journal of Food Microbiology 82: 59–70.

Figueroa-Gonzalez, I., G. Quijano, G. Ramırez and A. Cruz-Guerrero. 2011. Probiotics and prebiotics: perspectives and challenges. J. Sci. Food Agric. 91: 1341–1348.

Fontana, C., G. Vignolo and P.S. Cocconcelli. 2005. PCR-DGGE analysis for the identification of microbial populations from Argentinean dry fermented sausages. J. Microbiol. Methods 63: 254–263.

Goldin, B.R., S.L. Gorbach, M. Saxelin, S.Barakat, L. Gualtiere and S. Salminen. 1992. Survival of *Lactobacillus* species (strain GG) in human gastrointestinal tract. Dig. Dis. Sci. 37: 121–128.

Halos, L., M. Mavris, G. Vourch, R. Maillard, J. Barnouin, H.J. Boulouisand M. Vayssier-Taussat. 2006. Broad-range PCR-TTGE for the first-line detection of bacterial pathogen DNA in ticks. Vet. Res. 37: 245–53.

Han, M. and M.A. Robinson. 2003. PCR-SSCP Analysis of Polymorphism: A Simple and Sensitive Method for Detecting Differences Between Short Segments of DNA. Methods in Molecular Biology 226: 327–333.

Hayashi, K. 1991. PCR-SSCP: a simple and sensitive method for detection of mutations in the genomic DNA. Genome Res. 1: 34–38.

He, C., G. Lin and M.A. Smith. 2012. NMR identification of hexamethylenetetramine and its precursor in Titan tholins: Implications for Titan prebiotic chemistry. Icarus 220: 627–634.

Ho, C.S., C.W.K. Lam, M.H.M. Chan, R.C.K. Cheung, L.K. Law, L.C.W. Lit, K.F. Ng, M.W.M. Suen and H.L. Tai. 2003. Electrospray Ionisation Mass Spectrometry: Principles and Clinical Applications. Clin. Biochem. Rev. 24: 3–12.

Hoebregs, H. 1997. Fructans in foods and food products, ion-exchange chromatographic method: collaborative study, J. Assoc. Off. Anal. Chem. Int. 80: 1029–1037.

Hoefel, D., P.T. Monis, W.L. Grooby, S. Andrews and C.P. Saint. 2005. Culture-Independent Techniques for Rapid Detection of Bacteria Associated with Loss of Chloramine Residual in a Drinking Water System. Applied and Environmental Microbiology 71(11): 6479–6488.

Iwahana, H. and M. Itakura. 1998. Multiple Fluorescence-Based PCR-SSCP Analysis with Primer-, Post-, and Internal-Labeling. Methods in Molecular Medicine 16: 51–60.

Jan, G., A.S. Belzacq, D. Haouzi, A. Rouault, D. Metivier, G. Kroemer and C. Brenner. 2002. Propionibacteria induce apoptosis of colorectal carcinoma cells via short-chain fatty acids acting on mitochondria. Cell Death Differ. 9: 179–188.
Jany, J.L. and G. Barbier. 2008. Culture-Independent Methods for Identifying Microbial Communities in Cheese. Food Microbiology 25(7): 839–848.
Johansson, G.K., L. Ottova and J.A. Gustafsson. 1990. Shift from a mixed diet to a lactovegetarian diet: influence on some cancer-associated intestinal bacterial enzyme activities. Nutr. Cancer 14: 239–246.
Kochar, J., L. Djousse and J.M. Gaziano. 2007. Breakfast cereals and risk of type 2 diabetes in the Physicians' Health Study I. Obesity 15: 3039–3044.
Lee, L., S. Tin and S.T. Kelley. 2007. Culture-independent analysis of bacterial diversity in a child-care facility. BMC Microbiology 7: 27.
Lee, Y., K. and S. Salminen. 2009. Handbook of Probiotics and Prebiotics, America, John Wiley & Sons, Inc.
Liu, W., D.I. Smith, K.J. Rechtzigel, S.N. Thibodeau and C.D. James. 1998. Denaturing high performance liquid chromatography (DHPLC) used in the detection of germline and somatic mutations. Nucleic Acids Research 26(6): 1396–1400.
Lopez-Molina, D., M.D. Navarro-Martınez, F.R. Melgarejo, A.N.P. Hiner, S. Chazarra and J.N. Rodrıguez-Lopez. 2005. Molecular properties and prebiotic effect of inulin obtained from artichoke (*Cynara scolymus* L.). Phytochemistry 66: 1476–1484.
Lukow, T., P.F. Dunfield and W. Liesack. 2000. Use of the T-RFLP technique to assess spatial and temporal changes in the bacterial community structure within an agricultural soil planted with transgenic and non-transgenic potato plants. FEMS Microbiology Ecology 32: 241–247.
Macwana, S.R., S. Punj, J. Cooper, E. Schwenk and G.H. John. 2010. Identification and Isolation of an Azoreductase from Enterococcus faecium. Curr. Issues Mol. Biol. 12: 43–48.
Maina, N.H., M. Tenkanen, H. Maaheimo, R. Juvonen and L. Virkki. 2008. NMR spectroscopic analysis of exopolysaccharides produced by Leuconostoc citreum and Weissella confusa. Carbohydrate Research 343: 1446–1455.
Makras, L., G.V. Acker and L. De Vuyst. 2005. *Lactobacillus paracasei* Subsp. Paracasei 8700:2 Degrades Inulin-Type Fructans Exhibiting Different Degrees of Polymerization. Applied and Environmental Microbiology 71(11): 6531–6537.
Mandalari, G., C.N. Palop, K. Tuohy, G.R. Gibson, R.N. Bennett, K.W. Waldron, G. Bisignano, A. Narbad and C.B. Faulds. 2007. *In vitro* evaluation of the prebiotic activity of a pectic oligosaccharide-rich extract enzymatically derived from bergamot peel Appl. Microbiol. Biotechnol. 73: 1173–1179.
Marsh, T. and L. 1999. Terminal restriction fragment length polymorphism (T-RFLP): An emerging method for characterizing diversity among homologous populations of amplification products, 2, 3, Pages 323–327.
Marteau, P., P. Pochart, B. Flourie, P. Pellier, L. Santos, J.F. Desjeux and J.C. Rambaud. 1990. Effect of chronic ingestion of a fermented dairy product containing *Lactobacillus acidophilus* and *Bifidobacterium bifidum* on metabolic activities of the colonic flora in humans. Am. J. Clin. Nutr. 52: 685–688.
Masco, L., G. Huys, E. De Brandt, R. Temmerman and J. Swings. 2005. Culture-dependent and culture-independent qualitative analysis of probiotic products claimed to contain bifidobacteria. International Journal of Food Microbiology 102: 221–230.
Nakanishi, M., C. Yatome, N. Ishida and Y. Kitade. 2001. Putative ACP phosphodiesterase gene (acpd) encodes an azoreductase. J. Biol. Chem. 276: 46394–46399.
Oliveira, R.P.S., A.C.R. Florence, R.C. Silva, P. Perego, A. Converti, L.A. Gioielli and M.N. Oliveira. 2009. Effect of different prebiotics on the fermentation kinetics, probiotic survival and fatty acids profiles in nonfat symbiotic fermented milk. International Journal of Food Microbiology 128: 467–472.

Osborn, A.M., E.R.B. Moore and K.N. Timmis. 2000. An evaluation of terminal-restriction fragment length polymorphism (T-RFLP) analysis for the study of microbial community structure and dynamics. Environmental Microbiology 2: 39–50.

Otieno, D.O., J.F. Ashton and N.P. Shah. 2005. Stability of β-glucosidase Activity Produced by *Bifidobacterium* and *Lactobacillus* spp. in Fermented Soymilk During Processing and Storage. Journal of Food Science 70(4): 236–241.

Palframan, R., G.R. Gibson and R.A. Rastall. 2003. Development of a quantitative tool for the comparison of the prebiotic effect of dietary oligosaccharides. Letters in Applied Microbiology 37: 281–284.

Petzelbauer, I., R. Zeleny, A. Reiter, K.D. Kulbe and B. Nidetzky. 2006. Development of an Ultra-High-Temperature process for the enzymatic hydrolysis of lactose: II, Oligosaccharide formation by two thermostable b-glycosidases. Biotechnol. Bioeng. 69: 140–149.

Possemiers, S., K. Verth, S. Uyttendaele and W. Verstraete. 2004. PCR-DGGE-based quantification of stability of the microbial community in a simulator of the human intestinal microbial ecosystem. FEMS Microbiology Ecology 49: 495–507.

Prival, M.J., V.M. Davis, M.D. Peiperl and S.J. Bell. 1988. Evaluation of azo food dyes for mutagenicity and inhibition of mutagenicity by methods using *Salmonella typhimurium*, Mutat. Res. 206: 247–259.

Rafii, F., W. Franklin and C.E. Cerniglia. 1990. Azoreductase Activity of Anaerobic Bacteria Isolated from Human Intestinal Microflora. Applied and environmental microbiology 56(7): 2146–2151.

Rosenberg, E. and U. Gophna. 2011. Beneficial Microorganisms in Multicellular Life Forms. Springer Heidelberg Dordrecht, London, New York.

Saad, N., C. Delattre, M. Urdaci, J.M. Schmitter and P. Bressollier. 2012. An overview of the last advances in probiotic and prebiotic field. LWT—Food Science and Technology, In Press, 1–16.

Saengkanuk, A., S. Nuchadomrong, S. Jogloy, A. Patanothai and S. Srijaranai. 2011. A simplified spectrophotometric method for the determination of inulin in Jerusalem artichoke (*Helianthus tuberosus* L.) tubers. Eur. Food Res. Technol. 233: 609–616.

Sanders, M.E. and M.L. Marco. 2010. Food formats for effective delivery of probiotics. Annu. Rev. Food Sci. Technol. 1: 65–85.

Schrezenmeir, J. and M. de Vrese. 2001. Probiotics, prebiotics, and synbiotic: approaching a definition. Am. J. Clin. Nutr. 73: 361S–4S.

Shen, Z., C.H.D. Warren and D.S. Newburg. 2001. Resolution of structural isomers of sialylated oligosaccharides by capillary electrophoresis. J. Chromatogr 921: 315–321.

Shihata, A. and N.P. Shah. 2000. Proteolytic profiles of yogurt and probiotic bacteria International Dairy Journal 10: 401–408.

Šimonová, I., J. Karovičová, V. Mastihuba and Z. Kohajdová. 2010. HPLC Determination of Inulin in Plant Materials. Acta Chimica. Slovaca 3(2): 122–129.

Smalla, K., M. Oros-Sichler, A. Milling, H. Heuer, S. Baumgarte, R. Becker, G. Neuber, S. Kropf, A. Ulrich and C.C. Tebbe. 2007. Bacterial diversity of soils assessed by DGGE, T-RFLP and SSCP fingerprints of PCR-amplified 16S rRNA gene fragments: Do the different methods provide similar results? Journal of Microbiological Methods 69(3): 470–479.

Temmerman, R., I. Scheirlinck, G. Huys and J. Swings. 2003. Culture-Independent Analysis of Probiotic Products by Denaturing Gradient Gel Electrophoresis. Applied and environmental microbiology 69: 220–226

Thomas, D.W. and F.R. Greer. 2010. Committee on Nutrition. Probiotics and Prebiotics in Pediatrics. Pediatrics 126: 6.

Toward, R., S. Montandon, G. Walton and G.R. Gibson. 2012. Effect of prebiotics on the human gut microbiota of elderly persons 3(1): 57–60.

Ubeda, J.F., M. Fernandez-Gonzalez and A.I. Briones. 2009. Application of PCR-TTGE and PCR-RFLP for Intraspecific and Interspecific Characterization of the Genus Saccharomyces Using Actin Gene (ACT1) Primers. Curr. Microbiol. 58: 58–63.

Urashima, T., W.A. Bubb, M. Messer, Y. Tsuji and Y. Taneda. 1994. Studies of the neutral trisaccharides of goat (Capra hircus) colostrum and of the one- and two dimensional 1H and 13C NMR spectra of 6'-N acetylglucosaminyllactose. Carbohydr. Res. 262: 173–184.

Vasquez, A., S. Ahrne, B. Pettersson and G. Molin. 2001. Temporal temperature gradient gel electrophoresis (TTGE) as a tool for identification of *Lactobacillus casei, Lactobacillus paracasei, Lactobacillus zeae* and *Lactobacillus rhamnosus*. Letters in Applied Microbiology 32: 215–219.

Villarreal, J., V., T. Schwartz and U. Obst. 2010. Culture-independent techniques applied to food industry water surveillance: A case study. International Journal of Food Microbiology 141: 147–155.

Wichienchot, S., P. Thammarutwasik, A. Jongjareonrak, W. Chansuwan, P. Hmadhlu, T. Hongpattarakere, A. Itharat and B. Ooraikul. 2011. Extraction and analysis of prebiotics from selected plants from southern Thailand, Songklanakarin J. Sci. Technol. 33(5): 517–523.

Wolford, J.K., D. Blunt, C. Ballecer and M. Prochazka. 2000. High-throughput SNP detection by using DNA pooling and denaturing high performance liquid chromatography (DHPLC). Hum Genet. 107: 483–487.

Wollowski, I., G. Rechkemmer and B.L. Pool-Zobel. 2001. Protective role of probiotics and prebiotics in colon cancer. Am. J. Clin. Nutr. 73: 451–5.

Wong, J.M.W., R. de Souza, C.W.C. Kendall, A. Emam and D.J.A. Jenkins. 2006. Colonic Health: Fermentation and Short Chain Fatty Acids. Journal of Clinical Gastroenterology 40: 235–243.

Xie, J., T.C. Wehner and M.A. Conkling. 2002. PCR-based Single-strand Conformation Polymorphism (SSCP) Analysis to Clone Nine Aquaporin Genes in Cucumber. J. Amer. Soc. Hort. Sci. 127: 925–930.

Zenobi, R. and R. Knochenmuss. 1998. Ion Formation In Maldı Mass Spectrometry. Mass Spectrometry Reviews 17: 337–366.

5

Regulation and Guidelines of Probiotics and Prebiotics

Arturo Anadón, Víctor Castellano* and *María Rosa Martínez-Larrañaga*

Introduction

Probiotics are live microorganisms, generally bacteria but also yeast which, when ingested in adequate amount, interact with the gut microflora and host, having a positive effect on the health of an individual. These bacteria can help to maintain internal microbial balance and defend against harmful bacteria; three mechanisms of promoting human health have been described: (i) providing end-products of anaerobic fermentation of carbohydrates such as organic acids that can be absorbed by the host, these end-products being able to influence human mood, energy level and even cognitive abilities, (ii) successfully competing with pathogens, and (iii) stimulating host immune responses by producing specific polysaccharides (Saier and Mansour 2005). Probiotics should not cause disease in humans; they should be completely non-pathogenic and should not be able to evolve into pathogenic variants.

Probiotics are readily available to consumers and are commonly found as food probiotics (examples are yogurts, cheeses, milk-based beverages, fermented fish, meats, and vegetables, among others) and as food supplement probiotics (examples are tablets, capsules, pills, powders,

Departamento de Toxicología y Farmacología Facultad de Veterinaria Universidad Complutense de Madrid 28040-Madrid, Spain.
* Corresponding author

liquid concentrates in vials, and softgels among others). While probiotics can technically be any type of beneficial microorganism, certain bacteria are more commonly found in the market. Nowadays, the most widely used food-grade probiotics is lactic acid bacteria from the genera *Lactobacillus*, and *Bifidobacteria*, with some strains of *Enterococcus* and *Saccharomyces* species being amongst the exceptions. *Lactobacillus acidophilus* and *Bifidobacterium* bacteria can easily be added to traditionally fermented foods such as cheeses, yoghurts and other dairy products.

Prior to being categorized as probiotics, organisms need to follow a process of testing including strain identification by genotype and phenotype, functionalized characterization and safety assessment testing, and double-blind, placebo-controlled human trials to verify their health benefits and for that guidelines for the evaluation of probiotics in food have been proposed (FAO/WHO 2002).

The exact mechanisms of probiotics in the human body are complex; there is a delicate balance of gastrointestinal gut flora for which the interactions of the various bacteria, as well as their interaction with the rest of the body, are not entirely understood. There exists a wide variety of bacteria that are currently being categorized as probiotics, but much research remains to be done to understand the exact mechanisms of probiotics and to provide hard scientific evidence for their use in food and health industries.

A range of new probiotics and prebiotics is emerging and their market in food is growing rapidly. An example is the genetically modified organism (i.e., generic modification of the probiotic strain) which must be developed and evaluated carefully before it can be used in food, and whose objective is to improve desired functions. The research in this field indicates a promising future for the food and health industries.

The health benefits associated with probiotics vary widely depending on specific strains and circumstances. Certain types of probiotics have been shown to reduce diarrhea in infants as well as diarrhea caused by *Clostridium difficile*, common bacteria that proliferate when patients are given antibiotics. Other probiotics have improved eradication of *Helicobacter pylori*, which is known to cause peptic ulcers and gastritis. Probiotic strains administered in the vagina also showed vast reduction in urogenital infections. Studies have shown that probiotics, which normally pass through the gastrointestinal tract quickly, colonise in the gut for longer periods in the presence of pathogenic bacteria such as Salmonella and inhibit pathogenic activity; research supports the claim that probiotics help to maintain intestinal flora balance and fight infection (Namoto 2005).

Other kinds of products with health benefits are prebiotics. A prebiotic was defined as a non-digestive food ingredient that beneficially affects the host by selectively stimulating the growth and/or activity of one or a limited number of bacteria in the colon, and thus improving host health (Gibson and Roberfroid 1995). These authors revised this concept and proposed a new prebiotic definition as a selectively fermented ingredient that allows specific changes; both in the composition and/or activity in the gastrointestinal microbiota that confers benefits upon host well-being and health (Gibson et al. 2004, Roberfroid 2007). Although the concept of prebiotics has been developed over time, it is proposed that an ingredient must fulfill three fundamental aspects in order to be considered as an effective prebiotic: (i) resistance to digestion (i.e., not hydrolysed and absorbed in the upper part of the gastrointestinal tract; (ii) good fermentation by the large intestinal microbiota; and (iii) a selective effect on the microbiota that has associated health promoting effects (Macfarlane et al. 2006). Probiotics are found naturally in many foods and can also be isolated from plants or synthetized. They allow the selective growth of certain indigenous gut bacteria. The main candidates for prebiotic (recognized and emergent) status reported in the scientific literature are the following: lactulose, fructo-oligosaccharides (FOS), galacto-oligosaccharides (galactosa oligomers and some glucose/lactose/galactose units) (GOS), galacto-oligosaccharides (GOS)/transgalactoylated-oligosaccharides (GOS/TOS) inulins, isomalto-oligosaccharides, isomaltotetratose, lactulose, pyrodextrins (mixture of gluose-containing oligosaccharides), soya-oligosaccharides (SOS), genti-oligosaccharides, gluco-oligosaccharides, isomalto-oligosaccharides (IMO), lactosucrose, levans, pectic-oligosaccharides, resistant starch, sugar alcohols, and xylo-oligosaccharides (XOS) among others.

This chapter reviews deals with regulations and guidelines of probiotics and prebiotics published by several governments and formulated by different organizations respectively as well as the main procedures or approaches for their approval. The incorporation of novel strains of microbes into foods and/or medicinal products requires their authorization, for instance in the EU the authorization is based on assessment of safety, quality and efficacy.

Regulation and Guidelines of Probiotics

The current European Union legislation covers substances with a physiological effect, such as probiotic bacteria and prebiotic compounds. The new food probiotic are regulated by the novel foods regulation and the nutrition and health claims regulation. Microbial strains will have to pass one or both of those regulations. Any claims proposed for probiotic bacteria and prebiotic compounds to be used in food must be based on, and substantiated by, the generally accepted scientific data. The European Union regulations will prohibit any claims referring to the prevention, treatment or cure of a human disease for a food in contrast to that proposed by other countries such as Canada and the USA (Sanders et al. 2005). In USA, a probiotic product may be marketed and regulated to a generally health population as a food or dietary supplement and biological product (i.e., drugs) depending on the intended use of a probiotic. If a probiotic is intended for use as a drug an investigational New Drug Application must be submitted. Biological products require premarket review and approval by FDA while dietary supplements do not. The safety, purity and potency as well as efficacy of a biological product must be demonstrated for approval and dietary supplements need not demonstrate any of these to be marketed. Genetically modified probiotics are also subject to different legislations than naturally-occurring probiotic strains, and must comply with specific regulations for novel foods under the Food and Drugs Act and Regulations. There is great potential for the genetic modification of probiotic bacterial strains, but care must be taken in their development and distribution in the food market. In Canada, under the food provision of the Food and Drug Regulations, live bacterial cultures, including those represented as probiotics are food ingredients and can be added to food products. While there are currently no specific regulations regarding probiotic bacteria in foods, the general provisions of the Food and Drug Act and Regulations apply to food containing microorganisms including those microorganisms represented as probiotics. These provisions regulate the safety of foods and their ingredients, as well as the claims made on food labels and in advertising, including claims about probiotics.

One of the most difficult attempts facing those in the probiotic and prebiotic fields—substantiation domains of efficacy needed to support claims of health benefits. Hence, future food probiotic strains and prebiotic compounds will be carefully selected, identified and researched, and data generated to meet relevant requirements of these key European Union regulations. Probiotic products which claim specific nutritional, functional or therapeutic characteristics make ambiguous the borderlines between food, nutrition and health claims, dietary supplement, or human medicines, producing questions for regulators. Overall, probiotics consumed in foods and dietary supplements do not have to comply with more strict guidelines for probiotics than those requested for examplle for authorization of health claims that must be scientific substantiated.

Regulation of probiotics

Depending on the intended use of a probiotic, the legislative background requirements differ because it can be considered as a drug, novel food ingredient, nutritional and health claim or a dietary supplement, which sometimes are depending of countries and regulatory agencies approaches. The regulations of these products share may similarities, but overall the requirements for food differ significantly from those for drugs. Therefore, the statute's definitions of products merit close attention.

Drugs

Human drugs are regulated by the European Union under the Directive 2001/83/EC of the European Parliament and of the Council of 6 November 2001 on the Community code relating to medicinal products for human use and lays down harmonised rules for the authorisation, supervision and pharmacovigilance of medicinal products for human use within the Union (OJ No. L 311, 28.11.2004). A medicinal product may only be placed on the market in the European Economic Area (EEA) when a marketing authorization has been issued by the competent authority of a member state (or EEA country) for its own territory (national authorization) or when an authorization has been granted in accordance with Regulation (EC) No. 726/2004 of the European Parliament and of the Council of 31 March 2004 laying down Community procedures for the authorisation and supervision of medicinal products for human and veterinary use and establishing a European Medicines Agency (OJ No. L 136, 30.4.2004). In order to obtain a marketing authorisation, a summary of product characteristics in accordance with Article 11 of Directive 2001/83/EC must be included in the application. The summary of product characteristics sets out the agreed position of the medicinal product.

In the USA, drugs or medicinal products are regulated by the Federal Food, Drug and Cosmetic Act of 1938, and later amended, established limits for food additives, cosmetic, and drug safety, for human and veterinary use and requires drug manufacturers to demonstrate product safety to the regulatory agency U.S. FDA prior to marketing. The law also requires medicines to be labeled with adequate directions for safe use and prohibits false therapeutic claims. When a drug candidate is identified, the applicant for a medicinal product performs preclinical studies (i.e., *in vitro* and animal safety testing) to demonstrate that the product is reasonably safe for use in humans and later on the clinical trials are developed. Medicinal products could contain live microorganisms (e.g., bacteria or yeast) with an intended therapeutic effect in humans, may be used in disease prevention or treatment, intended local or regional action and includes "probiotic for clinical uses".

Novel food ingredients

To date probiotic foods are not governed under specific EU regulatory frameworks; although the Regulation (EC) No. 258/97 of the European

Parliament and of the Council of 27 January 1997 concerning novel foods and novel food ingredients (OJ No. L 043, 14.02.1997) may cover other more novel types of probiotic species that need to be discussed and assessed in the light of the Novel Food Guidelines (Jonas et al. 1996) and states safety rules for authorisation of novel food/ingredients. The principal objectives of the Regulation (EC) No. 258/97 are: (i) to protect the functioning of the internal market within the Community, and (ii) to protect public health. The basic criteria for the authorisation of novel foods is that they must not: (i) present a danger for the consumer; or (ii) mislead the consumer; or (iii) differ from foods or food ingredients that they are intended to replace, to an extent that their normal consumption would be nutritionally disadvantageous for the consumer. Commission Recommendation 97/618/EC of 29 July 1997 concerning the scientific aspects and the presentation of information necessary to support applications for the placing on the market of novel foods and novel food ingredients and the preparation of initial assessment reports under Regulation (EC) No 258/97 of the European Parliament and of the Council (OJ No. L 253, 16.9.1997) recommends instructions necessary to sustain an application to fulfill the safety assessment.

The Novel Food Regulation defines novel foods as foods and food ingredients that were not used for human consumption to a significant degree within the Community before 15 May 1997. 'Human consumption to a significant degree within the Community,' in this context, has been interpreted as being demonstrated by a food having been generally available within the Community (i.e., food/food ingredient that does not have a significant history of human consumption within the European Union prior to 15th May 1997).

Four novel food/food ingredient categories can be considered: (1) presenting a new or intentionally modified primary molecular structure; (2) consisting of micro-organisms, fungi or algae; (3) consisting of, or isolated from plants or isolated from animals; and (4) whose nutritional value, metabolism or level of undesirable substances has been significantly changed by the production process.

Actually there is an ongoing proposal for reviewing Regulation (EC) 258/97 in order to bring improvements on a number of important issues such as nanomaterials definition and labeling, a centralised and quicker authorisation procedure for novel foods and specific measures for traditional foods from third countries. In line with this, the European Commission is seeking feedback on how to create a more streamlined authorisation procedure (including the decision) which takes into account, for example, particular needs of traditional exotic food from third countries and which is adjusted to applications which cover several food uses.

For GM food and feed a specific regulation is in force in EU [Regulation (EC) No. 1829/2003 (OJ No. L 368, 18.10.2003) and Regulation (EC) No.

1830/2003 (OJ No. L 265, 18.10.2003)]. The additives and processing aids however fall outside the scope of this regulation. The case of a processing aid or additive consisting of live microorganisms thus remains uncertain. Microbial feed additives, however, are covered by Regulation (EC) No. 1831/2003 of the European Parliament and of the Council of 22 September 2003 on additives for use in animal nutrition (OJ No. L 268, 18.10.2003), and, in accordance with the guidelines of the FEEDAP Panel of EFSA, they are subjected to detailed efficacy and safety assessment, the latter with the intention of ensuring that they are safe to target animal species, users and consumers (Anadón et al. 2006, Anadón et al. 2010). The reader is referred to the paper of Roda et al. (2009) reporting a review on scientific risk assessment on GM foods.

Nutrition and health claims

Claims (Prebiotic and probiotic claims)

Prebiotic and probiotic foodstuffs with identifiable functions can be rightly considered as functional following the Consensus Document of the Scientific Concepts of Functional Foods in Europe (Diplock et al. 1999) where the following is stated: A food can be regarded as 'functional' if it is satisfactorily demonstrated to affect beneficially one or more target functions in the body, beyond adequate nutritional effects, in a way that is relevant to either an improved state of health and wellbeing and/or reduction of risk of disease. The term functional foods was first introduced in Japan in the mid-1980s although along the years they have been called by many different names in particular nutraceuticals, pharmafoods, medical foods, and a host of others, depending on the background and perspective of the researcher. Functional foods must remain foods and they must demonstrate their effects in amounts that can normally be expected to be consumed in the diet; they are not pills or capsules, but part of a normal food pattern.

European Union

Claims are governed by Regulation (EC) No. 1924/2006 of The European Parliament and the Council of 20 December 2006 on nutrition and health claims made on foods (OJ No. L 404, 30.12.2006) (Corrigendum OJ No. L 12, 18.1.2007). The Article 6 of this regulation indicates that health claims shall be based on and substantiated by generally accepted scientific evidence and the Article 13(1) indicates that health claims should be evaluated by the European Food Safety Authority (EFSA) after a scientific assessment of the highest possible standard. From 2008, hundreds of claims were adopted for over 200 foods/constituents. The claims were evaluated on a case-by-case basis (some were clustered into one opinion) respecting uniform scientific criteria to lead consistency. The opinions release by EFSA reflect the varying quality of the information submitted (characterization, dose, health relationship, scientific evidence) and the evaluation results were: (i) about one third was favourable with sufficient scientific evidence; (ii) half of the non favourable evaluations had insufficient information on the food/constituent, and (iii) probiotic claims on characterized strains with a non favourable outcome. The problems difficulties encountered with probiotic claims can be summarized as follows: identification and characterization of strains (phenotypic and genetic tests), impact of carrier or vehicle food matrix, stability (processes and storage), and clinical intervention studies often done in diseased subjects/populations. Referring to the claim application challenges the main strengths are: component should be well-defined, basis for effects should be sufficiently characterized, the clinical studies should be performed in target population, a clear rationale for extrapolation from specific populations should be presented, the cause-effects must be demonstrated, and the proposed health claim hypothesis and demonstration should clearly be carried through the whole application.

The EFSA has issued a scientific and technical guidance for the preparation and presentation of the application for authorization of a health claim (EFSA 2007) and its revision (EFSA 2011) under Regulation (EC) No. 1924/2006, and requested by the European Commission. This guidance applies to health claims related to the consumption of a food category, a food, or its constituents (including a nutrient or other substance, or a combination of nutrients/other substances); hereafter referred to as food/constituent.

The purpose of this guidance is to assist applicants in preparing and presenting their applications for authorization of health claims that fall under Article 14 of the Regulation (EC) No. 1924/2006 (i.e., reduction of disease risk claims and claims referring to children's development and health). This guidance will be updated at a later stage to cover applications for authorization of the health claims which fall under Article 18 of the Regulation (EC) No. 1924/2006. In other words, applications for inclusion of health claims in the Community list of permitted claims provided for in Article 13(3) based on newly developed scientific evidence and/or include a request for the protection of proprietary data. As specified in the Regulation (EC) No. 1924/2006, health claims should be substantiated by taking into account the totality of the available scientific data and by weighing the evidence, and subject to the specific conditions of use. Scientific substantiation is the main aspect to be taken into account to authorise health claims. Claims should be scientifically substantiated by taking into account the totality of the available scientific data, and by weighing the evidence, and should demonstrate the extent to which: (a) the claimed effect of the food/constituent is relevant for human health; (b) a cause and effect relationship is established between the consumption of the food/constituent and the claimed effect in humans (such as: the strength, consistency, specificity, dose-response, and biological plausibility of the relationship); (c) the quantity of the food/constituent and pattern of consumption required to obtain the claimed effect could reasonably be achieved as part of a balanced diet; and (d) the specific study group(s) in which the evidence was obtained is representative of the target population for which the claim is intended.

The Scientific Panel on Dietetic Products, Nutrition and Allergies (NDA) of EFSA prepared a draft opinion which was published for public consultation. After considering all comments received, the Panel adopted its opinion on 06 July 2007. In 2011, the NDA Panel was requested by EFSA to revise the opinion with regard to the forms to be used for the submission of an application for authorisation of health claims pursuant to Articles 13(5) and 14 of Regulation (EC) No 1924/2006, and for the modification of an existing authorisation in accordance with Article 19 of the same Regulation. The revision of the guidance, adopted by the NDA Panel on 13 May 2011, was of a purely administrative nature and concerned Parts 1 to 4, as well as the Appendices, of the guidance in order to simplify the presentation of an application (EFSA 2011).

In accordance with the requirements of the Regulation (EC) No. 1924/2006, the guidance imposes the layout of the submission dossier based on five parts and the correspondent content of the application.

Part 1. Administrative and Technical data: contains the specific requirements for the administrative and technical data, such as the application form, information related to the applicant and the nature of the application (including the national and international regulatory status of the health claim), health claim particulars, and the summary of the application.

Part 2. Food/Constituent Characteristics: contains information specific to the food/constituent and its characteristics (such as the composition, physical and chemical characteristics, manufacturing process, stability, and bioavailability data).

Part 3. Overall summary of Scientific Data: contains summaries (tabulated summaries of all pertinent studies identified and written summaries of data from pertinent human and non-human studies) and overall conclusions, which follow the scope and the outline of the body of scientific data identified under Part 4. All pertinent studies (human studies, animal studies, *in vitro* studies, other) identified should be included (published and unpublished); individual studies included in any review publication should be counted separately.

Part 4. Body of Pertinent Scientific Data Identified: contains all identified pertinent scientific data (published and unpublished, data in favour and not in favour) which form the basis for substantiation of the health claim.

Part 5. Annexes to the Application: comprises the glossary or abbreviation of terms quoted throughout the different Parts, copies/reprints of pertinent publications identified, full study reports of unpublished pertinent data, and scientific opinions of national/international regulatory bodies.

United States

The proposed use of a probiotic, whether as a drug or a dietary supplement, can determine how rigorously the product is regulated or whether the product is even legal. In accordance to the Food and Drug Administration (FDA) a drug is defined as an article intended for use in the diagnosis, cure, mitigation, treatment, or prevention of disease (FDA 2004). If a probiotic is intended for use as a drug, then it must do so through the drug status, which is similar to that of any new therapeutic agent. An Investigational New Drug application must be submitted and authorized by FDA before an investigational or biological product can be administered to humans. The probiotic drug must be proven safe and effective for its intended use before marketing (FDA 2009).

Historically the FDA prohibited health claims in food labeling prior to 1990. Under the Food, Drug, and Cosmetic Act prior 1990, all health claims were considered illegal drug claims. However as science has been progressing, evidence has proven links between diet and health. In 1990, the Nutrition Labeling and Education Act (NLEA) amended the Food, Drug, and Cosmetic Act to allow health claims for foods and dietary supplements under limited conditions. The Nutrition Labeling and Education (NLEA) Act of 1990 (NLEA) mandated nutritional labeling on most food products that are regulated by FDA. FDA promulgated additional regulations for the use of health and nutrient content claims. Most of these regulations went into effect in 1994. Certain nutrient information is mandatory, while other nutrients may be listed at the discretion of the manufacturer, unless the manufacturer makes a claim about the optional nutrient or indicates that the food product is fortified with an optional nutrient.

FDA Modernization Act of 1997 (FDAMA) further amended the Food, Drug, and Cosmetic Act to permit health claims based on an authoritative statement linking a nutrient to a disease made by a scientific body. Two new provisions of FDAMA permit distributors and manufacturers to use claims if based on current, published, authoritative statements from certain federal scientific bodies. The federal government agencies specifically identified as scientific bodies by FDAMA are: the National Academy of Sciences (NAS), the National Institutes of Health (NIH), and the Centers for Disease Control and Prevention (CDC).

When a probiotic is intended for use as a dietary supplement it is classified as foods and as such is formally authorized in the Dietary Supplement Health and Education Act (DSHEA) of 1994 by the FDA (Center for Food Safety and Applied Nutrition). A dietary supplement is a product taken by mouth that contains a dietary ingredient intended to supplement the diet. In contrast to medicines, dietary supplements do not need FDA approval before being marketed; the only requirement is manufacturers need to notify FDA before marketing a determined product. Under DSHEA, the dietary supplement manufacturer is responsible for ensuring that the dietary supplements that it manufactures or distributes are safe before marketing. However, the manufacturer is not required to demonstrate safety or efficacy before marketing a dietary supplement ingredient that was marketed before 1994. Therefore, manufacturers do not generally need to register with FDA, nor get FDA approval before producing or selling dietary supplements. In addition any representations or claims made about them are substantiated by adequate evidence to show that they are not false or misleading.

A health claim is defined in the USA as any claim made on the label or labeling that expressly or by implication characterizes the relationship of any substance to a disease or health-related condition. There are three different types of health-related claims that are not regulated as health claims which are called 'statements of nutritional support' and include: (1) description of general well-being from consumption of the food, (2) classical nutrient-deficiency disease and nutrition and (3) structure-function claims. In addition there are three different regulatory categories of health claims that may be used on a label or in labeling for a food: (1) pre-approved claims, (2) authoritative statements claims, and (3) qualified claims. Thus on that basis, a health claim must be pre-approved by the FDA or must be issued as authoritative statements by an agency of the US government with responsibility for dietary guidance or public health.

The law allows that in addition to nutrient content claims, manufacturers of dietary supplements may make structure/function or health claims for their products which were formally authorized in the DSHEA of 1994. Initially, such statements were regarded as being available for use only in the labeling of dietary supplements, not foods, but FDA extended the use of these claims to food in September 1997, in a Federal Register notice. Actually, FDA regulations allow a claim in the labeling of food that characterizes the relationship of any food substance to a disease or health-related condition if the claim is first approved by an FDA regulation, 21 of the Code of

Federal Regulation (CFR) at Section 101.14. Such claims are called 'health claims.' Examples include calcium to help prevent osteoporosis, folic acid to prevent neural tube defects, and consumption of soy protein to reduce the risk of cardiovascular disease (see 21 CFR 101.72, 101.79 and 101.82) (Anadón et al. 2010).

For a structure/function claim, FDA requires that manufacturers' substantiation is accepted by experts in the field and that the claim is truthful and not misleading. According to FDA, health claims describe a relationship between food, food components, or dietary supplement ingredients, and reducing risk of a disease or health-related conditions.

FDA has issued new guidance to industry allowing qualified health claims in the labeling of conventional foods and dietary supplements and four different health claims are allowed for dietary supplement products without an FDA approved regulation if certain legal requirements are met. The types of clinical studies must be rated on the basis of quality and strength of evidence (Anadón et al. 2010). Only data obtained from studies conducted in healthy populations are evaluated because health claims are usually directed at the general population or certain subgroups (e.g., elderly patients). The data supporting a health claim must be published and therefore apply to any product meeting the criteria for the claim (Saldanha 2008).

Generally Recognized As Safe (GRAS) and Qualified Presumption of Safety (QPS)

There are two comparative assessment procedures between the United States and the European Union. One for food supplements in the USA (FDA GRAS system) and the other for safety of bacterial dietary supplements in the EU (EFSA QPS system). In terms of safety for probiotic products when the product has a history of safe use in the target host, such as GRAS or its equivalents (i.e., the QPS) in the EU, then it is suggested that further animal and human toxicological studies may not be necessary.

Generally recognized as safe (GRAS) levels published in the USA

GRAS or Generally Recognized As Safe is a regulatory concept specific to the United States Federal Food, Drug, and Cosmetic Act. Under sections 201(s) and 409 of the Federal Food, Drug, and Cosmetic Act (the Act), any substance that is intentionally added to food is a food additive, that is subject to premarket review and approval by FDA, unless the substance is GRAS, among qualified experts (consensus), as having been adequately shown to be safe under the conditions of its intended use, or unless the use of the substance is otherwise excluded from the definition of a food additive. A GRAS substance is distinguished from a food additive on the basis of the common knowledge about the safety of the substance for its intended use and it is exempt from the food additive requirement. Thus, the difference between use of a food additive and use of a GRAS substance relates to the widespread awareness of the data and information about the substance (i.e., who has access to the data and information and who has reviewed that data and information). GRAS is related to the use of a food substance rather than the substance itself. As a practical matter, GRAS affirmation is only applicable for substances found safe based on common use in USA before 1958 and it is the Food, Drug, and Cosmetic Act section 201(s) that lays down that general recognition of safety must be established either through scientific procedures or experience based on common use in food before January 1, 1958. As a practical matter, GRAS affirmation based on scientific procedures works out to be nearly as involved as a new food additive petition.

Food ingredients, whose use is GRAS, are not required by law to receive FDA approval before marketing. These substances include live microorganisms added to food for uses other than as starter cultures. GRAS organisms are regarded as safe within their specific conditions of use and therefore do not need formal approval. Information relied on to establish the safety of the use of a GRAS substance can be deduced through publication in the scientific literature or can be empirically obtained through the long history of safe use (i.e., a substantial history of consumption in foods by a significant number of consumers). Evidently, that approach is not of use for products containing novel species/strain of microorganisms. Regardless of whether the use of a substance is a food additive or is GRAS, there must be evidence that the substance is safe under the conditions of its intended use. FDA has several lists of GRAS substances, consequently these lists are not all-inclusive and it is impracticable to list all substances that are used in food on the basis of the GRAS provision (open list). The use of a substance can be GRAS even if it is not listed by FDA. Finally, when an ingredient is GRAS for one use, is not necessarily GRAS for all uses.

The Qualified presumption of safety (QPS) concept published in the EU

The European Food Safety Authority (EFSA) has established a system based on the concept of qualified presumption of safety (QPS), presumption being defined in the European Union as 'a belief or assumption based on reasonable evidence' and qualified to allow certain restrictions to apply (EFSA 2005, Anadón et al. 2010). The QPS approach regarding microorganisms in food and feed is a system similar to the GRAS definition used in USA, but modified to take into account the different regulatory practices in Europe. This was considered necessary since issues of importance to Europe would not necessarily influence a GRAS listing. Consequently it was proposed that any 'generic listing' of a microorganism should be qualified, allowing the general safety of the organism/group of organisms to be concluded provided that certain specific criteria were met. It is intended to provide a mechanism to recognise and give weight to prior knowledge (whether gained through formal investigation or by experience of use).

QPS is a qualified approval system that would harmonize the safety assessment of microorganisms throughout the food chain without compromising the safety principles. However, certain issues could be addressed on a case-by-case basis, including the presence of virulence factors, toxic metabolites and antibiotic resistant determinants. QPS status would not be applied in case a microorganism commonly causes pathogenicity (e.g., species or subspecies with pathogenic potential). Therefore, an assessment on a case-by-case basis would always be required.

In the case of novel use of a microorganism that also has a traditional use in food or feed production, the Novel Food Regulation covers the safety assessment of the products [Regulation (EC) No. 258/97 of the European Parliament and of the Council of 27 January 1997 concerning novel foods and novel food ingredients (OJ No. L 43, 14.2.1997)].

To establish the QPS status for a given microorganism, several requirements should be fulfilled which rest on four pillars: Taxonomy, familiarity, pathogenicity and end use. The main requirements relating to taxonomy and familiarity of the candidate organisms for QPS assessment include: taxonomic entity (i.e., the taxon or taxonomic unit such a genus, species, subspecies or other grouping such as homofermentative lactobacilli should be considered). QPS should seek the highest possible taxonomic level that is practically possible by using the mechanism of 'qualification' to exclude undesirable strains. Familiarity reflects that enough is known about the proposed group of organisms to reach a decision on their safety. According to EFSA (2005), the questions relating to taxonomy and familiarity include: What evidence of taxonomic status is needed? What if a microorganism that has been granted QPS needs to be reclassified? Will the QPS status be retained? What taxonomic level is appropriate for QPS? Is a history of apparent safe use sufficient evidence of safety (and for all purposes)? Is lack of clinical data evidence of a lack of pathogenicity? and Should taxonomic units which include pathogenic strains be excluded from QPS? Figure 1 shows an example of a decision tree approach intended to be used by expert groups for the acceptance of a QPS micro-organism, but not for notification (EFSA 2005).

The term 'body of knowledge' (or familiarity) should be used following the recommendation of EFSA Scientific Colloquium (EFSA 2005). This term

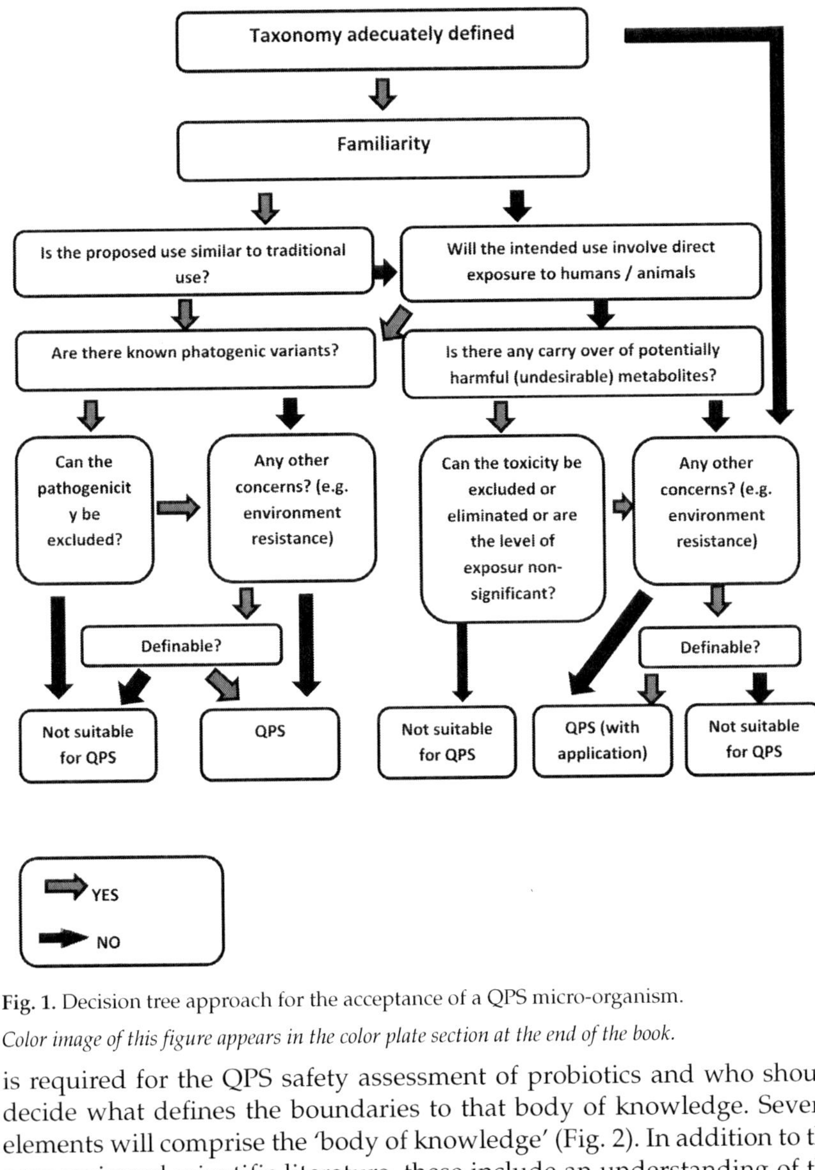

Fig. 1. Decision tree approach for the acceptance of a QPS micro-organism.

Color image of this figure appears in the color plate section at the end of the book.

is required for the QPS safety assessment of probiotics and who should decide what defines the boundaries to that body of knowledge. Several elements will comprise the 'body of knowledge' (Fig. 2). In addition to the peer-reviewed scientific literature, these include an understanding of the history of use of a microorganism, its industrial applications, its ecology, any clinical reports concerning the microorganisms and information from the scientific literature (Anadón et al. 2010). Pathogenicity, whether the grouping considered for QPS contains known pathogens. If so, whether enough is known about their virulence determinants or toxigenic potential to exclude pathogenic strains (example, the *Bacillus subtilis* group to exclude

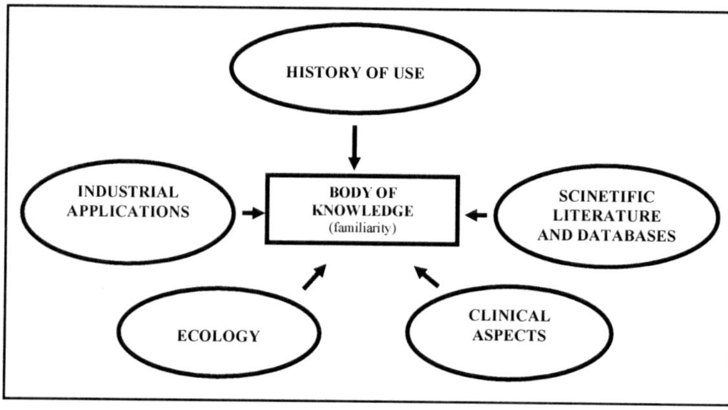

Fig. 2. Body of knowledge.

the occasional strain with a toxigenic potential, to fungal species to exclude mycotoxin producers or to distinguish between the virulent and avirulent forms of enterococci); and the end use, whether viable organisms enter the food chain or whether they are used to produce other products. As a principle, it is desirable to conclude on the safety of the taxonomic unit for all purposes. This is likely to be the case with many fungal species where safety might be established for a particular use (e.g., for plant protection purposes) but not enough is known about metabolic activities to allow extrapolation to other uses.

On that basis, EFSA will start with the evaluation of the four groups of microorganism used in food and feed production (e.g., lactic acid bacteria, *Bacillus* spp., yeasts, and commonly encountered filamentous fungi). Other groups could be added depending on the outcome of the inventory and the frequency of notifications (EFSA 2005). In conclusion, QPS guidelines at difference of GRAS are applied to: (i) microorganisms, (ii) a positive list will be available (no an open list as it is in GRAS), (iii) the inclusion of the microorganism is based on history of use and adverse effects, and (iv) define taxonomic unit (e.g., genus, species or strain).

Guidelines of probiotics

In order to standardize the requirements needed to make claims for probiotics, the Joint Food and Agriculture Organization of the United Nations and World Health Organization Expert Consultation on *Evaluation of Health and Nutrition Properties of Probiotics* met in Córdoba (Argentina) in October 2001 to create a set of guidelines (Joint FAO/WHO 2002) to set up a systematic approach for the evaluation of probiotics in food leading to the substantiation of health claims. As a consequence of this meeting a

Working Group was convened by FAO/WHO to generate guidelines and recommend criteria and methodology for the evaluation of probiotics, and to identify and define what data is needed to be available to accurately substantiate health claims. The aims of the Working Group were to identify and outline the minimum requirements needed for probiotic status. In this document no reference was made either to the term bio-therapeutic agents (i.e., microorganisms having therapeutic effects in humans) or non-food beneficial microorganisms. The GMOs were also excluded. The proposed guidelines recommended for the evaluation of probiotics for food are as following:

1) **Identification of the genus, species and strain of the probiotics.** The speciation of the bacteria must be established using a combination of phenotypic and genotypic tests [DNA-DNA hybridization or 16S rRNA gene sequence analysis for species identification, and DNA macrorestriction followed by pulsed field gel electrophoresis (PFGE), and randomly amplified polymorphic DNA analysis (RAPD) between others for strain identification]. The presence of extra-chromosomal genetic elements such as plasmids can contribute to strain typing and characterization.
   All strains are recommended to be deposited in an internationally recognized culture collection.
2) ***In vitro* testing to screen potential probiotics** (these tests are essentials to gain knowledge of strains and the mechanisms of probiotic effects). At this point it is essential to highlight that probiotics for human use will require substantiation of clinical efficacy with human trials and are recommended appropriate target-specific *in vitro* tests correlated with *in vivo* results. Table 1 summarizes the *in vitro* tests used in developing probiotics.
3) **Safety considerations: Requirements for proof that a probiotic strain is safe and without contamination in its delivery form.** The probiotics may theoretically be responsible for four types of side-effects (Marteau 2001): systemic infections, deleterious metabolic activities, excessive

Table 1. Available *in vitro* tests to study probiotic strains.

1. Resistance to gastric acidity
2. Bile acid resistance (colonization)
3. Adherence to mucus and/or human epithelial cells and cell lines
4. Antimicrobial activity against potentially pathogenic bacteria
5. Ability to reduce pathogen adhesion to surfaces
6. Bile salt hydrolase activity (as a probiotic marker)
7. Resistance to spermicides (applicable to probiotics for vaginal use)

immune stimulation in susceptible individuals, and gene transfer. While most of the species and genera are apparently safe, certain microorganisms may be problematic, particularly the enterococci (*E. faecium* and *E. faecalis*). These have emerged as opportunistic pathogens in hospital environments causing nosocomial infections such as endocarditis, bacteremia, and intra-abdominal, urinary tract and central nervous system infections, and may also harbor transmissible antibiotic resistance determinants (i.e., vancomycin resistant *Enterococcus* strains) and bacilli, especially those belonging to the *B. cereus* group that are known to produce enterotoxins and an emetic toxin (Anadón et al. 2006). With *Bifidobacterium* no cases of infections have been reported.

Some groups of bacteria are considered as GRAS, but the FAO/WHO working group recommends that strains of probiotics should be characterized at a minimum with the tests listed in Table 2.

The lack of infectivity in immunocompromised animal models can be of importance.

4) ***In vivo* studies using animals and humans**. In this requirement the FAO/WHO working group (2002) encourages the use of animal models to provide substantiation of *in vitro* effects and for determining the probiotic mechanism previous to human trials.

The principal outcome of efficacy studies on probiotics should be proven benefits in human trials, such as statistically and biologically significant improvement in condition, symptoms, signs, well-being or quality of life, reduced risk of disease or longer time to next occurrence, or faster recovery from illness. Each of them should have a proven correlation with the probiotic tested.

Probiotics have been tested in a variety of clinical conditions to prove efficacy and safety of products. Standard methods for clinical evaluations are comprised of Phase 1 (focused on safety), Phase 2 (efficacy), Phase 3 (effectiveness) and Phase 4 (surveillance). Phase 2 human trials, generally in the form of double blind, randomized, placebo-controlled (DBPC) or design, measure efficacy compared

Table 2. Safety assessment of the probiotic strain.

1. Determination of antimicrobial resistance patterns.
2. Assessment of certain metabolic activities (i.e., undesirable) (e.g., D-lactate production, and bile salt deconjugation).
3. Assessment of side effects during human studies.
4. Post-market epidemiological surveillance of adverse incidents in consumers
5. If the strain under evaluation belongs to a species known to be either a mammalian toxin producer, or to have hemolytic potential it must be tested for toxin production or haemolytic activity.

with placebo and adverse effects. Phase 3 is appropriate to compare probiotics with standard treatment of a specific condition and Phase 4, is a post-marketing surveillance system in human populations ingesting the probiotic microorganisms. In connection with the post-market surveillance system it can be stated that all stakeholders should develop some form of system to monitor the health outcome of long term probiotic administration, the potential side-effects and long term benefits should be recorded and documented and proper trace back system is a pre-requisite for surveillance of probiotic products.

The guidelines recommend that information accumulated to show that a strain(s) is a probiotic, including clinical trial evidence must be published in peer-reviewed scientific or medical journals. Furthermore, publication of negative results is encouraged as these contribute to the totality of the evidence to support probiotic efficacy.

5) **Health claims and labeling.** The consultation recommends that specific health claims should be permitted on the label and promotional material. In Table 3 is described the information needed on the label. It is also recommended that the product manufacturer takes responsibility to make sure that an independent third party reviews and evaluates the scientific evidence.

Table 3. Information to be described on the label.

1. Genus, species and strain designation. Strain designation should not mislead consumers about the functionality of the strain
2. Minimum viable numbers of each probiotic strain at the end of the shelf-life
3. The suggested serving size must deliver the effective dose of probiotics related to the health claim
4. Health claim(s)
5. Proper storage conditions
6. Corporate contact details for consumer information

Guidelines of prebiotics

Following the FAO Technical Meeting held in 2007 on prebiotics (FAO 2007, Anadón et al. 2010) the way to evaluate and substantiate a product as a prebiotic is indicated in flow-chart of Fig. 3. The steps to be accomplished are the following:

1) ***Product Specification/Characteristics of the Prebiotic.*** The component, to which the claim of being prebiotic is ascribed, must be characterized for any given product including: source and origin, purity, chemical composition and structure, vehicle, concentration, and the amount in which it is to be delivered to the host.

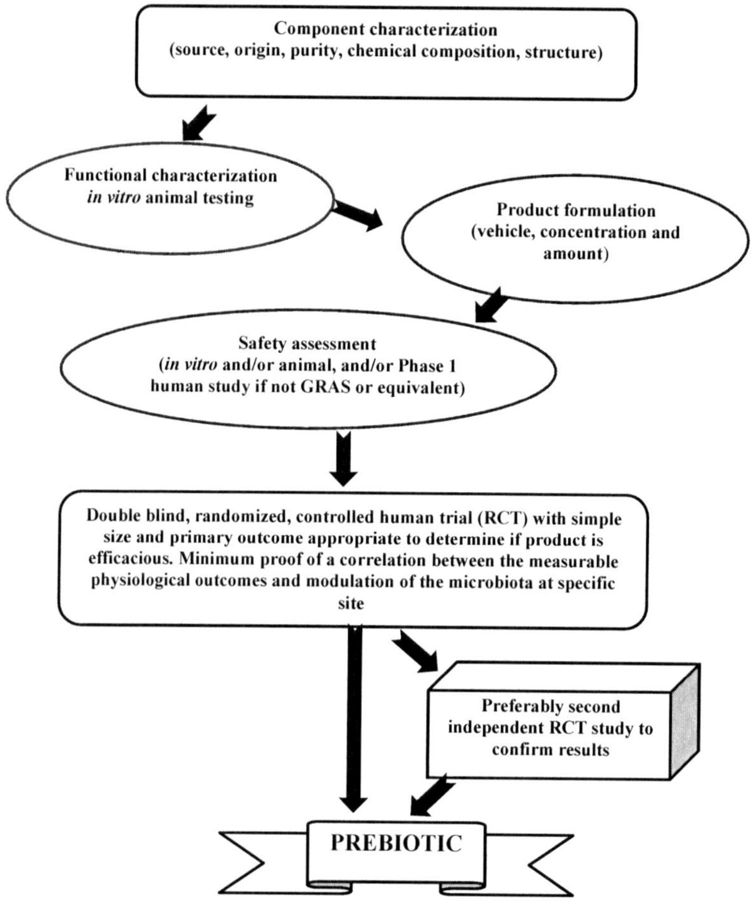

Fig. 3. Guidelines for the evaluation and substantiation of prebiotics.

2) ***Functionality.*** At least a correlation between the measurable physiological outcomes and modulation of the microbiota at a specific site (primarily the gastrointestinal tract, but potentially also other sites such as vagina and skin) should be evidenced. Also needed to be correlated is a specific function at a specific site with the physiological effect and its associated timeframe.

   Within a study, the target variable should change in a statistically significant way and the change should be biologically meaningful for the target group consistent with the claim to be supported. Substantiation of a claim should be based on studies with the final product type, tested in the target host. An appropriate sized randomized control trial (compared to a placebo or a standard control substance) is requested, more desirable with a second independent study.

Physiological indices to be tested, resulting from the administration of prebiotics could be: (i) satiety (e.g., estimation of carbohydrates, fats, and total energy intake); (ii) endocrine mechanisms regulating food intake and energy usage in the body; (iii) effects on absorption of nutrients (e.g., calcium, magnesium, trace elements, protein); (iv) reduced incidence or duration of infection; (v) blood lipid and classic endocrine parameters; (vi) bowel movement and regularity; (vii) biomarkers for cancer risk; and (viii) variations in innate and acquired immunity that are evidence of a health benefit.

3) ***Qualifications.*** For a specific prebiotic the qualifications can be summarized in: (i) component (chemical substance or a food grade component), (ii) health benefit (measurable and not due to the absorption of the component or due to the component acting alone, and over-riding any adverse effects) and (iii) modulation (changes in the composition or activities of the microbiota in the target host). For instance, a prebiotic can be a fiber but a fiber need not be a prebiotic. It is known that bifidogenic effects are not sufficient without demonstrated physiological health benefits. It has also been recognized that the determining incidents that happen within compartments of the intestine are often difficult. Specific site sampling or more sophisticated methods can reliably link microbiota modulation with health benefits, so that fecal analysis will be relevant.
4) ***Safety.*** It is recommended that the following issues are covered in any safety assessment of a prebiotic final product formulation.
   - When the product has a history of safe use in the target host, such as GRAS or its equivalents (i.e., QPS), then it is suggested that further animal and human toxicological studies may not be necessary.
   - Safe consumption levels with minimal symptoms and side effects should be established.
   - The product must not contain any contaminants and impurities. The contaminants should be identified and measured, and the impurities should be well characterized and submitted to toxicity evaluation if needed.
   - Based upon current knowledge, the prebiotic should not alter the microbiota in such a way as to have long-term detrimental effects on the host.

For functional ingredients, animal models can be used to ascertain the target organs and effects that are produced as a result of toxicity. The extent of testing necessary for a functional ingredient is increased in response to the lack of understanding of potential for toxicity because of inadequately characterized products. The following criteria must be met to derive a safe level of exposure without additional toxicology testing (Kruger and Mann

2003): (i) active component(s) and related substances are well-characterized and there is adequate understanding of the lack of potential for toxicity at the human dose levels recommended based upon existing data from the literature; (ii) impurities are well-characterized and there is an adequate understanding of the lack of potential for toxicity based upon existing data from the literature; and (iii) the manufacturing process is standardized and reproductive.

When the active component(s) or impurities are either not fully characterized, or there is not enough data available to evaluate the potential for toxicity, the following preclinical toxicological information is needed to assess the functional ingredient: toxicity studies *in vitro* and *in vivo*, including mutagenicity studies, reproductive and developmental toxicity studies, pharmacokinetics and special pharmacology studies and long-term feeding studies, following a tiered approach on a case-by-case basis. One element that must be considered in the design of animal studies for functional ingredients is the margin of safety between the no-observed-effect level (NOEL)/no-observed-adverse-effect level (NOAEL) determined in the animal studies and the dietary human exposure. For some experts, the margin of safety has the same meaning as the margin of exposure.

Conflict of Interest Statement

None of the authors of this paper has a financial or personal relationship with other people or organisations that could inappropriately influence or bias the content of the paper.

Acknowledgements

This review was supported by the Universidad Complutense de Madrid, Comunidad de Madrid, and Ministerio de Educación y Ciencia, Projects Refs. UCM-BSCH/GR35/10-A and S2009/AGR-1469, and Consolider-Ingenio 2010 Ref. CSD/2007/00063 (FUN-C-FOOD), Madrid, Spain.

References

Anadón, A., M.R. Martinez-Larrañaga and M.A. Martinez. 2006. Probiotics for animal nutrition in the European Union. Regulation and safety assessment. Reg. Toxicol. Pharmacol. 45: 91–95.

Anadón, A., M.R. Martínez-Larrañaga, V. Caballero and V. Castellano. 2010. Assessment of prebiotics and probiotics: An overview. Chapter 2. *In*: R. Ross Watson and V.R. Preedy [eds.]. Bioactive Foods in Promoting Health: Probiotics and Prebiotics. Academic Press, Oxford (ISBN: 978-0-12-374938-3), pp. 19–41.

Diplock, A.T., P.J. Aggett, M. Ashwell, F. Bornet, E.B. Fern and M.D. Robertfroid. 1999. Scientific concepts of functional foods in Europe Consensus document. Br. J. Nutr. 81(suppl): S1–S27.

FAO 2007. FAO Technical Meeting on Prebiotics, September 15–16, 2007.

FAO/WHO. 2002. Joint FAO/WHO Group report on drafting guidelines for the evaluation of probiotics in food, London, Ontario, Canada, April 30 and May 2002.

FDA 2004. Federal Food, Drug, and Cosmetic Act (FD&C Act). 21 U.S.C. 321. Sect 201 (2004) [updated 2009 Oct 14; cited 2010 Jun 24]. http://www.fda.gov/regulatoryinformation/legislation/federalfooddrugandcosmeticactfdcact/default.htm.

FDA 2009. Development & approval process (drugs) [updated 2009 Oct 14; cited 2010 Jun 24]. http://www.fda. gov/Drugs/DevelopmentApprovalProcess/default.htm.

EFSA 2005. EFSA Scientific Colloquium Summary Report. OPS Qualified Presumption of Safety of Microorganisms in Food and Feed, 13–14 December 2004, Brussels, Belgium. ISBN 92-9199-012-4 [http://www.efsa.eu.int/science/colloquium_series/no2_qps/948_en.html/].

EFSA 2007. Scientific and technical guidance for the preparation and presentation of the application for authorization of a health claim. Request no EFSA-Q-2007-066.

EFSA 2011. Scientific and technical guidance for the preparation and presentation of an application for authorisation of a health claim (revision 1). EFSA Journal 9(5): 2170.

Gibson, G.R. and M.B. Roberfroid. 1995. Dietary modulation of the human colonic microbiota: Introducing the concept of prebiotics. J. Nutrition 125: 1401–1412.

Gibson, G.R., H.M. Probert, J.V. Loo, R.A. Rastall and M.B. Roberfroid. 2004. Dietary modulation of the human colonic microbiota: updating the concept of prebiotics. Nutr. Res. Rev. 17: 259–275.

Jonas, D.A., E. Abtignac, J.M. Antoine, H.G. Classen, A. Huggett, I. Knudsen, J. Mahler, T. Ockhuizen, M. Smith, M. Teuber, R. Walker and P. De Vogel. 1996. The safety assessment of novel foods. Guidelines prepared by ILSI Europe novel foods task force. Food Chem. Toxicol. 34: 931–940.

Kruger, C.L. and S.W. Mann. 2003. Safety evaluation of functional ingredients. Food Chem. Toxicol. 41: 793–805.

Marteau, P. 2001. Safety aspects of probiotic products. Scand. J. Nutr. 45: 22–24.

Macfarlane, S., G.T. Macfarlane and J.H. Cummings. 2006. Prebiotics in the gastrointestinal tract. Alim. Pharmacol. Therapeut. 24: 701–714.

Namoto, K. 2005. Prevention of Infections by Probiotics. J. Biosci. Bioeng. 100(6): 584–592.

Roberfroid, M. 2007. Prebiotics: the concept revisited. J. Nutrition 137: 830–837.

Roda, L., A. Anadón, M.A. Recuerda, P. Díaz Peralta, F. González Botija, A. Martínez Cañellas, A. Lago, E. Alonso and M.R. Martínez-Larrañaga. 2009. Risk Analysis and GM Foods: Scientific Risk Assessment. E.F.F.L.R. 4: 235–250.

Saier, M.H. and N.M. Mansour. 2005. Probiotics and prebiotics in human. J. Mol. Microbiol. Biotechnol. 10: 22–25.

Saldanha, L.G. 2008. US Food and Drug Administration regulations governing label claims for food products, including probiotics. Clin. Infect. Dis. 46: S119–121.

Sanders, M.E., T. Tompkins, J. Heimbach and S. Kolida. 2005. Weight of evidence needed to substantiate a health effect for probiotics and prebiotics. Regulatory considerations in Canada, EU, and US. Eur. J. Nutrition 44: 303–310.

6

Probiotics and Prebiotics in Infant Nutrition

Dan W. Thomas and *Vivien A. Nguyen**

Introduction

It is well-accepted that resident microbes contribute fundamentally to infant and childhood health and immunity. Efforts to optimize the composition of enteric microflora have led to the study of probiotic and prebiotic supplementation in the prevention and treatment of disease. The purpose of this chapter is to review the rationale for, and evidence supporting the use of probiotics and prebiotics in infant nutrition.

Establishment of Enteric Microbiota and Development of Immunity

The gastrointestinal tract supports a diverse and dynamic microbial ecosystem. Interactions between the host and certain bacterial communities have evolved with time to foster mutualism. Enteric symbionts are metabolically and immunologically active, degrading indigestible compounds and providing essential nutrients, while defending against opportunistic pathogens and shaping the development of the intestinal architecture and mucosal immune system (Round and Mazmanian 2009).

Division of Gastroenterology & Nutrition, Children's Hospital Los Angeles, 4650 Sunset Boulevard, Los Angeles, CA 90027.
* Corresponding author

The amalgam of microorganisms that live within the intestine have the potential to exert both pro- and anti-inflammatory responses, and their composition is intimately linked to the health of the host. Dysbiosis, or perturbation of this balance, has been implicated in a number of disease processes, ranging from atopic (allergic rhinitis, atopic dermatitis, asthma) to autoimmune (inflammatory bowel disease, type 1 diabetes mellitus).

The process of bacterial colonization begins in utero, with exposure of fetal mucosal surfaces to amniotic fluid microflora, and progresses rapidly such that within days of birth, the bacterial cells contained within the lumen of the intestine far outnumber the infant's cells (Indrio and Neu 2011). A number of factors affect this sequence, including gestational age, exposure to antibiotics, hygiene of the environment, mode of delivery, and dietary source (Ganguli and Walker 2011). In premature infants, bacterial colonization is delayed and the proportion of potentially pathogenic microorganisms is high, likely due to antibiotic exposure and characteristics of the medical environment (Morowitz et al. 2010, Westerbeek et al. 2006). Establishment of the intestinal microbiota is similarly delayed in infants born via Cesarean section and differences in the composition of the flora have been shown to persist up to 6 months after birth (Gronlund et al. 1999). The enteric bacterial populations of breast-fed and formula-fed infants are markedly divergent. Breast-fed infants are primarily colonized with *Bifidobacterium*, with subpopulations of *Lactobacillus* and *Streptococcus*. In contrast, formula-fed infants are primarily colonized with *Bacteroides* and *Bifidobacterium*, but in fewer numbers compared to breast-fed infants, with subpopulations of *Clostridia*, *Staphylococcus*, and *Escherichea coli* (Pietzak 2004). The significance of these differences and their relevance to health and disease is uncertain, particularly as the intestinal microbiota remains mutable during the first 6 to 12 months of life (Magne et al. 2005, Schwiertz et al. 2003).

The composition of the infant microbiota and the temporal pattern in which it evolves, vary profoundly from individual to individual. In a ribosomal DNA microarray-based study tracing the development of intestinal flora in full term infants during the first year of life, Palmer et al. (2008) noted that "healthy" neonatal microbial profiles are heterogeneous, with population dynamics marked by abrupt shifts interspersed with intervals of remarkable stability. With time, these profiles converge to a distribution characteristic of the adult gastrointestinal tract, with a preponderance of *Bacteroides* and *Firmicutes* and a relative paucity of *Proteobacteria* and aerobic Gram-negative bacteria. Nonetheless, inter-individual differences persist into adulthood and the distinctive features of the microbiota are thought to influence host physiology and disease pathogenesis (Cilieborg 2012).

The enteric microflora plays an integral role in the development of intestinal mucosal defenses. Studies of gnotobiology, or the selective colonization of germ-free animals, have better elucidated the effect of the microbiota on the immune system. For instance, germ-free mice have defects in the morphology of the epithelium, in the development and maturation of gut-associated lymphoid tissues, and in the production of secretory IgA (Round and Mazmanian 2009). Sensing of commensal microflora through innate pattern recognition receptors induces a number of responses that maintain host-microbial homeostasis. Toll-like receptor-MyD88 signaling pathways encourage repair of damaged intestinal epithelium and production of antimicrobial peptides (Hooper et al. 2012). Commensals also assist in determining the composition of lamina propria T lymphocyte subsets, each with its own distinct effector functions. It has become evident that certain microbial communities are capable of eliciting either pro- or anti-inflammatory responses by enhancing T_H1 or T_H17 polarization versus $Foxp3^+$ regulatory T lymphocyte polarization, respectively. The balance of effector lymphocytes is therefore thought to have a profound impact on mucosal response to stressors that may elicit damage. Further, imbalances in T cell subsets may contribute to allergic and autoimmune diseases at sites distal to the intestine.

Given the impact of the enteric microflora on immune function and its dependence on environmental factors, it follows that an infant's intestinal microbial milieu and diet are integral determinants of health. Probiotics and prebiotics, of which human milk contains substantial quantities, are therefore thought to have the potential to be of benefit in supporting mucosal immune defenses and in protecting against disease.

Supplementation of Infant Nutrition with Probiotics and Prebiotics in the Prevention and Treatment of Diseases

Necrotizing Enterocolitis

Necrotizing enterocolitis (NEC), a potentially debilitating disease characterized by intestinal inflammation and necrosis, predominantly affects preterm infants after enteral nutrition has been initiated (Ganguli and Walker 2011, Thomas et al. 2010). Although the pathophysiology of NEC is incompletely understood, genetic susceptibility, dietary constituents, intestinal immaturity, imbalanced microvascular tone, altered acquisition of intestinal microflora, and dysregulated immunoreactivity are thought to be implicated (Ganguli and Walker 2011, Neu and Mihatsch 2012, Neu and Walker 2011). Interventions to favorably alter the intestinal microbiota, including supplementation of probiotics and prebiotics, could therefore be of benefit in prevention of this disease.

Deshpande et al. (2010) published a systematic review of 11 studies between 1997 and 2009, which included 2176 infants. In an effort to reduce heterogeneity, criteria for inclusion in the meta-analysis were as follows:

randomized controlled trial (RCT) involving preterm very-low-birth-weight (VLBW) infants; severe NEC (stage II or more, per Modified Bell criteria); initiation of enteral probiotic supplementation within 10 days of life and continuation for at least 7 days. A fixed-effects model estimated a lower risk of severe NEC in infants supplemented with probiotics (relative risk [RR]: 0.35; 95% confidence interval (CI):0.23–0.55; P<.00001). Risk of death from all causes mortality was found to be reduced in infants supplemented with probiotics (RR: 0.42; 95% CI: 0.29–0.62; P<0.00001); however, no significant difference in the risk for mortality as a result of NEC was demonstrated. The authors concluded that probiotics should be offered as routine therapy for preterm neonates; however, given the variability in strains, formulations, and dosages among the studies included in the meta-analysis, they acknowledged that prospective, observational studies or head-on trials of individual probiotic preparations are warranted.

In a 2011 Cochrane Collaboration review based on 16 RCTs that included 2942 neonates, enteral probiotic supplementation was found to significantly reduce the incidence of severe NEC (stage II or more) (RR: 0.35; 95% CI: 0.24–0.52) and mortality (RR: 0.40; 95% CI: 0.27–0.60) in preterm VLBW infants (AlFaleh et al. 2011). Reduction in the rate of sepsis trended towards significance; however, one study showed an increased risk of sepsis in neonates with birth weights <750g (Lin et al. 2008). Given the relative immaturity of immune system and intestinal function in preterm infants, probiotic translocation and subsequent sepsis are of concern. Instances of *Lactobacillus*, *Saccharomyces*, and *Bifidobacterium* sepsis have been documented previously; however, no episodes occurred in the included studies (Broughton et al. 1983, Land et al. 2005, Ohishi et al. 2010, Perapoch et al. 2000, Thompson et al. 2001). Infants received various preparations and dosages of probiotics, including *Lactobacillus* species, *Bifidobacterium* species, *Saccharomyces boulardii*, and *Streptococcus thermophilus*. The time of initiation and duration of therapy also differed among included studies. And a minority of studies assessed the duration of total parenteral nutrition and time to full enteral feeds, hospitalization days, weight gain, or long-term neurodevelopmental outcomes. Although the authors support a change in practice to provide enteral supplementation of probiotics to premature infants weight more than 1000 grams at birth, the heterogeneity of the included studies raises questions regarding optimal probiotic preparation, dosing, and protocol. Furthermore, given the potential influence that probiotics may have on host gene expression and composition of microflora, long-term outcome data are of particular interest.

Human milk, a natural prebiotic, has been shown to reduce the incidence of NEC (Corpeleijn et al. 2012, Lucas and Cole 1990, McGuire 2003, Menzen-Derr et al. 2009, Sisk et al. 2007). The prebiotic effects of human milk are attributed to the abundance of oligosaccharides, which selectively serve as a source of energy and nutrients for commensal bacteria (Bode 2009). Thus, a proposed strategy to prevent NEC is to supplement feedings with prebiotics, including inulin, galacto-oligosaccharides (GOS), fructo-oligosaccharides (FOS), acidic oligosaccharides, and lactulose (Sherman et al. 2009). These compounds have been shown to increase fecal *Bifidobacteria* counts, reduce stool pH and viscosity, and improve gastrointestinal motility (Neu and Mihatsch 2012). Thus far, the limited number of RCTs assessing the effect of enteral prebiotic supplementation have not shown a convincing benefit in the prevention of NEC. Westerbeek et al. (2011a,b, 2010) noted that the supplementation of non-human neutral and acidic oligosaccharides to preterm infants neither reduced intestinal permeability nor risk of serious infection . In a pilot study that included 28 preterm infants, Riskin et al. (2010) documented fewer episodes of late-onset sepsis and lower stage NEC in those receiving lactulose supplementation; however, the findings were not significant due to sample size. Modi et al. 2010 assessed the effect of oligosaccharides on enteral tolerance in preterm neonates and found a beneficial effect in those born at 26–28 weeks gestation (2.9%–9.9% improved tolerance; $p<0.001$); however, the incidence of NEC was not assessed and the findings could not be extrapolated to all premature infants.

In conclusion, there is evidence to support the use of probiotics to prevent NEC in preterm infants with VLBW. However, the effects of individual probiotic preparations on the incidence of NEC should be reconfirmed in adequately powered, high-quality, randomized controlled trials (Thomas et al. 2010). With regard to prebiotics, the theoretical benefits have not been substantiated with evidence demonstrating clinical efficacy and safety. Caution in the use of probiotics to prevent NEC must still be emphasized.

Infantile colic

Infantile colic is a common condition characterized by inconsolable crying and irritability in an otherwise healthy infant during the first 3 months of life. The diagnosis is clinical and incorporates the rule of three: unexplained paroxysmal fussing occurring for >3 hours per day, 3 days per week, and continuing for at least 3 weeks (Wessel et al. 1954). The pathogenesis of infantile colic is incompletely understood. Psychosocial factors, intolerance of certain nutritional constituents, and visceral hypersensitivity are thought to be involved (Gupta 2007). Aberrant bacterial colonization, with associated defects in barrier function and mucosal immune regulation, may also be implicated. For instance, Savino et al. (2009) studied the microbiota of colicky infants and found a preponderance of coliforms. And Rhoads et al. (2009) found that infants with colic had less diverse intestinal microflora as well as evidence of increased neutrophilic infiltration, as demonstrated by

2-fold higher fecal calprotectin levels. It has therefore been suggested that modulating the intestinal microflora of colicky infants with administration of probiotics, may be of benefit.

To date, no RCTs have been conducted with prevention of infantile colic as a primary endpoint, but a number of studies examining the effect of probiotic supplementation in the treatment of colic have been performed. In 2007, Savino and Pelle compared the effects of *Lactobacillus reuteri* ATCC 55730 supplementation to simethicone administration in 90 breastfed infants with colic. Median crying times were significantly reduced from 159 minutes per day at baseline to 41 minutes per day after 7 days of probiotic supplementation. And the response rates for treatment with *L. reuteri* versus simethicone were 95% and 7%, respectively. A subsequent study by the same authors confirmed the efficacy of *L. reuteri* DSM 17938 (a daughter strain of *L. reuteri* ATCC 55730) in reducing symptoms of colic in breastfed infants (Savino et al. 2010). Although these studies support the use of *L. reuteri* in treatment of colic in breasted infants, further RCTs are warranted prior to recommending the routine use of probiotics in the treatment of colic in both breastfed and formula-fed infants (Thomas et al. 2010).

The use of prebiotics in the treatment of infantile colic has not been studied in depth. Vivatvakin and Mahayosnond (2010) compared indicators of gastrointestinal comfort in infants fed a whey-predominant formula containing long-chain polyunsaturated fatty acids, GOS, and FOS, and infants fed a casein-predominant formula. Exclusively breastfed infants were also included as a reference. Infants supplemented with prebiotics had stool microbiota composition, stool consistency, and gastrointestinal transit times similar to those of breastfed infants. However, the study was not sufficiently powered to assess for differences in symptoms of colic. Infant formulas containing GOS and/or FOS are widely available, but more information is needed before the efficacy of adding prebiotics to infant formulas can be determined.

Diarrhea

Prevention of diarrhea acquired in child-care centers

Infants and children in child-care centers are exposed to a plethora of pathogens, resulting in frequent infectious diseases, such as upper respiratory tract infections and acute gastroenteritis. A number of RCTs conducted in child care centers demonstrated a reduction in the number of acute gastrointestinal tract infections in healthy infants supplemented with probiotics. Thibault et al. (2004) found that infants receiving a formula fermented with *Bifidobacterium breve C50* and *Streptococcus thermophilius 065* had no difference in incidence or duration of diarrheal episodes.

However, episodes were less severe in the fermented formula group, with fewer cases of dehydration (2.5% versus 6.1%, p=0.01), fewer medical consultations (46% versus 57%, p-0.003), and fewer oral rehydration therapy prescriptions (41.0% versus 51.0%, p-0.003). Chouraqui et al. 2004 noted a significant reduction in risk and duration of diarrhea in infants receiving *Bifidobacterium lactis* strain Bb 12. Long-term consumption of a milk-based formula containing *Bifidobacterium lactis* and *Streptococcus thermophilus* was associated with lower frequency of infection requiring antibiotic therapy (Saavedra et al. 2004). In 2005, Weizman et al. conducted a double-blind, placebo-controlled, randomized trial comparing the effect of 2 different species of probiotic bacteria (*Bifidobacterium lactis* and *Lactobacillus reuteri*) in preventing infectious illnesses in infants attending child care centers. The controls had more episodes of diarrhea compared with those fed *B. lactis* or *L. reuteri* (mean:0.31, CI:0.22–0.40 versus mean:0.13, CI:0.05–0.21 versus mean 0.02, CI:0.01–0.05, respectively). Further, the duration of the episodes was longer in controls than in those supplemented with probiotics. The effects were more pronounced with *L. reuteri*, which also improved other morbidity parameters. While the probiotics studied thus far appear to be safe, evidence of their efficacy in preventing diarrhea acquired in child-care centers is modest. Routine use of probiotics for this indication is not recommended, but there may circumstances in which probiotic supplementation in children cared for at long-term health facilities is beneficial (Thomas et al. 2010).

Prebiotics have not been extensively evaluated in the prevention of diarrhea in infants. In a prospective, randomized, double-blind, placebo-controlled trial, 206 healthy infants with family history of atopy were fed either prebiotic-supplemented (8g/L short chain GOS and long chain FOS) or placebo supplemented hypoallergenic formula during the first 6 months of life (Arslanoglu et al. 2007). During the study period, infants receiving prebiotics had fewer episodes of upper respiratory tract infections, otitis media, gastrointestinal infections, and urinary tract infections (p=0.01).

Prevention of nosocomial diarrhea

Hospitalized infants and children often acquire infectious diarrhea during their stay. Rotavirus is commonly implicated in this phenomenon and is often responsible for extended length of stay (Guandalini 2011). A handful of RCTs examining the effect of probiotic supplementation on occurrence of nosocomial diarrhea in infants and children have been performed. The first such study, performed in 1994 by Saavedra et al., found that *Bifidobacterium bifidum* and *S. thermophilus* supplementation significantly reduced the incidence of acute diarrhea and rotavirus shedding in infants admitted to a chronic medical care facility. Szajewska et al. (2001a) evaluated the efficacy of orally administered Lactobacillus GG in the prevention of nosocomial diarrhea in 81 children aged 1 to 36 months. *LGG* was found to reduce the risk of nosocomial diarrhea in comparison with placebo (6.7% versus 33.3%, respectively; RR:0.2; CI:0.06–0.6; number needed to treat [NNT]: 4). Further, the use of *LGG* compared with placebo significantly reduced the risk of rotavirus gastroenteritis (2.2% versus 16.7%; RR:0.13; CI:0.02-0.79; NNT: 7). A randomized, double-blind, placebo-controlled trial by Hojsak et al. (2010) echoed these findings. 742 hospitalized infants and children were allocated to receive either a milk product fermented with *LGG* or a pasteurized fermented milk product. A significant reduction in the risk of gastrointestinal infections was conferred by *LGG* supplementation (RR:0.4, CI:0.25-0.70, NNT 15). The effects of *LGG* and breast-feeding in the prevention of nosocomial rotavirus infection in 220 children aged 1 to 18 months, were also assessed by Mastretta et al. (2002). In contrast to the findings of Szajewska (2012a) and Hojsak et al. (2010), the authors found no difference in attack rate of rotavirus infection among patients who received probiotic versus those who received placebo (25.4% versus 30.2%, respectively). However, the attack rate of rotavirus infection among breast-fed infants was 10.6% (p=0.003). Although supplementation with LGG may be a reasonable means of reducing nosocomial rotavirus infection, the routine use of the pentavalent rotavirus vaccine presents a more effective preventative strategy (Thomas et al. 2010).

Prevention of antibiotic-associated diarrhea

Antibiotic therapy may disturb the balance of intestinal microflora, resulting in a range of symptoms, including diarrhea. In infants and children, antibiotic-associated diarrhea is estimated to occur in up to 40% of cases (Guandalini 2011). A 2011 Cochrane Collaboration review based on 16 RCTs that included 3432 children varying in age from 2 weeks to 17 years, assessed the capacity of probiotics to prevent this condition (Johnston et al. 2011). The studies utilized a range of probiotic strains, including *Bacillus, Bifidobacterium, Lactobacilli, Lactococcus, Leuconostoc cremoris, Saccharomyces,* and *Streptococcus,* alone or in combination. Probiotics were generally well tolerated, without serious adverse events. The incidence of antibiotic-associated diarrhea in the probiotic group was 9% compared to 18% in the control group, but statistically significant heterogeneity was detected. A similar trend was noted in relation to mean duration of diarrhea. Interestingly, a dose-dependence was noted, with an enhanced protective effect in children receiving high-dose *L. rhamnosus* or *Saccharomyces boulardii* (≥5 billion colony-forming units per day). The authors estimated that 7 children needed to be treated to prevent 1 case of antibiotic-associated diarrhea. Three previous meta-analyses addressing the use of probiotics in the prevention of antibiotic-associated diarrhea in children also suggested benefit (Johnston et al. 2006: RR: 0.43, CI:0.25–0.75; Szajewska et al. 2006: RR:0.44, CI:0.25–0.77; Johnston et al. 2007: RI:0.43, CI0.25–0.74). The current data suggest that probiotics may prevent the onset of antibiotic-associated diarrhea; however, further studies evaluating strain- and dose-specific efficacy are warranted.

It is unclear whether prebiotics are effective in the prevention of antibiotic-associated diarrhea. A randomized, double-blind, controlled clinical trial was carried out in 140 infants 1 to 2 years of age to assess the effects of a prebiotic-supplemented milk-based formula on the intestinal microbiota after a course of antibiotic therapy (Brunser et al. 2006). Prebiotics were found to increase fecal *Bifidobacteria* counts, but no significant differences in gastrointestinal symptoms were noted. The Working Group for Probiotics and Prebiotics of the European Society for Pediatric Gastroenterology, Hepatology, and Nutrition (ESPGHAN) recently conducted a multicenter RCT to determine the efficacy of administering a combination of inulin and FOS in the prevention of antibiotic associated diarrhea (Szajewska 2012b). There was no difference between the prebiotic and placebo groups in the proportions of participants with diarrhea (6.7% versus 10.6%, respectively; CI:0.16%–2.5%).

Treatment of acute infectious diarrhea

The use of probiotics in the treatment of acute infectious diarrhea in infants and children has been thoroughly investigated in numerous clinical trials, incorporating a range of probiotic strains, formulations, and dosages. A recent Cochrane Collaboration review examined 56 studies published between 1966 and 2010 (Allen et al. 2010). 6489 infants and young children were included. Probiotics reduced the duration of diarrhea (mean difference 24.76 hours, CI:15.9–33.6 hours), the risk of diarrhea lasting ≥ 4 days (RR:0.41, CI:0.32–0.53), and the stool frequency on day 2 of illness (mean difference 0.8, CI:0.45–1.14). Differences in effect size between studies were noted; however, the heterogeneity was not due to probiotic strain, the number of different strains, the viability of the organisms, low versus high dose preparations, the causes or severity of diarrhea, or whether the studies were done in developed or developing countries. These findings suggest that mechanisms common to the majority of probiotics, such as competition for available niches, acidification of luminal contents, and elicitation of immune responses, are effective against a wide range of enteric pathogens. With the exception of mild hypersensitivity to *E. coli* Nissle reported in one participant, no serious adverse events were attributed to probiotics in the included studies.

A number of preceding systematic reviews provide further insight. Szajewska et al. 2001b assessed the effect of probiotics in acute diarrhea lasting ≥ 3 days in infants and children. Probiotics reduced the duration of diarrhea by 18.2 hours and *LGG* was thought to be particularly effective in rotavirus diarrhea. Chmielewska et al. (2008) reviewed studies assessing the effects of *Lactobacillus reuteri* strain ATTCC 55730 and found that the probiotic reduced the duration of acute infectious diarrhea by 22 hours. The effects of *Saccharomyces boulardii* were evaluated in a meta-analysis of 7 randomized controlled trials that included 944 otherwise healthy infants and children with acute gastroenteritis (Sjazewska et al. 2009). The duration of diarrhea was reduced by 1.08 days in children who received the yeast compared to those who received the placebo. With regard to dose, Wolvers et al. (2010) found that probiotic supplementation in the range of 10^{10}–10^{11} colony-forming units per day was associated with enhanced efficacy. Dose-dependence was significant in watery diarrhea and viral gastroenteritis but not in invasive bacterial diarrhea. Further, the protective effects were greater when probiotics were administered early in the course of the illness.

In summary, probiotics appear to be safe and efficacious in reducing the frequency and duration of diarrhea in acute viral gastroenteritis in otherwise healthy children. The benefit is dependent upon strain, dose, and early initiation of therapy. However, further studies to guide the use of probiotic regimens in specific patient populations should be considered (Thomas et al. 2010).

Atopic diseases

The prevalence of atopic diseases, including food hypersensitivity, eczema, allergic rhinoconjunctivitis, and asthma, has increased over time, particularly in developed countries. Though genetic predisposition plays an integral role in the development of this spectrum of disorders, diet and the acquisition of intestinal microbiota in infancy are also thought to shape the process by which the immune system learns to differentiate between self and non-self (Penders et al. 2007). In utero, in a simplified way of stating it, the balance between cell-mediated immune responses (T_H1) and humoral immune responses (T_H2) is biased towards the latter in an effort to prevent rejection of the fetus by the mother. T_H2 responses continue to predominate in early infancy. However, exposure to pathogenic and commensal microflora elicits T_H1 and regulatory T cell responses, which counterbalance the T_H2 milieu. Numerous studies have documented differences in the microbiota of children with atopy compared to non-allergic children. The most consistent finding in these studies was a reduction in the proportion of *Bifidobacteria* in the feces of infants with eczema and atopic sensitization, but not in the feces of children with asthma (Bjorksten et al. 2001, Kalliomaki et al. 2001, Murray et al. 2005). These findings suggest that alterations in enteric microflora during infancy may contribute to the pathogenesis of atopic diseases. Probiotics and prebiotics have therefore been studied in the prevention of food hypersensitivity and allergic disease in infants.

In a Finnish study, 159 pregnant women were randomly allocated to receive either placebo or 10^{10} colony-forming units of *LGG* daily for 4 weeks prior to delivery (Kalliomaki et al. 2003). Probiotic supplementation was continued postnatally for 6 months in both mother and infant. At the 2 year time-point, atopic dermatitis was diagnosed in 23% of children who had received *LGG* versus 46% of children who had not (RR:0.51, CI:0.32–0.84, $p<0.01$). The number of mother-infant pairs required to be treated with *LGG* to prevent 1 case of atopic dermatitis was 4.5. The prevalence of atopic dermatitis had not changed at the 4 year time-point; however, only 67% of the original cohort was available for inclusion in the analysis. These findings were not confirmed in a subsequent clinical trial by Taylor et al. (2007), in which 231 neonates at risk for atopy were supplemented with either *Lactobacillus acidophilus* or placebo for the first 6 months of life. At 6 months, atopic dermatitis rates were similar in the probiotic and placebo groups (25.8% versus 22.7%, respectively). And at 12 months, the rate of sensitization was significantly higher in the probiotic group. A Cochrane Collaboration review examined the effect of probiotic supplementation on allergic disease and food hypersensitivity in infants (Osborn and Sinn 2007a). Probiotics were not found to confer a significant protective effect in allergy, food hypersensitivity, asthma, or allergic rhinitis. However, meta-analysis of 5 studies reporting the outcomes of 1477 infants demonstrated a reduction in atopic dermatitis (RR:0.82, CI:0.7–0.95). Due to the heterogeneity between studies, the authors concluded that there is insufficient evidence to recommend the routine supplementation of probiotics to either pregnant women or infants to prevent allergic disease or food hypersensitivity in childhood.

A limited number of randomized controlled trials have evaluated the use of prebiotics in prevention of atopic disease in infants. Osborn and Sinn (2007b) performed a meta-analysis of 7 studies reporting allergic disease outcome for 432 infants. No significant difference was noted in the incidence of atopic dermatitis between infants receiving prebiotics and those receiving placebo. However, there was marked heterogeneity between studies, potentially attributable to differences in risk factors, prebiotic formulation, and assessment of atopic dermatitis. The authors concluded that there is insufficient evidence to determine the role of prebiotic supplementation of infant formula for prevention of allergic disease and food hypersensitivity. A subsequent meta-analysis by van der Aa et al. (2010) led to similar conclusions.

The concomitant administration of probiotics and prebiotics for prevention of allergic diseases in infants was evaluated by Kukkonen et al. (2007). 1223 pregnant women carrying children at risk for allergy were randomized to receive either probiotics or placebo 2–4 weeks before delivery. Their infants received the same probiotic plus GOS or a placebo for 6 months. Combined probiotic and prebiotic supplementation showed no effect on the cumulative incidence of allergic diseases by age 2 years, but significantly reduced the risk of atopic dermatitis. Confirmatory studies are therefore warranted.

Effects of Probiotics and Prebiotics on Growth in Infancy

In 2011, the ESPGHAN Committee on Nutrition systematically reviewed published evidence relating to the safety and efficacy of the administration of formula supplemented with probiotics or prebiotics compared with unsupplemented formula. The authors stratified the impact of probiotics on growth by strain, with results as follows:

Bifidobacterium lactis

Two trials assessing the impact of *B. lactis* on weight and growth were carried out in infants of HIV-positive mothers. Urban et al. (2008) noted that infants who received chemically acidified formula supplemented with *B. lactis* had more rapid head growth; however, no differences were noted in terms of weight gain or linear growth velocity. Similarly, Velaphi et al. (2008) did not note a difference in weight for age, length for age, head circumference for age, or weight for length among treatment groups. A third study, evaluating the effect of *B. lactis*-supplemented formula on growth revealed no differences between infants receiving probiotic-supplemented formula and infants receiving unsupplemented formula (Weizman and Alsheikh 2006).

Bifidobacterium bifidum, Streptococcus thermophilus, and *Lactobacillus helveticus*

Langhendries et al. (1995) assessed the impact of an infant formula fermented by *S. thermophilus* and *L. helveticus* and supplemented with *B. bifidum* on growth. The authors reported normal growth, without significant differences between infants in the probiotic-supplemented formula and unsupplemented formula groups; however, the study was insufficiently powered to evaluate these effects.

Bifidobacterium longum BL999 and *Lactobacillus rhamnosus* LPR

Chouraqui et al. (2008) evaluated the safety and efficacy of a number of formulas, containing probiotics and/or prebiotics, including BL999 and LPR. Weight gain was equivalent among the study and control groups.

LGG

Vendt et al. (2006) found that infants supplemented with LGG for 6 months had higher length and weight standard deviation scores than infants who received regular formula.

L. reuteri ATCC 55730

Weizman et al. (2006) demonstrated no significant difference in growth parameters between infants who received *L. reuteri* compared with infants who received unsupplemented formula.

The ESPGHAN Committee on Nutrition noted that interpreting studies on the effects of probiotic supplementation in infant formula on growth parameters was limited by the number, size, and follow-up periods of available studies. However, in general, the aforementioned probiotic strains were thought to support normal growth in infants.

Of note, a single study assessed growth and neurodevelopmental outcomes in very low-birth-weight preterm infants supplemented with oral probiotics for the prevention of NEC (Sari et al. 2012). 221 infants completed the trial protocol and174 infants were evaluated in follow-up. There was no significant difference in either parameter between infants supplemented with probiotics and infants supplemented with placebo.

The effects of prebiotic supplementation on infant growth have been evaluated in 2 meta-analyses. In 2009, Rao et al. identified 10 publications that assessed these effects. Individually, the RCTs did not demonstrate significant differences in growth parameters between study and control groups; however, when pooled, administration of formula supplemented with prebiotics had a significant effect on weight gain. A Cochrane Collaboration review by Osborn et al. (2007a, 2007b) analyzed data from 3 RCTs reporting growth parameters in term infants. The authors noted a significant increase in weight gain in infants fed a prebiotic formula; however, there were no significant differences in linear growth or in head circumference between infants supplemented with prebiotics and infants receiving control formula. Given these findings, the ESPGHAN Committee on Nutrition concluded that supplementation of infant formula with prebiotics does not adversely affect growth and may be associated with modest improvements therein.

Safety of Probiotics and Prebiotics in Infants and Children

The safety of probiotics and prebiotics is discussed at length in Chapter 23. In relation to infants and children, the ESPGHAN Committee on Nutrition recommended that more studies be performed to establish the safety and efficacy of these products in the pediatric population (Braegger et al. 2011).

The committee also cited the need for a centralized mechanism of oversight to ensure probiotic microorganism safety, identity, and genetic stability. Similarly, the American Academy of Pediatrics Committee on Nutrition concluded that the addition of probiotics to powdered infant formulas did not appear to be harmful in healthy term infants. However, the authors raised concerns regarding the use of probiotics in infants and children who are immunocompromised, chronically debilitated, or seriously ill with indwelling medical devices (Thomas et al. 2010).

Acknowledgements

We would like to thank our colleagues and our administrative secretary, Ms. Claudia Tejeda, in the Division of Gastroenterology and Nutrition at Children's Hospital Los Angeles for their assistance in the preparation and completion of this monograph. And to our family and friends, for their unwavering support, our utmost gratitude.

References

AlFaleh, K., J. Anabrees, D. Bassler and T. Al-Kharfi. 2011. Probiotics for prevention of necrotizing enterocolitis in preterm infants. Cochrane Database of Systematic Reviews. Issue 3. Art. No.:CD005496.

Allen, S.J., E.G. Martinez, G.V. Gregorio and L.F. Dans. 2010. Probiotics for treating acute infectious diarrhea. Cochrane Database of Systematic Reviews. Issue 11. Art. No.:CD003048. DOI:10.1002/14651858.CD003048.pub3.

Arslanoglu, S., G.E. Moro et al. 2007. Early supplementation of prebiotic oligosaccharides protects formula-fed infants against infections during the first 6 months of life. J. Nutr. 137: 2420–2424.

Bjorksten, B., E. Sepp et al. 2001. Allergy development and the intestinal microflora during the first year of life. Journal of Allergy and Clinical Immunology 108: 516–520.

Bode, L. 2009. Human milk oligosaccharides: prebiotics and beyond. Nutr Rev. S183–91.

Braegger, C., A. Chmielewska et al. 2011. Supplementation of infant formula with probiotics and/or prebiotics: a systematic review and comment by the ESPGHAN committee on nutrition. JPGN 52: 238–250.

Broughton, R.A., W.C. Gruber, A.A. Haffar and C.J. Baker. 1983. Neonatal meningitis due to *Lactobacillus*. Pediatr. Infect. Dis. 2: 3892–384.

Brunser, O., M. Gotteland et al. 2006. Effect of a milk formula with prebiotics on the intestinal microbiota of infants after an antibiotic treatment. Pediatr. Res. 59: 451–456.

Chmielewska, A., M. Ruszcynski et al. 2008. *Lactobacillus reuteri* strain ATCC 55730 for the treatment of acute infectious diarrhea in children: a meta-analysis of randomized controlled trials. Meta-analysis 10: 32–36.

Chouraqui, J.P., D. Grathwohl et al. 2008. Assessment of the safety, tolerance, and protective effect against diarrhea of infant formulas containing mixtures of probiotics or probiotics and prebiotics in a randomized controlled trial. Am. J. Clin. Nutr. 87: 1365–1373.

Chouraqui, J.P. 2004. Acidified milk formula supplemented with *Bifidobacterium lactis*: impact on infant diarrhea in residential care settings. JPGN 38: 288–292.

Cilieborg, M.S., M. Boye and P.T. Sangild. 2012. Bacterial colonization and gut development in preterm neonates. Early Hum. Dev. 88: S41–S49.

Corpeleijn, W.E., S.M.P. Kouwenhoven et al. 2012. Intake of own mother's milk during the first days of life is associated with decreased morbidity and mortality in very low birth weight infants during the first 60 days of life. Neonatology 102: 276–281.

Deshpande, G., S. Rao, S. Patole and M. Bulsara. 2010. Updated meta-analysis of probiotics for preventing necrotizing enterocolitis in preterm neonates. Pediatrics 125: 921–930.

Ganguli, K. and W.A. Walker. 2011. Probiotics in the prevention of necrotizing enterocolitis. J. Clin. Gastroenterol. 45: S133–138.

Gronlund, M.M., O.P. Lehtonen, E. Eerola and P. Kero. 1999. Fecal microflora in healthy infants born by different methods of delivery: permanent changes in intestinal flora after Cesarean delivery. J. Pediatr. Gastroenterol. Nutr. 29: 19–25.

Guandalini, S. 2011. Probiotics for prevention and treatment of diarrhea. J. Clin. Gastroenterol. 45: S149–153.

Gupta, S.K. 2007. Update on infantile colic and management options. Curr. Opin. Investig. Drugs 8: 921–926.

Hojsak, I. et al. 2010. *Lactobacillus* GG in the prevention of nosocomial gastrointestinal and respiratory tract infections. Pediatrics 125: e1171–e1177.

Hooper, L.V., D.R. Littman and A.J. Macpherson. 2012. Interactions between the microbiota and the immune system. Science 336: 1269–1273.

Indrio, F. and J. Neu. 2011. The intestinal microbiome of infants and the use of probiotics. Current Opinion in Pediatrics 23: 145–150.

Johnston, B.C., J.Z. Goldenberg et al. 2011. Probiotics for the prevention of pediatric antibiotic-associated diarrhea. Cochrane Database of Systematic Reviews. Issue 11. Art. No.:CD004827. DOI:10.1002/14651858.CD004827.pub3.

Johnston, B.C., A.L. Supina et al. 2007. Probiotics for the prevention of pediatric antibiotic-associated diarrhea. Cochrane Database of Systematic Reviews. Issue 2. DOI:10.1002/14651858.CD004827.pub2.

Johnston, B.C., A.L. Supina et al. 2006. Probiotics for pediatric antibiotic-associated diarrhea: a meta-analysis of randomized placebo-controlled trials. CMAJ 175: 377–383.

Kalliomaki, M., S. Salminen et al. 2003. Probiotics and prevention of atopic disease: 4 year follow-up of a randomized placebo-controlled trial. Lancet. 361: 1869–1871.

Kalliomaki, M., P. Kirjavainen et al. 2001. Distinct patterns of neonatal gut microflora in infants in whom atopy was and was not developing. Journal of Allergy and Clinical Immunology 107: 129–134.

Kukkonen, K., E. Savilahti et al. 2007. Probiotics and prebiotic galacto-oligosaccharides in the prevention of allergic diseases: a randomized, double blind, placebo-controlled trial. J. Allergy Clin. Immunol. 119: 192–198.

Land, M.H., K. Rouster-Stevens et al. 2005. *Lactobacillus* sepsis associated with probiotic therapy. Pediatrics 115: 179–181.

Langhendries, J.P., J. Detry et al. 1995. Effect of a fermented infant formula containing viable *Bifidobacteria* on the fecal flora composition and pH of healthy full-term infants. JPGN 21: 177–181.

Lin, H.C., C.H. Hsu, H.L. Chen et al. 2008. Oral probiotics prevent necrotizing enterocolitis in very low birth weight preterm infants: a multicenter, randomized, controlled trial. Pediatrics 122: 693–700.

Lucas, A. and T.J. Cole. 1990. Breast milk and neonatal necrotizing enterocolitis. Lancet. 336: 1516–1523.

Magne¸ F., A. Suau, P. Pochart and J.F. Desjeux. 2005. Fecal microbial community in preterm infants. J. Pediatr. Gastroenterol. Nutr. 41: 386–392.

Mastretta, E., P. Longo et al. 2002. Effect of *Lactobacillus* GG and breast-feeding in the prevention of rotavirus nosocomial infection. JPGN 35: 527–531.

McGuire, W. and M.Y. Anthony. 2003. Donor human milk versus formula for preventing necrotising enterocolitis in preterm infants: systematic review. Arch. Dis. Child Fetal Neonatal Ed. 88: F11–14.

Menzen-Derr, J. et al. 2009. Role of human milk in extremely low birth weight infants' risk of necrotizing enterocolitis or death. J. Perinatol. 27: 427–433.

Modi, N., S. Uthaya et al. 2010. A randomized, double-blind, controlled trial of the effect of prebiotic oligosaccharides on enteral tolerance in preterm infants (ISRCTN77444690). Pediatr. Res. 68: 440–445.

Morowitz, M.J., V. Poroyko, M. Caplan, J. Alverdy and D.C. Liu. 2010. Redefining the role of intestinal microbes in the pathogenesis of necrotizing enterocolitis. Pediatrics 125: 777–785.

Murray, C.S., G. Tannock et al. 2005. A fecal microbiota in sensitized wheezy and non-sensitized non-wheezy children: a nested case-control study. Clinical and Experimental Allergy 35: 741–745.

Neu, J. and W. Mihatsch. 2012. Recent developments in necrotizing enterocolitis. JPEN. 36: 30S–35S.

Neu, J. and W.A. Walker. 2011. Necrotizing enterocolitis. NEJM. 364: 255–264.

Ohishi, A., S. Takahashi et al. 2010. *Bifidobacterium* septicemia associated with postoperative probiotic therapy in a neonate with omphalocele. J. Pediatr. 56:79–681.

Osborn, D.A. and J.K.H. Sinn. 2007a. Probiotics in infants for prevention of allergic disease and food hypersensitivity. Cochrane Database of Systematic Reviews. Issue 4. Art. No.:6475. DOI:10.1002/14651858.CD006475.pub2.

Osborn, D.A. and J.K.H. Sinn. 2007b. Prebiotics in infants for prevention of allergic disease and food hypersensitivity. Cochrane Database of Systematic Reviews. Issue 4. Art. No.:CD06474. DOI:10.1002/14651858.CD006474.

Palmer, C., E.M. Bik, D.B. DiGiulio, D.A. Relman and P.O. Brown. 2008. Development of the human infant intestinal microbiota. PLOS Biology 5: 1556–1573.

Penders, J., E.E. Stobberingh et al. 2007. The role of the intestinal microbiota in the development of atopic disorders. Allergy 62: 1223–1236.

Perapoch, J., A.M. Planes et al. 2000. Fungemia with *Saccharomyces cerevisiae* in two newborns, only one of whom had been treated with ultra-levure. 2000. Eur J. Clin. Microbiol. Infect. Dis. 19: 468–470.

Pietzak, M. 2004. Bacterial colonization of the neonatal gut. J. Pediatr. Gastroenterol. Nutr. 38: 389–391.

Rao, S., R. Srinivasjois and S. Patole. 2009. Prebiotic supplementation in full-term neonates: a systematic review of randomized controlled trials. Arch. Pediatr. Adolesc. Med. 163: 755–764.

Riskin, A., O. Hochwald et al. 2010. The effects of lactulose supplementation to enteral feedings in premature infants: a pilot study. J. Pediatr. 156: 209–14.

Round, J.L. and S.K. Mazmanian. 2009. The gut microbiota shapes intestinal immune responses during health and disease. Nat. Rev. Immunol. 9: 313–323.

Rhoads, J.M., N.H. Fatheree et al. 2009. Altered fecal microflora and increased fecal calprotectin in infants with colic. J. Pediatr. 155: 823–828.

Saavedra, J.M., A. Abi-Hanna et al. 2004. Long-term consumption of infant formulas containing live probiotic bacteria: tolerance and safety 79: 261–267.

Saavedra, J.M. and N.A. Bauman. 1994. Feeding of *Bifidobacterium bifidum* and *Streptococcus thermophilus* to infants in hospital for prevention of diarrhea and shedding of rotavirus. Lancet 344: 1046–1049.

Sari, F.N., Z. Eras et al. 2012. Do oral probiotics affect growth and neurodevelopmental outcomes in very low-birth-weight preterm infants? Am. J. Perinatol. 29: 579–586.

Savino, F., L. Cordisco et al. 2010. *Lactobacillus reuteri* DSM 17938 in infantile colic: a randomized double blind, placebo controlled trial. Pediatrics 126: e526-e533.

Savino, F., L. Cordisco et al. 2009. Molecular identification of coliform bacteria from colicky breastfed infants. Acta Pediatr. 981582–1588.

Savino, F. and E. Pelle. 2007. *Lactobacillus reuteri* (American type culture collection strain 55730) versus simethicone in the treatment of infant colic: a prospective randomized study. Pediatrics 110:124–130.

Schwiertz, A., B. Gruhl, M. Lobnitz, P. Michel, M. Radke and M. Blaut. 2003. Development of the intestinal bacterial composition in hospitalized preterm infants in comparison with breastfed, full-term infants. Pediatr. Res. 54: 393–399.

Sherman, P.M. and M. Cabana et al. 2009. Potential roles and clinical utility of prebiotics in newborns, infants, and children: proceedings from a global prebiotic summit meeting. J. Pediatr. 155: S61–S70.

Sisk, P.M. et al. 2007. Early human milk feeding is associated with a lower risk of necrotizing enterocolitis in very low birth weight infants. J. Perinatol. 27: 428–433.

Szajewska, H. 2012a. Inulin and fructo-oligosaccharides for the prevention of antibiotic-associated diarrhea in children: report by the ESPGHAN Working Group on Probiotics and Prebiotics. JPGN 54: 828–829.

Szajewska, H. 2012b. Supplementation of infant formula with probiotics/prebiotics: lessons learned with regard to documenting outcomes. J. Clin. Gastroenterol. 46: S67–S68.

Szajewska, H. and A. Skorka et al. 2009. *Saccharomyces boulardii* for treating acute gastroenteritis in children: updated meta-analysis of randomized controlled trials. Alimentary Pharmacology and Therapeutics 30: 960–961.

Szajewska, H., M. Ruszczynski et al. 2006. Probiotics in the prevention of antibiotic-associated diarrhea in children: a meta-analysis of randomized controlled trials. J. Pediatr. 149: 367–372.

Szajewska, H., M. Kotowska et al. 2001a. Efficacy of *Lactobacillus* GG in prevention of nosocomial diarrhea in infants. J. Pediatr. 138: 361–365.

Szajewska, H. and J.Z. Mrukowicz. 2001b. Probiotics in the treatment and prevention of acute infectious diarrhea in infants and children: a systematic review of published, randomized, double-blind, placebo controlled trials. JPGN 33: S17–S25.

Taylor, A.L., J.A. et al. 2007. Probiotic supplementation for the first 6 months of life fails to reduce the risk of atopic dermatitis and increases the risk of allergen sensitization in high-risk children: a randomized controlled trial. J. Allergy Clin. Immunol. 119: 184–191.

Thibault, H., C. Aubert-Jacquin et al. 2004. Effects of long-term consumption of a fermented infant formula (with *Bifidobacterium breve* c50 and *Streptococcus thermophilus* O65) on acute diarrhea in healthy infants. JPGN 39: 147–152.

Thomas, D.W. and F.R. 2010. Greer, and the American Academy of Pediatrics Committee on Nutrition. Clinical report—probiotics and prebiotics in pediatrics. Pediatrics 126: 1217–1231.

Thompson, C., Y.S. McCarter, P.J. Krause and V.C. Herson. 2001. *Lactobacillus acidophilus* sepsis in a neonate. J. Perinatol. 21: 258–260.

Urban, M.F., K.D. Bolton et al. 2008. Growth of infants born to HIV-infected women, when fed a biologically acidified starter formula with and without probiotics. S. Afr J. Clin. Nutr. 21: 28–32.

van der Aa, L.B. and H.S. Heymans et al. 2010. Probiotics and prebiotics in atopic dermatitis: review of the theoretical background and clinical evidence. Pediatr. Allergy. Immunol. 21: e355–367.

Velaphi, S.C., P.A. Cooper et al. 2008. Growth and metabolism of infants born to women infected with human immunodeficiency virus and fed acidified whey-adapted starter formulas 24: 203–211.

Vendt, N., H. Grunberg et al. 2006. Growth during the first 6 months of life in infants using formula enriched with *Lactobacillus rhamnosus* GG: double-blind, randomized trial. J. Hum. Nutr. Diet. 19: 51–58.

Vivatvakin, B., A. Mahayosnond et al. 2010. Effect of a whey-predmoninant starter formula containing LCPUFAs and oligosaccharides (FOS/GOS) on gastrointestinal comfort in infants. Asia Pac. J. Clin. Nutr. 19: 473–480.

Weizman, Z. and A. Alsheikh. 2006. Safety and tolerance of a probiotic formula in early infancy comparing two probiotic agents: a pilot study.

Weizman, Z., G. Asli et al. 2005. Effects of a probiotic infant formula on infections in child care centers: comparison of two probiotic agents. Pediatrics 115: 5–9.

Wessel, M.A., J.C. Cobb et al. 1954. Paroxysmal fussing infancy, sometimes called 'colic.' Pediatrics 14: 421–435.

Westerbeek, E.A., R.L. Hensgens et al. 2011a. The effect of neutral and acidic oligosaccharides on stool viscosity, stool frequency, and stool pH in preterm infants. Acta Paediatr. 100: 1426–31.

Westerbeek, E.A., A. van den Berg et al. 2011b. The effect of enteral supplementation of a prebiotic mixture of non-human milk galacto-, fructo-, and acidic oligosaccharides on intestinal permeability in preterm infants. Br. J. Nutr. 105: 268–274.

Westerbeek, E. A. and J.P. van den Berg. 2010. Neutral and acidic oligosaccharides in preterm infants: a randomized, double-blind, placebo-controlled trial. Am. J. Clin. Nutr. 91: 679–86.

Westerbeek, E.A., A. van den Berg, H.N. Lafeber, J. Knol, W.P. Fetter and R.M. van Elburg. 2006. The intestinal bacterial colonization in preterm infants: a review of the literature. Clin. Nutr. 25: 361–368.

Wolvers, D., J.M. Antoine et al. 2010. Guidance for substantiating the evidence for beneficial effects of probiotics: prevention and management of infections by probiotics. J. Nutr. 140: 698S–712S.

7

Probiotics and Prebiotics in Lipid Metabolism

Derya Koçak Yanık,[1,] Ayşe Binnur Erbağcı[2] and Fahrettin Göğüş[1]*

Introduction

Nowadays, most consumers are aware of a strong relationship between diet and health. Although the primary role of diet is to provide nutrients, foods are no longer considered only in terms of taste and nutritional needs. The use of foods to improve health is an increasingly accepted idea. The *World Health Organization* (WHO) delineated that unhealthy diets such as those high in fat, salt and free sugar, and low in complex carbohydrates, fruits and vegetables, lead to increased risk of cardiovascular diseases (WHO 2003). Lipids have an important role in humans and abnormalities in lipid metabolism can cause serious disorders such as obesity, diabetes, etc. More recent findings show that elevated fasting triglyceride levels are associated with a greater risk of cardiovascular diseases. The WHO has predicted that by 2030, cardiovascular diseases will remain the leading causes of death, affecting approximately 23.6 million people around the World (WHO 2009). In view of this, there is extensive interest in the dietary management of

[1] Gaziantep Üniversitesi Gıda Mühendisliği Bölümü 27310 Gaziantep, Turkey.
Emails: dkocak@gantep.edu.tr; fahret@gantep.edu.tr
[2] Gaziantep Üniversitesi Tıp Fakültesi Tıbbi Biyokimya ABD, 27310 Gaziantep, Turkey.
Email: aerbagci@gantep.edu.tr
* Corresponding author

triglyceride levels. Drug therapy is largely used for this purpose, but most people are affected by unwanted side effects of such treatments. Current dietary strategies for prevention of cardiovascular disorders include low-fat/low-saturated fat diet (Taylor and Williams 1998). Although such diets are an effective therapy, they are difficult to maintain on a long-term and their efficacy diminishes over time (Pereira and Gibson 2002). There is a growing interest in alternative agents which have preventative and therapeutic potential. Probiotic and prebiotic foods fall into this category. Probiotics are live microorganisms that promote health benefits upon consumption, while prebiotics are nondigestible food ingredients that selectively stimulate the growth of beneficial microorganisms in the gastrointestinal tract. Probiotics and/or prebiotics could be used as alternative supplements to exert health benefits. Past *in vivo* studies showed that the administration of probiotics and/or prebiotics are effective in improving lipid profiles, including the reduction of serum/plasma total cholesterol, LDL-cholesterol and triglycerides or increment of HDL-cholesterol. However, other past studies have also shown that probiotics and prebiotics had insignificant effects on lipid profiles (Ooi and Liong 2010). It is still unclear which mechanisms are used by probiotics or prebiotics to bring about improved lipid metabolism. Nowadays, combining probiotics and prebiotics into "synbiotics" is a new approach to further enhance their effects.

Probiotics Foods and Lipid Metabolism

Use of probiotics goes back to ancient times. The first record of the influence of certain dairy products on blood lipids dates back more than 40 years (Pereira and Gibson 2002). The digestion and absorption of lipids are complex metabolic phenomena occurring mainly in the small intestine and some probiotics are able to interfere with this metabolism. Although the mechanisms of action are poorly understood for probiotics, some specific probiotics have a beneficial impact on lipid metabolism and some others are not active, and there is as yet no good agreement about the reason of these differences (Rabot et al. 2010b).

Several mechanisms have been proposed in order to explain the positive effect of probiotics on lipid metabolism in the past *in vivo* and *in vitro* studies. There are still a limited number of human studies which obviously show these mechanisms. Therefore it seems to be difficult to get a coherent picture about how probiotics work. Most suggested mechanisms underline the effect on serum cholesterol level. Fuller and Gibson (1998) suggested that food-derived indigestible carbohydrates are fermented to produce short-chain fatty acids (SCFAs) in the gut by probiotic bacteria. These SCFAs cause a decrease in the systemic levels of blood lipids by inhibiting hepatic cholesterol synthesis and/or redistributing cholesterol from plasma to the liver. Acetate, propionate, butyrate, and lactate are the main SCFAs which play a role on the lipid metabolism. Each SCFA has a different metabolic feature. For example butyrate is used as an energy substrate for colonocytes, acetate is potentially used as a cholesterol or fatty acid precursor and propionate is a gluconeogenic substrate in the liver, but propionate may also counteract *de novo* lipogenesis from acetate or glucose in the same tissue (Delzenne and Cani 2011).

Begley et al. (2006) suggested another mechanism which includes enzymatic deconjugation of bile acids by bile-salt hydrolase of probiotics. Bile is released into the duodenum upon ingestion of food. It consists of cholesterol, phospholipids, conjugated bile acids, bile pigments and electrolytes. Once deconjugated, bile acids are less soluble and absorbed by the intestines, leading to their elimination in the feces. Cholesterol is used to synthesize new bile acids in a homeostatic response, resulting in lowering of serum cholesterol (Begley et al. 2006). The hypocholesterolemic effect of the probiotics has also been attributed to their ability to bind cholesterol in the small intestines. Cholesterol was also removed by probiotics by incorporation into the cellular membranes during growth. Cholesterol can also be converted in the intestines to coprostanol, which is directly excreted in feces. This decreases the amount of cholesterol being absorbed, leading to a reduced concentration in the physiological cholesterol pool (Ooi and Liong 2010).

In another study, the gut microbiota has been suggested to alter fat storage through the regulation of FIAF (fasting-induced adipose factor). FIAF, produced by brown and white fat, liver and intestine, inhibits Lipoprotein lipase (LPL), regulating fatty acid oxidation in both muscle and adipose tissue. LPL promotes release of fatty acids from circulating chylomicrons and Very-low-density lipoprotein which results in their storage as triglycerides in the adipose tissue. FIAF inhibition of LPL therefore reduces fat storage (Conterno et al. 2011).

The understanding that probiotic foods can have beneficial effects on health has recently enhanced their commercial value. Yoghurt is the oldest and most widely used health promoting food to increase the number of advantageous bacteria in the intestinal tract. Nowadays, a wide range of probiotic foods such as cheese, fermented meats, ice-cream, desserts, etc. have received market interest. Some probiotic foods and their effect on lipid metabolism reported in the literature are summarized as follows:

The effect of fermented milk containing *L. acidophilus* on serum cholesterol in hypercholesterolemic humans was investigated by Anderson et al. (1999). They reported that daily consumption of 200 g of yogurt containing *L. acidophilus* after each dinner contributed to a significant reduction in serum cholesterol concentration. In another study, the effect of a low-fat yogurt containing *B. longum* was evaluated by Xiao et al.

(2003). Results from this study showed a significant decline in serum total cholesterol, LDL-cholesterol and triglycerides after 4-weeks. It has been reported that a Danish fermented milk product, Gaio®, that is produced through the action of a bacterial culture containing a strain of *Enterococcus faecium* and two strains of *Streptococcus thermophilus* was effective in reducing plasma cholesterol and LDL-cholesterol at relatively modest levels of intake. No change was observed in HDL-cholesterol or plasma triacylglycerol levels (Agerbaek et al. 1995). However, in another study with a similar milk product fermented with same bacterial culture, no significant effect on serum cholesterol levels was reported (Sessions et al. 1998). Bukowska et al. (1998) evaluated the effect of a Swedish food product, Pro viva, which contains *Lactobacillus plantarum*, on cholesterol and triglyceride levels of 30 healthy males. Total and LDL-cholesterol level in the study group reduced after 6 weeks of supplementation of the diet with Pro viva. Triglyceride, HDL-cholesterol, glucose levels and BMI remained unchanged. The effect of the daily consumption of 50 g of probiotic sausage containing *L. paracasei* LTH 2579 on immunity and blood serum lipids was investigated in healthy volunteers during several weeks (Jahreis et al. 2002). There was no significant influence of the probiotics on serum concentration of different cholesterol fractions and triacylglycerides.

Post studies have revealed controversial results. There is as yet no convincing evidence in humans that probiotics have positive effect on lipid metabolism. Thus, more studies are needed to understand the effect of probiotic foods on lipid metabolism in humans.

Prebiotic Foods and Lipid Metabolism

Prebiotics consist mainly of oligosaccharides, sugar molecules of three to six chains and soluble fiber. Prebiotic carbohydrates are found naturally in some fruits, vegetables, whole grains and legumes (Crittenden and Payne 2008). Fructo-oligosaccharides, inulins, isomalto-oligosaccharides, lactilol, lactosucrose, lactulose, oligofructose, pyrodextrins, soligosaccharides, transgalacto-oligosaccharides, and xylo-oligosaccharides are oligosaccharides classified as prebiotics which are added to processed foods and supplements (Sekhon and Jairath 2010). Nowadays, various prebiotic products are widely available in the market. While the effect of probiotics on lipid metabolism has been well-documented, prebiotics have also received much attention in lipid metabolism studies. Prebiotics have been proposed to modify serum triglyceride levels and cholesterol. This hypothesis has been supported by several mechanisms and some of the possible mechanisms have been reviewed by Pereira and Gibson (2002). One mechanism for a lipid-lowering is increasing the synthesis of fermentation byproducts (e.g., propionate) which reach the liver by the portal vein and potentially modulate the hepatic cholesterol synthesis (Chen et al. 1984, Levrat et al. 1994). Another commonly accepted mechanism for the hypo-triacylglycerolemic effect of oligofructose and inulin is a reduction in the hepatic de novo fatty acid and triacylglycerol synthesis (Fiordaliso et al. 1995, Kok et al. 1996).

In another study, Roberfroid (2000) reviewed the possible effect of inulin-type fructans on the modulation of triacylglycerol metabolism. Two effects were discussed in his study. The first effect is the modification of glucose or insulin concentrations, because dietary modulation of lipogenesis is often linked to such physiologic changes. Indeed, the induction of lipogenic enzymes by glucose, occurring via an increased gene transcription, is potentiated by insulin. The second effect is the production in the large bowel of short-chain carboxylic acids, which results in increase in the portal concentration of both acetate and propionate.

For prebiotic substances, the majority of studies have been done with the fructo-oligosaccharides inulin and oligofructose. Some relevant human studies reported in the literature are summarized in Table 1 which is modified from Delzenne et al. (2008).

Synbiotic Foods and Lipid Metabolism

Nowadays it is common to find synbiotic foods such as yogurts and fermented milks with added prebiotics in the market. A synbiotic can be defined as a mixture of probiotics and prebiotics that beneficially affects the host by improving the survival and implantation of live microbial dietary supplements in the gastrointestinal tract (Tuohy et al. 2003). Synbiotics have the potential to enhance the effect of probiotics and prebiotics and they are considered as a promising area for the development of new functional foods.

Human studies about the effectiveness of synbiotics are few in number. The effect of consumption of synbiotic milk containing 107-108 CFU/g of *Lactobacillus acidophilus* and 2.5% (g/100 g) of fructo-oligosaccharides on blood serum lipids was investigated by Schaafsma et al. (1998). Authors observed a significant decline in total cholesterol, LDL-cholesterol and LDL/HDL ratio. Kiessling et al. (2002) evaluated the hypocholesterolemic effect of a synbiotic yoghurt (containing *L. acidophilus* 145, *B. longum* 913 and oligofructose) on twenty-nine women. The authors found that the daily consumption of 300 g synbiotic yoghurt over 21 weeks significantly increased ($P < 0.002$) serum HDL cholesterol, leading to an improved ratio of LDL/HDL. In another study, Greany et al. (2008) evaluated the effect of synbiotic products (*L. acidophilus* & *B. longum* & fructooligo-saccharides) on 55 normocholesterolemic volunteers. There was no significant improvement in lipid profiles.

Table 1. Effects of some prebiotics on lipid metabolism in humans

Prebiotics	Dose (g/day)/ Duration (week)	Effects	References
FOS	8/2	Decrease in Total cholesterol and LDL cholestrol	Yamashita et al. 1984
FOS	8/5	Decrease in Total cholesterol	Hidaka et al. 1991
Inulin	9/-	Decrease in TAG and Total cholesterol	Canzi et al. 1995
FOS	20/4	No significant effect	Luo et al. 1996
Inulin	14/4	No significant effect	Pedersen et al. 1997
Inulin, FOS	17/-	No significant effect	Ellegard et al. 1997
Inulin	18/6	Decrease in Total cholesterol and LDL cholesterol	Davidson et al. 1998
Inulin	10/8	Reduce in TAG	Jackson et al. 1999
Inulin	9/4	Reduce in TAG and Total cholesterol	Brighenti et al. 1999
FOS	15/3	No significant effect	Alles et al. 1999
Inulin, FOS	15/3	No significant effect	Havenaar et al. 1999
FOS	20/4	No significant effect	Luo et al. 2000
Inulin	20/3	Reduce in TAG	Causey et al. 2000
Inulin	7/4	Reduce in TAG and Total cholesterol	Balcazar-Munoz et al. 2003
Inulin	10/3	Reduce in TAG	Letexier et al. 2003
FOS	10.6/8	No significant effect	Giacco et al. 2004
FOS	16/8	No significant effect	Daubioul et al. 2005
FOS, Inulin	10/24	No significant effect	Forcheron and Beylot 2007
GOS	5.5/20	No significant effect	Vulevic et al. 2008

FOS: Fructo- oligosaccharides, GOS: Galacto-oligosaccharides, TAG: Triacylglycerol

Impact of Prebiotics and Probiotics on Atherosclerosis

Most of the biological systems regulating food and energy intake in human is not adapted to the sedentary life style, high calorie, high fat (HF) diet, and long life expectancy which are related to the era of science, and industrialization. These homeostatic systems formerly preventing starvation and sustaining the species, now pose a threat to the health. Atherosclerosis has become more prevalent since the second half of the 21st century, and it is more common in well-developed countries and urban societies than developing countries, and rural societies respectively (Onat et al. 2012). Atherosclerosis is associated with a cluster of metabolic disorders such as obesity, insulin resistance, metabolic syndrome, fatty liver disease, and Diabetes Mellitus (DM), because they are all at the interface of nutrition and systemic low-grade inflammation (Delzenne et al. 2011). The most common forms of atherosclerosis, coronary artery disease (CAD), and cerebrovascular disease are implicated in one third of the deaths world-wide (Onat et al. 2012). Apart from obesity and visceral fat accumulation (discussed in Chapter 11) major health concern for lipid and lipoprotein metabolism is associated with atherosclerosis. Atherosclerosis develops as a consequence of dislipidemia, inflammation, oxidative stress, thrombogenesis, endothelial damage, and endoplasmic reticulum stress (Hotamisligil 2010, Hotamisligil and Erbay 2008, Polonsky 2012, Rıfai and Warnick 2006, Tabas 2010). Among these factors impaired plasma lipids and lipoproteins, low grade inflammation, lipid peroxidation, increased thrombogenesis, and increased blood pressure raise intriguing biological issues linking prebiotics, and probiotics to the atherosclerosis.

Plasma Lipids and Lipoproteins

Serum cholesterol levels in young adulthood have been shown to be proportional to the rate of atherogenesis or development of CAD later in life (NECP-ATPIII 2002, Anderson et al. 1987, Stamler et al. 2000). The robust relationship between total cholesterol and CAD found in epidemiological studies implies that an elevated low-density lipoprotein (LDL) is a powerful risk factor, as most of the total cholesterol is contained in LDL. Subsequent controlled clinical trials of LDL cholesterol lowering studies have corroborated LDL as the most abundant and clearly evident atherogenic lipoprotein (Gordon 2000). The Third Report of the Expert Panel on Detection, Evaluation, and Treatment of High Blood Cholesterol in Adults (Adult Treatment Panel III, or ATP III) on behalf of the National Cholesterol Education Program (NCEP) recommended total cholesterol, LDL cholesterol, high-density lipoprotein (HDL) cholesterol, and triglycerides as the preferred initial test, and continued identification of LDL cholesterol lowering as the primary goal of cholesterol-lowering therapy. NCEP- ATP III defined total cholesterol levels <200 mg/dL as desirable, LDL cholesterol levels <100 mg/dL as optimal, triacylglycerol level <150 mg/dL as normal, associated with a relatively low risk for CAD in populations. A low HDL cholesterol level which is strongly associated with increased risk for CAD was set as <40 mg/dL, both in men and women (NECP-ATPIII 2002).

Molecular mechanisms of probiotics and prebiotics preventing dyslipidemia

Among the beneficial effects attributed to probiotics and prebiotics, improvement of plasma lipids and lipoproteins is of particular interest (Backhed et al. 2004). An efficient member of the cholesterol-lowering drugs, namely statins inhibit 3-hydroxy-3 methylglutaryl-CoA (HMG-CoA) reductase activity. However, expression of HMG-CoA reductase after probiotics treatment was conflicting in animal studies. Some studies showed unchanged (Edwards and Moore 2003), or decreased expression of HMG-CoA reductase (Park et al. 2007). Even germ free (GF) mice on a HF diet exhibited increased cholesterol synthesis through overexpression of liver sterol regulatory element-binding protein (SREBP) 2, and up-regulation of HMG-CoA synthase 1, HMG-CoA reductase, farnesyl diphosphate synthase, squalene epoxydase, and 7-dehydrocholesterol reductase mRNAs compared to conventionally raised/HF mice despite marginally reduced plasma total and HDL cholesterols (Rabot et al. 2010a, Wostmann and Wiech 1961). The authors claimed an increase in cholesterol excretion by the liver, supported by elevated hepatic expression of membrane transporters ABCG5 and ABCG8 (McGillicuddy et al. 2009, Rabot et al. 2010a), and increased fecal cholesterol. In the enterocytes, these transporters excrete cholesterol in feces via reverse cholesterol transport, and bacterial lipopolysaccarides (LPS) are suggested to have an impact on their expression (McGillicuddy et al. 2009, Rabot et al. 2010a).

Alternative arguments for altered cholesterol metabolism associated with probiotics are reduced serum cholesterol levels through incorporation of cholesterol into the cellular membrane inhibiting formation of the intestinal cholesterol micelles, and increased production of bile salt hydrolase which catalyzes the hydrolysis of conjugated bile salt into free bile acids (Kim et al. 2004, Liong and Shah 2005). Park et al. suggest increased cholesterol absorption, and excretion, despite reduced total serum cholesterol (25%) and non-HDL cholesterol (42%) levels in rats. In this study supplementation with lactic acid bacteria increased cholesterol absorption, excretion, and formation of secondary bile acids through stimulating LDL receptor, cholesterol 7α-hydroxylase, and bile acid deconjugating/dehydroxylating enzyme expressions respectively (Park et al. 2007). Cholesterol 7α-hydroxylase is the rate limiting step in bile acid synthesis, and formation of secondary bile acids contributes to their solubility, thus both steps are important in elimination of cholesterol from the body (Rıfai and Warnick 2006). An association of bacterial endotoxin activity with dyslipidemia, insulin resistance, obesity, and chronic inflammation in humans has been shown, suggesting contribution of the inflammatory pathways (Lassenius et al. 2011). Finally, fermentation products of dietary fiber have been postulated to suppress cholesterol synthesis in the liver through improved glucose tolerance, decreased free fatty acid availability, or anti-inflammatory pathways (Jenkins et al. 1997, Nishina and Freedland 1990, Venter and Vorster 1989).

Animal studies

Lactic acid bacteria predominantly from the genera *Lactobacillus*, *Bifidobacterium*, certain *Bacillus* and *Enterococcus* species are the main probiotic bacteria associated with cholesterol lowering effects (Ali et al. 2004, Agerholm-Larsen et al. 2000a, Fabian and Elmadfa 2006). These bacteria constitute a significant proportion of probiotic cultures in nutritional supplements, pharmaceuticals and functional foods (Del Piano et al. 2006, Guo et al. 2011). Consumption of probiotic-fermented foods, mainly milk and dairy products, have been demonstrated to have cholesterol-lowering effect in rats (Lee do et al. 2009), hamsters (Chiu and Lu 2006) and pigs (Park et al. 2008). Rossi et al. have shown that *Enterococcus faecium* pure culture in an *in vitro* model, and a soy product fermented with *Enterococcus faecium* and *Lactobacillus helveticus,* in animal tests and clinical trials exhibited significant hypocholesterolemic effects (Rossi et al. 1999, 2003). Cholesterol lowering effect of *Enterococcus faecium* was not verified *in vivo*, but raised HDL cholesterol and decreased triglycerides levels were observed in hypercholesterolemic rabbits following an atherogenic diet plus *Enterococcus faecium* compared to their counterperts that consumed an atherogenic diet only. However, neither the atherosclerotic lesion area in the aortic arch nor the extent of atherosclerosis in the thoracic and abdominal aorta was reduced after 60 days of the study period (Cavallini et al. 2009). Supplementation with *Lactobacillus acidophilus* and *Lactobacillus casei* decreased plasma total cholesterol, triglycerides, LDL cholesterol, and free fatty acids in a diabetic rat model (Yadav et al. 2007). Another study showed similar fasting serum triglyceride values in both GF and conventionalized mice (Yusof et al. 2008).

High-fat diet is well-known to induce expression of hepatic HMG-CoA reductase mRNA, and activity, leading to increased cholesterol storage (Chan et al. 2008, Lin and Yin 2008). HF fed GF mice consumed less calories, exhibited reduced plasma total cholesterol and HDL cholesterol concentrations when compared with the conventionally raised/HF controls. However, a moderate accumulation of hepatic cholesterol, along with pronounced SREBP2 proteins and up-regulation of cholesterol biosynthesis genes (2.8-fold increase in hepatic HMG-CoA reductase mRNA expression) was observed, despite increased fecal cholesterol excretion. In association with lower plasma free fatty acid, and triacylglycerol concentrations, GF mice also revealed down-regulation of liver peroxisome proliferator-activated receptor gamma (PPAR-γ), and stearoyl-CoA desaturase 1 mRNAs (Rabot et al. 2010a).

Human studies

Human clinical trials that have evaluated effect of probiotic, prebiotic and synbiotic intake on biomarkers of lipid metabolism have yielded contradictory results, with some studies finding no effect (Lewis and Burnmeister 2005, Roos De et al. 1999, Simons et al. 2006), while others have identified a significant cholesterol-lowering effect (Ataie-Jafari et al. 2009, Anderson and Gilliland 1999, Xiao et al. 2003). Supplementation with *Lactobacillus fermentum* capsules for ten weeks has not caused a significant change over time or between treatments in total cholesterol, HDL cholesterol or triacylglycerol levels in subjects with total cholesterol levels ≥154 mg/dL (Simons et al. 2006). Studies with *Lactobacillus acidophilus* as the probiotic, or *Lactobacillus acidophilus+Bifidobacterium longum* and oligofructose as the synbiotics revealed a poor (3%) or no reduction in LDL cholesterol levels (Anderson and Gilliland 1999, Kiessling et al. 2002), but increased HDL cholesterol by 11,5 mg/dL in hypercholesterolemic subjects improving the ratio of LDL: HDL cholesterol (Kiessling et al. 2002). Supplementation with the probiotic bacteria *Lactobacillus plantarum* did not cause any significant changes in plasma concentrations of total cholesterol, triacylglycerol, and lipoprotein (a) in a group of heavy smokers (Naruszewicz et al. 2002). Bukowska et al. (1998) investigated role of both a fermentable oat fraction and *Lactobacillus plantarum* supplementation on LDL-cholesterol and fibrinogen, and found significant reductions. Kawase et al. (2000) showed a significant increase in HDL cholesterol and a decrease in triacylglycerol concentrations in healthy volunteers after supplementation of the diet with fermented milk containing *Lactobacillus casei* and *Streptococcus thermophilus*.

A meta-analysis study including 485 participants from 13 trials with high, borderline high and normal cholesterol levels, treated with probiotics, revealed a modest decrease in circulating levels of total cholesterol, and LDL cholesterol with pooled mean net changes –6.40 mg/dL, and –4.90 mg/dL, but no significant difference in HDL cholesterol, and triglycerides levels compared to controls (Guo et al. 2011). Another meta-analysis of six intervention studies evaluated effect of yoghurt consumption fermented with the Causido (R) culture (a strain of *Enterococcus faecium*), observed similar reductions in plasma total cholesterol and, LDL cholesterol levels with pooled net changes of –8.51 mg/dL, and –7.74 mg/dL respectively (Agerholm-Larsen et al. 2000b).

Supplementation with prebiotics such as oligofructose, galactosyl-oligosaccharides, and inulin revealed either no effect on plasma lipids or a decrease in triacylglycerol levels (Alliet et al. 2007, Giacco et al. 2004, Letexier et al. 2003, Luo et al. 2000). A meta-analysis study conducted to quantify the effects of dietary inulin-type fructans in the human showed an association with decreased serum triacylglycerols which was consistent across conditions like gender, amount fed, duration of the study, background diet, overweight, hyperlipidemia, or diabetes (Brighenti 2007). Because inulin-type fructans are not absorbed, and, at least in humans, they have no effect on postprandial blood glucose (Brighenti et al. 1999, Causey et al. 2000, Giacco et al. 2004) but can inconsistently either raise (Causey et al. 2000) or suppress (Giacco et al. 2004) insulin response, it is likely that their effects are mediated by events related to colonic fermentation (Brighenti 2007).

Insulin Resistance and Diabetes Mellitus

Diabetes is a complex, heterogeneous disorder defined as fasting blood glucose of 126 mg/dL or greater. Type 1 DM is due to selective autoimmune destruction of the pancreatic beta cell, leading to insulin deficiency, while insulin resistance is essential in the pathogenesis of type 2 DM along with impaired beta-cell function. Metabolic syndrome is a clinical phenotype, which includes insulin resistance, high waist measurement, hypertension, hypertriglyceridemia, and low HDL cholesterol levels (Rıfai and Warnick 2006, Scott et al. 2011). Risk for all forms of cardiovascular disease, including CAD is substantially increased with DM, and metabolic syndrome. Both hyperglycemia and insulin resistance have been proposed to have an implication in the pathogenesis of macrovascular complications, by several moleculer mechanisms including dislipidemia (Polonsky 2012, NECP-ATPIII 2002, Rıfai and Warnick 2006).

Effect of intestinal microflora on insulin sensitivity has been shown on GF mice fed on a HF diet. GF animals were resistant to diet-induced insulin resistance with improved glucose tolerance, reduced fasting and non-fasting insulinemia, and increased phospho-Akt (Ser-473) in adipose tissue (Rabot et al. 2010a). Backhed et al. (2004) found statistically significant elevations in fasting glucose and insulin levels, an insulin-resistant state, as defined by glucose and insulin tolerance tests, and increased fat content after conventionalization. In addition, several studies have demonstrated that antibiotic administration improves oral glucose tolerance in obese and HF diet-induced insulin resistant mice (Cani et al. 2007b, Membrez et al. 2008). In two different diabetes models in rats, supplementation with *Lactobacillius acidophilus* and *Lactobacillius casei* decreased plasma glucose, glycosylated haemoglobin, and insulin levels (Yadav et al. 2007), and decreased incremental peaks and delayed reduction of insulin secretion during oral glucose tolerance test respectively (Yadav et al. 2008).

These data suggest that not only intestinal microflora is associated with insulin resistance but modulation or reduction of the gut microbiota can be a candidate strategy in managing insulin resistance as well. However, human studies are contradictory. Supplementation with *Lactobacillus fermentum* capsules for ten weeks has not caused a significant change over time or between treatments in fasting blood glucose levels in hypercholesterolemic subjects (Simons et al. 2006).

Inflammation

Dysfunction of immune and nutrient sensing homeostatic systems underlies many chronic metabolic diseases; including atherosclerosis, type 2 DM and obesity (Erbağcı et al. 2001, 2002, Hotamisligil and Erbay 2008, Vrieze et al. 2010). Although inflammation has been shown to contribute metabolic dysregulation at several points, modulation of insulin signalling is the most crucial, as it is a dominant metabolic pathway in energy homeostasis. Antigenic components of bacteria trigger inflammatory signaling pathways and pro-inflammatory cytokine production. Organelle stress due to nutrient overload and processing defects induce other inflammatory pathways. Both conditions lead to the serine phosphorylation of Insulin receptor substrate 1 impeding the insulin signalling pathway and, leading to insulin resistant states (Hotamisligil 2006, Hotamisligil and Erbay 2008).

Main component of the gram negative bacteria wall, LPS have been postulated to be the source of endotoxaemia and inflammation associated with the gut microbiota (Cani and Delzenne 2007, Cani and Delzenne 2011, Delzenne et al. 2011, Nakamura and Omaye 2012). Modulation of intestinal microflora, e.g., by antibiotic treatment or dietary intervention with oligofructose, inhibited inflammation, reduced glucose intolerance, and decreased body weight gain in mice (Cani et al. 2007a, 2007b, Membrez et al. 2008). Additionally, a high-fat diet decreased bifidobacterium genus, and increased plasma LPS. It is suggested that transport of LPS to the circulation may be the prominent determinent of metabolic endotoxaemia rather than quantity of LPS containing bacteria in the gut. LPS are internalized into the enterocytes and transported to the lympathic circulation by chylomicrons along with dietary fats. Thus, increased lipid content of the diet enhances absorption of LPS in normal humans (Deopurkar et al. 2010, Ghoshal et al. 2009, Ji et al. 2011). Decreased gut barrier integrity, and decreased intestinal degradation of LPSs due to low alkaline phosphatase activity are also proposed to take part in enhanced transport of LPS (Cani and Delzenne 2011, Delzenne et al. 2011). In prebiotic treated animals LPS absorption decreases through an improvement of the expression and activity of proteins involved in gut barrier-function; Zonula-occludens 1 and Occludin (Delzenne et al. 2011). In addition, treatment with prebiotics improves endocannabinoid system responsiveness of the gut, consequently decreasing gut permeability, metabolic endotoxemia of LPS. Nevertheless, those effects are not verified in human studies (Cani et al. 2006, 2007b, 2009a,b, Cani and Delzenne 2009a,b, Muccioli et al. 2010). Inflammatory action of LPS are mediated by Toll-like receptor 4 (TLR4), an innate immune receptor localized on the surface of various cells. LPS-induced activation of TLR4 leads to increased production of proinflammatory cytokines and chemokines (Erridge 2011, Nakamura and Omaye 2012).

Dietary macronutrients can act as ligands of TLR4 (Dandona et al. 2010, Dasu et al. 2008, Lee et al. 2001, Wong et al. 2009). Saturated fatty acids in the diet have a structural similarity to lipid A derived from LPS, and could be recognized by pathogen sensing systems, subsequently leading to inflammation. Overnutrition, particularly saturated fatty acids stimulate inflammation through TLR4 activation (Nijhuis et al. 2009, Wong et al. 2009). Free saturated fatty acids appear to contribute to the reduced levels of GLUT4 found in type 2 DM through blocking activation of PPARγ (Armoni et al. 2005). On the contrary, Ω-3 polyunsaturated fatty acids eicosapentaenoic acid and decosahexaenoic acid exhibit anti-inflammatory and anti-diabetic properties by up regulating adiponection, and β-oxidation of free fatty acids, through activation of PPARγ (Fedor and Kelley 2009, Kalupahana et al. 2010, Kelley and Adkins 2012). PPAR-γ has suppressive effect on the inflammatory response. As a transcription factor it activates the PON1 gene increasing synthesis and release of paraoxanase 1 from the liver, which could contribute to prevention of atherosclerosis (Hamblin et al. 2009). Conjugated linoleic acid (CLA) is another agonist of redox-sensitive transcription factors PPARγ and NF-κB (Nakamura and Omaye 2009, Nakamura et al. 2012). CLA is produced from linoleic acid by *Lactobacillus rhamnosus, Lactobacillus acidophilus, Lactobacillus plantarum, Lactobacillus paracasei,* and *Lactobacillus casei* in mice (Ewaschuk et al. 2006, Kishino et al. 2002), and has health promoting properties including anti-oxidant, anti-inflammatory, anti-atherogenic, and anti-obesity effects (Bassaganya-Riera and Hontecillas 2010, Bougnoux et al. 2010, Kennedy et al. 2010, Nakamura and Omaye 2008, Nakamura and Omaye 2009, Schoeller et al. 2009).

Contribution of the gut microbiota and HF diet to development of atherosclerosis was further demonstrated by changing the gut microbiota of atherosclerotic prone apolipoprotein E (ApoE) –/– mice. Feeding with prebiotics significantly reduced development of atherosclerotic lesions as compared to the mice fed a control diet (Rault-Nania et al. 2006). Compared to TLR4 knock-out mice, wilde type mice fed a HF diet developed vascular inflammation (higher thoracic aorta IKBA-phosphorylation, ICAM, IL-6) and vascular insulin resistance (Kim et al. 2007, Timmers et al. 2008). Finally,

TLR4 or Myd88 deficiency attenuates the HF diet induced atherosclerosis, chemokine secretion and macrophage infiltration in ApoE deficient mice (Björkbacka et al. 2004a,b, Björkbacka 2006, Michelsen et al. 2004, Michelsen and Arditi 2006).

Another hypothesis linking gut microbiota to meta-inflammation could be butyrate bioavailability (Vice et al. 2005). Butyrate is an essential energy source for colon epithelial cells, and has anti-inflammatory, anti-diabetic properties (Gao et al 2009, Saemann et al. 2000, Segain et al. 2000, Vinolo et al. 2009). Consumption of non-digestible carbohydrates stimulates growth of particular butyrate-producing bacteria (*Roseburia/Eubacterium rectale* species and *Faecalibacterium prausnitzii*-cluster of *Firmicutes*) and raise plasma levels of butyrate (Louis et al. 2007, Mahowald et al. 2009).

Increased blood pressure

Increased blood pressure is among the important etiologic factors of the endothelial damage contributing to atherosclerosis. Studies about probiotics effect on blood pressure usually have given favorable results. Supplementation with *Lactobacillus plantarum* resulted in a significant reduction in systolic blood pressure in a group of heavy smokers, which was more evident in subjects with higher systolic blood pressure at baseline (Naruszewicz et al. 2002). Nakajima et al. have shown blood pressure-lowering effect of a diet supplemented with *Lactobacillus casei* in hypertensive patients (Nakajima et al. 1995). Healthy volunteers who consumed fermented milk containing *Lactobacillus casei*, and *Streptococcus thermophilus* have experienced significantly decreased systolic blood pressure (Kawase et al. 2000). Also, an indirect effect of probiotics on blood pressure has been suggested through improved insulin sensitivity, and decreased leptin levels, which takes part in modulation of neuropeptide Y and angiotensinogen release (Kazumi et al. 1999, Naruszewicz et al. 2002). However, supplementation with *Lactobacillus fermentum* capsules for ten weeks has not caused a significant change over time or between treatments in systolic or diastolic blood pressure (Simons et al. 2006).

Oxidative stress

Only limited number of studies has investigated impact of gut microbiota, or effect of prebiotic/probiotics supplementation on oxidative status. A significant decrease in lipid peroxidation marker F2-isoprostanes (31%) was observed in chronic cigarette smokers, after treatment with *Lactobacillus plantarum*, but production of reactive oxygen species by resting and stimulated monocytes was not influenced (Naruszewicz et al. 2002).

Further, changing gut microbiota by *lactobacilli* fermented goat milk feeding decreased conjugated diene level in plasma lipoprotein fraction, diminished the level of oxidized LDL and suppressed production of 8-isoprostanes (Kullisaar et al. 2003). In two different diabetes rat models, supplementation with *Lactobacillus acidophilus* and *Lactobacillus casei* revealed decreased thiobarbituric acid-reactive substances, increased reduced-glutathione in liver (Yadav et al. 2007), and decreased oxidative damage in pancreatic tissues by inhibiting lipid peroxidation, nitric oxide formation and by enhancing superoxide dismutase, catalase and glutathione peroxidase activities (Yadav et al. 2008). It can be concluded that the present data is consistent about preventive or remedial effect of prebiotics on lipid peroxidation and anti-oxidant capacity. However, molecular mechanisms linking oxidative status to prebiotics need clarifying.

Plaque progression

An obligate role for gut flora in choline, trimethylamine N-oxide (TMAO), and betaine formation from dietary lipid phosphatidylcholine has been shown. These metabolites were identified predictors of cardiovascular disease (CVD) in a metabolomics study, and further demonstrated to have dose-dependent associations with the presence of CVD, and multiple individual CVD phenotypes including peripheral artery disease, CAD, and history of myocardial infarction in an independent large clinical cohort (Wang et al. 2011). Additionally, dietary supplementation of mice with phosphatidylcholine metabolites promoted up-regulation of multiple macrophage scavenger receptors, augmented macrophage cholesterol accumulation and foam cell formation, while suppression of intestinal microflora inhibited development of atherosclerosis. The study revealed a significant positive correlation between plasma levels of TMAO and atherosclerotic plaque size. However, plasma cholesterol, triglycerides, lipoproteins, glucose levels, and hepatic triglyceride content in the mice failed to show significant increases that could account for the enhanced atherosclerosis (Wang et al. 2011). Multiple members of Flavin monooxygenases which participate in TMAO formation from dietary phosphatidylcholin were significantly correlated with aortic lesion development and HDL cholesterol concentrations (Wang et al. 2011).

Thrombogenesis

Increased thrombogenesis is implicated in acute complications of atherosclerosis such as stroke and myocardial infarction. Although studies evaluating thrombogenesis markers are limited, anti-inflammatory state favoured by prebiotics and probiotics is expected to suppress their formation as thrombogenesis is a part of acute phase reaction. Concordantly, supplementation with *Lactobacillus plantarum* resulted in a significant decrease in plasma fibrinogen and IL-6 concentrations by 21% and 41% respectively (Naruszewicz et al. 2002).

References

Agerbaek, M., L.U. Gerdes and B. Richelsen. 1995. Hypocholesterolemic effect of a new fermented milk product in healthy middle-aged men. Eur. J. Clin. Nutrit. 49: 346–352.

Agerholm-Larsen, L., M.L. Bell, G.K. Grunwald and A. Astrup. 2000b. The effect of a probiotic milk product on plasma cholesterol: a meta-analysis of short-term intervention studies. Eur. J. Clin. Nutr. 54: 856–860.

Agerholm-Larsen, L., A. Raben, N. Haulrik, A.S. Hansen, M. Manders and A. Astrup. 2000a. Effect of 8 week intake of probiotic milk products on risk factors for cardiovascular diseases. Eur. J. Clin. Nutr. 54: 288–297.

Alles, M.S., N.M. De Roos, J.C. Bakx, E. van de Lisdenk, P.L. Zock and G.A. Hautvast. 1999. Consumption of fructooligosaccharides does not favorably affect blood glucose and serum lipid concentrations in patients with type 2 diabetes. Am. J. Clin. Nutr. 69: 64–69.

Ali, A.A., M.T. Velasquez, C.T. Hansen, A.I. Mohamed and S.J. Bhathena. 2004. Effects of soybean isoflavones, probiotics, and their interactions on lipid metabolism and endocrine system in an animal model of obesity and diabetes. J. Nutr. Biochem. 15: 583–590.

Alliet, P., P. Scholtens, M. Raes, K. Hensen, H. Jongen, J.L. Rummens, G. Boehm and Y. Vandenplas. 2007. Effect of prebiotic galacto-oligosaccharide, long-chain fructo-oligosaccharide infant formula on serum cholesterol and triacylglycerol levels. Nutrition 23: 719–723.

Anderson, J.W. and S.E. Gilliland. 1999. Effect of fermented milk (yoghurt) containing *Lactobacillus acidophilus* L1 on serum cholesterol in hypercholesterolemic humans. J. Am. Coll. Nutr. 18: 43–50.

Anderson, K.M., W.P. Castelli and D. Levy. 1987. Cholesterol and mortality: 30 years of follow-up from the Framingham Study. JAMA 257: 2176–2180.

Armoni, M., C. Harel, F. Bar-Yoseph, S. Milo and E. Karnieli. 2005. Free fatty acids repress the GLUT4 gene expression in cardiac muscle via novel response elements. J. Biol. Chem. 80: 34786–34795.

Ataie-Jafari, A., B. Larijani and H.A. Majd. 2009. Cholesterol-lowering effect of probiotic yogurt in comparison with ordinary yogurt in mildly to moderately hypercholesterolemic subjects. Ann. Nutr. Metab. 54: 22–27.

Backhed, F., H. Ding, T. Wang, L.V. Hooper, G.Y. Koh, A. Nagy, C.F. Semenkovich and J.I. Gordon. 2004. The gut microbiota as an environmental factor that regulates fat storage. PNAS 104: 15718–15723.

Balcazar-Munoz, B.R., E. Martinez-Abundis and M. Gonzalez-Ortiz. 2003. Effect of oral inulin administration on lipid profile and insulin sensitivity in dyslipidemic obese subjects. Rev. Med. Chil. 131: 597–604.

Bassaganya-Riera, J. and R. Hontecillas. 2010. Dietary conjugated linoleic acid and n-3 polyunsaturated fatty acids in inflammatory bowel disease. Curr. Opin. Clin. Nutr. 13: 569–573.

Begley, M., C. Hill and C.G.M. Gahan. 2006. BileSalt Hydrolase Activity in Probiotics. Appl. Environ. Microbiol. 72: 1729–1738.

Bjorkbacka, H. 2006. Multiple roles of Toll-like receptor signaling in atherosclerosis. Curr. Opin. Lipidol. 17: 527–533.

Björkbacka, H., K.A. Fitzgerald, F. Huet, X. Li, J.A. Gregory, M.A. Lee, C.M. Ordija, N.E. Dowley, D.T. Golenbock and M.W. Freeman. 2004a. The induction of macrophage gene expression by LPS predominantly utilizes Myd88-independent signaling cascades. Physiol. Genomics 19: 319–330.

Björkbacka, H., V.V. Kunjathoor, K.J. Moore, S. Koehn, C.M. Ordija, M.A. Lee, T. Means, K. Halmen, A.D. Luster, D.T. Golenbock and M.W. Freeman. 2004b. Reduced atherosclerosis in MyD88-null mice links elevated serum cholesterol levels to activation of innate immunity signaling pathways. Nat. Med. 10: 416–421.

Bougnoux, P., N. Hajiaji, K. Maheo, C. Couet and S. Chevalier. 2010. Fatty acids and breast cancer: sensitization to treatments and prevention of metastatic re-growth. Prog. Lipid. Res. 49: 76–86.

Brighenti, F., M.C. Casiraghi, E. Canzi and A. Ferrari. 1999. Effect of consumption of a ready-to-eat breakfast cereal containing inulin on the intestinal milieu and blood lipids in healthy male volunteers. Eur. J. Clin. Nutr. 53: 726–33.

Brighenti, F. 2007. Dietary Fructans and Serum Triacylglycerols: A Meta-Analysis of Randomized Controlled Trials. The Journal of Nutrition 137: 2552–2556.

Bukowska, H., J. Pieczul-Mroz, M. Jastrzebska, K. Chelstowski and M. Naruszewicz. 1998. Decrease in fibrinogen and LDL-cholesterol levels upon supplementation of diet with *Lactobacillus plantarum* in subjects with moderately elevated cholesterol. Atherosclerosis 137: 437–438.

Cani, P.D., C. Knauf, M.A. Iglesias, D.J. Drucker, N.M. Delzenne and R. Burcelin. 2006. Improvement of glucose tolerance and hepatic insulin sensitivity by oligofructose requires a functional glucagon-like Peptide 1 receptor. Diabetes 55: 1484–1490.

Cani, P.D., J. Amar, M.A. Iglesias, M. Poggi, C. Knauf, D. Bastelica, A.M. Neyrinck, F. Fava, K.M. Tuohy, C. Chabo, A. Waget, E. Delmee, B. Cousin, T. Sulpice, B. Chamontin, J. Ferrieres, J.F. Tanti, G.R. Gibson, L. Casteilla, N.M. Delzenne, M.C. Alessi and R. Burcelin. 2007a. Metabolic endotoxemia initiates obesity and insulin resistance. Diabetes 56: 1761–1772.

Cani, P.D., A.M. Neyrinck, F. Fava, C. Knauf, R.G. Burcelin, K.M. Tuohy, G.R. Gibson and N.M. Delzenne. 2007b. Selective increases of bifidobacteria in gut microflora improve high-fat-diet-induced diabetes in mice through a mechanism associated with endotoxaemia. Diabetologia 50: 2374–2383.

Cani, P.D., E. Lecourt, E.M. Dewulf, F.M. Sohet, B.D. Pachikian, D. Naslain, B.F. De, A.M. Neyrinck and N.M. Delzenne. 2009a. Gut microbiota fermentation of prebiotics increases satietogenic and incretin gut peptide production with consequences for appetite sensation and glucose response after a meal. Am. J. Clin. Nutr. 90: 1236–1243.

Cani, P.D., S. Possemiers, W.T. Van de, Y. Guiot, A. Everard, O. Rottier, L. Geurts, D. Naslain, A. Neyrinck, D.M. Lambert, G.G. Muccioli and N.M. Delzenne. 2009b. Changes in gut microbiota control inflammation in obese mice through a mechanism involving GLP-2-driven improvement of gut permeability. Gut. 58: 1091–1103.

Cani, P.D. and N.M. Delzenne. 2007. Gut microflora as a target for energy and metabolic homeostasis. Curr. Opin. Clin. Nutr. Metab. Care. 10: 729–734.

Cani, P.D. and N.M. Delzenne. 2009a. The role of the gut microbiota in energy metabolism and metabolic disease. Curr. Pharm. Des. 15: 1546–1558.

Cani, P.D. and N.M. Delzenne. 2009b. Interplay between obesity and associated metabolic disorders: new insights into the gut microbiota. Curr. Opin. Pharmacol. 9: 737–743.

Cani, P.D. and N.M. Delzenne. 2011. The gut microbiome as therapeutic target. Pharmacol. Ther. 130: 202–212.

Canzi, E., F. Brighenti, M.C. Casiraghi, E. Del Puppo and A. Ferrari. 1995. Prolonged consumption of inulin in ready-to-eat breakfast cereals: effects on intestinal eco-system, bowel habits and lipid metabolism. *In:* Proceedings of COST Action 92 workshop-Dietary fibre and fermentation in the colon., Finland, pp. 280–284.

Causey, J.L., J.M. Feirtag, D.D. Gallaher, B.C. Tungland and J.L. Slavin. 2000. Effects of dietary inulin on serum lipids, blood glucose and the gastrointestinal environment in hypercholesterolemic men. Nutr. Res. 20: 191–201.

Cavallini, D.C., R. Bedani, L.Q. Bomdespacho, R.C. Vendramini and E.A. Rossi. 2009. Effects of probiotic bacteria, isoflavones and simvastatin on lipid profile and atherosclerosis in cholesterol-fed rabbits: a randomized double-blind study. Lipids Health Dis. 8: 1.

Chan, M.Y., Y. Zhao and C.K. Heng. 2008. Sequential responses to high-fat and high-calorie feeding in an obese mouse model. Obesity 16: 972–978.

Chen, W.-J.L. and J.W. Anderson and D. Jennings. 1984. Propionate may mediate the hypocholesterolemic effects of certain soluble plant fibers in cholesterol-fed rats. Proceed. Soc. Exp. Biol. Med. 175: 215–218.

Chiu, C.H. and T.Y. Lu. 2006. The effects of *Lactobacillus*-fermented milk on lipid metabolism in hamsters fed on high-cholesterol diet. Appl. Microbiol. Biotechnol. 71: 238–245.

Conterno, L., F. Fava, R. Viola and K.M. Tuohy. 2011. Obesity and the gut microbiota: does up-regulating colonic fermentation protect against obesity and metabolic disease? Genes Nutr. 6: 241–260.

Crittenden, R. and M.J. Payne. 2008. Nutrition News. Facts and functions of prebiotics, probiotics and synbiotics. *In:* Lee Y.K. and S. Salminen [eds.]. Handbook of Probiotics and Prebiotics. Wiley-Interscience, New York, USA, pp. 535–582.

Dandona, P., H. Ghanim, A. Chaudhuri, S. Dhindsa and S.S. Kim. 2010. Macronutrient intake induces oxidative and inflammatory stress: potential relevance to atherosclerosis and insulinresistence. Exp. Mol. Med. 42: 245–253.

Dasu, M.R., S. Devaraj, L. Zhao, D.H. Hwang and I. Jialal. 2008. High glucose induced Toll like receptor expression in human monocytes. Diabetes. 57: 3090–3098.

Daubioul, C.A., Y. Horsmans, P. Lambert, E. Danse and N.M. Delzenne. 2005. Effects of oligofructose on glucose and lipid metabolism in patients with nonalcoholic steatohepatitis: results of a pilot study. Eur. J. Clin. Nutr. 59: 723–726.

Davidson, M.H., K.C. Maki, C. Synecki, S.A. Torri and K.B. Drennan. 1998. Effects of dietary lipids in men and women with hypercholestrolemia. Nutr. Res. 18: 503–517.

Del Piano, M., L. Morelli, G.P. Strozzi, S. Allesina, M. Barba, F. Deidda, P. Lorenzini, M. Ballare, F. Montino, M. Orsello, M. Sartori, E. Garello, S. Carmagnola, M. Pagliarulo and L. Capurso. 2006. Probiotics: from research to consumer. Dig. Liver. Dis. 38: 248–255.

Delzenne, M.N., D.P. Cani and M.A. 2008. Neyrinck. Prebiotics and Lipid metabolism. *In:* J. Versalovic and M. Wilson [eds.]. Therapeutic Microbiology: Probiotics and Related Strategies. ASM Press, Washington, DC, pp. 183–192.

Delzenne, N.M. and P.D. Cani. 2011. Interaction Between Obesity and the Gut Microbiota: Relevance in Nutrition. Annu. Rev. Nutr. 31: 15–31.

Delzenne, N.M., A.M. Neyrinck and P.D. Cani. 2011. Modulation of the gut microbiota by nutrients with prebiotic poperties: consequences for host health in the context of obesity and metabolic syndrome. Microbial Cell Factories. 10(Suppl 1):S10 doi:10.1186/1475-2859-10-S1-S10.

Deopurkar, R., H. Ghanim, J. Friedman, S. Abuaysheh, C.L. Sia, P. Mohanty, P. Viswanathan, A. Chaudhuri and P. Dandona. 2010. Differential effects of cream, glucose, and orange juice on inflammation, endotoxin, and the expression of Toll-like receptor-4 and suppressor of cytokine signaling-3. Diabetes Care 33: 991–997.

Edwards, J.E. and R.A. Moore. 2003. Statins in hypercholesterolaemia: a dose-specific meta-analysis of lipid changes in randomized, double blind trials. B.M.C. Fam. Pract. 4: 1–19.

Ellegard, L., H. Andersson and I. Bosaeus. 1997. Inulin and oligofructose do not in-fluence the absorption of cholesterol, or the excretion of cholesterol, Ca, Mg, Zn, Fe, or bile acids but increases energy ex-cretion in ileostomy subjects. Eur. J. Clin. Nutrit. 51: 1–5.

Erbağcı, A.B., M. Tarakçıoğlu, Y. Coşkun, E. Sivaslı and E.S. Namıduru. 2001 Mediators of inflammation in children with type1 Diabetes Mellitus. Clin. Biochem. 38: 645–650.

Erbağcı, A.B., M. Tarakçıoğlu, M. Aksoy, R. Kocabaş, M. Nacak, Ş. Aynacıoğlu and C. Sivrikoz. 2002. Diagnostic value of CRP and Lp(a) in coronary heart disease. Acta. Cardiol. 57: 197–204.

Erridge, C. 2011. The capacity of food stuffs to induce innate immune activation of human monocytes *in vitro* is dependent on food content of stimulants of Toll-like receptors 2 and 4. Br. J. Nutr. 105: 15–23.

Ewaschuk, J.B., J.W. Walker, H. Diaz and K.L. Madsen. 2006. Bioproduction of conjugated linoleic acid by probiotic bacteria occurs in vitro and in vivo in mice. J. Nutr. 136: 1483–1487.

Fabian, E. and I. Elmadfa. 2006. Influence of daily consuption of probiotic and conventional yoghurt on the plasma lipid profile in young healthy women. Ann. Nutr. Metab. 50: 387–393.

Fedor, D. and D.S. Kelley. 2009. Prevention of insulin resistance by n-3 polyunsaturated fatty acids. Curr. Opin. Clin. Nutr. 12: 138–146.

Fiordaliso, M., N. Kok, J.P. Desaher, F. Goethals, D. Deboyser, M. Roberfroid and N. Delzenne. 1995. Dietary oligofructose lowers triglycerides, phospholipids and cholesterol in serum and very low density lipoproteins of rats. Lipids 30: 163–167.

Forcheron, F. and M. Beylot. 2007. Long-term administration of inulin-type fructans has no significant lidid-lowering effect in normolipidemic humans. Metabolism 56: 1093–1098.

Fuller, R. and G. Gibson. 1998. Probiotics and prebiotics. *In:* Encyclopaedia of Food Microbiology. 1633–1639, Academic Press.

Gao, Z., J. Yin, J. Zhang, R.E. Ward, R.J. Martin, M. Lefevre, W.T. Cefalu and J. Ye. 2009. Butyrate improves insulin sensitivity and increases energy expenditure in mice. Diabetes 58: 1509–1517.

Ghoshal, S., J. Witta, J. Zhong, W. de Williers and E. Eckhardt. 2009. Chylomicrons promote intestinal absorption of lipopolysaccharides. J. Lipid Res. 50: 90–97.

Giacco, R., G. Clemente, D. Luongo, G. Lasorella, I. Fiume, F. Brouns, F. Bornet, L. Patti, P. Cipriano, A.A. Rivellese and G. Riccardi. 2004. Effects of short-chain fructo-oligosaccharides on glucose and lipid metabolism in mild hypercholesterolaemic individuals. Clin. Nutr. 23: 331–340.

Gordon, D.J. 2000. Cholesterol lowering reduces mortality: the statins. *In:* S.M. Grundy (ed.). cholesterol-lowering therapy: evaluation of clinical trial evidence. Marcel Dekker Inc. 299–311.

Greany, K.A., M.J.L. Bonorden, J.M. Halmiton-Reeves, M.H. McMullen, K.E. Wangen, W.R. Phipps, J. Feirtag, W. Thomas and M.S. Kurzer. 2008. Probiotic Capsules Do Not Lower Plasma Lipid in Young Women and Men. Eur. J. Clin. Nutr. 62: 232–237.

Guo, Z., X.M. Liu, Q.X. Zhang, Z. Shen, F.W. Tian, H. Zhang, Z.H. Sun, H.P. Zhang and W. Chen. 2011. Influence of consumption of probiotics on the plasma lipid profile: A meta-analysis of randomised controlled trials. Nutrition, Metabolism & Cardiovascular Diseases 21: 844–850.

Hamblin M., L. Chang, Y. Fan, J. Zhang and Y.E. Chen. 2009. PPARs and the Cardiovascular System. Antioxid. Redox Signal 11: 1415–52.

Havenaar, R., S. Bonnin-Marol, W. Van Dokkum, S. Petitet and G. Schaafsma. 1999. Inulin: fermentation and microbial ecology in the intestinal track. Food Rev. Int. 15: 109–120.

Hidaka, H., Y. Tashiro and T. Eida. 1991. Proliferation of bifidobacteria by oligosaccharides and their useful effect on human health. Bifidobacteria Microflora 10: 65–79.

Hotamisligil, G.S. 2006. Inflammation and metabolic disorders. Nature 444: 860–867.

Hotamisligil, G.S. 2010. Endoplasmic reticulum stress and atherosclerosis. Nat. Med. 6: 396–399.

Hotamisligil, G.S. and E. Erbay. 2008. Nutrient sensing and inflammation in metabolic diseases. Nat. Rev. Immunol. 8: 923.

Jackson, K.G., G.R.J. Taylor, A.M. Clohessy and C.M. Williams. 1999. The effect of the daily intake of inulin on fasting lipid insulin and glucose consentrations in middle aged men and women. Br. J. Nutr. 82: 23–30.

Jahreis, G., H. Vogelsang, G. Kiessling, R. Schubert, C. Bunte and W.P. Hammes. 2002. Influence of probiotic sausage (Lactobacillus paracasei) on blood lipids and immunological parameters of healthy volunteers. Food Res. Int. 35: 133–138.

Ji, Y., Y. Sakata and P. Tso. 2011. Nutrient-induced inflammation in the intestine. Curr. Opin. Clin. Nutr. 14: 315–321.

Jenkins, D.J., A.R. Leeds, M.A. Gassull, B. Cochet and G.M. Alberti. 1997. Decrease in postprandial insulin and glucose concentrations by guar and pectin. Ann. Intern. Med. 86: 20–23.

Kalupahana, N.S., K.J. Claycombe, S.J. Newman, T. Stewart, N. Siriwardhana, N. Mathhan, A.H. Lichtenstein and N. Moustaid-Moussa. 2010. Eicosapentaenoic acid prevents and reverses insulin resistance in high-fat diet-induced obese mice via modulation of adipose tissue inflammation. J. Nutr. 140: 1915–1922.

Kawase, M., H. Hashimoto, M. Hosoda, H. Morita and A. Hosono. 2000. Effect of administration of fermented milk containing whey protein concentrate to rats and healthy men on serum lipids and blood pressure. J. Dairy. Sci. 83: 255–263.

Kazumi, T., A. Kawaguchi, J. Katoh, M. Iwahashi and G. Yoshino. 1999. Fasting insulin and leptin serum levels are associated with systolic blood pressure independent of percentage body fat and body mass index. J. Hypertens 17: 1451–1455.

Kelley, D.S. and Y. Adkins. 2012. Similarities and differences between the effects of EPA and DHA on markers of atherosclerosis in human subjects. P. Nutr. Soc. 28: 1–10.

Kennedy, A., K. Martinez, S. Schmidt, S. Mandrup, K. Lapoint and M.K. McIntosh. 2010. Antiobesity mechanisms of action of conjugated linoleic acid. J. Nutr. Biochem. 21: 171–179.

Kiessling, G., J. Schneider and G. Jahreis. 2002. Long-Term Consumption of Fermented Dairy Products over 6 Months Increases HDL Cholesterol. Eur. J. Clin. Nutr. 56: 843–849.

Kim, F., M. Pham, I. Luttrell, D.D. Bannerman, J. Tupper, J. Thaler, T.R. Hawn, E.W. Raines and M.W. Schwartz. 2007. Toll-like receptor-4 mediates vascular inflammation and insulin resistance in diet-induced obesity. Circ. Res. 100: 1589–1596.

Kim, G.B., S.H. Yi and B.H. Lee. 2004. Purification and characterization of three different types of bile salt hydrolases from *Bifidobacterium* strains. J. Dairy. Sci. 87: 258–266.

Kishino, S., J. Ogawa, Y. Omura, K. Matsumura and S. Shimuzu. 2002. Conjugated linoleic acid production from linoleic acid by lactic acid bacteria. JAOCS 79: 159–163.

Kok, N., M. Roberfroid, A. Robert and N. Delzenne. 1996. Involvement of lipo-genesis in the lower VLDL secretion induced by oligofructose in rats. Br. J. Nutrit. 76: 881–890.

Kullisaar, T., E. Songisepp, M. Mikelsaar, K. Zilmer, T. Vihalemm and M. Zilmer. 2003. Antioxidative probiotic fermented goats' milk decreases oxidative stress-mediated atherogenicity in human subjects. Br. J. Nutr. 90: 449–456.

Lassenius, M.I., K.H. Pietilainen, K. Kaartinen, P.J. Pussinen, J. Syrjanen, C. Forsblom, I. Porsti, A. Rissanen, J. Kaprio, J. Mustonen, P.H. Groop and M. Lehto. 2011. Bacterial endotoxin activity in human serum is associated with dyslipidemia, insulin resistance, obesity, and chronic inflammation. Diabetes Care 34: 1809–1815.

Lee do, K., S. Jang, E.H. Baek, M.J. Kim, K.S. Lee, H.S. Shin, M.J. Chung, J.E. Kim, K.O. Lee and N.J. Ha. 2009. Lactic acid bacteria affect serum cholesterol levels, harmful fecal enzyme activity, and fecal water content. Lipids Health. Dis. 11: 8–21.

Lee, J.Y., K.H. Sohn, S.H. Rhee and D. Hwang. 2001. Saturated fatty acids, but not unsaturated fatty acids, induce the expression of cyclooxygenase-2 mediated through Toll-like receptor 4. J. Biol. Chem. 276: 16683–16689.

Letexier, D., F. Diraison and M. Beylot. 2003. Addition of inulin to a moderately high-carbohydrate diet reduces hepatic lipogenesis and plasma triacylglycerol concentrations in human subjects. Am. J. Clin. Nutr. 77: 559–564.

Levrat, M.A., M.L. Favier, C. Moundras, C. Remesy, C. Demigne and C. Morand. 1994. Role of dietary propionic acid and bile acid excretion in the hypocholesterolemic effects of oligosaccharides in rats. J. Nutrit. 124: 531–538.

Lewis, S.J. and S. Burmeister. 2005. A double-blind placebo-controlled study of the effects of *Lactobacillus acidophilus* on plasma lipids. Eur. J. Clin. Nutr. 59: 776–780.

Lin, C.C. and M.C. Yin. 2008. Effects of cysteine-containing compounds on biosynthesis of triacylglycerol and cholesterol and anti-oxidative protection in liver from mice consuming a high-fat diet. Br. J. Nutr. 99: 37–43.

Liong, M.T. and N.P. Shah. 2005. Acid and bile tolerance and the cholesterol removal ability of *Lactobacilli* strains. J. Dairy. Sci. 88: 55–66.

Louis, P., K.P. Scott, S.H. Duncan and H.J. Flint. 2007. Understanding the effects of diet on bacterial metabolism in the large intestine. J. Appl. Microbiol. 102: 1197–1208.

Luo, J., S.W. Rizkalla, C. Alamowitch, A. Boussarri, A. Blayo, J.L. Barry, A. Laffitte, F. Guyon, F.R. Bornet and G. Slama. 1996. Chronic consumption of short-chain fructooligosaccharides by healthy subjects decreased basal hepatic glucose production but had no effect on insulin-stimulated glucose metabolism. Am. J. Clin. Nutr. 63: 939–945.

Luo, J., M. Van Yperselle, S.W. Rizkalla, F. Rossi, F.R. Bornet and G. Slama. 2000. Chronic consumption of short-chain fructooligosaccharides does not affect basal hepatic glucose production or insulin resistance in type 2 diabetics. J. Nutr. 130: 1572–1577.

Mahowald, M.A., F.E. Rey, H. Seedorf, P.J. Turnbaugh, R.S. Fulton, A. Wollam, N. Shah, C. Wang, V. Magrini, R.K. Wilson, B.L. Cantarel, P.M. Coutinho, B. Henrissat, L.W. Crock, A. Russell, N.C. Verberkmoes, R.L. Hettich and J.I. Gordon. 2009. Characterizing a model human gut microbiota composed of members of its two dominant bacterial phyla. Proc. Natl. Acad. Sci. 106: 5859–5864.

McGillicuddy, F.C., M.M. de la Llera, C.C. Hinkle, M.R. Joshi, E.H. Chiquoine, J.T. Millheimer, G.H. Rothblat and M.P. Reilly. 2009. Inflammation impairs reverse cholesterol transport *in vivo*. Circulation 119: 1135–1145.

Membrez, M., F. Blancher, M. Jaquet, R. Bibiloni, P.D. Cani, R.G. Burcelin, I. Corthesy, K. Mace and C.J. Chou. 2008. Gut microbiota modulation with norfloxacin and ampicillin enhances glucose tolerance in mice. FASEB J. 22: 2416–2426.

Michelsen, K.S. and M. Arditi. 2006. Toll-like receptor signaling and atherosclerosis. Curr. Opin. Hematol. 13: 163–168.

Michelsen, K.S., M.H. Wong, P.K. Shah, W. Zhang, J. Yano, T.M. Doherty, S. Akira, T.B. Rajavashisth and M. Arditi. 2004. Lack of Toll-like receptor 4 or myeloid differentiation factor 88 reduces atherosclerosis and alters plaque phenotype in mice deficient in apolipoprotein E. Proc. Natl. Acad. Sci. 101: 10679–10684.

Muccioli, G.G., D. Naslain, F. Backhed, C.S. Reigstad, D.M. Lambert, N.M. Delzenne and P.D. Cani. 2010. The endocannabinoid system links gut microbiota to adipogenesis. Mol. Syst. Biol. 6: 392.

Nakajima, K., Y. Hata, Y. Osono, M. Hamura, S. Kobayashi and M. Watanuki. 1995. Antihypertensive effect of extract of *Lactobacillus casei* in patients with hypertension. J. Clin. Biochem. Nutr. 18: 181–187.

Nakamura, Y.K. and S.T. Omaye. 2009. Conjugated linoleic acid isomers' roles in the regulation of PPARγ and NF-κBDNA binding and subsequent expression of antioxidant enzymes in human umbilical vein endothelial cells. Nutrition 25: 800–811.

Nakamura, Y.K., M.A. Dubick and S.T. Omaye. 2012. Modulation of oxidative stress by gamma-glutamylcysteine (GGC) and conjugated linoleic acid (CLA) in human umbilical vein endothelial cells. Food Chem. Tox. 50: 1854–1859.

Nakamura, Y.K. and S.T. Omaye. 2012. Metabolic diseases and pro- and prebiotics: Mechanistic insights. Nutrition & Metabolism 9: 60.

Nakamura, Y.K. and S.T. Omaye. 2008. Conjugated linoleic acid modulation of risk factors associated with atherosclerosis. Nutr. Metab. 5: 22.

[NECP-ATPIII], National Cholesterol Education Program (Third Report) 2002. Detection, Evaluation, and Treatment of High Blood Cholesterol in Adults (Adult Treatment Panel III). National Heart, Lung, and Blood Institute National Institutes of Health NIH Publication. No. 02-5215.

Naruszewicz, M., M.L. Johansson, D.Z. Downar and H. Bukowska. 2002. Effect of *Lactobacillus plantarum* 299v on cardiovascular disease risk factors in smokers. Am. J. Clin. Nutr. 76: 1249–1255.

Nijhuis, J., S.S. Rensen, Y. Slaats, F.M.H. van Dielen, W.A. Buurman and J.W.M. Greve. 2009. Neutrophil activation in morbid obesity, chronic activation of acute inflammation. Obesity 17: 2014–2018.

Nishina, P.M. and R.A. Freedland. 1990. Effects of propionate on lipid biosynthesis in isolated rat hepatocytes. J. Nutr. 120: 668–672.

Onat, A., M. Aydın, B. Köroğlu, E. Örnek, S. Altay, E. Çelik and A. Karagöz. 2012. TEKHARF 2011 taraması: Ölümler ve uzun vadeli takipte performans. Arch. Turk. Soc. Cardiol. 40: 117–121.

Ooi, L.-G. and M.-T. Liong. 2010. Cholesterol-Lowering Effects of Probiotics and Prebiotics: A Review of *in vivo* and *in vitro* Findings. Int. J. Mol. Sci. 11: 2499–2522.

Park, Y.H., J.G. Kim, Y.W. Shin, S.H. Kim and K.Y. Whang. 2007. Effect of dietary inclusion of Lactobacillus acidophilus ATCC 43121 on cholesterol metabolism in rats. J. Microbiol. Biotechnol. 17: 655–662.

Park, Y.H., J.G. Kim, Y.W. Shin, H.S. Kim, Y.J. Kim, T. Chun, S.H. Kim and K.Y. Whang. 2008. Effects of *Lactobacillus acidophilus* 43121 and a mixture of *Lactobacillus casei* and *Bifidobacterium longum* on the serum cholesterol level and fecal sterol excretion in hypercholesterolemia-induced pigs. Biosci. Biotechnol. Biochem. 72: 595–600.

Pedersen, A., B. Sandstrom and M.M.J. VanAmelsvoort. 1997. The effect of ingestion of inulin on blood lipids and gastrointestional symptoms in healthy females. Br. J. Nutr. 78: 215–222.

Pereira, D.I.A. and G.R. Gibson. 2002. Effects of Consumption of Probiotics and Prebiotics on Serum Lipid Levels in Humans. Crit. Rev. Biochem. Mol. Biol. 37: 259–281.

Polonsky, K.S. 2012. The Past 200 Years in Diabetes. N. Engl. J. Med. 367: 1332–13340.

Rabot, S., M. Membrez, A. Bruneau, P. Gerard, T. Harach, M. Moser, F. Raymond, R. Mansourian and C.J. Chou. 2010a. Germ-free C57BL/6J mice are resistant to high-fat-diet-induced insulin resistance and have altered cholesterol metabolism. The FASEB Journal 24: 4948–4959.

Rabot, S., J. Rafter, G.T. Rijkers, B. Watzl and J.-M. Antoine. 2010b. Guidance for Substantiating the Evidence for Beneficial Effects of Probiotics: Impact of Probiotics on Digestive System Metabolism. J. Nutr. 140: 677–689.

Rault-Nania, M.H., E. Gueux, C. Demougeot, C. Demigne, E. Rock and A. Mazur. 2006. Inulin attenuates atherosclerosis in apolipoprotein E deficient mice. Br. J. Nutr. 96: 840–844.

Rıfai N. and R. Warnick. 2006. Lipids, lipoproteins, apolipoproteins, and other cardiovascular risk factors. *In:* C.A. Burtis and D.E. Bruns [eds.]. Tietz Textbook of Clinical Chemistry and Molecular Diagnostics. Elsevier, New York, USA, pp. 903–982.

Roberfroid, M.B. 2000. Prebiotics and probiotics: are they functional foods? Am. J. Clin. Nutr. 71: 1682–1687.

Roos De, N.M., G. Schouten and M.B. Katan. 1999. Yogurt enriched with *Lactobacillus acidophilus* does not lower blood lipids in healthy men and women with normal to borderline high serum cholesterol levels. Eur. J. Clin. Nutr. 53: 277–280.

Rossi, E.A., R.C. Vendramini, I.Z. Carlos, I.C. Pey and G.F. Valdez. 1999. Development of a novel fermented soymilk product with potential probiotic properties. Eur. Food. Res. Technol. 209: 305–307.

Rossi, E.A., C.R. Vendramini, I.Z. Carlos, M.G. Oliveira and G.F. Valdez. 2003. Efeito de um novo produto fermentado de soja sobre lípides séricos de homens adultos normocolesterolêmicos. Arch. Latinoam. Nutr. 53: 47–51.

Saemann, M.D., G.A. Bohmig, C.H. Osterreicher, H. Burtscher, O. Parolini, C. Diakos, J. Stöckl, W.H. Hörl and G.J. Zlabinger. 2000. Antiinflammatory effects of sodium butyrate on human monocytes: potent inhibition of IL-12 and up-regulation of IL-10 production. FASEB J. 14: 2380–2382.

Schaafsma, G., W.J.A. Meuling, W. van Dokkum and C. Bouley. 1998. Effects of A Milk Product, Fermented by Lactobacillus acidophilus and with Fructo-Oligosaccharides Added, on Blood Lipids in Male Volunteers. Eur. J. Clin. Nutr. 52: 436–440.

Schoeller, D.A., A.C. Watras and L.D. Whigham. 2009. A meta-analysis of the effects of conjugated linoleic acid on fat-free mass in humans. Appl. Physiol. Nutr. Me. 34: 975–978.

Scott, R., M. Donoghoe, G.F. Watts, R. O'Brien, C. Pardy, M.R. Taskinen, T.M. Davis, P.G. Colman, P. Manning, G. Fulcher and A.C. Keech. 2011. FIELD Study Investigators. Impact of metabolic syndrome and its components on cardiovascular disease event rates in 4900 patients with type 2 diabetes assigned to placebo in the FIELD randomised trial. Cardiovasc. Diabetol. 21; 10: 102.

Segain, J.P., D. Raingeard de la Blétière, A. Bourreille, V. Leray, N. Gervois, C. Rosales, L. Ferrier, C. Bonnet, H.M. Blottière and J.P. Galmiche. 2000. Butyrate inhibits inflammatory responses through NFkappaB inhibition: implications for Crohn's disease. Gut. 47: 397–403.

Sekhon, B.S. and S. Jairath. 2010. Prebiotics, probiotics and synbiotics: an overview. J. Pharm. Educ. Res. 1: 13–36.

Sessions, V.A., J.A. Lovegrove, T.S., Dean, C.M. Williams, T.A.B. Sanders, I.A. Macdonald and A.M. Salter. 1998. The effect of a new fermented milk product on plasma cholesterol and apolipoprotein B concentrations in middle-aged men and women. *In:* M.J. Sadler [ed.]. Functional Foods the consumer, the products and the evidence. Royal Society of Chemistry, UK, pp. 15–19.

Simons, L.A., S.G. Amansec and P. Conway. 2006. Effect of *Lactobacillus* fermentum on serum lipids in subjects with elevated serum cholesterol. Nutrition, Metabolism & Cardiovascular Diseases 16: 531–535.

Stamler, J., M.L. Daviglus, D.B. Garside, A.R. Dyer, P. Greenland and J.D. Neaton. 2000. Relationship of baseline serum cholesterol levels in 3 large cohorts of younger men to long-term coronary, cardiovascular, and all-cause mortality and to longevity. JAMA 284: 311–318.

Tabas, I. 2010. The Role of Endoplasmic Reticulum Stress in the Progression of Atherosclerosis. Circ. Res. 107: 839–850.

Taylor, G.R.J. and C.M. Williams. 1998. Effects of probiotics and prebiotics on blood lipids. Br. J. Nutrit. 80: 225–230.

Timmers, L., J.P. Sluijter, J.K. van Keulen, I.E. Hoefer, M.G. Nederhoff, M.J. Goumans, P.A. Doevendans, C.J. van Echteld, J.A. Joles, P.H. Quax, J.J. Piek, G. Pasterkamp and D.P. de Kleijn. 2008. Toll-like receptor 4 mediates maladaptive left ventricular remodeling and impairs cardiac function after myocardialinfarction. Circ. Res. 102: 257–264.

Tuohy K.M., M.P. Hollie, W.S. Chris and R.G. Glann. 2003. Using probiotics and prebiotics to improve gut health. DDT 8: 692–700.

Venter, C.S. and H.H. Vorster. 1989. Possible metabolic consequences of fermentation in the colon for humans. Med. Hypotheses 29: 161–166.

Vice, E., J.D. Privette, R.C. Hickner and H.A. Barakat. 2005. Ketone body metabolism in lean and obese women. Metabolism 54: 1542–1545.

Vinolo, M.A., E. Hatanaka, R.H. Lambertucci, P. Newsholme and R. Curi. 2009. Effects of short chain fatty acids on effector mechanisms of neutrophils. Cell. Biochem. Funct. 27: 48–55.

Vrieze, A., F. Holleman, E.G. Zoetendal, W.M. de Vos, J.B.L. Hoekstra and M. Nieuwdorp. 2010. The environment within: how gut microbiota may influence metabolism and body composition Diabetologia 53: 606–613.

Vulevic, J., A. Drakoularakou, P. Yaqoob, G. Tzortzis and G.R. Gibson. 2008. Modulation of the Fecal Microflora Profile and Immune Function by a Novel Trans Galactooligosaccharide Mixture (B-GOS) in Healthy Elderly Volunteers. Am. J. Clin. Nutr. 88: 1438–1446.

Wang, Z., E. Klipfell, B.J. Bennett, R. Koeth, B.S. Levison, B. DuGar, A.E. Feldstein, E.B. Britt, X. Fu, Y.M. Chung, Y. Wu, P. Schauer, J.D. Smith, H. Allayee, W.H.W. Tang, J.A. DiDonato,

A.J. Lusis and S.L. Hazen. 2011. Gut flora metabolism of phosphatidylcholine promotes cardiovascular disease. Nature 472: 57–63.

WHO, World Health Organization. 2003. Diet, Nutrition and Prevention of Chronic Diseases; Report of a Joint WHO/FAO Expert Consultation, Geneva, Switzerland.

WHO, World Health Organization. 2009. Cardiovascular Disease; Fact sheet N°317, Geneva, Switzerland.

Wong, S.W., M.J. Kwon, A.M.K. Choi, H.P. Kim, K. Nakahara and D.H. Hwang. 2009. Fatty acids modulate Toll-like receptor 4 activation through regulation of receptor dimerization and recruitment into lipid rafts in a reactive oxygen species-dependent manner. J. Biol. Chem. 284: 27384–27392.

Wostmann, B.S. and N.L. Wiech. 1961. Total serum and liver cholesterol in germfree and conventional male rats. Am. J. Physiol. 201: 1027–1029.

Xiao, J.Z., S. Kondo, N. Takahashi, K. Miyaji, K. Oshida, A. Hiramatsu, K. Iwatsuki, S. Kokubo and A. Hosono. 2003. Effects of milk products fermented by *Bifidobacterium longum* on blood lipids in rats and healthy adult male volunteers. J. Dairy Sci. 86: 2452–2461.

Yadav, H., S. Jain and P.R. Sinha. 2007. Antidiabetic effect of probiotic dahi containing *Lactobacillus acidophilus* and *Lactobacillus casei* in high fructose fed rats. Nutrition 23: 62–68.

Yadav, H., S. Jain and P.R. Sinha. 2008. Oral administration of dahi containing probiotic *Lactobacillus acidophilus* and *Lactobacillus casei* delayed the progression of streptozotocin induced diabetes in rats. J. Dairy. Res. 75: 189–195.

Yamashita, K., M. Itakura and K. Kawai. 1984. Effects of fructo-oligosaccharides on blood glucose and serum lipids in diabetic subjects. Nutr. Res. 4: 961–966.

Yusof, H.M., E.A. Miles and P. Calder. 2008. Influence of very long-chain n-3 fatty acids on plasma markers of inflammation in middle-aged men. Prostaglandins Leukot Essent Fatty Acids 78: 219–228.

8

Pro- and Prebiotics for Elderly

Lyra A., Tiihonen K., Lehtinen M.J.* and *Ouwehand A.C.*

Introduction

The average life expectancy continues to increase in Western societies. Together with a low nativity, this causes the proportion of seniors to increase presenting a financial challenge to health care systems. However, it also presents an interesting opportunity for products aimed at health maintenance. Ageing predisposes us to a natural degeneration in gastrointestinal (GI) function, epithelial barrier integrity, GI microbiota composition and immune system function (adaptive and innate) elevating the risk of infections. The increased risk of infections together with an increased risk for both chronic and acute inflammatory responses may enhance the aforementioned phenomenon, further elevating the risk of infection and systemic diseases. These age-related effects can potentially be held back with the use of pro- and prebiotics together with healthy life-style and diet choices.

Effects of Ageing on Gut Function

Ageing has relatively little effect on the GI tract function but involves multiple small changes which may increase the incidence of common

DuPont Nutrition & Health, Danisco Sweeteners Oy, Sokeritehtaantie 20, 02460 Kantvik, Finland.
* Corresponding author

complaints such as constipation, diverticulosis and malnutrition among the elderly (Table 1) (Drozdowski and Thomson 2006, Morley 2007, Salles 2007). Weakening of cell-mediated immune responses (Pawelec 2012) and intrinsic ageing of epithelial stem cells (Kirkwood 2004) may increase infections as well as cancers in stomach and colon. Some physiological changes in the ageing gut may be difficult to differentiate from disease or medication induced alterations. Thus, clinically significant abnormalities effecting GI function, such as reduction in food intake, should be evaluated and not attributed only to ageing.

The age-related physiological changes in the GI tract are most pronounced in the proximal and distal parts of the GI tract where the skeletal muscle has an important functional role (Bitar et al. 2011). With ageing, poor dentition as well as decreased sensations of taste, smell and thirst may contribute to decreased food intake and concomitant malnutrition and dehydration (Bhutto and Morley 2008). Saliva protects teeth and lubricates the mouth and esophagus, facilitating chewing and swallowing. Although decreased salivary secretion is common in older people it is mainly due to medication rather than ageing itself. Ageing also impairs esophagus motility contributing to dysphagia and GI reflux.

Gastric and small intestinal motor and sensory functions are not substantially affected with increasing age. The delayed gastric emptying rate could increase satiety, reduce food intake, and thus increase the risk of malnutrition (Benelam 2009). In addition, altered secretion and sensitivity of satiety hormones, such as cholecystokinin (CCK) and leptin (Di Francesco

Table 1. Age-related physiological changes in the gastrointestinal tract and their effects on nutritional status and the gastrointestinal health.

Organ/Sense	Changes with ageing	Effect
Taste, smell and vision	Impaired sensations	Decreased appetite, dehydration
Mouth	Tooth decay, decreased saliva secretion	Malnutrition, dysphagia
Esophagus	Decreased peristalsis, increased gastrointestinal reflux	Dysphagia, aspiration pneumonia
Stomach	Decreased emptying, reduced gastric acidity	Prolonged satiety, malabsorption of nutrients
Pancreas, liver	Minor decrease in pancreatic secretions, delayed drug metabolism in liver	Decreased digestibility, increased drug toxicity
Small intestine	Increased CCK secretion, decreased absorption of vitamins and minerals	Decreased appetite, nutritional deficiencies
Colon, rectum	Decreased motility, altered colonic wall resistance, decreased rectal sensations	Constipation, diverticula, fecal incontinence

et al. 2007), are likely to induce pronounced satiety which may contribute to the inability to compensate for underfeeding periods. In the gut, malnourishment can lead to damage of the epithelial cells causing decreased local immunity and further reduced absorption of nutrients. Moreover the secretion of digestive juices may be slightly reduced, which impairs the digestion of some nutrients such as lactose (Di Stefano et al. 2001).

The small intestinal motor and sensory functions are well preserved although the increased prevalence of small intestinal bacterial overgrowth is proposed to be related to decreased motility of the small intestine. The disorders of the gastric secretion and small intestinal epithelium may interfere with the absorption of minerals (calcium, magnesium, iron) and vitamins (D, B_{12}, C) (Holt 2007). Decreased calcium absorption in the elderly can be enhanced by the concomitant supply of vitamin D (Lips 2012). On the other hand, the decreased absorption of B_{12} vitamin and folate are affected more by the atropic gastritis and medication (e.g., proton pump inhibitors) than ageing itself. It is also notable that the increased drug absorption due to decreased gut motility and delayed drug elimination due to decreased liver metabolism may increase the drug level in blood circulation and thus the risk of side effects from medication.

The changes in the lower GI tract contribute to constipation, diverticulosis and fecal incontinence (Bhutto and Morley 2008). Constipation in the elderly is a common problem with several potential causes. The role of neurodegeneration in constipation has recently been reviewed by Wade and Cowen (Wade and Cowen 2004), Camilleri et al. (Camilleri et al. 2008) and Wiskur et al. (Wiskur and Greenwood-Van Meerveld 2010). Not only altered rectal sensations but also weakening of GI neuromuscular and pelvic floor functions may contribute to constipation as well as faecal incontinence and diverticulosis (Bhutto and Morley 2008, Bouras and Tangalos 2009). Due to the increasing prevalence of constipation the use of laxatives also increases with ageing (Bouras and Tangalos 2009).

Although ageing has relatively little effect on the overall GI function, the impaired adaptation to epithelial damages and stress as well as medication and diseases, may cause malnutrition and GI disorders in the elderly.

Effects of Ageing on Gut Microbiota

The human GI microbiota goes through transitional stages throughout life, with the most profound changes detected before 3 years of age or after 100 years of age (Biagi et al. 2010, Mariat et al. 2009, Yatsunenko et al. 2012). Commonly encountered effects on the GI microbiota due to ageing have been brought forward although the elderly subject groups studied vary by age range, health status and life-style (Table 2).

Table 2. Characteristics of the gastrointestinal microbiota of elderly.

Reference	Method for microbiota analysis	Sample	n	Age range (years)	Study set-up	Outcome on microbiota analyses	Firmicutes/ Bacteroidetes -ratio	Comment
(Benno et al. 1989)	Culturing		30		Comparison of rural long-living (median 84 y) and urban (median 64 y) Japanese populations.	Tendency for long life associated with bifidobacteria and less lecithinase-positive clostridia; Elderly from urban areas associated with higher counts of bacilli and *Clostridium perfringens*.	NA	Alterations discussed to be due to dietary intake of fibre.
(Hopkins and Macfarlane 2002)	Cellular fatty acid profile analysis (MIDI)	fecal	15	21–34/ 67–88	Healthy young adults, healthy elderly and elderly suffering from *Clostridium difficile* associated diarrhea (CDAD).	Diversity reduced with age. *Bacteroides* diversity increased, *Bifidobacterium* diversity decreased and enterobacteria more abundant with age. CDAD increased diversity among clostridia and lactobacilli and reduced counts of *Bacteroides*, *Prevotella* and bifidobacteria.	NA	Only few subjects. Culture-dependent method.
(Hayashi et al. 2003)	16S rRNA gene Sanger sequencing (27 to 520 *Escherichia coli* 16S rRNA gene position) and terminal restriction fragment length polymorphism (T-RFLP)	fecal	6	74–94	Elderly in long-term residential care.	In comparison to young adults, the elderly gut microbiota frequently harbored Gammaproteobacteria and was abundant in *Clostridium* cluster XIVa.	NA	Age affecting the microbiota in a characteristic manner shown by novel phylotypes clustering together.

Table 2. contd....

Table 2. contd.

Reference	Method for microbiota analysis	Sample	n	Age range (years)	Study set-up	Outcome on microbiota analyses	Firmicutes/ Bacteroidetes -ratio	Comment
(Bartosch et al. 2004)	16S rRNA gene qPCR	fecal	94	63–103	Comparison of healthy elderly in the community, hospitalized elderly and hospitalized elderly receiving antibiotics.	Hospitalization reduced total fecal bacterial count, *Bacteroides-Prevotella*, *Bifidobacterium* spp., *Desulfovibrio* spp., *Clostridium clostridioforme*, and *Faecalibacterium prausnitzii* whereas enterobacteria increased.	NA	Presenting results as proportional values of total bacteria presented a more uniform outcome for the different groups. An array of different antibiotics consumed. Collection and processing of samples efficient and uniform between groups. DNA extracted from fresh fecal samples.
(Mueller et al. 2006)	16S rRNA fluorescence *in situ* hybridization	fecal	145	20–50 y and >60 y	Comparison of two age groups from four different European countries (France, Germany, Italy, and Sweden).	High enterobacteria counts were characteristic for the elderly in all four countries. *Bifidobacterium* spp. were not significantly higher among elderly within countries, although a trend for increase with ageing was observed. However, the location had a significant effect on bifidobacterial abundance.	NA	Shows the importance of location (which effects the lifestyle and diet) over age on the characteristics of the gastrointestinal microbiota.

(Tiihonen et al. 2008)	16S rRNA gene qPCR	fecal	55	68–88 (and 14 younger controls)	Elderly NSAID users compared with elderly non-NSAID users and younger controls.	Ageing reduced aerobic bacteria counts. NSAID use reduced *Clostridium coccoides –Eubacterium rectale* counts and butyrate and other SCFA bacterial metabolite levels.	NA	Effect on NSAIDs also analyed. SCFAs in addition to microbes quantified from feces.
(Mariat et al. 2009)	16S rRNA gene qPCR	fecal	62	< 10 mo/25–45 y/70–90 y	Infants, young adults and elderly.	*Bacteroidetes* and *E. coli* associated with elderly. *Firmicutes/Bacteroidetes*–ratio was reduced among infants and elderly in comparison to young adults.	Decreased with ageing.	Whole lifespan analysis. Similar characteristics found in infants and elderly of harboring an undeveloped or declining immune system and possibly a less diverse diet.
(Rajilic-Stojanovic et al. 2009)	16S rRNA gene microarray (HITChip)	fecal	5	average 71 (and 5 younger controls)	Comparison of young adults and elderly. Timely follow-up with 3 samples over at least 2 months.	Overall microbiota clustered according to age group. Alterations in 20 family/genus level groups most profound. *Bacilli* abundant among elderly and *Bacteroidetes* abundant among younger controls.	NA (*Bacteroidetes* decreased).	Study aimed to test novel methodology and thus low subject count. With the 15 samples from 3 subjects within the two age groups a clear distinction seen with a method covering the overall microbiome.

Table 2. contd....

Table 2. contd.

Reference	Method for microbiota analysis	Sample	n	Age range (years)	Study set-up	Outcome on microbiota analyses.	Firmicutes/ Bacteroidetes -ratio	Comment
(Biagi et al. 2010)	16S rRNA gene V1 and V6 targetting microarray (HITChip) and qPCR	fecal	63	25–104	Comparison of young adults (~30 y), elderly (~70 y) and centenarians (~100 y).	Age related changes become profound only at extreme high age (> 100 y): decrease in diversity, *F. prausnitzii* and relatives and *Bifidobacterium* spp. *Eubacterium limosum* highly associated with centenarians. The overall gut microbiota of centenarians grouped apart from young and elderly with hierarchical clustering, whereas that of elderly and young adults overlapped.	No significant change.	Adding extremely high age group to analyses gave new perspective to the field. Two methods applied to microbiota analysis.
(Claesson et al. 2011)	16S rRNA gene V4 pyrosequencing	fecal	161	65–96 (and 9 younger controls)	Elderly compared with younger controls and previously published core microbiota of younger adults. Timely follow-up for a subset of subjects.	Core microbiome of elderly differs from that of younger adults. Among elderly subjects there is less uniformity between gut microbiotas and uncommon phyla are more prevalent. *Bacteroidetes* dominant microbiotas and high abundance of *Proteobacteria* typical for elderly.	Decreased among antibiotic users.	High subject count and thorough microbiome analysis. The applied wide age range for elderly putatively adds to the high variability encountered.

(Mäkivuokko et al. 2010)	%G+C profiling, Sanger sequencing (28 to 928 *E. coli* 16S rRNA gene position)	fecal	18	70–85 (and 14 younger controls)	Elderly NSAID users compared with elderly non-NSAID users and younger controls. (Subjects from Tiihonen et al. 2008)	*Bacteroidetes* and lactobacilli increased and *Firmicutes* including *Roseburia* and *Ruminococcus* decreased with age. NSAID use lowered the overall bacterial count, *Collinsella* levels within *Actinobacteria* and lactobacilli within Firmicutes.	Decreased with ageing.	Methodology allows less biased detection of high GC and less prominent species. Effect of NSAID use had a significant effect on lactobacilli and Actinobacteria (*Collinsella*) within the elderly microbiota.
(Mikelsaar et al. 2010)	PCR, cultivation	fecal	42	>65 (72±5.0)	Lactobacilli species analyzed in relation to health.	Lactobacilli found among elderly individuals varied highly. Positive associations with white blood cell count (*Lactobacillus reuteri*) and negative associations with blood glucose level (*Lactobacillus fermentum*) and oxidized lipoprotein (lactobacilli) detected.	NA	Methodology allowed for detection of species specific associations detected.
(Claesson et al. 2012)	16S rRNA gene V4 pyrosequencing	fecal	178	64–102 (and 13 younger controls)	Elderly subjects from among community residents, day-hospital visitors, temporarily hospitalized and long-term residential care.	Elderly in long-term care had less diverse microbiota. Diet, health and microbiota are associated among elderly even with the effect of residence taken into account. Proportion of *Bacteroidetes* and the genera *Parabacteroides, Eubacterium, Anaerotruncus, Lactonifactor* and *Coprobacillus* were increased	No correlation with BMI among elderly.	High subject count with thorough analysis of the microbiome, nutritional and health status, diet, fecal metabolites and immune markers. Reasons the importance

Table 2. contd....

Table 2. contd.

Reference	Method for microbiota analysis	Sample	n	Age range (years)	Study set-up	Outcome on microbiota analyses.	Firmicutes/ Bacteroidetes -ratio	Comment
						among elderly in long-term care. *Coprococcus* and *Roseburia* associated with elderly living among the community.		of nutritional and associated microbial factors in healthy ageing.
(O'Sullivan et al. 2012)	Culturing 16S rRNA gene V4 pyrosequencing	fecal	185	≥65	Effect of antibiotics on the gastrointestinal microbiota of elderly.	Culturable bifidobacteria reduced 7-fold after antibiotic therapy. *Bifidobacterium* spp. reduced and all nine detected genera affected with antibiotic treatment. Changes proceed even post-therapy. Nucleic acid inhibitor antibiotics had a particularly profound impact.	Decreased with antibiotics among community residents	Analyses the long term effects on antibiotics with a large subject group and underlines the lengthy effect antibiotics have on the gastrointestinal microbiota.
(Yatsunenko et al. 2012)	16S rRNA gene V4 pyrosequencing and metagenomics	fecal	531	0–70 (531 subjects in total)	Children and adults up to 70 y of age from Amazonas of Venezuela, rural Malawi and metropolitan US areas.	Geographical origin and age has an effect on the gastrointestinal microbiota: age related changes seen in childhood as increased variation between individuals.	NA	A wide age range analyzed although limited to only 70 y of age: shows that successions in the microbiota are likely to happen in a wide timely span during life and thus narrow age limits in sampling may distort outcomes.

Bifidobacteria, often associated with health, have been linked with longer life expectancy as seen in the rural population of Yuzurihara (Japan), who consume a fiber rich diet compared with the urban elderly from Tokyo (Benno et al. 1989). Indeed, diminished bifidobacteria have been suggested to correlate to several phenomena putatively shaping the gut microbiota of ageing individuals such as *Clostridium difficile* associated diarrhoea (CDAD) (Hopkins and Macfarlane 2002), hospitalization (Bartosch et al. 2004), extremely high age (Biagi et al. 2010) and antibiotic treatment (O'Sullivan et al. 2012) among elderly. Nevertheless, not all studies have shown a significant decrease in bifidobacteria due to ageing (Mueller et al. 2006).

Claesson and colleagues have characterized the elderly microbiota in two large cohorts by pyrosequencing the 16S rRNA gene V4 variable region (Claesson et al. 2011, 2012). Overall, the elderly subjects had less *Firmicutes* and more *Bacteroidetes*. However, the number of younger subjects in the analysis was substantially smaller than that of the elderly (9 and 161, respectively), this difference should therefore be interpreted with caution (Claesson et al. 2011). The sequence data from the samples of the elderly were also analyzed together with previously published data from studies on adults and differences in the core microbes were detected that were in accordance with the predominance of *Bacteroidetes* among elderly subjects' fecal microbiota. For the analysis, a core microbiome was defined as a phylotypes present in at least 50% of the subjects. Moreover, substantial timely variation was seen among a subgroup of subjects followed over a three-month time period, although to a lesser extent than variation observed between subjects. Increased *Bacteroidetes* spp. with ageing has been detected in other cohorts as well (Hopkins and Macfarlane 2002, Mariat et al. 2009, Mäkivuokko et al. 2010).

Thereafter, Claesson and colleagues (Claesson et al. 2012) published another thorough analysis of the microbial composition of 178 elderly subjects (newly recruited) with additional data gathered regarding residence, diet, health status and metabolomic analysis (for a subgroup of 29 subjects) emphasizing the importance of diet. On community level, the fecal microbiotas were separated according to (1) residence: with community residents abundant in *Firmicutes* and long-term care residents abundant in *Bacteroidetes* and (2) diet: with a healthy diet (low to moderate in fat and high in fiber) being associated to a higher diversity in the GI microbiota. Residence impacts the GI microbiota comparatively slowly taking up to a year for the alterations to be enforced and has less impact than diet (Claesson et al. 2012). The effects of residence and wellbeing (frailty) were also linked with microbial short chain fatty acid (SCFA) metabolites (Claesson et al. 2012).

The microbiotas sub-typed weakly into two enterotypes *Bacteroides* and *Prevotella* (Claesson et al. 2012) noted to have been previously associated with a protein-rich or carbohydrate-rich diet, respectively (Wu et al. 2011). Stronger associations could be seen when the sequence data was analysed group-wise instead of subject-wise revealing six co-abundance groups (COG) (*Bacteroidetes, Prevtella, Ruminococcus, Oscillibacter, Allistipes* and *Ordiobacter*) which may allow more reliable interpretations on the effect of alterations if a subjects microbiota changes from one COG group to another.

Taken together, the elderly microbiome has its unique characteristics often presented as low levels of bifidobacteria, increased *Bacteroidetes* spp. and high variability. Lifestyle, diet, health and medication may, however, have a greater impact on the GI microbiome than ageing as such.

Effects of Ageing on Immune Function

The immune system senses commensal and pathogenic microbes by receptors that bind to microbe associated molecular patterns (MAMPs), like lipopolysaccharide (LPS). Important receptors for MAMPs are Toll-like receptors (TLRs) that bind to MAMPs and activate innate and adaptive immune responses (Medzhitov et al. 1997). All the human immune system cells express some or all of the ten, currently described, TLRs at different levels. When immune cells encounter microbes, the TLRs trigger responses in innate immune system cells, like natural killer (NK) cells, macrophages, neutrophils, dendritic cells (DC) and epithelial cells. These cells respond to microbes by trying to eradicate them and by eventually triggering adaptive T- and B-cell immune responses. Effector T-cells attack the infective agent by destroying the infected cells and by stimulating the function of the innate immune cells. B-cells on the other hand mature into plasma cells that secrete antibodies against the invading pathogen.

Immunological ageing of the adaptive immune system starts in the mid-twenties when involution of the thymus and thus reduction of naïve T-cell output starts (Miller 1961). In adulthood, this does not have major clinical importance, since the naïve T-cell repertoire together with memory T-cells is adequate to respond to infections. In ageing individuals, one of the hallmarks of immunosenescence is the lower number and/or proportions of peripheral blood naïve $CD8^+$ cells and increased number of memory/effector $CD8^+$ cells (Pawelec 2012). These changes in $CD8^+$ T-cells are associated especially with cytomegalovirus (CMV) seropositivity in humans and it has been shown that many of the memory/effector cells are against CMV antigens (Hadrup et al. 2006). Interestingly, it was shown in the National Health and Nutrition Examination Survey study (USA; 14000 adults followed over 10 years) that CMV negativity might provide survival advantage for humans (Simanek et al. 2011). In addition to T-cells, B-cell function and numbers also decline in old age (Ademokun et al. 2010).

The decline in innate immune system function coincides with general age-related deterioration of physiological functions. One important change seems to be decreased sensitivity of TLR signaling in elderly that has implications for the functions of most of the immune system cells (Shaw et al. 2011). Neutrophils show decreased chemotaxis, phagocytic activity, and declined superoxide generation due to changes in intracellular signaling (Butcher et al. 2001, Shaw et al. 2010). The number of monocytes and NK-cells increase with age, but their signaling efficiency, cytokine production, and up regulation of co-stimulatory molecules is suboptimal, leading to a net decrease in function (Della Bella et al. 2007, Mocchegiani et al. 2009, van Duin et al. 2007a, van Duin et al. 2007b). These changes in immune system function lead to susceptibility to infections and poor vaccine responses in old age (Pawelec et al. 2005, Siegrist and Aspinall 2009), and to a low-grade inflammation—a condition coined inflammaging—that is associated with age related metabolic changes predisposing to for example frailty and type 2 diabetes (Franceschi et al. 2007).

Probiotics in Elderly

Probiotics and elderly gut function

Abnormal bowel function may present symptoms of bloating, diarrhoea, constipation or recurrent abdominal pain related to alterations in bowel function (Longstreth et al. 2006). As discussed above, of these symptoms, constipation is commonly encountered among elderly, although it is not necessarily brought about through ageing itself, but rather as a secondary phenomenon due to age related decline in health and altered life style (including medication) (McCrea et al. 2008).

Certain probiotic strains have the capacity to alleviate constipation: *Bifidobacterium lactis* strains HN019 and DN-173010 have been shown to reduce the intestinal transit time (Waller et al. 2011) or increase weekly defecation frequency (Yang et al. 2008), respectively, from levels representative of functional constipation to those comparable to normal gut function. However, neither of these strains have been tested for constipation on elderly participants, although *B. lactis* HN019 effectively elevates fecal bifidobacteria counts and reduces enterobacteria in the intestine of 60 to 87 year old subjects (Ahmed et al. 2007). With elderly subjects, Zaharoni and colleagues (Zaharoni et al. 2011) conducted a large intervention trial with 243 over 65 year-old subjects hospitalized for orthopedic rehabilitation showing a reduction in both need for laxatives and days with diarrhea due to 45 day VSL#3 (a commercial probiotic mixture) supplementation. Several probiotic supplements, including VSL#3 (Guandalini et al. 2010) and *B. lactis*

HN019 (Waller et al. 2011), have shown alleviation of other functional bowel symptoms, including pain and bloating (Parkes et al. 2010).

Regarding bowel function studies targeting elderly subjects, health care routines and the risk of decline in health over a long term intervention need to be considered. A reduction in defecation frequency may not easily reduce laxative use among elderly home residents, as laxative dosing for each individual may have been accustomed over a long period of time and thus may not respond as efficiently as in community-dwelling elderly and the younger adult population. Thus, proving a clinically significant response may be challenging even with altered defecation frequency detected (Ouwehand et al. 2002). Moreover, the use of laxatives may mask the alleviation of constipation by other means, especially over short intervention times (An et al. 2010). With longer intervention periods, on the other hand, the putative beneficial effects of probiotics may be overcome due to declining health among the elderly subjects. In a small pilot study with *Lactobacillus rhamnosus* LB21 and *Lactococcus* La1 supplemented yoghurt, a significant decrease in body weight was observed, but no effect on defecation frequency was seen over a 6 month period (Carlsson et al. 2009). Enteral feeding may allow for efficient delivery of probiotics, at least as to compliance, but the enterally fed are a challenging target group regarding outcomes on gut function. Enteral feeding of *Lactobacillus johnsonii* La1 as a single strain supplement in fermented milk did not affect bowel function among hospitalized elderly during a 12-week intervention although positive effects were seen in the nutritional and immunological status of the subjects (Fukushima et al. 2007). In the above examples, of course, the negative result on bowel function outcomes may also be explained by the small subject groups and strain and dose selection regarding the supplement.

Another important target of probiotic research among elderly consumers is the reduction of the incidence and duration of antibiotic associated diarrhea (AAD; including CDAD) which poses a major health threat among the elderly. Although AAD is not a true functional disorder, an ability to retain balanced gut function (Zaharoni et al. 2011) and reduce fecal levels of *C. difficile* (Lahtinen et al. 2012) among elderly with a probiotic could reduce the risk of AAD.

To conclude, not many sufficiently powered clinical trials on probiotics for elderly are available, but it is likely that effects seen in the younger adult populations are valid for at least most elderly subjects as the gut microbiota shows most drastic age related changes only after 100-years of age (Biagi et al. 2010). For the ageing population, probiotics have potential to alleviate constipation and to reduce the risk of potentially detrimental imbalances in the gut microbiota.

Probiotics and elderly immune function

It has been extensively shown that probiotics interact with TLRs and other pattern recognition receptors on immune system cells and thus directly influence their functions. Furthermore, specific probiotic strains may induce beneficial changes in gut microbiota that have an impact on immune status. For example, a clinical trial in an elderly population showed that consumption of *B. longum* 2C and 46 induced changes in the bifidobacteria population that correlate with TNF-α and IL-10 levels in plasma (Ouwehand et al. 2008).

Clinical trials so far have quite clearly shown that probiotic (e.g., *L. rhamnosus* HN001, *B. lactis* HN019 or *Lactobacillus acidophilus* NCFM) consumption enhances *ex vivo* cytotoxicity of NK cells against model tumor cells and phagocytic activity of neutrophils and monocytes against *Escherichia coli* (Gill et al. 2001a, Gill et al. 2001b, Ibrahim et al. 2010, Sheih et al. 2001). Intriguingly, opposite effect on phagocytosis was shown in a study where consumption of *L. acidophilus* La1 decreased phagocytic activity of monocytes and neutrophils in elderly (Schiffrin et al. 2009). The difference between the studies may simply indicate strain specificity, but alternatively it may be a consequence of decreased TLR4 (LPS receptor) stimulation on phagocytes, as it was shown in the same study that LPS, sCD14, and LBP levels were decreased in the blood samples of the probiotic group.

In a recent study, ingestion of *Lactobacillus delbrueckii* subsp. *bulgaricus* 8481 decreased inflammatory marker IL-8 levels in the elderly (Moro-Garcia et al. 2012), adding to evidence from adult studies that probiotics may counteract inflammation and related metabolic disorders. In addition it was shown that intake of this probiotic could improve hallmarks of ageing: the number of recent thymus emigrant T-cells ($CD8^{+}CD31^{+}$) was increased, the number of senescent effector/memory type T-cells ($CD8^{+}CD28^{null}$) was decreased, and importantly CMV reactivation was prevented in probiotic group, indicating that probiotic consumption could counteract markers of immunosenescence.

Vaccine adjuvants stimulate TLRs or mimic their activation that improves vaccination responses. Probiotics share similar TLR activating properties and it has been shown that *Lactobacillus casei* DN-114 001 and *Lactobacillus plantarum* CECT7315/7316 improve vaccination responses against influenza in elderly (Boge et al. 2009, Bosch et al. 2012). On the other hand a large study involving 737 healthy aged (>65 yrs) volunteers did not show any improvements in vaccination response upon *L. casei* Shirota consumption nor reduction in influenza-like symptoms (Van Puyenbroeck et al. 2012), indicating perhaps strain specificity of the effect. Further evidence on clinical benefits and immune system stimulation by probiotics were observed in common cold studies where duration but not the rate of infections decreased upon *L. casei* DN-114 001 (Guillemard et al. 2010, Turchet et al. 2003) or *L. acidophilus* La1 consumption (Fukushima et al. 2007). In addition, study with elderly consuming *L. delbrueckii* ssp. *bulgaricus* OLL1073R-1 and *Streptococcus thermophilus* OLS3059 showed reduction in the rate of respiratory tract infections (Makino et al. 2010).

Although some clinical trials have been conducted, there is still lack of research on many aspects of immune response to probiotics in elderly. In contrast to modifications in microbiota composition, it is likely that results from clinical studies with other age groups cannot be directly extrapolated to elderly due to changes in immune systems upon ageing. Just like TLR response to pathogens changes, it was recently shown that cells of the elderly and young respond differently to probiotics *in vitro* (You and Yaqoob 2012).

Prebiotics in Elderly

Unlike probiotics, prebiotics do not currently have a generally accepted definition. Although various definitions exist, the understanding of the concept is the same. A recent definition of the prebiotic concept is the one given by Roberfroid (Roberfroid et al. 2010) "The selective stimulation of growth and/or activity(ies) of one or a limited number of microbial genus(era)/species in the gut microbiota that confer(s) health benefits to the host".

The main target of prebiotics is thus the GI microbiota. In elderly subjects, this is very appropriate as it has been commonly documented that their microbiota may be different from younger adults. The often referred reduction in fecal *Bifidobacterium* levels in elderly is, however, not always observed (Mueller et al. 2006). The intestine is the body's main site of antibody production. It is therefore not surprising that functional foods such as prebiotics may influence the immune system; this is likely to happen primarily through modulation of the microbiota activity and/or composition. Also immune function is reduced in elderly as compared to younger adults and therefore an appropriate target for prebiotics.

Gut benefits

Although the difference in fecal microbiota composition between elderly and younger adults is not always reported to be the same, reduction in bifidobacteria is an often recurring topic. Many prebiotics have been selected for their so-called bifidogenic activity and may indeed increase fecal *Bifidobacterium* levels; although this depends on the starting level of the bifidobacteria (Tuohy et al. 2001). Changing the microbiota on it's own such as increasing levels of bifidobacteria is not necessarily a health benefit and should thus always be correlated with other health benefits. It is also important to look beyond bifidobacteria as actually the majority of microbiota members are still unknown (Björklund et al. 2012). The effect of a prebiotic on potential intestinal pathogens has been little investigated, especially in the elderly. It can be speculated that, in particular colonic pathogens maybe affected by prebiotics. It is less likely that prebiotics affect pathogens in the small intestine, although a direct interaction between prebiotics and the pathogen cannot be excluded (Fig. 1). Indeed, reduction in CDAD has been reported (Lewis et al. 2005). Fermentation of the prebiotic in the intestine will lead to the formation of SCFAs. This may reduce colonization by (potential) pathogens (Fig. 1) and explain the anti-*C. difficile* effect described above. Fermentation of a prebiotic, instead of proteinacious substrates and the formation of SCFAs is also likely to explain the reduction in carcinogenic potential of intestinal contents and

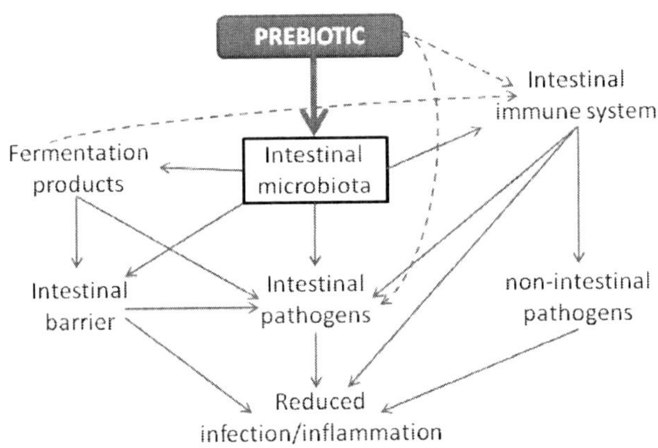

Fig. 1. Potential pathways by which prebiotics can influence health; note, this is not exclusive for elderly.

to explain improved mineral absorption. For prebiotics, a typical health benefit for seniors would be improved bowel function. While this has been a well documented benefit in adults, this has been reported very little for elderly. However, lactitol in combination with *L. acidophilus* NCFM was able to improve bowel function in healthy elderly (Ouwehand et al. 2009).

Immune benefits

The reduced immune function in elderly is referred to as immunosenesence. As discussed above, for probiotics ample evidence exists that specific strains are able to modulate immune function in elderly. For prebiotics, this evidence is much scarcer (Lomax and Calder 2009). On one hand relatively few studies have been done with prebiotics on this topic in general and in elderly in particular. On the other hand, the studies that have been done failed to show any substantial change in immune markers. Changes in serum cytokine levels have been documented, but it is difficult to translate such changes into a health benefit. Some changes in adaptive immunity have, however been reported, such as improved response to vaccination (Langkamp-Henken et al. 2006) and lower infection levels (Bunout et al. 2004).

These changes in immunity may explain the observed reductions in respiratory tract infections (Bunout et al. 2004) and intestinal infections (diarrhea) (Langkamp-Henken et al. 2006, Lewis et al. 2005).

Although effects on respiratory tract infections may seem surprising, it is in line with observations for probiotics, as discussed above. It strengthens the idea that changing the GI microbiota leads to modulation of the immune system, with benefits beyond the gut (Fig. 1).

Nutritional aspects

Prebiotics may be useful in elderly due to the age-related decrease in bowel function, reduced immune function and nutrient metabolism (Macfarlane and Macfarlane 2011). Various prebiotics have been observed to improve bowel function in the elderly (Tiihonen et al. 2010). The fermentation of prebiotics by colonic bacteria increases the bacterial biomass leading to an increased fecal output. During the fermentation, SCFAs, butyrate, propionate and acetate are formed, which are efficiently absorbed to epithelial cells and circulation.

Fermentation of prebiotics by the colonic bacteria produces SCFAs which not only supply energy to the host but also provide energy to the gut epithelial and immune cells. Butyrate, especially, is the main source of energy in colonic mucosa and thus prebiotics that increase its availability may have important health implications. Prebiotic fermentation also decreases pH which increases mineral (i.e., calcium and magnesium) solubility and thus the mineral bioavailability and utilization (Legette et al. 2012). Decreased colonic pH can inhibit the conversion of primary bile acids to carcinogenic secondary bile acids.

Colonic fermentation metabolites may also modulate gut morphology, gene expression and also lipid metabolism. Recently, polydextrose fermentation metabolites have been shown to regulate the transcription factors involved in energy metabolism and induction of apoptosis in colon cancer cells (Putaala et al. 2011). Propionate has been shown to exhibit hypocholesterolemic effects due to its action in lipidogenesis in liver.

By promoting the growth of potential beneficial microbes and/or reducing adverse microbial metabolism, prebiotics may improve the gut barrier function and thus attenuate the risk for systemic low-grade inflammation which is an underlying condition in many age-related diseases.

Conclusions

The importance of an overall healthy life-style and diet cannot be overcome with supplements, but many of these age-related impairments in gut and immune physiology and functions can be targeted with pre- and/or probiotics for better resilience against internal and external threats. As discussed above, evidence exists that pro-and/or prebiotics may contribute to the alleviation of constipation, enhance immunity, suppress immunosenescence, increase resilience against intestinal and respiratory pathogens, reduce carcinogenic potential of colonic digesta and improve availability of certain nutrients. To better understand the potential pro- and prebiotics may have on the health status of elderly, further research is needed in this growing part of the population in Western societies as most of the clinical studies thus far have been conducted with adult populations (under ~65 years of age).

References

Ademokun, A., Y.C. Wu and D. Dunn-Walters. 2010. The ageing B cell population: composition and function. Biogerontology 11: 125–137.

Ahmed, M., J. Prasad, H. Gill, L. Stevenson and P. Gopal. 2007. Impact of consumption of different levels of Bifidobacterium lactis HN019 on the intestinal microflora of elderly human subjects. The journal of nutrition, health & aging 11: 26–31.

An, H.M., E.H. Baek, S. Jang, K. Lee do, M.J. Kim, J.R. Kim, K.O. Lee, J.G. Park and N.J. Ha. 2010. Efficacy of Lactic Acid Bacteria (LAB) supplement in management of constipation among nursing home residents. Nutrition journal 9: 5.

Bartosch, S., A. Fite, G.T. Macfarlane and M.E. McMurdo. 2004. Characterization of bacterial communities in feces from healthy elderly volunteers and hospitalized elderly patients by using real-time PCR and effects of antibiotic treatment on the fecal microbiota. Applied and environmental microbiology 70: 3575–3581.

Benelam, B. 2009. Satiety and the anorexia of ageing. British journal of community nursing 14: 332–335.

Benno, Y., K. Endo, T. Mizutani, Y. Namba, T. Komori and T. Mitsuoka. 1989. Comparison of fecal microflora of elderly persons in rural and urban areas of Japan. Applied and environmental microbiology 55: 1100–1105.

Bhutto, A. and J.E. Morley. 2008. The clinical significance of gastrointestinal changes with aging. Current opinion in clinical nutrition and metabolic care 11: 651–660.

Biagi, E., L. Nylund, M. Candela, R. Ostan, L. Bucci, E. Pini, J. Nikkila, D. Monti, R. Satokari, C. Franceschi, P. Brigidi and W. De Vos. 2010. Through ageing, and beyond: gut microbiota and inflammatory status in seniors and centenarians. PloS one 5: e10667.

Bitar, K., B. Greenwood-Van Meerveld, R. Saad and J.W. Wiley. 2011. Aging and gastrointestinal neuromuscular function: insights from within and outside the gut. Neurogastroenterology and motility: the official journal of the European Gastrointestinal Motility Society 23: 490–501.

Björklund, M., A.C. Ouwehand, S.D. Forssten, J. Nikkilä, K. Tiihonen, N. Rautonen and S.J. Lahtinen. 2012. Gut microbiota of healthy elderly NSAID users is selectively modified with the administration of Lactobacillus acidophilus NCFM and lactitol. Age (Dordr) 34: 987–999.

Boge, T., M. Remigy, S. Vaudaine, J. Tanguy, R. Bourdet-Sicard and S. van der Werf. 2009. A probiotic fermented dairy drink improves antibody response to influenza vaccination in the elderly in two randomised controlled trials. Vaccine 27: 5677–5684.

Bosch, M., M. Mendez, M. Perez, A. Farran, M.C. Fuentes and J. Cune. 2012. Lactobacillus plantarum CECT7315 and CECT7316 stimulate immunoglobulin production after influenza vaccination in elderly. Nutricion hospitalaria: organo oficial de la Sociedad Espanola de Nutricion Parenteral y Enteral. 27: 504–509.

Bouras, E.P. and E.G. Tangalos. 2009. Chronic constipation in the elderly. Gastroenterology clinics of North America 38: 463–480.

Bunout, D., G. Barrera, S. Hirsch, V. Gattas, M.P. de la Maza, F. Haschke, P. Steenhout, P. Klassen, C. Hager, M. Avendano, M. Petermann and C. Munoz. 2004. Effects of a nutritional supplement on the immune response and cytokine production in free-living Chilean elderly. JPEN. Journal of parenteral and enteral nutrition 28: 348–354.

Butcher, S.K., H. Chahal, L. Nayak, A. Sinclair, N.V. Henriquez, E. Sapey, D. O'Mahony and J.M. Lord. 2001. Senescence in innate immune responses: reduced neutrophil phagocytic capacity and CD16 expression in elderly humans. Journal of leukocyte biology 70: 881–886.

Camilleri, M., T. Cowen and T.R. Koch. 2008. Enteric neurodegeneration in ageing. Neurogastroenterology and motility : the official journal of the European Gastrointestinal Motility Society 20: 185–196.

Carlsson, M., Y. Gustafson, L. Haglin and S. Eriksson. 2009. The feasibility of serving liquid yoghurt supplemented with probiotic bacteria, Lactobacillus rhamnosus LB 21, and Lactococcus lactis L1A—a pilot study among old people with dementia in a residential care facility. The journal of nutrition, health & aging 13: 813–819.

Claesson, M.J., S. Cusack, O. O'Sullivan, R. Greene-Diniz, H. de Weerd, E. Flannery, J.R. Marchesi, D. Falush, T. Dinan, G. Fitzgerald, C. Stanton, D. van Sinderen, M. O'Connor, N. Harnedy, K. O'Connor, C. Henry, D. O'Mahony, A.P. Fitzgerald, F. Shanahan, C. Twomey, C. Hill, R.P. Ross and P.W. O'Toole. 2011. Composition, variability, and temporal stability of the intestinal microbiota of the elderly. Proceedings of the National Academy of Sciences of the United States of America 108 Suppl. 1: 4586–4591.

Claesson, M.J., I.B. Jeffery, S. Conde, S.E. Power, E.M. O'Connor, S. Cusack, H.M. Harris, M. Coakley, B. Lakshminarayanan, O. O'Sullivan, G.F. Fitzgerald, J. Deane, M. O'Connor, N. Harnedy, K. O'Connor, D. O'Mahony, D. van Sinderen, M. Wallace, L. Brennan, C. Stanton, J.R. Marchesi, A.P. Fitzgerald, F. Shanahan, C. Hill, R.P. Ross and P.W. O'Toole. 2012. Gut microbiota composition correlates with diet and health in the elderly. Nature 488: 178–184.

Della Bella, S., L. Bierti, P. Presicce, R. Arienti, M. Valenti, M. Saresella, C. Vergani and M.L. Villa. 2007. Peripheral blood dendritic cells and monocytes are differently regulated in the elderly. Clinical immunology (Orlando, Fla.) 122: 220–228.

Di Francesco, V., F. Fantin, F. Omizzolo, L. Residori, L. Bissoli, O. Bosello and M. Zamboni. 2007. The anorexia of aging. Dig. Dis. 25: 129–137.

Di Stefano, M., G. Veneto, S. Malservisi, A. Strocchi and G.R. Corazza. 2001. Lactose malabsorption and intolerance in the elderly. Scandinavian journal of gastroenterology 36: 1274–1278.

Drozdowski, L. and A.B. Thomson. 2006. Aging and the intestine. World journal of gastroenterology : WJG. 12: 7578–7584.

Franceschi, C., M. Capri, D. Monti, S. Giunta, F. Olivieri, F. Sevini, M.P. Panourgia, L. Invidia, L. Celani, M. Scurti, E. Cevenini, G.C. Castellani and S. Salvioli. 2007. Inflammaging and anti-inflammaging: a systemic perspective on aging and longevity emerged from studies in humans. Mechanisms of ageing and development 128: 92–105.

Fukushima, Y., S. Miyaguchi, T. Yamano, T. Kaburagi, H. Iino, K. Ushida and K. Sato. 2007. Improvement of nutritional status and incidence of infection in hospitalised, enterally fed elderly by feeding of fermented milk containing probiotic Lactobacillus johnsonii La1 (NCC533). The British journal of nutrition 98: 969–977.

Gill, H.S., M.L. Cross, K.J. Rutherfurd and P.K. Gopal. 2001a. Dietary probiotic supplementation to enhance cellular immunity in the elderly. British journal of biomedical science 58: 94–96.

Gill, H.S., K.J. Rutherfurd and M.L. Cross. 2001b. Dietary probiotic supplementation enhances natural killer cell activity in the elderly: an investigation of age-related immunological changes. Journal of clinical immunology 21: 264–271.

Guandalini, S., G. Magazzu, A. Chiaro, V. La Balestra, G. Di Nardo, S. Gopalan, A. Sibal, C. Romano, R.B. Canani, P. Lionetti and M. Setty. 2010. VSL#3 improves symptoms in children with irritable bowel syndrome: a multicenter, randomized, placebo-controlled, double-blind, crossover study. Journal of pediatric gastroenterology and nutrition 51: 24–30.

Guillemard, E., F. Tondu, F. Lacoin and J. Schrezenmeir. 2010. Consumption of a fermented dairy product containing the probiotic Lactobacillus casei DN-114001 reduces the duration of respiratory infections in the elderly in a randomised controlled trial. The British journal of nutrition 103: 58–68.

Hadrup, S.R., J. Strindhall, T. Kollgaard, T. Seremet, B. Johansson, G. Pawelec, P. thor Straten and A. Wikby. 2006. Longitudinal studies of clonally expanded CD8 T cells reveal a repertoire shrinkage predicting mortality and an increased number of dysfunctional cytomegalovirus-specific T cells in the very elderly. Journal of immunology (Baltimore, Md. (1950) 176: 2645–2653.

Hayashi, H., M. Sakamoto, M. Kitahara and Y. Benno. 2003. Molecular analysis of fecal microbiota in elderly individuals using 16S rDNA library and T-RFLP. Microbiology and immunology 47: 557–570.

Holt, P.R. 2007. Intestinal malabsorption in the elderly. Dig. Dis. 25: 144–150.

Hopkins, M.J. and G.T. Macfarlane. 2002. Changes in predominant bacterial populations in human faeces with age and with Clostridium difficile infection. Journal of medical microbiology 51: 448–454.

Ibrahim, F., S. Ruvio, L. Granlund, S. Salminen, M. Viitanen and A.C. Ouwehand. 2010. Probiotics and immunosenescence: cheese as a carrier. FEMS immunology and medical microbiology 59: 53–59.

Kirkwood, T.B. 2004. Intrinsic ageing of gut epithelial stem cells. Mechanisms of ageing and development 125: 911–915.

Lahtinen, S.J., S. Forssten, J. Aakko, L. Granlund, N. Rautonen, S. Salminen, M. Viitanen and A.C. Ouwehand. 2012. Probiotic cheese containing Lactobacillus rhamnosus HN001 and Lactobacillus acidophilus NCFM(R) modifies subpopulations of fecal lactobacilli and Clostridium difficile in the elderly. Age (Dordr) 34: 133–143.

Langkamp-Henken, B., S.M. Wood, K.A. Herlinger-Garcia, D.J. Thomas, J.K. Stechmiller, B.S. Bender, E.M. Gardner, S.J. DeMichele, J.P. Schaller and D.M. Murasko. 2006. Nutritional formula improved immune profiles of seniors living in nursing homes. Journal of the American Geriatrics Society 54: 1861–1870.

Legette, L.L., W. Lee, B.R. Martin, J.A. Story, J.K. Campbell and C.M. Weaver. 2012. Prebiotics enhance magnesium absorption and inulin-based fibers exert chronic effects on calcium utilization in a postmenopausal rodent model. Journal of food science 77: H88–94.

Lewis, S., S. Burmeister, S. Cohen, J. Brazier and A. Awasthi. 2005. Failure of dietary oligofructose to prevent antibiotic-associated diarrhoea. Alimentary pharmacology & therapeutics 21: 469–477.

Lips, P. 2012. Interaction between vitamin D and calcium. Scandinavian journal of clinical and laboratory investigation. Supplementum 243: 60–64.
Lomax, A.R. and P.C. Calder. 2009. Prebiotics, immune function, infection and inflammation: a review of the evidence. The British journal of nutrition 101: 633–658.
Longstreth, G.F., W.G. Thompson, W.D. Chey, L.A. Houghton, F. Mearin and R.C. Spiller. 2006. Functional bowel disorders. Gastroenterology 130: 1480–1491.
Macfarlane, G.T. and S. Macfarlane. 2011. Fermentation in the human large intestine: its physiologic consequences and the potential contribution of prebiotics. Journal of clinical gastroenterology 45 Suppl.: S120–127.
Makino, S., S. Ikegami, A. Kume, H. Horiuchi, H. Sasaki and N. Orii. 2010. Reducing the risk of infection in the elderly by dietary intake of yoghurt fermented with Lactobacillus delbrueckii ssp. bulgaricus OLL1073R-1. The British journal of nutrition 104: 998–1006.
Mäkivuokko, H., K. Tiihonen, S. Tynkkynen, L. Paulin and N. Rautonen. 2010. The effect of age and non-steroidal anti-inflammatory drugs on human intestinal microbiota composition. The British journal of nutrition 103: 227–234.
Mariat, D., O. Firmesse, F. Levenez, V. Guimaraes, H. Sokol, J. Dore, G. Corthier and J.P. Furet. 2009. The Firmicutes/Bacteroidetes ratio of the human microbiota changes with age. BMC microbiology 9: 123.
McCrea, G.L., C. Miaskowski, N.A. Stotts, L. Macera and M.G. Varma. 2008. Pathophysiology of constipation in the older adult. World journal of gastroenterology: WJG 14: 2631–2638.
Medzhitov, R., P. Preston-Hurlburt and C.A. Janeway, Jr. 1997. A human homologue of the Drosophila Toll protein signals activation of adaptive immunity. Nature 388: 394–397.
Mikelsaar, M., J. Stsepetova, P. Hutt, H. Kolk, E. Sepp, K. Loivukene, K. Zilmer and M. Zilmer. 2010. Intestinal Lactobacillus sp. is associated with some cellular and metabolic characteristics of blood in elderly people. Anaerobe 16: 240–246.
Miller, J.F. 1961. Immunological function of the thymus. Lancet 2: 748–749.
Mocchegiani, E., R. Giacconi, C. Cipriano and M. Malavolta. 2009. NK and NKT cells in aging and longevity: role of zinc and metallothioneins. Journal of clinical immunology 29: 416–425.
Morley, J.E. 2007. The aging gut: physiology. Clinics in geriatric medicine 23: 757–767, v–vi.
Moro-Garcia, M.A., R. Alonso-Arias, M. Baltadjieva, C. Fernandez Benitez, M.A. Fernandez Barrial, E. Diaz Ruisanchez, R. Alonso Santos, M. Alvarez Sanchez, J. Saavedra Mijan and C. Lopez-Larrea. 2012. Oral supplementation with Lactobacillus delbrueckii subsp. bulgaricus 8481 enhances systemic immunity in elderly subjects. Age (Dordr). 2012/05/31, 10.1007/s11357-012-9434-6.
Mueller, S., K. Saunier, C. Hanisch, E. Norin, L. Alm, T. Midtvedt, A. Cresci, S. Silvi, C. Orpianesi, M.C. Verdenelli, T. Clavel, C. Koebnick, H.J. Zunft, J. Dore and M. Blaut. 2006. Differences in fecal microbiota in different European study populations in relation to age, gender, and country: a cross-sectional study. Applied and environmental microbiology 72: 1027–1033.
O'Sullivan, O., M. Coakley, B. Lakshminarayanan, S. Conde, M.J. Claesson, S. Cusack, A.P. Fitzgerald, P.W. O'Toole, C. Stanton and R.P. Ross. 2012. Alterations in intestinal microbiota of elderly Irish subjects post-antibiotic therapy. The Journal of antimicrobial chemotherapy. 2012/09/06, 10.1093/jac/dks348.
Ouwehand, A.C., N. Bergsma, R. Parhiala, S. Lahtinen, M. Gueimonde, H. Finne-Soveri, T. Strandberg, K. Pitkälä and S. Salminen. 2008. Bifidobacterium microbiota and parameters of immune function in elderly subjects. FEMS immunology and medical microbiology 53: 18–25.
Ouwehand, A.C., H. Lågström, T. Suomalainen and S. Salminen. 2002. Effect of probiotics on constipation, fecal azoreductase activity and fecal mucin content in the elderly. Annals of nutrition & metabolism 46: 159–162.
Ouwehand, A.C., K. Tiihonen, M. Saarinen, H. Putaala and N. Rautonen. 2009. Influence of a combination of Lactobacillus acidophilus NCFM and lactitol on healthy elderly: intestinal and immune parameters. The British journal of nutrition 101: 367–375.

Parkes, G.C., J.D. Sanderson and K. Whelan. 2010. Treating irritable bowel syndrome with probiotics: the evidence. The Proceedings of the Nutrition Society 69: 187–194.
Pawelec, G. 2012. Hallmarks of human "immunosenescence": adaptation or dysregulation? Immunity & ageing : I & A. 9: 15.
Pawelec, G., A. Akbar, C. Caruso, R. Solana, B. Grubeck-Loebenstein and A. Wikby. 2005. Human immunosenescence: is it infectious? Immunological reviews 205: 257–268.
Putaala, H., H. Mäkivuokko, K. Tiihonen and N. Rautonen. 2011. Simulated colon fiber metabolome regulates genes involved in cell cycle, apoptosis, and energy metabolism in human colon cancer cells. Molecular and cellular biochemistry 357: 235–245.
Rajilic-Stojanovic, M., H.G. Heilig, D. Molenaar, K. Kajander, A. Surakka, H. Smidt and W.M. de Vos. 2009. Development and application of the human intestinal tract chip, a phylogenetic microarray: analysis of universally conserved phylotypes in the abundant microbiota of young and elderly adults. Environmental microbiology 11: 1736–1751.
Roberfroid, M., G.R. Gibson, L. Hoyles, A.L. McCartney, R. Rastall, I. Rowland, D. Wolvers, B. Watzl, H. Szajewska, B. Stahl, F. Guarner, F. Respondek, K. Whelan, V. Coxam, M.J. Davicco, L. Leotoing, Y. Wittrant, N.M. Delzenne, P.D. Cani, A.M. Neyrinck and A. Meheust. 2010. Prebiotic effects: metabolic and health benefits. The British journal of nutrition 104 Suppl. 2: S1–63.
Salles, N. 2007. Basic mechanisms of the aging gastrointestinal tract. Dig. Dis. 25: 112–117.
Schiffrin, E.J., A. Parlesak, C. Bode, J.C. Bode, M.A. van't Hof, D. Grathwohl and Y. Guigoz. 2009. Probiotic yogurt in the elderly with intestinal bacterial overgrowth: endotoxaemia and innate immune functions. The British journal of nutrition 101: 961–966.
Shaw, A.C., S. Joshi, H. Greenwood, A. Panda and J.M. Lord. 2010. Aging of the innate immune system. Current opinion in immunology 22: 507–513.
Shaw, A.C., A. Panda, S.R. Joshi, F. Qian, H.G. Allore and R.R. Montgomery. 2011. Dysregulation of human Toll-like receptor function in aging. Ageing research reviews 10: 346–353.
Sheih, Y.H., B.L. Chiang, L.H. Wang, C.K. Liao and H.S. Gill. 2001. Systemic immunity-enhancing effects in healthy subjects following dietary consumption of the lactic acid bacterium Lactobacillus rhamnosus HN001. Journal of the American College of Nutrition 20: 149–156.
Siegrist, C.A. and R. Aspinall. 2009. B-cell responses to vaccination at the extremes of age. Nature reviews. Immunology 9: 185–194.
Simanek, A.M., J.B. Dowd, G. Pawelec, D. Melzer, A. Dutta and A.E. Aiello. 2011. Seropositivity to cytomegalovirus, inflammation, all-cause and cardiovascular disease-related mortality in the United States. PloS one 6: e16103.
Tiihonen, K., A.C. Ouwehand and N. Rautonen. 2010. Human intestinal microbiota and healthy ageing. Ageing research reviews 9: 107–116.
Tiihonen, K., S. Tynkkynen, A. Ouwehand, T. Ahlroos and N. Rautonen. 2008. The effect of ageing with and without non-steroidal anti-inflammatory drugs on gastrointestinal microbiology and immunology. The British journal of nutrition 100: 130–137.
Tuohy, K.M., S. Kolida, A.M. Lustenberger and G.R. Gibson. 2001. The prebiotic effects of biscuits containing partially hydrolysed guar gum and fructo-oligosaccharides—a human volunteer study. The British journal of nutrition 86: 341–348.
Turchet, P., M. Laurenzano, S. Auboiron and J.M. Antoine. 2003. Effect of fermented milk containing the probiotic Lactobacillus casei DN-114001 on winter infections in free-living elderly subjects: a randomised, controlled pilot study. The journal of nutrition, health & aging 7: 75–77.
van Duin, D., H.G. Allore, S. Mohanty, S. Ginter, F.K. Newman, R.B. Belshe, R. Medzhitov and A.C. Shaw. 2007a. Prevaccine determination of the expression of costimulatory B7 molecules in activated monocytes predicts influenza vaccine responses in young and older adults. The Journal of infectious diseases 195: 1590–1597.
van Duin, D., S. Mohanty, V. Thomas, S. Ginter, R.R. Montgomery, E. Fikrig, H.G. Allore, R. Medzhitov and A.C. Shaw. 2007b. Age-associated defect in human TLR-1/2 function. Journal of immunology (Baltimore, Md. : 1950) 178: 970–975.

Van Puyenbroeck, K., N. Hens, S. Coenen, B. Michiels, C. Beunckens, G. Molenberghs, P. Van Royen and V. Verhoeven. 2012. Efficacy of daily intake of Lactobacillus casei Shirota on respiratory symptoms and influenza vaccination immune response: a randomized, double-blind, placebo-controlled trial in healthy elderly nursing home residents. The American journal of clinical nutrition 95: 1165–1171.

Wade, P.R. and T. Cowen. 2004. Neurodegeneration: a key factor in the ageing gut. Neurogastroenterology and motility : the official journal of the European Gastrointestinal Motility Society 16 Suppl. 1: 19–23.

Waller, P.A., P.K. Gopal, G.J. Leyer, A.C. Ouwehand, C. Reifer, M.E. Stewart and L.E. Miller. 2011. Dose-response effect of Bifidobacterium lactis HN019 on whole gut transit time and functional gastrointestinal symptoms in adults. Scandinavian journal of gastroenterology 46: 1057–1064.

Wiskur, B. and B. Greenwood-Van Meerveld. 2010. The aging colon: the role of enteric neurodegeneration in constipation. Current gastroenterology reports 12: 507–512.

Wu, G.D., J. Chen, C. Hoffmann, K. Bittinger, Y.Y. Chen, S.A. Keilbaugh, M. Bewtra, D. Knights, W.A. Walters, R. Knight, R. Sinha, E. Gilroy, K. Gupta, R. Baldassano, L. Nessel, H. Li, F.D. Bushman and J.D. Lewis. 2011. Linking long-term dietary patterns with gut microbial enterotypes. Science 334: 105–108.

Yang, Y.X., M. He, G. Hu, J. Wei, P. Pages, X.H. Yang and S. Bourdu-Naturel. 2008. Effect of a fermented milk containing Bifidobacterium lactis DN-173010 on Chinese constipated women. World journal of gastroenterology: WJG 14: 6237–6243.

Yatsunenko, T., F.E. Rey, M.J. Manary, I. Trehan, M.G. Dominguez-Bello, M. Contreras, M. Magris, G. Hidalgo, R.N. Baldassano, A.P. Anokhin, A.C. Heath, B. Warner, J. Reeder, J. Kuczynski, J.G. Caporaso, C.A. Lozupone, C. Lauber, J.C. Clemente, D. Knights, R. Knight and J.I. Gordon. 2012. Human gut microbiome viewed across age and geography. Nature 486: 222–227.

You, J. and P. Yaqoob. 2012. Evidence of immunomodulatory effects of a novel probiotic, Bifidobacterium longum bv. infantis CCUG 52486. FEMS immunology and medical microbiology. 2012/08/14, 10.1111/j.1574-695X.2012.01014.x.

Zaharoni, H., E. Rimon, H. Vardi, M. Friger, A. Bolotin and D.R. Shahar. 2011. Probiotics improve bowel movements in hospitalized elderly patients—the PROAGE study. The journal of nutrition, health & aging 15: 215–220.

9

Probiotics and Prebiotics in Animal Nutrition

Mattarelli Paola,[1,*] *Modesto Monica,*[1] *Sgorbati Barbara*[1]
and *Tacconi Stefano*[2]

Introduction

There has been a close relationship between microbes residing in the gastrointestinal tract (GIT) and the animal host during the long course of evolution (Ley et al. 2008). Nowadays the microbiota within the GIT of mammals can be considered a metabolically active organ: culture independent studies of the human microbiota recently identified a complex symbiotic environment with a wide biodiversity with more than 1,000 bacterial phylotypes representing more than 7,000 strains and with a high number of cells that can reach 10^{14} (Backhed et al. 2005, Murphy et al. 2009). Under normal circumstances, commensal bacteria are an essential health asset with a nutritional function and a protective influence on the intestinal structure and homeostasis. The intestinal microbiota promotes supply, digestion and absorption of nutrients, improves growth performance,

[1] Department of Agricultural Sciences, Bologna University, Via Fanin 42, 40127 Bologna, Italy.
Emails: paola.mattarelli@unibo.it; monica.modesto@unibo.it; barbara.sgorbati@unibo.it
[2] Aptuit Srl, Via Alessandro Fleming 4, 37135 Verona, Italy.
Email: stefano.tacconi4@tin.it
* Corresponding author

prevents pathogen colonization, and shapes and maintains normal mucosal immunity. Although the intestinal microbiota is complex and the role of most of the bacteria in providing benefit to the host is not clear, bacterial species of the genera *Lactobacillus* and *Bifidobacterium* have been shown to supply beneficial host effects because of their metabolic function and end products. Regulating the homeostasis, which is maintained by the microbiota, by enhancing its beneficial components, it could be possible to treat various intestinal disorders and maintain host well-being (O'Hara and Shanahan 2007).

Early GIT colonization and immune system development

The succession of microbiota colonization has been mostly studied in mammals. Animal's GIT is sterile before birth. The newborn GIT are rapidly colonized by pioneering microbes: newborn has a fully developed but naive mucosal immune system that undergoes rapid changes in response to enteric antigens which vary greatly in their potential danger (Taschuk and Griebel 2012). Studies with gnotobiotic animal models revealed that an absence of microbial stimulation results in severe developmental and immunological consequences. The gut microbiome has been implicated in many different aspects of homeostasis and neonatal mucosal immune development, including angiogenesis (Stappenbeck et al. 2002), quorum sensing and biofilm production (Zimmermann et al. 2006), and host defense peptide secretion (Bevins and Salzman 2011). Based on these considerations it can be deducted that the establishment of the GIT beneficial microbial population is of outmost importance in animal health and immune development for a lifetime of good health. In mammals microbial colonization starts with facultative bacteria, such as *Enterobacteriaceae*, *Enterococcus*, and *Streptococcus*, which are followed by anaerobic bacteria such as *Bifidobacterium*, *Bacteroides*, *Clostridium*, and *Eubacterium* (Fig. 1).

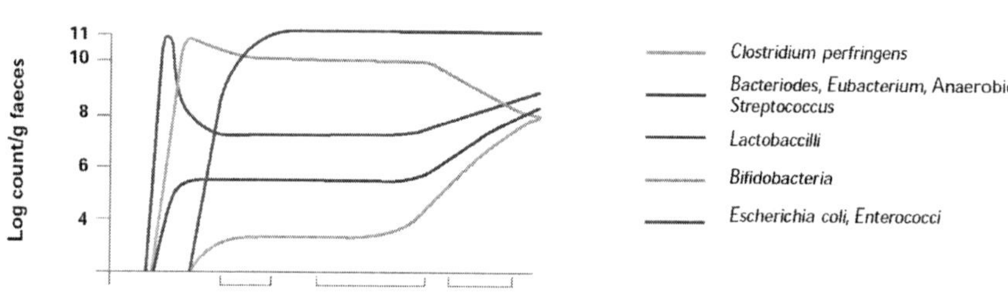

Fig. 1. Changes in gut microbiota groups with ageing in humans (adapted from Mitsuoka 1992).

Color image of this figure appears in the color plate section at the end of the book.

The origin of these intestinal microbes has attracted continuous attention. It has been hypothesized that they are acquired during transit through the birth canal and immediately after birth from the surroundings and from diet. For mammals, breast milk can be an important source of prebiotic compounds which can stimulate the growth of beneficial microbial groups: for example, in humans bifidobacteria becomes the dominant microorganism in the intestine of breast-fed infants within a week after birth and remains so throughout until weaning (Nicholson et al. 2012). The same situation, even if there are few studies in this field, could be hypothesized for other non human breast-fed mammals. Further studies of early microbiota colonization in mammal and non mammal vertebrates will certainly enrich the knowledge on this important topic. Profound changes occur in the intestinal ecosystem when young mammals are weaned from their mother's milk or liquid suckling diet onto solid food. Once solid food is digested, obligate anaerobes increase in number and diversity, especially in the hindgut and the intestine is subjected to profound modifications. The effects are especially evident in young fast growing animals, such as pigs, or during specific periods when the microbial community is subject to large change, such as weaning, and decrease during aging. The age effect is consistent with the capacity of the normal gut microbiota to resist the change as the animal grows. Moreover, access to beneficial microorganisms has been suggested to be one of the selective advantages of social behavior in animals (Ley et al. 2008). In particular, the close proximity of individuals in livestock or poultry farms could facilitate the host–host transmission of microbiota. Therefore, in high population density, it is important to maintain a "healthy" microbiota for the homeostasis of the whole organisms.

GIT microbiota and host nutrition

The GIT microbiota plays an important role in nutrient digestion and absorption: it is involved in the fermentation of nondigestible dietary fiber and related nutrients (resistant starch or oligosaccharides), the anaerobic metabolism of peptides and proteins, the biotransformation of conjugated bile acids, the degradation of oxalate-based complexes, and the synthesis of aminoacids, enzymes, volatile fatty acids (VFA) and some vitamins (e.g., B12, folic acid and K) (Cani and Delzenne 2007, D'Aimmo et al. 2012). Intestinal bacteria themselves are rich sources of protein (as much as 60–65%), which can also be used by hosts. In pigs, up to 30% of energy for maintenance could be retained due to microbial biodegradation, particularly in the large intestine.

Gut microbiota composition is involved in the regulation of energy homeostasis. Backhed et al. (2004) found that young conventionally reared mice have a 40% higher body fat content and 47% higher gonadal fat content than germ-free mice. Strikingly, this phenomenon was associated with a lower food intake in mice with the conventional microbiota than in their germ-free counterparts. In the same line, the authors demonstrated that germ-free mice colonized with the gut microbiota derived from the conventional mice produces a 60% increase in body fat content and insulin resistance within 14 days despite reduced food intake.

Microbiota and pathogens

The mucosal surface of the GIT tract represents a major entry point and ecological niche for many microbial pathogens and the presence of different strains and species of probiotics can help to prevent their invasion through different mechanisms: immune system stimulation (described above), competitive exclusion, consumption of nutrient sources, and production of antimicrobial substances. Beneficial gut microbes also stimulate the host to produce various antimicrobial compounds.

In competitive exclusion probiotic can displace the incoming pathogens by competition adhesion to the GIT mucus sites, by coaggregation mechanisms and by regulation of intestinal motility and mucus secretion (Schachtsiek et al. 2004).

The capacity to control the proliferation of pathogen microorganisms could be done also through modulation of the intestinal environment by probiotics which compete for the occupancy of a common biotope (e.g., access to nutrients) (Oelschlaeger 2010). Iron, for example, being essential for most bacteria, is a limiting nutrient and probiotics can compete for its availability. *Lactobacillus* can render iron unavailable for pathogenic microorganisms, either by binding ferric hydroxide on its surface or by secreting siderophores that chelate and transport iron. Some probiotics are also able to influence the composition and equilibrium of the gut microbiota. For example, probiotic consumption using a mixture of probiotics (VSL#3) was shown to increase the total number of intestinal bacteria and to restore the diversity of the GIT bacterial microbiota (Kuhbacher et al. 2006).

One of the important effects of GIT microbiota, especially related to bifidobacteria and lactobacilli, is to counteract the load of gram negative bacteria and in this way decrease the concentration of lipopolysaccharide (LPS) on the intestinal mucosa. The decrease of LPS and CD14/signaling on the immune cells leads to a decrease of proinflammatory cytokine product and in this way reduces the inflammatory tone which has been recognized to be one of the causes for obese and diabetes 2 type metabolism (Cani and Delzenne 2007).

The above considerations are especially related to homoeothermic animals. On the other hand the exact role of gut microbiota in nutrition of heterothermic animals, e.g., fish or insects, is difficult to assess because of the complex and variable ecology of their GIT microbiota. Despite recent conventional and gnotobiotic studies that indicate the possible involvement of GIT bacteria in several physiological and nutritional functions in these animals, more emphasis and/or thorough research is required in order to establish the nutritional importance of their gut microbiota.

Another important mechanism for counteracting the pathogens is the secretion of active molecules (e.g., VFA, hydrogen peroxide and bacteriocins) by probiotics. The lower pH produced by the organic acids and the hydrogen peroxide produced by probiotics can control growth and/or survival of pathogen microorganisms. The bacteriocins are secreted peptides or proteins that generally kill closely related bacteria by permeabilizing their membranes or by interfering with essential enzyme. Many of them are produced by *Lactobacillus* probiotic strains such as lactacin B, lactacin F, nisin, reuterin, etc. (Wohlgemuth et al. 2010).

The control of pathogens colonization is very important not only for reducing infections but also because intestinal pathogens produce toxins and other classes of substances, i.e., mucinases, adhesins and invasins, which interfere with epithelial metabolism. All together, the pathogenic phenotype is likely to directly trigger uncontrolled pathological inflammation. Increasing evidence indicates that changes in gut microbiota, with an increase of pathogenic bacteria and a decrease of health-promoting bacteria, such as bifidobacteria and lactobacilli, play an important role in promoting and maintaining intestinal inflammation in inflammation bowel diseases (Andoh and Fujiyama 2006).

Finally the new concept that probiotics could also counteract eukaryotic pathogens is emerging. Approaches with probiotics could help to reduce the risks of infestation by specific parasites (*Cryptosporidium*, *Giardia*, *Eimeria*, worms) or to complement classical antiparasite treatments (Travers et al. 2011).

Probiotics, Prebiotics and Synbiotics

Probiotics

The more widely accepted definition for the term "probiotic" is "live microorganisms which, when administered in adequate amounts, confer a health benefit on the host" (FAO/WHO 2002). This definition implies that a health effect must be demonstrated for the probiotic. The beneficial modes of action include: regulation of intestinal microbial homeostasis, stabilization of the gastrointestinal barrier function, expression of bacteriocins, enzymatic activity inducing absorption and nutrition, immunomodulatory effects, inhibition of procarcinogenic enzymes and interference with the ability of pathogens to colonize and infect the mucosa (Gaggia et al. 2010). The most used probiotics are *Bifidobacterium* and *Lactobacillus* and most literature concern their probiotic activity but also studies of other probiotic genera especially for animal feeding (*Enterococcus, Bacillus, Saccharomyces*, etc.) are now in progress. The expected health-promoting characteristics and safety criteria of probiotics are shown in Fig. 2.

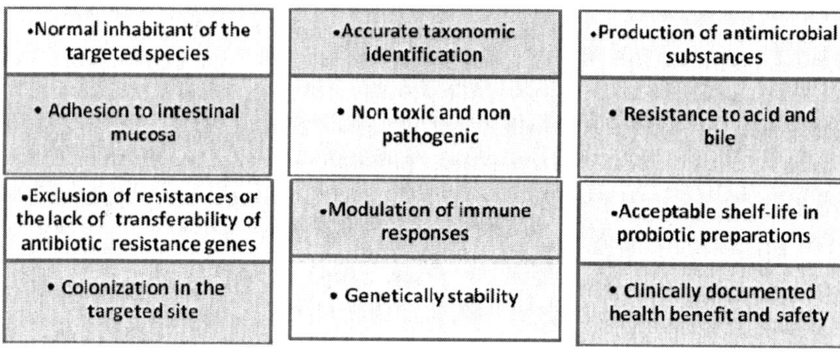

Fig. 2. Characteristics and safety criteria of probiotics.

Regulations on probiotic

Significant progress in legislation for the safety evaluation of probiotics has been made in the USA, Canada, and Europe (EFSA 2005a, HC 2006, FAO/WHO 2002); however, no unique standard is available. In the USA, specific utilization of microorganisms for human consumption should possess "GRAS" status ("Generally Regarded As Safe") regulated by the Food and Drug Administration. In Europe, the European Food Safety Authority (EFSA) has introduced the concept of Qualified Presumption of Safety (QPS) similar in purpose to the GRAS approach. The QPS concept provides a generic assessment system for use within EFSA that in principle can be applied to all requests received for the safety assessments of microorganisms deliberately introduced into the food chain (EFSA 2005b). According to recent evaluation (Wassenaar and Klein 2008), QPS system appears more flexible because it takes into account additional criteria to evaluate the safety of bacterial additives such as a history of safe use in the food industry and the acquisition of antibiotic resistance or virulence determinants. EFSA has published a list of microorganism, which possess a known history of safety, proposed for QPS status (EFSA 2007).

The studies on the efficacy of probiotics and prebiotics in animals and man have often produced contrasting results: these can derive from the heterogeneity of the experimental protocol utilized. There is no standardization concerning doses, time and way of administration, animal condition, etc. Recently the recommended guidelines for the design of probiotic studies to substantiate health claims has been published (Fig. 3). These guidelines are a very important milestone to compare data from different experiments and to provide the basis for more refined hypothesis-driven clinical trials.

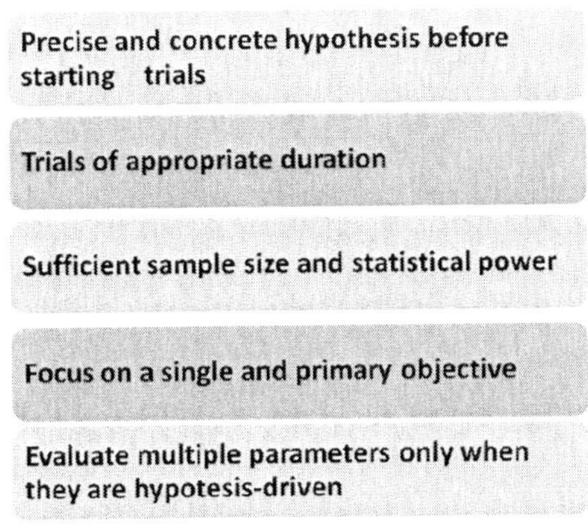

Fig. 3. Guidelines for the design of probiotic/prebiotic studies to substantiate health claims (from Gibson et al. 2011).

Probiotic utilization

Both the food industry and pharmaceutical manufacturers have started adding probiotic cultures to animal feed and to pharmaceuticals: and in the last 20 years a growing number of probiotic adjuncts have become available to the human or animal "consumer". Very close attention must be paid to describing the identity of strains that are candidate probiotics as such strains require standardized and accurate classical and molecular procedures during their identification. Following strain identification, there is a screening step that classifies the strains on the basis of their specific health-promoting effects (vitamin production, i.e., folic acid, induction of antinflammatory cytokine, bacteriocin production, etc.), their host colonization properties (survival in gastric environment, intestinal adhesion, etc.) and industrial features (freezing and freeze-drying survival, oxygen sensibility, growth performance, etc.). Recently, microencapsulation technologies have enabled the introduction of viable probiotic bacteria into industrial preparations, with the objective of enhancing the survival of probiotic bacteria during their exposure to the adverse conditions of the gastro-intestinal tract (Chávarri et al. 2010). Health-promoting properties are known to be strain-dependent, and specific strains have now been demonstrated to have beneficial properties (Gaggia et al. 2010).

Theoretically, host colonization should be facilitated by choosing probiotic strains of human origin for human consumption, and those of animal origin for animal consumption. However, numerous studies have shown that the effect of probiotic treatment reaches its maximum during the administration of the probiotic independently of the host-specificity of the strain used, and the presence of probiotic bacteria is not maintained after the cessation of probiotic consumption. Therefore the use by humans of probiotics of animal origin, and vice versa, can produce a positive response in the host. An example of this is the worldwide use of *Bifidobacterium animalis* subsp. *lactis*, which appears to be safe and to have probiotic properties for the human host (Holmes et al. 2012), despite its animal origin. In addition, it has been recognized that functionality of multistrain and multispecies probiotics could be more effective than that of monostrain probiotics. The advantages of administering multistrain and multispecies probiotics include the enhanced capability of colonizing the gastrointestinal tract and to combine the different mechanisms of action of each strain in a synergistic way (Timmerman et al. 2004).

Most used probiotic genera for animal feed

The most widely used probiotics in feed and pharmaceutical preparation for animals are *Lactobacillus, Enterococcus, Bacillus* and *Saccharomyces.* Competitive exclusion (CE) mixed cultures have been developed for animal breeding (Schneitz 2005), but most of the products being developed so far are preparations of unknown bacterial composition posing the risk of containing pathogenic bacteria or viruses. From the point of view of the risks associated with undefined preparations, a defined CE product consisting of various well-characterized bacterial strains that meet the European regulatory demands has being developed (Callaway et al. 2008). Other probiotics used are, *Bifidobacterium, Escherichia coli* Nissle 1917, *Pediococccus* and *Streptococcus.* In man the most used probiotics are *Bifidobacterium* and *Lactobacillus* (Biavati and Mattarelli 2012). The use of enterococci as probiotics, which are currently used in animal feed, remains a controversial issue because of the emergence of the increased association of enterococci with human diseases. The concern that enterococci antimicrobial resistance genes or genes encoding virulence factors could be transferred to other bacteria in the gastrointestinal tract contributes to safety concerns of entetroccci: no members of this genus have been proposed for QPS status (EFSA 2007).

Prebiotic

A prebiotic was first defined in 1995 by Gibson and Roberfroid as "a non-digestible food ingredient that beneficially affects the host by selectively stimulating the growth and/or activity of one or a limited number of bacteria in the colon, and thus improves host health." This definition was updated in 2010 into "a selectively fermented ingredient that allows specific changes, both in the composition and/or activity in the gastrointestinal microbiota that confers benefits upon host wellbeing and health" (Gibson et al. 2010).

Most identified prebiotics with the exception of inulin, which is a mixture of fructo-oligo and polysaccharides, are mixtures of indigestible oligosaccharides, consisting of 3–10 carbohydrate monomers. In the last two decades prebiotics have been normally utilized for human and animal feed consumption; dietary carbohydrates such as fibers are candidate prebiotics, but most promising are nondigestible oligosaccharides (NDOs).

Currently, the target genera are lactobacilli and bifidobacteria; however, prebiotic success has primarily been achieved with bifidobacteria. This may be due to the fact that more bifidobacteria usually reside in the human colon than lactobacilli and they exhibit a preference for oligosaccharides (Brownawell et al. 2012).

The majority of the effects claimed by the prebiotics are described in Fig. 4.

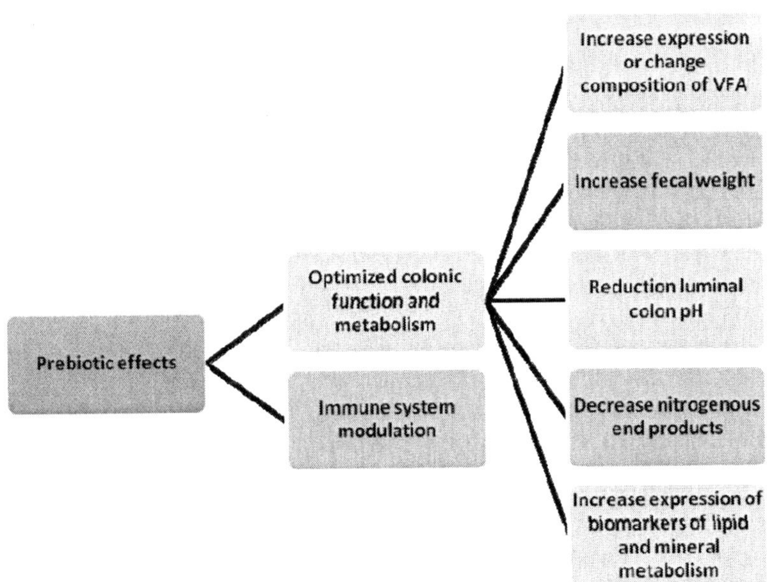

Fig. 4. Major effects of prebiotics.

All these effects on colonic microbiota and on the biochemistry and histology of the host bowel support the logic of the use of prebiotics for promoting health benefits. Candidate prebiotics must fulfill the cited criteria which are to be proven by in vitro and finally *in vivo* tests.

Synbiotic

The recent development of commercial prebiotic oligosaccharides and probiotic bacteria has led to a new concept of symbiotic which combines probiotics and prebiotics.

Moreover, compared with the use of individual components, application of synbiotics seems to confer a beneficial modulation of the composition of the gut microbiota by increasing the levels of purportedly beneficial bacteria, such as lactobacilli and bifidobacteria and reducing the levels of the other less desirable species, such as coliforms and enterococci (Modesto et al. 2011).

Animal Feeding and Antibiotic

Antibiotics at sub-therapeutic levels as growth promoter (antibiotic growth promoter, AGP) have been used over many decades. Despite the positive impact on improving growth performance and reducing diarrhea and mortality of AGP, there are concerns that their use in animal feeds can result in the development of antibiotic resistant bacterial strains and antibiotic residue problems in animal products. It is also perceived that these antibiotic-resistant strains can be transferred to humans, which may impair the effectiveness of certain antibiotics in the treatment of human diseases. In the United States, legislation has been introduced that seeks to restrict the use of certain antimicrobial drugs for subtherapeutic or nontherapeutic purposes in food-producing animals. Most U.S. livestock and poultry producers are opposed to such restrictions because of concerns about animal welfare and food safety, as well as concerns about possible increases in production costs, among other reasons. The European Union in January 2006 completely banned the use of AGP for the possibility of generation of resistant pathogen strains. However, the discontinuation of using AGP may reduce performance and increase the therapeutic use of antibiotics because enteric diseases in particular are difficult to control with no use of in-feed antibiotics (Kil and Stein 2010). In Europe the negative consequences for animal health and welfare and for food safety on a multi-national scale have become particularly evident in the food chain. For example the incidence of *C. jejuni*, one of the most common pathogens in livestock, has increased from 2006 to date (EFSA 2009). This is in agreement with the idea that the use of antibiotics resulted, besides the improvement of nutrient absorption enhancing feed intake and weight gain, in the inhibition of pathogens, widespread at the primary production level: in fact , an increase in the amount of prescription medication used in livestock and poultry production has been observed, Therefore the research of a rational development of new alternative strategies for good animal performance together with low or absence of pathogens in the livestock food chain has to be intensified.

Probiotic and Prebiotic Application in Different Animals

Appling probiotics and prebiotics in farm or pet animals seeks to restore or beneficially alter the microbiota present in young, stressed or antibiotic treated animals so that they can better resist gastrointestinal disease, especially infectious disease. Specific nutritional stresses may include the change from milk to solid feed, from high fiber to high protein diets in young animals, and are associated with imbalances in the microbiota, which can lead to increased disease susceptibility, for example, increased incidences in diarrhea in post weaning piglets. The following paragraphs give a general overview of the probiotics and prebiotics used in animal nutrition, which is important not only for humans working with animals but also for meat consumption.

Mammals

Strategy to protect and maintain mammal health

When breastfeeding is not possible, in animal breeding the inclusion of probiotics in milk formula has been suggested. Specific probiotics in milk formula focus on aiding healthy gut microbiota development and may constitute a new model of preventing pathogen action through competitive exclusion and aggregation with pathogens (Collado et al. 2007). Thus, the inclusion of probiotics in powdered infant formula may enhance their resemblance to breast milk (Callaway et al. 2008).

Cattle

Feeding practices and farm management of calve production cope with major stress events, such as transportation, marketing, dietary changes and exposure to a variety of infectious agents. Moreover, in intensive rearing of calves, the possibility of acquiring natural, autochthonous microbiota is strongly diminished with a high incidence of intestinal and respiratory diseases in veal calves. Feeding probiotics such as lactic acid bacteria,

Propionibacterium, Bifidobacterium, Bacillus or yeasts, to young pre-ruminants has been shown to improve digestion with beneficial repercussions on gut health; in particular, the rate, severity and length of diarrheal episodes are reduced. Probiotics generally target the intestine, because the rumen is not yet developed. With regard to animal performance, improved weight gain and rumen development have been reported in young calves with several bacterial and yeast strains supplementation (Adams et al. 2008).

In adult ruminants, probiotics have mostly been selected to target the rumen compartment, which is the main site of feed digestion. The most common marketed products for ruminants are live yeast (*Saccharomyces cerevisiae*) preparations. A growing interest for using probiotics is to reduce pathogen infections and digestive carriage by adult ruminants of human pathogens, such as *Escherichia coli* O157 or *Salmonella*. Certain strains of *Lactobacillus acidophilus* have shown to decrease numbers of *E. coli* O157 in feedlot cattle feces (Tabe et al. 2008). The use of prebiotics in cattle has been limited due to the ability of ruminants to degrade most prebiotics; however enhancements in rumen-protective technologies may allow these compounds to be used in feedlot and dairy cattle, considering also that several classes of nondigestible oligosaccharides are found in plant cell wall in nature including feeds normally used for livestock feeding (Callaway et al. 2008).

Pig

In pig production, from birth to post-weaning, piglets are subjected to major stressful events, making them highly sensitive to digestive disorders. Piglets are very sensitive to gut colonisation by pathogenic bacteria (*E. coli, Clostridium difficile, Clostridium perfringens, Salmonella, Listeria*), parasites (*Isospora, Cryptosporidium*) or viruses (*Coronavirus, Rotavirus*), which are responsible for growth reduction and diarrhea. At this time, the development of both innate and adaptive immunity at the mucosal surface is critical in preventing the potential harmful effects of intestinal pathogenic agents. Probiotics are therefore recommended during this period and numerous studies have shown a beneficial role of probiotic administration in piglets, improving the number of beneficial bacteria and decreasing the load of pathogens; moreover, they display a major role in stimulating the immune cell response, showing high IgM and IgA activities towards pathogens in comparison to control, and increasing defensive tools against pathogenic invasion (Casey et al. 2007, Konstantinov et al. 2008). In contrast, some authors reported an enhancement of the course of infection or a partial alleviation of diarrhea.

Different types of chemically defined or undefined dietary compounds are added to the diet of pigs to test their influence on gastrointestinal microbiota or on health status improvement during challenge with pathogens. Incorporations of prebiotic oligosaccharides into pig feeds have resulted in mixed but generally no significant effects regarding beneficial modulation of microbial populations in various intestinal segments and feces of swine. Based on the results of current studies, a beneficial effect of synbiotic applications could be suggested as their administration showed significant improvements of growth performance parameters in suckling (Modesto et al. 2011) and in growing pigs (Piva et al. 2005).

Horses

Horses are sensitive to environmental stress (during the feed change, transport, competition, and weaning) and, for example, may develop colic or laminitis in response to sudden diet changes or a carbohydrate overload (Hudson et al. 2001). Abrupt changes in the diet of horses were associated with drastic modifications in their microbial population in the large intestine, and microbial population in the colon seemed to be more sensitive than the population in the cecum (de Fombelle et al. 2001). Therefore it is essential to develop feeding practices, which can efficiently supply the horse with required energy and are also able to prevent nutrition related diseases by achieving a balance in the gut microbial population. There is a lack of studies on application of pro and prebiotics in horses. Some results are positive such as the beneficial effects in preventing digestive disorders associated with starch intake after supplementing the diet of the horse with short chain FOS (Respondek et al. 2011). Probiotic yeast seems to be more effective in respect to probiotic bacteria: *S boulardii* efficacy for decreasing the duration and severity of clinical signs in horses with enterocolitis was showed by Desrochers et al. (2005).

Poultry

Newly hatched broiler chickens of the modern poultry husbandry do not come into contact with the mother hens. This lack of contact is believed to result in a delayed development of the intestinal microbiota with all related consequences and broilers at very young age are particularly susceptible to pathogen colonization. In this respect, probiotics from the animals guts could be of great interest because they offer biological alternatives to protect and improve health which should find acceptance by both the producers and consumers. In poultry, benefits of probiotic supplementation (live yeast or bacteria) are reported in broilers' performance and health, with evidence of increased resistance of chickens to *Salmonella, E. coli, Clostridium perfringens, Campylobacter jejuni* infections (Higgins et al. 2008, La Ragione et al. 2004). Probiotics can increase feed efficiency and productivity of laying hens (Yörük et al. 2004), and an improvement in egg quality (decreased yolk cholesterol level, improved shell thickness, egg weight) has also been reported (Xu et al. 2003). Some studies about prebiotic in poultry indicate their usefulness in controlling or reducing the growth of *C. perfringens* which is very important to the poultry industry because it is one of the most important causes of necrotic enteritis (Hofacre et al. 2005). The new interesting prebiotic fucosyllacrose seems to favour the coaggragation with pathogens instead of mucosal lining of the poultry intestine eliminating the pathogen contaminant (Lee et al. 2012).

Howewer reviewing the results of different *in vivo* studies, the effect of prebiotics on gut health, performance, and reduction of pathogen shedding appears variable, depending on the type and on the dose of prebiotic used. The inclusion criteria of the supplement is not consistent among authors and high dosage of prebiotics showed negative effects on gut system, causing diarrhea and consequently decreasing growth performance (Biggs 2007). Several studies have revealed that synbiotic treatment was more efficacious in reducing pathogen carriage and infections in poultry than an individual prebiotic or probiotic treatment.

Fish

The study of the GIT microbiota in fish is still in its infancy and in the future the new high-throughput sequencing methods could help in discovering its biodiversity. The early exposure of the intestine to live bacteria and subsequent colonization is very important for the development of gut barrier like in homoeothermic animals. The developmental stage of fish, gut structure, the surrounding environment like ambient water temperature, rearing and farming conditions are critical factors that affect the initial colonization and the subsequent establishment process. When different types of chemicals, antibiotics, pollutants like pesticides, herbicides and insecticides enter into the digestive tract of an aquatic animal, they can drastically affect the composition of dominant GIT microbiota and may lead to the elimination of individual species from the whole microbial community (Navarrete et al. 2009). Unlike the microbiota in warm-blooded animals, fish GIT microbiota seems to be highly variable, depending on seasonal and day-to-day fluctuations (Pelletier et al. 2007). The intensively farmed marine or fresh water fishes are often affected by numerous viruses, bacteria, fungi and parasites causing infectious diseases, and thereby leading to heavy losses in aquaculture production. These problems arise particularly during the larval stage, the most critical period of rearing.

The use of antimicrobial agents causes environmental concerns, and their effectiveness in preventing or controlling fish diseases has been questioned, given extensive documentation on the evolution of drug resistance by pathogenic bacteria. The use of probiotics in animal nutrition has recently begun to be applied in aquaculture as health protective agents (Defoirdt et al. 2011). Numerous microorganisms have been used as probiotics such as *Aeromona* spp., *Vibrio* spp., *Lactobacillus* spp., etc., to improve growth or survival of larval aquatic species. It has been suggested that the efficacy of probiotics is highest in the host species from which they are isolated, therefore candidate aquatic probiotics for larviculture are isolated from healthy adults and larvae. However, even probiotics isolated from man such as *Lactobacillus rhamnosus* enhanced survival of rainbow trout challenged with a virulent strain of *Aeromonas salmonicida*, and some probiotics used for human beings and terrestrial animals have given promising results in aquaculture species (Nikoskelainen et al. 2003).

Prebiotics are found to stimulate the growth of species of intestinal bacteria in fish. Furthermore, dietary supplementation of prebiotics like mannan oligosaccharides leads to improved growth and immunity and enhancement of digestive enzymes like protease and amylase, respectively (Xu et al. 2009). However, there are some concerns associated with the use of prebiotics in aquaculture practices. Several pathogens as well as opportunistic bacteria can utilize a wide range of carbohydrates and can eventually pose health hazards by proliferating inside the gut by metabolizing the prebiotics (Merrifield et al. 2010). Similarly, another major concern for prebiotics is that high concentrations of prebiotics can be harmful as evidenced from the damaging effect of inulin at a high concentration on the enterocytes of *Salvelinus alpinus* (Olsen et al. 2001). However, the growth enhancement and health improvement of fish/shell-fish by promoting the growth of certain microbes in the GI tract through prebiotics and/or probiotics is a beneficial and rational strategy but their use in aquaculture is still in its infancy. Therefore, the fish intestinal microbiota might be a key pool of potential probiotics for cultured fish species.

Honey-Bee

Honey bees (*Apis mellifera*) as pollinators in agriculture play a critical role in the economy for global food production. Recently, honey bee populations in the United States, Canada, and Europe have suffered an unexplained increase in annual losses due to a phenomenon known as colony collapse disorder.

Several members of the *A. mellifera* microbiota (*Acetobacteraceae*, *Bifidobacterium*, *Lactobacillus*, and *Simonsiella*) produce short chain fatty acids such as lactic or acetic acid as waste products during the metabolism of carbohydrates (Vasquez et al. 2012). Assimilation of these compounds could supplement bee nutrition, just as short chain fatty acids produced by rumen microbes supply nearly all the energy requirements of ruminant mammals. Short chain fatty acids can be absorbed through the rectal wall in insects, and the majority of the pollen and bacterial biomass within an adult *A. mellifera* is contained inside the rectum (Bradley 2008). Overwintering *Apis* may obtain additional nutrition from these rectal bacteria, as consumed food is stored for longer periods of time within the rectum during winter months (Lindstrom et al. 2008). Recently in the crop of honey bee, which is a central organ in the honeybee's food production between the oesophagus and ventriculus and is used for collection and transport of nectar to the hive, 13 bacterial species within the genera *Lactobacillus* and *Bifidobacterium* (L-B) have been found (Vasquez et al. 2012). These bacteria play a key role in the production of honey and beebread, long term stored food for both adult honeybees and larvae. Both *in vitro* and *in vivo* studies showed that the L-B microbiota in *A. mellifera* inhibit *Paenibacillus larvae* that is the cause of the brood disease American foulbrood (Forsgren et al. 2010).

The probiotic L-B have evolved in synergy with bees and play an important role in defending their hosts: they exert a protective role in bacterial brood diseases such as American and European fullbroad. Any beneficial effect from these bacterial groups may be undermined where prophylactic use of antibiotics is practiced. It is important to discover the mechanisms of action and functional analysis of L-B against pathogens and food spoiling microbes, and how they can be used to resolve ongoing honeybee colony losses, in which L-B may be the important missing link.

Concluding Remarks

There is a strong requirement for natural alternatives to prevent the proliferation of pathogenic bacteria and to modulate indigenous GIT microbiota so that the health, immune status and performance of animals and humans could be improved. Characterization of the GIT microbiota is essential in providing insights for the understanding of its role in animal health and disease. A key issue is to identify and understand the species present in the gut microbiota of the different animals and to functionally characterize their gut microbiota. The application of probiotics and prebiotics for animal welfare is very promising, even in the perspective of a natural approach for animal feeding. They are very important in the animal production system: infact after the ban of antibiotics as growth promoters they can be a valid support both for prevention of infections, for growth performance and for food safety. Moreover it has to be considered that probiotics are not an alternative to conventional medicine; but they are supplements and adjuncts to it in the prevention or the treatment of diseases, not substitutes.

Isolation, characterization, and risk assessment of intestinal strains are essential parts in the development of a safe probiotic feed additive. The selection of specific probiotic strains and prebiotics and their combinations for the use in feed additives requires a critical evaluation, especially according to different regulations in different countries. Natural or syntethized/extracted prebiotics in the diet is of particular importance for animal species when they are closely tied to optimal animal growth and health.

Today, the molecular mechanisms underlying the beneficial effects of probiotics and prebiotics have been clarified by a multitude of metagenomic analysis. However, further deeper investigations are needed using validated defined protocols (specific probiotics and prebiotics and experimental models), as well as extended clinical investigations for studying for a specific strain in a specific indication. Gnotobiotics, whose genotype and microbial status can be clearly defined and whose diet and environmental conditions can be easily controlled, are invaluable tools to go forward in this direction.

The full application of prebiotic in animal livestock and poultry is at its beginning because its cost/benefits ratio has not been fully established. Moreover, the costs are very high, mainly for the probiotic preparation, which has to contain high concentration of viable bacteria which allow the maximum viability of the bacterial species utilized at the moment of administration in animal feed.

Although probiotic and prebiotic research and application will be costly, it has the great potential to reduce the risk of pathogenic infections and improve animal health and food safety.

References

Adams, M.C., J. Luo, D. Rayward, S. King, R. Gibson and G.H. Moghaddam. 2008. Selection of a novel direct-fed microbial to enhance weight gain in intensively reared calves. An. Feed Sci. Technol. 145: 41–52.

Andoh, A. and Y. Fujiyama. 2006. Therapeutic approaches targeting intestinal microflora in inflammatory bowel disease. World J. Gastroenterol. 12: 4452–4460.

Backhed, F., H. Ding, T. Wang, L.V. Hooper, G.Y. Koh, A. Nagy, C.F. Semenkovich and J.I. Gordon. 2004. The gut microbiota as an environmental factor that regulates fat storage. Proc. Natl. Acad. Sci. USA 101: 15718–15723.

Backhed, F., R.E. Ley, J.L. Sonnenburg, D.A. Peterson and J.I. Gordon. 2005. Host-bacterial mutualism in the human intestine. Science 307: 1915–1920.

Bevins, C.L. and N.H. Salzman. 2011. Paneth cells, antimicrobial peptides and maintenance of intestinal homeostasis. Nat. Rev. Microbiol. 9: 356–368.

Biavati, B. and P. Mattarelli. 2012. Genus *Bifidobacterium*. *In*: M. Goodfellow, P. Kampfer, H.-J. Busse, K.-I. Suzuki, W. Ludwig and W.B. Whitman [eds.]. Bergey's Manual of Systematic Bacteriology 2nd Edition. The Actinobacteria. Vol. 5. Springer, New York, pp. 171–206.

Biggs, P., C.M. Parson and G.C. Fahey. 2007. The effects of several oligosaccharides on growth performance, nutrient digestibilities, and cecal microbial populations in young chicks. Poult. Sci. 86: 2327–2336.

Brownawell, A.M., W. Caers, G. Gibson, C.W.C. Kendall, K.D. Lewis, Y. Ringel and J.L. Slavin. 2012. Prebiotics and the health benefits of fiber: current regulatory status, future research, and goals. J. Nutr. 142: 962–974.

Callaway, T.R., T.S. Edrington, R.C. Anderson, J.A. Byrd and D.J. Nisbet. 2008. Gastrointestinal microbial ecology and the safety of our food supply as related to *Salmonella*. J. Anim. Sci. 86: E163–E172.

Cani, P.D. and N.M. Delzenne. 2007. Gut microflora as a target for energy and metabolic homeostasis. Curr. Opin. Clin. Nutr. Metab. Care 10: 729–734.

Casey, P.G., G.E. Gardiner, G. Casey, B. Bradshaw, P.G. Lawlor, P.B. Lynch, F.C. Leonard, C. Stanton, R.P. Ross, G.F. Fitzgerald and C. Hill. 2007. A five-strain probiotic combination reduces pathogen shedding and alleviates disease signs in pigs challenged with *Salmonella enterica* Serovar Typhimurium. Appl. Environmen. Microbiol. 73: 1858–1863.

Chávarri, M., I. Marañón, R. Ares, F.C. Ibáñez, F. Marzo and M.C. Villarán. 2010. Microencapsulation of a probiotic and prebiotic in alginate chitosan capsules improves survival in simulated gastro-intestinal conditions. Int. J. Food Microbiol. 142: 185–189.

Collado, M.C., C. Grzeskowiak and S Salminen. 2007. Probiotic strains and their combination inhibit *in vitro* adhesion of pathogens to pig intestinal mucosa. Curr. Microbiol. 55: 260–265.

D'Aimmo, M.R., P. Mattarelli, B. Biavati, N.G. Carlsson and T. Andlid. 2012. The potential of bifidobacteria as a source of natural folate. J. Appl. Microbiol. 112: 975–984.

de Fombelle, A., V. Julliand, C. Drogoul and E. Jacotot. 2001. Feeding and microbial disorders in horses. 1. Effects of an abrupt incorporation of two levels of barley in a hay diet on microbial profile and activities. J. Equine Vet. Sci. 21: 439–445.

Defoirdt, T., P. Sorgeloos and P. Bossier. 2011. Alternatives to antibiotics for the control of bacterial disease in aquaculture. Curr. Opin. Microbiol. 14: 251–258.

Desrochers, A.M., B.A. Dolente, M.F. Roy, R. Boston and S. Carlisle. 2005. Efficacy of *Saccharomyces boulardii* for treatment of horses with acute enterocolitis. J. Am. Vet. Med. Assoc. 227: 954–959.

EFSA. 2005a. European Food Safety Authority. Opinion of the scientific committee on a request from EFSA related to a generic approach to the safety assessment by EFSA of microorganisms used in food/feed and the production of food/feed additives. EFSA Journal 226: 1–12.

EFSA. 2005b. Opinion of the scientific panel on additives and products or substances used in animal feed on the updating of the criteria used in the assessment of bacteria for resistance to antibiotics of human or veterinary importance. EFSA Journal 223: 1–12.

EFSA. 2007. European Food Safety Authority. Introduction of a Qualified Presumption of Safety (QPS) approach for assessment of selected microorganisms referred to EFSA Opinion of the Scientific Committee (Question No EFSA-Q-2005-293) Adopted on 19 November 2007. EFSA Journal 587: 1–16.

EFSA. 2009. European Food Safety Authority. The community summary report on trends and sources of zoonoses and zoonotic agents in the European Union in 2007. The EFSA Journal 223: 1–320. European Food Safety Authority. Community Summary Report on Trends and Sources of Zoonoses and Zoonotic Agents in the European Union in 2007. (http://www.efsa.europa.eu/cs/BlobServer/Report/zoonoses_report_2007,3.pdf?ssbinary=true.

FAO/WHO 2002. Joint FAO/WHO (Food and Agriculture Organization/World Health Organization) working group report on drafting guidelines for the evaluation of probiotics in food. London, Ontario, Canada.guidelines for the evaluation of probiotics in food. Joint working group report on drafting. London, Ontario 1–11.

Forsgren, E., T.C. Olofsson, A. Vásquez and I. Fries. 2010. Novel lactic acid bacteria inhibiting *Paenibacillus larvae* in honey bee larvae. Apidologie 41: 99–108.

Gaggìa, F., P. Mattarelli and B. Biavati. 2010. Probiotics and prebiotics in animal feeding for safe food production. Int. J. Food Microbiol. 141: S15–S28.

Gibson, G.R. and M.B. Roberfroid. 1995. Dietary modulation of the human colonic microbiota: introducing the concept of prebiotics. J. Nutr. 125: 1401–1412.

Gibson, G.R., K.P. Scott, R.A. Rastall, K.M. Tuohy, A. Hotchkiss, A. Dubert-Ferrandon, M. Gareau, E.F. Murphy, D. Saulnier, G. Loh, S. Macfarlane, N. Delzenne, Y. Ringel, G. Kozianowski, R. Dickmann, I. Lenoir-Wijnkoop, C. Walker and R. Buddington. 2010. Dietary prebiotics: current status and new definition. Food Sci. Tec. Bull.: Functional Foods 7: 1–19.

Gibson, G.R., R.J. Brummer, E. Isolauri, H. Lochs, L. Morelli, T. Ockhuizen, I.R. Rowland, J. Schrezenmeir, C. Stanton and K. Verbeke. 2011. The design of probiotic studies to substantiate health claims. Gut Microbes 2: 299–305.

HC. 2006. Health Canada. Evidence for Safety and Efficacy of Finished Natural Health Products. [online]. Ottawa (ON): Natural Health Products Directorate, Health Canada. [Accessed 2008 May 28]. Available from: http://www.hc-sc.gc.ca/dhpmps/prodnatur/legislation/docs/efe-paie_e.html.

Higgins, S.E., J.P. Higgins, A.D. Wolfenden, S.N. Henderson, A. Torres-Rodriguez, G. Tellez and B. Hargis. 2008. Evaluation of a *Lactobacillus*-based probiotic culture for the reduction of *Salmonella enteritidis* in neonatal broiler chicks. Poul. Sci. 87: 27-

Hofacre, C.L., G.F. Mathis and M.A. Quiroz. 2005. Natural alternatives to prevent necrotic enteritis. Int. Poult. Prod. 13: 7–9.

Holmes, E., J. Kinross, G.R. Gibson, R. Burcelin, W. Jia, S. Pettersson and J.K. Nicholson. 2012. Therapeutic modulation of microbiota-host metabolic interactions. Sci. Transl. Med. 4: 137rv6.

Hudson, J.M., N.D. Cohen, P.G. Gibbs and J.A. Thompson. 2001. Feeding practices associated with colic in horses. J. Amer. Vet. Med. Assoc. 219: 1419–1425.

Kil, D.Y. and H.H. Stein. 2010. Invited Review. Management and feeding strategies to ameliorate the impact of removing antibiotic growth promoters from diets fed to weanling pigs. Can. J. Anim. Sci. 90: 447–460.

Konstantinov, S.R., H. Smidt, A.D.L. Akkermans, L. Casini, P. Trevisi, M. Mazzoni, S. De Filippi, P. Bosi and W.M. De Vos. 2008. Feeding of *Lactobacillus sobrius* reduces *Escherichia coli* F4 levels in the gut and promotes growth of infected piglets. FEMS Microbiol. Ecol. 66: 599–607.

Kuhbacher, T., S.J. Ott, U. Helwig, T. Mimura, F. Rizzelo, B. Kleessen, P. Gionchetti, M. Blaut, M. Campieri, U.R. Fölsch, M.A. Kamm and S. Schreiber. 2006. Bacterial and fungal microbiota in relation to probiotic therapy (VSL #3) in pouchitis. Gut. 55: 833–841.

La Ragione, R.M., A. Narbad, M.J. Gasson and M.J. Woodward. 2004. *In vivo* characterization of *Lactobacillus johnsonii* FI9785 for use as a defined competitive exclusion agent against bacterial pathogens in poultry. Lett. Appl. Microbiol. 38: 197–205.

Lee, W.-H., P. Pathanibul, J. Quarterman,1 J.-H. Jo, N.S. Han, M.J. Miller, Y.-S. Jin and J.-H. Seo. 2012. Whole cell biosynthesis of a functional oligosaccharide, 2′-fucosyllactose, using engineered *Escherichia coli*. Microb. Cell Fact. 11: 48.

Ley, R.E., M. Hamady, C. Lozupone, P.J. Turnbaugh, R.R. Ramey, J.S. Bircher, M.L. Schlegel, T.A. Tucker, M.D. Schrenzel, R. Knight and J.I. Gordon. 2008. Evolution of mammals and their gut microbes. Science 320: 1647–1651.

Lindström A., S. Korpela and I. Fries. 2008. Horizontal transmission of *Paenibacillus larvae* spores between honey bee (*Apis mellifera*) colonies through robbing, Apidologie 39: 515–522.

Merrifield, D.L., A. Dimitroglou, A. Foey, S.J. Davies, R.T.M. Baker, J. Bogwald, M. Castex and E. Ringo. 2010. The current status and future focus of probiotic and prebiotic applications for salmonids. Aquaculture 302: 1–2: 1–18.

Mitsuoka, T. 1992. Intestinal flora and aging. Nutr. Rev. 50: 438–446.

Modesto, M., I. Stefanini, M.R. D'Aimmo, L. Nissen, D. Tabanelli, M. Mazzoni, P. Bosi, G.P. Strozzi and B. Biavati. 2011. Strategies to augment non-immune system based defence mechanisms against gastrointestinal diseases in pigs. NJAS-Wageningen Journal of Life Sciences.

Murphy, W.J., D.M. Larkin, A.E. van der Wind, G. Bourque, G. Tesler, L. Auvil, J.E. Beever, B.P. Chowdhary, F. Galibert, L. Gatzke, C. Hitte, S.N. Meyers, D. Milan, E.A. Ostrander, G. Pape, H.G. Parker, T. Raudsepp, M.B. Rogatcheva, L.B. Schook, L.C. Skow, M. Welge, J.E. Womack, S.J. O'brien, P.A. Pevzner and H.A. Lewin. 2009. Dynamics of mammalian chromosome evolution inferred from multispecies comparative maps. Science 309: 613–618.

Navarrete, P., R.T. Espejo and J. Romero. 2009. Molecular analysis of microbiota along the digestive tract of juvenile Atlantic salmon (*Salmo salar* L.). Microb. Ecol. 57: 550–561.

Nicholson, J.K., E. Holmes, J. Kinross, R. Burcelin, G. Gibson, W. Jia and S. Pettersson. 2012. Host-gut microbiota metabolic interactions science 8: 1262–1267.

Nikoskelainen, S., A.C. Ouwehand, G. Bylund, S. Salminen and E.M. Lilius. 2003. Immune enhancement in rainbow trout (*Oncorhynchus mykiss*) by potential probiotic bacteria (*Lactobacillus rhamnosus*). Fish Shellfish Immunol. 15: 443–452.

Oelschlaeger, T.A. 2010. Mechanisms of probiotic actions—a review. Int. J. Med. Microbiol. 300: 57–62.

O'Hara, A.M. and F. Shanahan. 2007. Gut microbiota: mining for therapeutic potential. Clin. Gastroenterol. Hepatol. 5: 274–284.

Olsen, R.E., R. Myklebust, H. Kryvi, T.M. Mayhew and E. Ring. 2001. Damaging effect of dietary inulin on intestinal enterocytes in Arctic charr (*Salvelinus alpinus* L.). Aquacult. Res. 32: 931–934.

Pelletier, M.C., M. Boussaha, D.G. Douet, A. Lautraite and P. Tailliez. 2007. Diversity of lactic acid bacteria associated with fish and the fish farm environment, established by amplified rRNA gene restriction analysis. Appl. Environ. Microbiol. 73: 2947–2955.

Piva, A., G. Casadei, P.P. Gatta, J.B. Luchansky and G. Biagi. 2005. Effect of lactitol, lactic acid bacteria, or their combinations (synbiotic) on intestinal proteolysis *in vitro*, and on feed efficiency in weaned pigs. Can. J. Anim. Sci. 85: 345–353.

Respondek, F., K. Myers, T.L. Smith, A. Wagner and R.J. Geor. 2011. Dietary supplementation with short-chain fructo-oligosaccharides improves insulin sensitivity in obese horses. J. Anim. Sci. 89: 77–83.

Schachtsiek, M., W.P. Hammes and C. Hertel. 2004. Characterization of *Lactobacillus coryniformis* DSM 20001T surface protein Cpf mediating coaggregation with and aggregation among pathogens. Appl. Environ. Microbiol. 70: 7078–7085.

Schneitz, C. 2005. Competitive exclusion in poultry—30 years of research. Food Control 16: 657–667.

Stappenbeck, T.S., L.V. Hooper and J.I. Gordon. 2002. Developmental regulation of intestinal angiogenesis by indigenous microbes via Paneth cells. Proc. Natl. Acad. Sci. USA 24: 15451–15458.

Tabe, E.S., J. Oloya, D.K. Doetkott, M.L. Bauer, P.S. Gibbs and M.L Khaitsa. 2008. Comparative effect of direct-fed microbials on fecal shedding of *Escherichia coli* O157:H7 and *Salmonella* in naturally infected feedlot cattle. J. Food Protect 71: 539–544.

Taschuk, R. and P.J. Griebel. 2012. Commensal microbiome effects on mucosal immune system development in the ruminant gastrointestinal tract. Anim. Health Res. Rev. 13: 129–141.

Timmerman, H.M., C.J.M. Koning, L. Mulder, F.M. Rombouts and A.C. Beynen. 2004. Monostrain, multistrain and multispecies probiotics --a comparison of functionality and efficacy. Int. J. Food Microbiol. 96: 219–33.

Travers, M.A., I. Florent, L. Kohl and P. Grellier. 2011. Probiotics for the Control of Parasites: An Overview. J. Parasitol. Res. Volume 2011, Article ID 610769, 11 pages. doi:10.1155/2011/610769.

Vasquez, A., E. Forsgren, I. Fries, R.J. Paxton, E. Flaberg, L. Szekely and T.C. Olofsson. 2012. Symbionts as major modulators of insect health: lactic acid bacteria and honeybees. PLoS ONE 7: e33188.

Wassenaar, T.M. and G. Klein. 2008. Safety aspects and implications of regulation of probiotic bacteria in food and food supplements. J. Food Protect. 71: 1734–1741.
Wohlgemuth, S., G. Loh and M. Blaut. 2010. Recent developments and perspectives in the investigation of probiotic effects. Int. J. Med. Microbiol. 300: 3–10.
Xu, Q., Y. Chao and Q. Wan. 2009. Health benefit application of functional oligosaccharides. Carbohydr. Polym. 77: 435–441.
Xu, Z.R., C.H. Hu, M.S. Xia, X.A. Zhan and M.Q. Wang. 2003. Effects of dietary froctooligosaccharide on digestive enzyme activities, intestinal microflora and morphology of male broilers. Poult. Sci. 82: 1030–1036.
Yörük, M.A., M. Gül, A. Hayirli and M. Macit. 2004. The effects of supplementation of humate and probiotic on egg production and quality parameters during the late laying period in hens. Poult. Sci. 83: 84–88.
Zimmermann, S., C. Wagner, W. Müller, W.G. Brenner, F. Hug, B. Prior, U. Obst and G.M. Hänsch. 2006. Induction of neutrophil chemotaxis by the quorum-sensing molecule N-(3-oxododecanoyl)-L-homoserine lactone. Infect. Immun. 74: 5687–5692.

10

Interactions of Probiotics and Prebiotics with Minerals

Corrie M. Whisner[1] and *Connie M. Weaver*[2,*]

Introduction

Osteoporosis is a major public health concern for men and women over the age of 50, characterized by structural deterioration and low bone mass (NOF 2011) leading to hip and spine fractures, as well as loss of function, independence, and increased risk of death with mortality rates between 20–24% one year after fracture (Leibson et al. 2002). This debilitating disease is highly influenced by genetic and environmental factors with diet as an especially important modifiable lifestyle factor that helps maximize and prolong skeletal health.

Despite daily recommendations, calcium intakes remain inadequate in key population groups, including adolescent girls and elderly males and females (FNB-IOM 2010). Calcium intakes from food and supplements from the NHANES 2003–2006 averaged between 918 and 1,296 mg/day in Americans, one year and older (Bailey et al. 2010). Consumption of milk, a primary source of dietary calcium has suffered significant decline in recent times with only 48% of the American adolescents reported to consume milk, as against 76% consumption in 1977–78 (USDA 2010). This decrease

[1] Cornell University, Division of Nutritional Sciences, 215 Savage Hall, Ithaca, New York 14853.
[2] Purdue University, Nutrition Science, Stone Hall, 700 W State St, West Lafayette, IN 47907.
* Corresponding author

in milk consumption can pose a threat to bone health, as it provides a highly bioavailable source of calcium. Humans absorb about one-third of calcium consumed which may decrease if calcium intake comes from less bioavailable sources, such as plants which often contain phytic and oxalic acids thought to hinder mineral absorption.

The rise of functional foods has brought about a new approach to maximize bone health where the addition of specialized compounds to food items can beneficially affect one or more target functions in the body to improve health beyond that possible with adequate nutritional intakes alone (Diplock et al. 1999). Bioactive compounds, such as prebiotics, probiotics and synbiotics, may specifically help offset mineral deficits in the diet by increasing absorption and retention in the body. Research has supported the effects of these bioactive compounds on calcium, which will be the primary focus of this chapter, but evidence also suggests positive effects on other minerals, including iron, magnesium and phosphorus.

Calcium Metabolism

Calcium homeostasis is a complex and tightly regulated process occurring through the coordinated actions of intestine, kidney, and bone. This three-organ system is activated when calcium intakes are low and ionic calcium concentrations decrease in the plasma. Calcium-sensing receptors in the parathyroid glands sense the drop in calcium, thereby signaling the release of parathyroid hormone (PTH). Elevated levels of PTH increase renal reabsorption and bone resorption in order to return extracellular calcium levels to within the small range of 8.5–10.5 mg/dL (Goldstein 1990). PTH also stimulates the activity of renal 1α-hydroxylase (1-OHase) to convert the inactive form of vitamin D, 25-hydroxyvitamin D (25-OH-D) to the active metabolite 1α-dihydroxyvitamin D ($1,25(OH)_2D$) in the kidney. The active metabolite of vitamin D increases calcium absorption in the small intestine and promotes bone resorption through osteoclastic activity on the bone surface.

This system is self-regulated through a negative feedback loop. As plasma levels of calcium increase, the calcium-sensing receptors are no longer stimulated to release PTH. Reversing the system, elevated or hypercalcemic levels stimulate the secretion of calcitonin from the thyroid gland which inhibits bone resorption.

Calcium Intake

Calcium intakes in most diets come predominantly from dairy products which accounted for 72% of calcium consumed in the U.S., in 2000 (HHS-

ARS 2005). Generally, females are less likely to achieve recommended calcium intakes compared to males (Ervin et al. 2004). NHANES data from 2003–2006 indicated that young girls 9–13 and 14–18 years, women 51–70 years of age, and both women and men over the age of 70 were at the greatest risk of not meeting recommended intakes of calcium for their respective ages (FNB-IOM 2010, Bailey et al. 2010). In fact, it has been estimated that only 6% of girls and 28% of boys between the ages of 9 and 13 consumed greater than the recommended 1300 mg of calcium each day while the percentages of girls and boys aged 14–18 years were 9% and 31%, respectively (Moshfegh et al. 2005).

Maintaining an adequate calcium intake during childhood is crucial for establishing peak bone mass in order to reduce the risk of osteoporosis later in life. Early work by Matkovic et al. made this connection between dietary calcium intake and peak bone mass when an association between low calcium intakes and decreased bone mineral density (BMD) was seen in adolescent girls (Matkovic et al. 1979). This relationship has been further illustrated by clinical trials showing that increased calcium intake through diet and supplements resulted in greater bone mineral accrual rates compared to control treatments (Bonjour et al. 1997, Lloyd et al. 1993, Nowson et al. 1997). One long-term calcium supplementation trial has suggested the presence of a catch-up phenomenon where, regardless of calcium intake levels, BMD of those consuming habitually low intakes will eventually rise to that of children on high intakes during late adolescence (Matkovic et al. 2005).

Controlled metabolic studies have continued to explain the importance of calcium intake. In a study of adolescent girls, calcium intake explained 12.3% of skeletal calcium retention (Braun et al. 2007). In general, dietary calcium results in increased bone size (Lee et al. 1996) but these effects differ with different skeletal locations, pubertal stage, and habitual calcium intakes. More interestingly, the greatest bone mineral increases have been seen in cortical bone of prepubertal children consuming habitually low calcium intakes (Bonjour et al. 1997). Because it can be difficult to increase calcium intakes among those with already low intakes, functional foods which improve the bioavailability of calcium already in the diet may be a more feasible approach to improving bone health as convincing evidence exists for both prebiotics and probiotics.

Calcium Absorption in Small Intestine

Calcium is absorbed in its ionized form (Ca^{2+}) after being released from insoluble calcium salts. Calcium release from salts occurs in the presence of stomach acid, after which, calcium is absorbed in the small intestine both transcellularly and paracellularly. Proposed methods of absorption include facilitated diffusion, vesicular transport, transcaltachia and regulated paracellular transport (Fleet and Schoch 2010). Together these methods of absorption account for an approximate gross absorption efficiency of 30%.

The facilitated diffusion model is saturable and occurs trancellularly (Bronner et al. 1986) in the duodenum and jejunum. This method of calcium transport requires both energy and calcium-binding proteins (Calbindin D9k, TRPV6, and PMCA1b) as well as the active vitamin D metabolite, 1,25 $(OH)_2D$. Vitamin D regulates TRPV6, a luminal protein that binds and internalizes calcium into the enterocyte. Calbindin D9k then transports calcium across the cytosol to the basolateral membrane where PMCA1b actively shuttles calcium into the plasma. Active calcium transport is highly efficient and occurs when calcium intakes are low and extracellular calcium concentrations fall below the tightly regulated range. This results in the conversion of 25-OH-D to $1,25(OH)_2D$ through increased PTH secretion. Active vitamin D binds the vitamin D receptor leading to transcriptional regulation of calcium binding proteins and more efficient absorption. Regardless of the increase in calcium transport proteins, the active transport mechanism is unable to compensate for habitually low calcium intakes (Bronner 2009) and its effectiveness decreases with age following the pubertal growth spurt (Pansu et al. 1983). In addition to the actions of calbindin D, transcellular calcium absorption may also occur through vesicular transport. In this method, $1,25(OH)_2D$ increases the presence of intestinal lysosomes in which calcium can be accumulated and transported to the basolateral membrane and into the blood. A transcaltachia model which involves rapid transepithelial calcium transport may also exist for transcellular transport. In this model, the movement of calcium is activated by 1,25 $(OH)_2D$ and utilizes a basolateral membrane receptor which may either be MARRS (Nemere et al. 2004) or localized VDR on the intestinal cell membrane (Norman et al. 2002). Lastly, calcium transport is also believed to occur through paracellular fluxes in the jejunum and ileum (Sheikh et al. 1990, Karbach 1992). This mechanism is also influenced by vitamin D with 1,25 $(OH)_2D$ inducing ion movement through tight junctions which may be mediated by increased levels of claudin 2 and 12 in intestinal cells which improve calcium permeability (Fujita et al. 2008).

Calcium Absorption in Large Intestine

The large intestine plays a small role in calcium absorption but the proportion of absorption in the lower gut is believed to increase in the presence of prebiotics and probiotics. Using a dual isotope technique, Barger-Lux and colleagues found that only 4.2% of the total amount of calcium absorbed after 26 hours occurred in the large intestine, accounting for an absolute size of 6.8 mg/day (Barger-Lux et al. 1989). Gastrointestinal transit times can vary greatly with travel from mouth to cecum ranging between 71–114 minutes in healthy young adults (Haboubi et al. 1988). Magee and Dalley have reported that transit to the ileum for an average-sized meal occurs in 4 hours and they suggested that most of the delay in movement occurs in the cecum and colon (Magee and Dalley 1986). Such variation in transit time may also influence the amount of calcium that is absorbed in the large intestine, especially in the presence of prebiotics and probiotics.

Prebiotics

As defined by the International Scientific Association for Probiotics and Prebiotics (ISAPP), "a dietary prebiotic is a selectively fermented ingredient that results in specific changes, in the composition and/or activity of the gastrointestinal microbiota, thus conferring benefit(s) upon host health" (Fig. 1) (Gibson et al. 2010). While host benefits are many, ranging from improved intestinal health and immunity to decreased risk of cancer, the benefit of greatest importance to bone health is enhanced mineral absorption (Roberfroid et al. 2010).

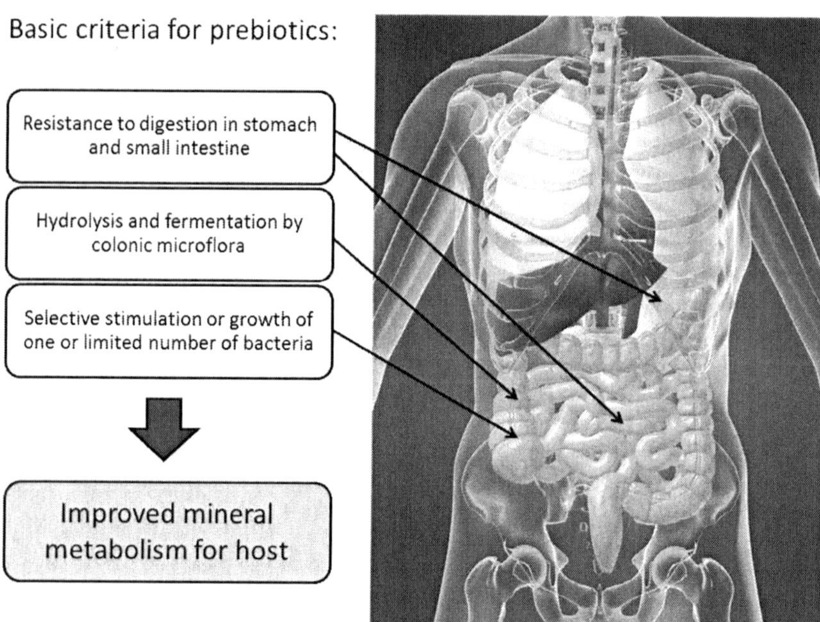

Fig. 1. Prebiotics are special carbohydrates that must meet three criteria before they can have beneficial effects for the host.

Poorly digested carbohydrates classified as non-digestible oligosaccharides (NDOs) are currently regarded as the most promising form of prebiotics that improve mineral metabolism. These compounds include galactooligosaccharides (GOS) and fructooligosaccharides (FOS) (Roberfroid et al. 2010). Oligosaccharides generally vary in chain length, with a degree of polymerization between 4 and 10, but other short chain disaccharides and longer chain polysaccharides exist. Prebiotic disaccharides include milk sugar derivatives, lactulose and lactitol, while polysaccharides include long-chain fructooligosaccharides and resistant starches.

Fructooligosaccharides (Yazawa and Tamura 1982, Yazawa et al. 1978, Gibson et al. 1995) and galactooligosaccharides (Rowland and Tanaka 1993) have been associated with stimulated growth and proliferation of bifidobacteria in the colon, a microbe thought to have beneficial health effects. Fermentation of NDOs is also thought to enhance production of short-chain fatty acids (SCFAs) altering the pH of colonic contents and increasing the solubility of minerals to increase their absorption (Bongers and van den Heuvel 2003).

There is convincing evidence that fructooligosaccharides stimulate calcium absorption and also enhance bone density and strength through mechanisms in the colon. Ohta et al. (1995) found that calcium and magnesium absorption correlated with colon and rectal length and transit time. Involvement of the lower gut in prebiotic effects was noted when the same group found no increase in calcium absorption when rats were cecectomized (Ohta et al. 1994). Animal studies have also suggested that the effects of NDO on mineral absorption are dose-dependent (Levrat et al. 1991, Brommage et al. 1993), and increases in mineral absorption have been associated with improved bone health. Zafar et al. (2004a) found that bone mineral density (BMD) of the femur was increased when rats were fed a mixture of short- and long-chain inulin-type fructans (ITF-mix). Increases in tibial calcium content (Lobo et al. 2009, Chonan et al. 1995) and improved microarchitectural properties including greater trabecular number and thickness (Scholz-Ahrens et al. 2002) have also been noted in rats.

Research in humans has shown equally promising results. A one-year intervention in adolescent boys and girls demonstrated that the effects of inulin-type fructans persist long-term. Calcium absorption, bone mineral density and bone mineral content increased significantly after daily supplementation with 8 g ITF-mix (mixture of long and short-chain inulin-type fructans). Similar to animal studies, a study in young adults identified the colon as the primary site for fructan-induced calcium absorption, accounting for 69.6% of the increase in total absorption (Abrams et al. 2007a).

Products with Established Prebiotic Effects

Fructooligosaccharides

Fructooligosaccharides (FOS) are NDOs comprised of fructose residues with varying degrees of polymerization (DP). Those with a DP of 3–6 are classified as short chain fructooligosaccharides while oligofructose has a DP of 4 and chicory inulin a DP of 12. Longer NDOs include long-chain FOS (lcFOS) and high molecular weight inulin (DP 25) (Roberfroid et al. 2010). FOS chains are made of fructosyl units connected by β(2-1) fructosyl-fructose linkages which are not digested by human enzymes allowing these polymers to be hydrolyzed and fermented by bacteria in the colon. Significant and selective growth of fecal bifidobacteria has been seen in humans fed 10 g of FOS for 7 days (Bouhnik et al. 2004). FOS are known to increase calcium absorption, improve BMD in growing rats, and decrease the loss of bone mineral in postmenopausal rat models by improving mineral solubility and increasing the surface area available for absorption in the large intestine (Roberfroid et al. 2010).

Galactans

Galactooligosaccharides (GOS) make up the third largest component of breast milk (5–10 g/l) after lactose and lipids, making them a unique oligosaccharide that contributes greatly to the protective microbiome that persists in the intestines of breast fed infants (Bode 2006, Kunz et al. 2000, German et al. 2008, Coppa et al. 1993). GOS inhibits pathogen growth and promotes the growth and proliferation of beneficial microbes such as bifidobacteria and lactobacillus in the large intestine (Fanaro et al. 2005). Galactans can also be commercially prepared through the enzymatic conversion of lactose with beta-D-galactosidase resulting in lactose connected to a chain of galactose monomers ranging between 2 and 8 monomers in length. These polymers vary in chain length and linkages (β(1-4), β(1-2), and β(1-6)) making them indigestible by human enzymes; thus, giving them their prebiotic effect. Previous studies have show positive effects of GOS on calcium absorption in postmenopausal women (van den Heuvel et al. 2000) and on the calcium content of rat bones (Chonan et al. 1995).

Lactose and Lactulose

Milk derivatives, including lactose and lactulose, have prebiotic effects. Consumption of calcium and lactose has been associated with improved bone mineral content and strength in vitamin D deficient rats (Schaafsma et al. 1988). While humans possess enzymes in the small intestine to digest lactose, it is possible for this disaccharide to evade digestion under the condition of lactase deficiency. Griessen et al. (1989) found that lactase-deficient individuals absorb more calcium from lactose-containing milk than do individuals with normal lactase activity. The increased absorption among maldigesters may be the result of up regulated calcium absorption in response to the lower intakes common in lactose intolerant individuals; however, this effect may also be explained by colonic fermentation and β-galactosidase activity of colonic microflora. However, other studies find no benefit of lactose on calcium absorption (Tremaine et al. 1986, Cochet et al. 1983). Lactulose is also believed to be a fermentable substrate for bacteria leading to improved mineral absorption. A crossover study (van den Heuvel et al. 1999b) providing 0, 5, and 10 g lactulose to 12 postmenopausal women found a positive linear trend between lactulose intake and fractional calcium absorption.

Prebiotic Combinations

Other carbohydrates that influence the microflora of the colon and lead to positive health effects include prebiotic mixtures which are unique combinations of NDO with varying degrees of polymerization. Inulin-type fructan mixtures (ITF-mix) are among the most cited for their beneficial effects on mineral bioavailability and bone outcomes (Abrams et al. 2005, Abrams et al. 2007a,b, Griffin et al. 2002, Griffin et al. 2003, Holloway et al. 2007). Combining short- and long-chain NDO is thought to have synergistic effects along the entire length of the colon (Coxam 2005) with short-chain NDO, such as oligofructose, acting proximally and longer chains acting more distally.

Prebiotic Effects on Mineral Metabolism and Bone in Animals

Experimental animal models have shown positive prebiotic effects on mineral metabolism, including calcium, magnesium, iron and zinc absorption, and also helped elucidate the mechanisms by which NDO elicit their response. The primary mechanism reported in rats has been decreased pH followed by bacterial fermentation in the cecum and colon. In addition to improved mineral absorption, dietary supplementation with NDO in rats has been associated with improved bone mineral content (BMC) during growth and reduced losses of BMC and BMD after ovariectomy.

Mineral Absorption

In animals, supplementation with prebiotics increases the availability and absorption of calcium (Ohta et al. 1995,Ohta et al. 1998, Chonan and Watanuki 1996, Weaver et al. 2011, Wang et al. 2010, Lobo et al. 2006), magnesium (Ohta et al. 1994, Ohta et al. 1995, Delzenne et al. 1995, Lopez et al. 2000, Weaver et al. 2011, Rondon et al. 2008, Wang et al. 2010, Lobo et al. 2006), iron (Asvarujanon et al. 2005, Wang et al. 2010), zinc (Coudray et al. 2006), and copper (Coudray et al. 2006). Given the relationship between inadequate calcium consumption and osteoporosis risk, the potential for improving calcium absorption and retention becomes extremely important. Early animal studies have found prebiotics to have a dose-dependent effect on calcium absorption. ITF given in a range of 0–20% in the diet (Levrat et al. 1991) and lactulose at 5 and 10% (Brommage et al. 1993) resulted in greater absorption with increasing amounts of NDO. In general, NDO have positive effects on calcium balance. However, results vary with differences in animal age, experimental conditions, and chosen outcome measures.

While effects in rats can often be seen in 2–3 weeks, many of the differences seen in calcium absorption responses to NDO have been attributed to the duration of treatment. A few studies have found treatment effects on calcium absorption and retention in as few as one to three days when rats received daily NDO doses of 5 g/100 kg (Brommage et al. 1993, Morohashi et al. 1998) or 50 g/kg body weight (Ohta et al. 1995). A study in young, growing rats, aimed to determine the difference between short (13 days) and longer (40 days) periods of inulin (10% by weight of diet) supplementation at varying calcium intakes (0.25%, 0.50%, and 0.75%) (Coudray et al. 2005b). After 13 days, inulin increased apparent calcium absorption on all calcium intakes compared to controls; while, longer supplementation with inulin only provided greater improvements to calcium absorption in rats on the low calcium diet (0.25%).

In contrast, studies with duration greater than 40 days have found NDO effects on calcium absorption and retention to persist long-term. In growing Wistar rats, treatment with oligofructose or inulin for 3 months resulted in increased intestinal calcium absorption compared with controls (Nzeusseu et al. 2006). Increased absorption in these rats may have been the result of morphological changes in cecal tissue where, compared to control-fed animals, cecal wall weight increased equally for both FOS treatments. Furthermore, inulin and oligofructose treatment led to a 4- and 2-fold increase, respectively, in the important calcium transport protein, Calbindin D9K.

Other studies have found improved effects of NDO with age as well as NDO mixtures. In a study in adult male rats, comparing types of fructans with varying chain length and branching, alone or in combination, it was found that, although calcium absorption and balance increased with all treatments, only the oligofructose-inulin combination produced significant increases (Coudray et al. 2003). This enhanced response was likely due to the synergistic response of combining short and long-chain fructans together. Differing degrees of polymerization seen in these two fibers allowed them to be fermented at different rates, thereby, enhancing absorption throughout the entire length of the colon. Another study by Coudray et al. (2005a) assessed the effects of inulin on calcium and magnesium absorption in rats two, five, 10, and 20 months old. Mineral absorption decreased with age; calcium and magnesium absorption were lowest in 10 and 20 month old rats. However, consumption of 7.5% inulin by weight, resulted in increased calcium absorption in all age groups compared to controls. Additionally, the level of calcium absorption in older rats on inulin was numerically greater than levels seen in younger inulin-treated rats.

Similar to young and adult animals, NDO have positive effects during menopause and post-menopausal states. Chonan et al. (1995) found that treatment with GOS for 20 days improved the apparent calcium absorption of OVX rats compared to sham-controls. A study in ovariectomized rats consuming 55 g/kg of an inulin and FOS mixture for 21 days found improvements in calcium balance compared to control rats (Zafar et al. 2004a). Calcium absorption was significantly increased while bone resorption, in relation to bone formation rate, was significantly decreased. Another study found that difructose anhydride III (DFAIII), a nondigestible disaccharide, increases apparent calcium absorption in OVX rats fed either a control or vitamin D-deficient diet (Mitamura and Hara 2006). DFAIII had a greater effect on calcium absorption in OVX, vitamin D deficient rats, where bone turnover is high compared to sham counterparts. Accounting for other nutrients important to bone, Scholz-Ahrens et al. (2002) found oligofructose (50 g/kg) to be most effective at increasing calcium absorption in rats consuming a high calcium diet (10 g/kg) for 16 weeks.

Bone effects

The effects on calcium absorption and retention are important outcomes for measuring the efficacy of NDO, however, balance studies used to assess these outcomes only provide acute information regarding calcium metabolism. To understand the long-term effects on bone it is necessary to measure mineral accretion in bone by assessing the mineral content and density as well as the architectural structure of bone. Similar to absorption and retention, bone outcomes are also dependent on other factors such as animal sex, age, hormonal status, mineral content of the diet, length of intervention, methodologies, and skeletal location.

In growing rat models, NDO have generally been associated with increases in tibial and femoral calcium content (Takahara et al. 2000, Richardson et al. 2002, Lobo et al. 2009, Weaver et al. 2011) but some studies have shown no effect on bone calcium content (Coudray et al. 2003, Zafar et al. 2004b).

Regardless of calcium content, bones from growing rats have experienced other positive structural changes in response to NDO. Weaver et al. (2011) carried out a study in 75 4-week old Sprague-Dawley rats to evaluate the effect of diets containing 0, 2, 4, 6, or 8% GOS by weight on calcium metabolism and bone parameters. Treatment with GOS increased tibial bone breaking strength, total and trabecular volumetric bone mineral density (vBMD) of the distal femur, and area and vBMD of the proximal tibia. Results from this study suggested that trabecular-rich bones benefited the most from GOS supplementation and regression analysis revealed that GOS improved vBMD through increased bifidobacteria content, trophic effects in cecal tissue, and decreased cecal pH.

Similar bone outcomes have been noted in OVX rat models as well. Scholz-Ahrens et al. (2002) found that 25, 50 and 100 g/kg of oligofructose led to similar reductions in structural bone loss in ovariectomized rats, regardless of the dose; however, changes in trabecular architecture varied by treatment. When calcium was provided at recommended levels (5 g/kg), 25 g/kg oligofructose resulted in increased trabecular thickness, while 50 and 100 g/kg doses increased the circumference of trabeculae. In rats fed high calcium (10 g/kg) diets, trabecular thickness did not differ between groups with and without oligofructose, though consumption of oligofructose reduced the loss of bone area through increased trabecular number, trabecular area, and cortical thickness. Results from this study suggested that oligofructose benefits weight-bearing sites as results were only seen in appendicular skeletal sites when calcium was adequately supplied. Lumbar vertebrae did experience an increase in calcium content when rats were given the high calcium diet supplemented with 100 g/kg oligofructose compared to those on 5 g/kg calcium and 25 or 50 g/kg oligofructose.

In addition to microarchitectural changes, increases in bone strength have been seen in both OVX and growing rat models, evidenced by increased load in 3-point bone breaking (Mathey et al. 2004, Lobo et al. 2006, Demigné et al. 2008, Lobo et al. 2009). Despite the increase in breaking strength, complementary increases in BMD were not always seen (Lobo et al. 2006, Demigné et al. 2008). While changes in bone mineral composition and structure are thought to result from the bacterial fermentation of NDO, bacteria also interact with other compounds to improve bone. Soy isoflavones which have beneficial effects on bone are cleaved by colonic microbes to create equol, a potentially more potent estrogen metabolite

(Yuan et al. 2007). Studies in OVX mice and rats have found that diets containing FOS and isoflavones result in greater femoral bone mineral density compared to treatment groups receiving either isoflavone or FOS alone (Ohta et al. 2002, Hooshmand et al. 2010). Conversely, a study investigating the effects of a mixture of ITF and soy found no additive effect on BMD but it did help retain trabecular microarchitectural properties of the tibia (Devareddy et al. 2006).

Other bioactive compounds have been mixed with NDO to enhance or synergistically affect bone. A study by Arjmandi et al. (2010) compared the effects of FOS to a variety of different mixtures of FOS and bioactive compounds (dried plum fractions (purees, skins, juice, extract), whole raisins, dates, and figs, and nutrition supplement β-hydroxy-β-methylbutyrate) on restoring bone mineral density in OVX rats. Among 15 different treatment groups, OVX rats fed diets containing 5% FOS + 7.5% dried plum experienced significant improvements in BMD of the femur and fourth lumbar vertebrae compared to OVX controls. Conversely, results from Zafar et al. (2004b) suggested that isoflavones enhanced calcium absorption independently of NDO and that addition of ITF to rat diets actually decreased the conversion of isoflavones to equol.

Prebiotic Effects on Mineral Metabolism and Bone in Humans

The effects of NDO on calcium metabolism in humans are often inconsistent. Studies have provided evidence for increases in calcium absorption and retention (Abrams et al. 2005, Coudray et al. 1997, Griffin et al. 2002, Griffin et al. 2003, van den Heuvel et al. 1999a, van den Heuvel et al. 1999b, van den Heuvel et al. 2000) while others have shown no prebiotic effect (Martin et al. 2010, Tahiri et al. 2003, van den Heuvel et al. 1998, van den Heuvel et al. 2009). In order to fully understand the benefits of NDO on calcium absorption and retention in bone, a number of factors must be considered. Conditions that may influence the effect of NDOs on bone health include life-stage, mineral status, NDO dose, composition and matrix of food-items in which NDO is given, and length of treatment.

It is more challenging to study the effects of prebiotics in humans because measuring the colonic component of absorption can be difficult. In healthy, adult men, supplementation with 15 g/day of inulin, oligofructose or galactooligosaccharide had no effect on calcium or iron absorption (van den Heuvel et al. 1998), while daily supplementation with 40 g of inulin resulted in significant increases in apparent calcium absorption (Coudray et al. 1997). The contradictory results seen here may have been due to differences in the measurement of calcium absorption. The study by Coudray et al. (1997) using 40 g NDO used chemical balance techniques which measure net calcium absorption, including that which occurs in the colon, while van den Heuvel et al. (1998) used fractional absorption to measure the ratio of dual stable isotopes in 24 h urine collections. It is possible that van den Heuvel et al. missed the colonic component related to prebiotic consumption because the effects of colonic absorption may not appear in urine until after 24 hours (van den Heuvel et al. 1998, van den Heuvel et al. 1999a). In addition to these methodological issues, contradictory responses may have been a dose response effect as Coudray used more than twice as many grams of NDO than van den Heuvel. In a similar study, van den Heuvel's group found a significant increase in fractional calcium absorption in adolescent males given 15 g oligofructose each day when they extended the measurement of calcium isotopes in 24-h urine to a 36-h collection (van den Heuvel et al. 1999a). Nonetheless, a study in adolescent girls aged 11–13 y showed no differences in calcium absorption or retention measured in urine collected over 4 days after girls were given 9 g/d oligofructose-enriched inulin for 3 weeks (Martin et al. 2010). This lack of effect was partially attributed to calcium intakes of 1500 mg exceeding the recommended 1300 mg/day but small sample size (n=14) may have also hindered the ability to see an effect.

The effects of prebiotics on calcium absorption and bone outcomes also seem to vary by age. One study has been done in babies to assess prebiotic effects on mineral metabolism. This study was done in 6–12 month infants fed formula supplemented with short-chain inulin (0.75, 1, and 1.25 g/d) and resulted in improved iron and magnesium retention with no effect on calcium, copper and zinc metabolism (Yap et al. 2005). Treatment with inulin also resulted in significant decreases in luminal pH but no increase in short-chain fatty acids (SCFAs), a change commonly associated with prebiotic-induced mineral absorption. Although calcium and magnesium absorption are thought to occur via the same mechanism, the lack of effect on calcium seen in this study may have been due to limited SCFA production. One study found that SCFAs have a unique effect on calcium absorption, where, in the presence of SCFAs, calcium solubility was greatly increased in the large intestine but not from acidic pH alone (Mineo et al. 2001). Conversely, magnesium solubility and absorption have been shown to increase as a result of reduced colonic pH (Heijnen et al. 1993) which may help explain the differences in mineral absorption seen in the above infant study. Other infant studies have shown positive effects of NDO on fecal bacteria which were associated with improved mineral absorption. Fanaro et al. (2008) showed that the addition of 5 g GOS per liter of baby formula significantly increased the number of colony-forming units in 4–6 month infants compared to babies fed a control formula.

In young girls at or near menarche, a modest intake of ITF-mix for 3 weeks resulted in a 30% improvement in calcium absorption compared to placebo and oligofructose treatments (Griffin et al. 2002). Furthermore, this effect was observed in girls with low habitual calcium intakes. Van den Heuvel et al. have demonstrated in boys that NDO are able to stimulate true fractional calcium absorption in boys but not girls. Absorption was increased by 10% in 14–16 year old boys (n=12) fed 15 g of oligofructose daily for 9 days (van den Heuvel et al. 1999a) but in adolescent girls (n=14) with low calcium intakes, short-chain FOS treatment (10 g/d) for 36 days had no effect on calcium absorption (van den Heuvel et al. 2009). It is unclear why such variable results exist during this life-stage and differences could be the result of differences in NDO types, length of treatment, and calcium status prior to intervention.

To begin to understand the long-term effects of NDO on bone health, Abrams et al. completed a 1-year clinical trial on pre-pubertal girls and boys who received 8 g/d of ITF-mix (mixture of short and long chain ITF). Using dual isotope techniques, they found that calcium absorption in the fructan group was significantly higher after 8 weeks than in the control group. This effect persisted throughout the entire intervention resulting in significant increases in whole body BMC and BMD after 1 year of treatment (Fig. 2) (Abrams et al. 2005). Supplementation with ITF-mix in this cohort also suggested that the effects of dietary factors such as prebiotics may be modulated by genetic factors through vitamin D receptor (VDR) gene polymorphisms. An interaction of fructan supplementation with the Fok1

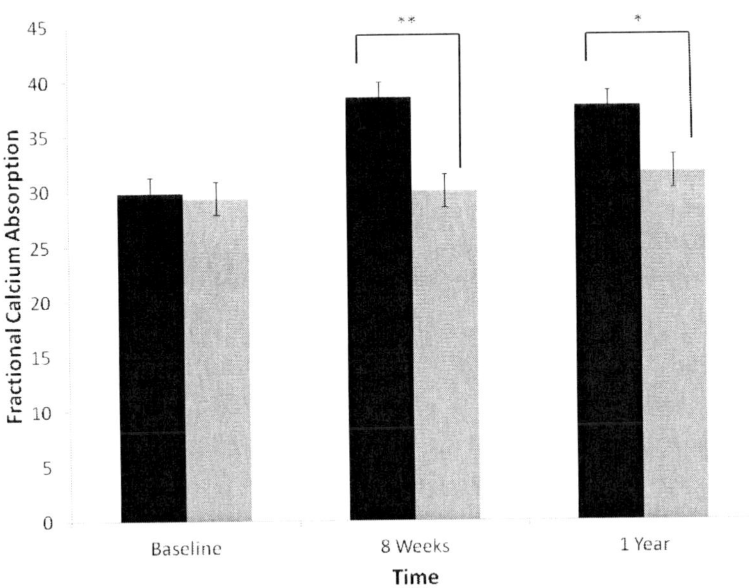

Fig. 2. Daily treatment with Synergy 1® improves fractional calcium absorption over 8 weeks which persists after 1 year of treatment; * $P<0.05$, ** $P<0.001$. Data from S.A. Abrams et al. 2005. Am. J. Clin. Nutr. 82: 471–476.

gene was a significant determinant of responders at 8 weeks with sex, tanner stage and ethnicity as covariates (Abrams et al. 2005). In those children classified as responders (3% increase in calcium absorption from treatment), calcium accretion was greater after one year compared to non-responders and those on the control treatment (Abrams et al. 2007b). However, this could reflect regression to the mean.

Studies done in adults have continued to show contradictory results with huge variation in intervention length, treatment vehicle/matrix as well as NDO dose. Two studies have shown no effect of NDO in adults. In a 5-day intervention with 15 g inulin plus 210 mg calcium added to 100 g cheese, 23 year old women experienced no change in blood levels of calcium (Teuri et al. 1999). This study may have been too short to elicit a response from prebiotics. Additionally, ionized calcium and PTH concentration in the blood were assessed which are not preferred methods for measuring calcium absorption. The second study that found no prebiotic response investigated the response of 15 g FOS, GOS, or placebo after 21 days of treatment in young men (van den Heuvel et al. 1998). This particular study may have failed to see an effect because urine was only collected for 24 hours, and it has been shown that calcium absorption is not complete until after 24 hours (Barger-Lux et al. 1989). Another study by Lopez-Huertas et al. found that 1.1 g ITF given one time in milk increased calcium absorption compared to regular milk, but this effect did not reach statistical significance (Lopez-Huertas et al. 2006).

More promising results have come from studies in which high doses were used or "responder" vs. "non-responder" effects were assessed. During a 28-day intervention where 9 young men received 40 g/d of chicory root inulin, apparent calcium absorption was increased (Coudray et al. 1997). In another study, Abrams et al. (2007a) aimed to assess the relatedness of the mechanism for inulin-type fructan-induced calcium absorption and the colon. They found that, in those who respond to ITF, the increase in calcium absorption occurred primarily in the large intestine. In responders (absorption increase > 3%), supplementation with 8 g/d for 8 weeks resulted in an increase in calcium absorption from 22.7% to 31.0%.

Studies in postmenopausal women have found ITF to have a positive influence on the absorption of copper (Ducros et al. 2005), iron (Kim et al. 2004), magnesium (Tahiri et al. 2003), and calcium (Kim et al. 2004, Holloway et al. 2007, Adolphi et al. 2009). While one study found positive effects on magnesium metabolism, no significant effect of short-chain FOS on calcium absorption was seen (scFOS vs. placebo, 36.6 ± 8.5 vs. 35.6 ± 9.4) (Tahiri et al. 2003). However, this study did find a trend toward significance in women more than 6 years postmenopausal (37.4 ± 9.7) compared to those who were postmenopausal only 2–6 years (35.7 ± 7.9). In a randomized, double-blind, parallel investigation, 8 g of chicory fructan fiber increased calcium absorption by 42% while the maltodextrin control actually decreased absorption 29% after 3 months of treatment (Kim et al. 2004). This study also found a slight decrease in urinary excretion of DPD. Finally, using the gold standard in clinical trials, a randomized, double-blind, placebo-controlled cross over design with dual isotopes, Holloway et al. showed that daily consumption of 10 g ITF-mix for 6 weeks resulted in significant increases (7%) in fractional calcium absorption in women with initially low BMD (Holloway et al. 2007).

In addition to the effects of ITF on calcium absorption, other types of non-digestible carbohydrates have shown prebiotic effects on mineral metabolism. Van den Heuvel et al. found that daily consumption for 9 days with 10 g lactulose (van den Heuvel et al. 1999b) and 20 g trans-galactooligosaccharide (van den Heuvel et al. 2000) improved true calcium absorption in postmenopausal women. Lactulose was administered as 5 or 10 g in 100 ml of water at breakfast which resulted in a dose-response increase in calcium absorption; although, the effect from 5 g was not significant (van den Heuvel et al. 1999b). The intervention testing GOS was provided as two 200 mg yogurt drinks per day with 10 g of GOS added to each drink. The increase in calcium absorption observed from GOS treatment was 16% greater than the placebo with added sucrose (van den Heuvel et al. 2000). Additionally, no complementary increase in urinary calcium excretion was seen in the GOS group, suggesting that treatment with this NDO results in decreased resorption or increased calcium deposition in bone of postmenopausal women.

Prebiotic Effects in the Gut

A number of factors influence the quantity and diversity of bacterial communities throughout the gastrointestinal tract including pH, nutrient availability, health and age of the host, bacterial adhesion, inter-bacterial species relationships, mucin secretion and transit time (Collins and Gibson 1999). Of these factors, access to nutrient substrates seems to play an essential role in microbe viability. Non-digestible oligosaccharides capable of evading digestion in the small intestine, such as raffinose, inulin-type fructans (ITF), and galactans, are selectively fermented by bifidobacteria and lactobacilli in the colon (Hudson and Marsh 1995). Prebiotic effects also seem to vary with location in the large intestine. Short-chain prebiotics such as oligofructose seem to be proximally fermented while carbohydrate components with longer chains elicit their response in the distal colon (Roberfroid et al. 2010). Mixing both long and short-chain components results in a synergistic effect along the entire colon allowing for enhanced absorption in proximal, transverse, and distal portions (Coxam et al. 2005).

Prebiotic mechanisms

The positive effects of prebiotics on mineral absorption have been associated with increases in mineral solubility resulting from morphological, physiological and molecular changes in the intestine (Scholz-Ahrens et al. 2001). It is believed, these intestinal changes are mediated by intestinal flora rather than as direct substrate effects. Microbial involvement became evident when germ-free and bacterially colonized rats were given fructooligosaccharide and architectural changes were assessed in the intestinal mucosa (Kleessen et al. 2003). Treatment with FOS resulted in increased intestinal villi and deeper mucosal crypts in bacterial colonized rats but not in the germ-free group. A similar study concluded that actions of intestinal bacteria were necessary for galactooligosaccharide to elicit a response when rats fed neomycin antibiotics experienced no change in calcium and magnesium absorption compared to rats fed only GOS (Chonan et al. 2001). These studies suggest that microflora present in the large intestine play a crucial role in prebiotic-induced mineral absorption.

It has been well established that prebiotics have an impact on intestinal physiology (Macfarlane et al. 2006). An adult human intestine is home to 100 trillion microorganisms comprised of health-promoting and pathogenic strains but is mostly comprised of Cytophaga-Flavobacterium-Bacteroides and the Firmicutes (Backhed et al. 2005). Prebiotics cause this intricate concentration of bacteria to shift, favoring organisms that selectively ferment undigested material. More specifically, studies have noted significant proliferative effects on bifidobacteria and lactobacillus species in animals (Tzortzis et al. 2005, Rodriquez-Cabezas et al. 2010) and humans (Ben et al. 2008, Bouhnik et al. 1997). Recent advancements in molecular techniques have aided the discovery of new strains associated with prebiotic fermentation and improved mineral absorption, such as *Roseburia*, *Eubacterium*, and *Faecalibacterium* (Roberfroid et al. 2010).

Prebiotic effects may also be localized in the intestine with a number of studies indicating the importance of the cecum in mineral absorption. Previous work has suggested that prebiotics stimulate cecal absorption of magnesium (Levrat et al. 1991, Ohta et al. 1997, Younes et al. 2001, Weaver et al. 2011); however, there are studies in which no effects were seen (Ohta et al. 1994, Demigne et al. 1989). Similarly, a number of studies indicated that the cecum is an important site for calcium absorption (Demigne et al. 1989, Levrat et al. 1991, Ohta et al. 1994, Chonan and Watanuki 1995) while two studies found no relationship (Brommage et al. 1993, Ohta et al. 1997). Treatment with FOS was able to increase iron absorption and prevent post-gastrectomy anemia in gastrectomized rats (Sakai et al. 2000) but inulin did not influence fractional iron absorption in women with low iron status (Petry et al. 2012). Minerals such as iron, zinc, and copper have been studied less frequently and require further investigation.

Most of the research linking intestinal location and morphology with prebiotic effects has been in animals. Given the coprophagic behavior of rats and their highly developed cecum, it is difficult to extrapolate these mechanistic findings to humans. However, there are several theories that may explain the effects of prebiotics on mineral absorption in the lower gut of both humans and animals.

Colonic fermentation and production of short chain fatty acids

One of the possible mechanisms for prebiotic-related improvements on mineral bioavailability is through the selective fermentation of prebiotics by microbiota (Roberfroid 1998). Upon reaching the colon, prebiotics are hydrolyzed and fermented by the resident microflora resulting in the production of short chain fatty acids (SCFAs) and other organic acids (Fig. 3). The accumulation of these acidic compounds decreases the pH of the luminal contents which aids in the solubilization of calcium to its

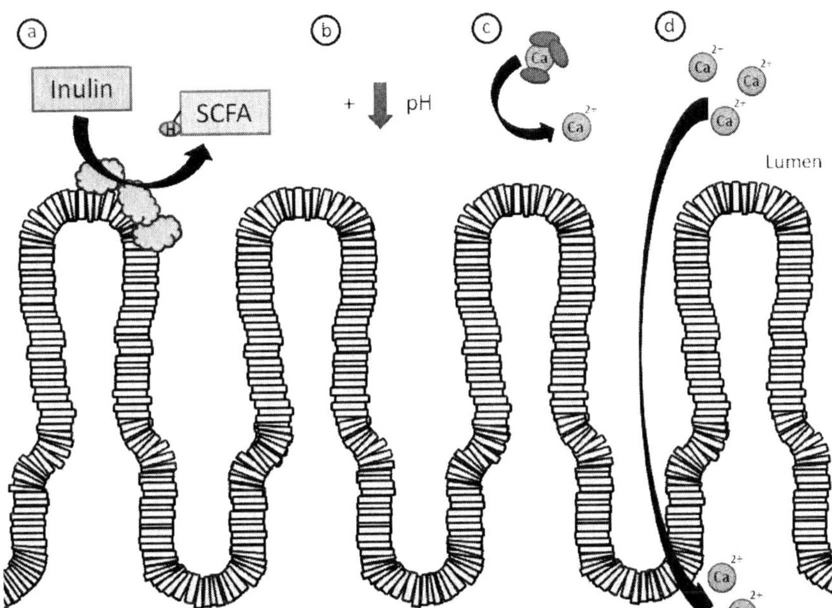

Fig. 3. Fermentation of prebiotics enhances calcium uptake at the enterocyte through the production of short chain fatty acids (SCFAs). a) Prebiotics such as inulin are fermented by saccharolytic bacteria residing in the colon to form SCFAs such as butyrate and acetate b) Accumulated SCFAs decreases the luminal pH c) This newly acidic environment ionizes calcium, freeing it from compounds to which it is bound d) Ionized calcium is absorbed more easily across the intestinal wall.

Color image of this figure appears in the color plate section at the end of the book.

ionized form. This improved solubility is thought to lead to increases in passive absorption (Ohta et al. 1995) while also enhancing transcellular absorption through hydrogen ion exchange across the cell membrane (Lutz and Scharrer 1991). It has also been speculated that the stimulation of paracellular transport creates an osmotic effect that pulls water into the intestinal lumen allowing for improved mineral dissolution and a resultant improvement in mineral availability in the lower gut (Bongers and van den Heuvel 2003). A recent study in women with low iron status found that supplementation with inulin 3 times per day, reaching a total amount of 20 g/d, resulted in significant increases in lactate and decreased pH (Petry et al. 2012). While there was no effect on total fecal SCFAs, bifidobacteria content of the feces did increase with inulin compared to control.

Morphological changes

Another way prebiotics contribute to improvements in mineral absorption is through trophic changes (Fig. 4) in the colon that increase the surface area available for absorption (Raschka and Daniel 2005). Production of

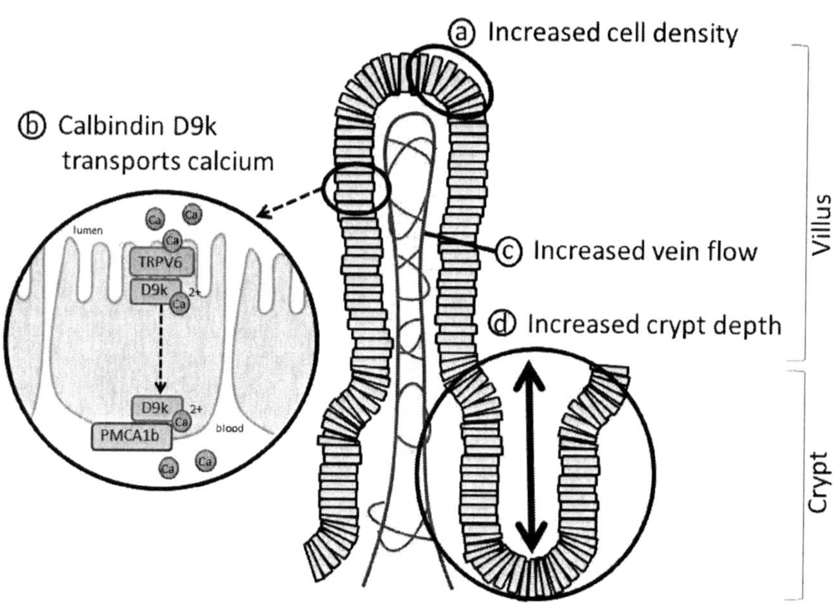

Fig. 4. Prebiotic-induced mineral absorption is mediated at the cellular level. a) epithelial cell density increases, b) calbindin D9k increases the active transport of calcium across the enterocyte, c) cecal vein flow increases resulting in increased surface area which allows for greater absorption, and d) crypt depth increases.

Color image of this figure appears in the color plate section at the end of the book.

SCFAs such as butyrate, acetate and propionate have been associated with cell proliferation as it acts as an important energy source for mucosal cells (Scheppach et al. 1995). Cellular proliferation results in increases in crypt depth, epithelial cell density, and improved cecal vein flow. Rats fed GOS diets had increased crypt depth and cell density in the proximal and distal colon which also correlated with apparent Ca absorption (Perez-Conesa et al. 2007). Raschka and colleagues reported that this increased surface area was associated with a two-fold increase in net calcium transport in the cecum and distal colon of rats fed prebiotics compared to controls (Raschka and Daniel 2005).

Regulation of transport proteins

In the presence of prebiotics, intestinal epithelial cells may increase active calcium transport across cells by increasing the presence of Calbindin D9k in the cecum and colorectum (Fig. 4). FOS treatment in rats significantly increased the relative concentration of Calbindin D9k in the cecum and colorectum compared to control animals (Ohta et al. 1998). The effects of FOS on Calbindin D9k were later shown to be independent of vitamin D pathways (Takasaki et al. 2000). More recent work from the same group further elucidated this mechanism by showing that FOS-induced calcium absorption through Calbindin D9k is mediated transcriptionally by VDR and cdx-2 (Fukushima et al. 2005).

Probiotics and Synbiotics

Probiotics have been defined by the FAO/WHO as "live microorganisms which when administered in adequate amounts confer a health benefit on the host" (FAO-WHO 2001). The accumulation of beneficial bacteria in the lower gut can also create an environment which is unfavorable for harmful bacterial growth (Parracho et al. 2007). Lactobacillus and bifidobacteria are among the organisms that naturally reside in the gut which are commonly accepted as probiotics. These bacterial species are also consumed in the diet, generally in the form of yogurt and other fermented dairy products.

Currently, few have studied the effects of probiotics on mineral metabolism or bone health. This lack of research may stem from their sensitivity to environmental factors, such as heat and moisture, making prolonged shelf-life difficult (Douglas and Sanders 2008). Probiotics may influence bone health through the production of metabolites such as vitamins D, C, K and B-6 (Weber 1999) which influence calcium metabolism. A secondary analysis of infant rhesus monkeys given *Lactobacillus reuteri* supplements showed no effect on calcium, iron or zinc metabolism (Kelleher et al. 2002). Similarly, adult men and women supplemented with vitamin D and/or probiotic in a randomized double-blind placebo-controlled trial experienced no effect on serum PTH or bone turnover markers (Hill et al. 2009). Soy milk fermented with *Lactobacillus acidophilus* also had no effect on calcium absorption in osteopenic postmenopausal women (Cheung et al. 2011).

Innovative work with Caco-2 cells has found that probiotic bacteria allow for higher bioavailability of certain minerals. Using a transwell cell culture system, the bioavailability of selenium and zinc found in commercial supplements were compared to organic selenium and zinc that had been internalized by *Lactobacillus buchneri* Lb26 and *Bifidobacterium lactis* Bb1, respectively (Mogna et al. 2012). Results from this work suggest that the two minerals internalized by probiotics diffused into the basolateral compartment at concentrations six to 65 times higher than inorganic and organic forms commonly found in commercial supplements. Such an effect merits further investigation into using probiotics as nutraceutical supplements.

Much of the work with probiotics has been done in conjunction with other compounds. These mixtures, commonly known as synbiotics, are combinations of prebiotics and probiotics. Reaching an optimum combination of probiotics and prebiotics is thought to produce better results as seen in both animal and human studies; however further work is needed to confirm this assumption. In postmenopausal women, Adolphi et al. showed that consumption of fermented milk at bedtime reduced the excretion of deoxypyridinoline (DPD), a marker of bone resorption (Adolphi et al. 2009). The addition of calcium from milk minerals, ITF and caseinophosphopeptides to the fermented milk also resulted in increased calcium and phosphorus excretion which the authors attributed to stimulated intestinal calcium absorption.

Work by Perez-Conesa (Perez-Conesa et al. 2006) compared the effects of probiotics, prebiotics, and synbiotics in rats using follow-up formulas to administer *Bifidobacterium bifidum* and *Bifidobacterium longum* (probiotics), galactooligosaccharides at 12, 50 and 100 g/kg (prebiotic) or bifidobacteria and galactooligosaccharides (synbiotics). Weanling rats consumed these seven treatments for 30 days and mineral balance was assessed for calcium, magnesium and phosphorus at three time intervals during the treatment period. Probiotic and prebiotic treatments increased calcium, magnesium and phosphorus bioavailability and mineral absorption after 8–10 days of treatment. Mineral absorption and retention decreased at 18–20 and 28–30 days. While all treatments were effective at improving mineral bioavailability, prebiotic (100 g/kg) and synbiotic (50 and 100 g/kg) treatments were most beneficial. Femur and tibia mineral content were also increased in these rats (Perez-Conesa et al. 2007).

A similar study in rats found that administration of Bifidobacterium cultures with lactulose promoted the absorption of calcium and improved bone strength (Igarashi et al. 1994). In comparison to the control diet containing whey protein, diets supplemented with *Bifidobacterium longum* alone and *Bifidobacterium longum* plus lactulose resulted in greater femur breaking force. The combined probiotic-lactulose diet had the greatest effect on gastrointestinal properties, resulting in significantly greater bifidobacteria and acetic acid in the feces and decreased cecal pH compared to control and Bifidobacteria diets.

A recent study investigating the effects of *Bifidobacterium longum* in combination with yacon flour, which has many fructooligosaccharides, found significantly greater tibia calcium, phosphorus and magnesium content in rats fed the *Bifidobacterium longum* and yacon flour + *Bifidobacterium longum* diets (Rodrigues 2012). Compared to control, fracture strength was also increased in the yacon flour, *Bifidobacterium longum*, and yacon flour + *Bifidobacterium longum* diets.

Probiotic Effects in the Gut

Given the limited information for probiotic effects on mineral metabolism, it is difficult to identify a specific mechanism of action in the lower gut. Available literature suggests that supplementing probiotics in the presence of prebiotics may mediate intestinal changes which promote mineral absorption and protect bone. Work in rats has shown that administering synbiotics instead of probiotics alone, leads to the production of short chain fatty acids (Igarashi et al. 1994) which has been associated with improved calcium and magnesium absorption and bone health in rats (Weaver et al. 2011). Probiotics may still promote SCFA production, although the effect may be less profound in the absence of prebiotics, but these compounds may act through additional mechanisms including the degradation of phytic acid and inflammatory responses.

Degradation of phytic acid

Minerals are often complexed with phytates commonly found in grain products, nuts and seeds. As a result, absorption of important minerals is reduced. Probiotics with phytase activity may ameliorate this effect by breaking down phytates that pass through the gut. *Mitsuokella jalaludinii*, a phytase-producing bacteria found in cattle had positive effects on broiler chickens consuming phytate-rich diets (Lan et al. 2002). Chickens consuming four levels of this bacteria in a phytate-rich diet were compared to chickens consuming positive (low-phytate diet) and negative (phytate-rich diet with no bacterial addition) control diets. While the phytate-rich diet led to significantly decreased tibia ash, the addition of phytase-producing cultures prevented the loss of bone.

Morphological changes in the intestine

Similar to prebiotics, probiotics and synbiotics have also been studied for their potential to improve intestinal surface area and luminal properties. Perez-Conesa (Perez-Conesa et al. 2007) compared the effects of follow-up formula supplemented with probiotics (*Bifidobacterium bifidum* and *Bifidobacterium longum*), prebiotics (galactooligosaccharides at 12, 50 and 100 g/kg) and synbiotics (bifidobacteria and galactooligosaccharide) in weanling rats on luminal pH, crypt depth and cell density. Those rats fed symbiotic formulas experienced lower cecal and colonic pH when compared to the control diet. Crypt depth and cell density in the cecum were not affected by any of the diets; however, crypt depth was improved by all seven diets in the proximal and distal colon. These morphological changes were associated with increased calcium and magnesium absorption in the distal colon and proximal and distal colon, respectively. Increases in femur and tibia calcium content were also seen in these rats.

A similar study looking at the effects of yacon flour and *Bifidobacterium longum* alone and in combination found that cecal anaerobe and content weight were highest in rats fed the yacon flour diet compared to all other diets (Rodrigues 2012). However, when yacon flour + *Bifidobacterium longum* was given to rats, the mineral concentration (calcium, magnesium and phosphorus) of bones increased more than with individual yacon flour and *Bifidobacterium longum* diets.

Anti-inflammatory effect and stimulated enterocyte uptake

Probiotics are also thought to have bone-preserving properties through their anti-inflammatory effect. *Enterrococcus faecium* further improved the bone-sparing effects of methotrexate but did not when bacteria were given alone (Rovensky et al. 2004). Milk fermented with *Lactobacillus helveticus* reduced parathyroid levels and increased serum calcium in postmenopausal women (Narva et al. 2004a) and increased osteoblast bone formation *in vitro* (Narva et al. 2004b). These effects are likely the result of specific peptides formed during milk fermentation rather than direct bacterial effects (Narva et al. 2004b).

Conclusions

Previous research has identified a relationship between NDO consumption and improved calcium metabolism but more work is needed to begin to understand the role of probiotics in mineral metabolism. For prebiotics, there is a specific need to understand the differences between NDO types and also to identify other carbohydrates with potential prebiotic effects. Additionally, identifying an optimal dose to see effects would provide further understanding of the relationship between NDO and bone health. Probiotics, on the other hand, require both animal studies and randomized clinical trials to begin to understand their effects on mineral absorption and retention. Research to date suggests their potential but it is difficult to identify strains which are most beneficial. Studying the effects of both prebiotics and probiotics, alone or together, in important populations such as adolescents and postmenopausal women, is equally important as these bioactive compounds may help make up for deficits in calcium intake to maximize bone mineral accretion and prevent the onset of osteoporosis.

References

Abrams, S.A., I.J. Griffin, K.M. Hawthorne, L. Liang, S.K. Gunn, G. Darlington and K.J. Ellis. 2005. A combination of prebiotic short- and long-chain inulin-type fructans enhances calcium absorption and bone mineralization in young adolescents. Am. J. Clin. Nutr. 82: 471–476.

Abrams, S.A., K.M. Hawthorn, O. Aliu, P.D. Hicks, Z. Chen and I.J. Griffin. 2007a. An inulin-type fructan enhances calcium absorption primarily via an effect on colonic absorption in humans. J. Nutr. 137: 2208–2212.

Abrams, S.A., I.J. Griffin and K.M. Hawthorn. 2007b. Young adolescents who respond to an inulin-type fructan substantially increase total absorbed calcium and daily calcium accretion to the skeleton. J. Nutr. 137: 2524S–2526S.

Adolphi, B., K.E. Scholz-Ahrens, M. de Vrese, Y. Açil, C. Laue and J. Schrezenmeir. 2009. Short-term effect of bedtime consumption of fermented milk supplemented with calcium, inulin-type fructans and caseinphosphopeptides on bone metabolism in healthy, postmenopausal women. Eur. J. Nutr. 48: 45–53.

Arjmandi, B.H., C.D. Johnsom, S.C. Campbell, S. Hooshmand, S.C. Chai and M.P. Akhter. 2010. Combining fructooligosaccharide and dried plum has the greatest effect on restoring bone mineral density among select functional foods and bioactive compounds. J. Med. Food 13: 312–319.

Asvarujanon, P., S. Ishizuka and H. Hara. 2005. Promotive effects of non-digestible disaccharides on rat mineral absorption depend on the type of saccharide. Nutrition 21: 1025–1035.

Backhed, F., R.E. Hey, J.L. Sonnenburg, D.A. Peterson and J.I. Gordon. 2005. Host-bacterial mutualism in the human intestine. Science 307: 1915–1920.

Bailey, R.L., K.W. Dodd, T.A. Goldman, J.J. Gahche, J.T. Dwyer, A.J. Moshfegh, C.T. Sempos and M.F. Picciano. 2010. Estimation of total usual calcium and vitamin D intakes in the United States. J. Nutr. 140: 817–822.

Barger-Lux, M.J., R.P. Heaney and R.R. Recker. 1989. Time course of calcium absorption in humans: Evidence for a colonic component. Calcif Tissue Intl. 44: 308–311.

Ben, X., J. Li, Z. Feng, S. Shi, Y. Lu, R. Chen and X. Zhou. 2008. Low level of galacto-oligosaccharide in infant formula stimulates growth of intestinal Bifidobacteria and Lactobacilli. World J. Gastroenterol. 14: 6564–6568.

Bode, L. 2006. Recent advances on structure, metabolism, and function of human milk oligosaccharides. J. Nutr. 136: 2127–2130.

Bongers, A. and E.G.H.M. van den Heuvel. 2003. Prebiotics and the bioavailability of minerals and trace elements. Food Rev. Int. 19: 397–422.

Bonjour, J.P., A.L. Carrie, S. Ferrari, H. Clavien, D. Slosman, G. Theintz and R. Rizzoli. 1997. Calcium-enriched foods and bone mass growth in prepubertal girls: A randomized double blind, placebo-controlled trial. J. Clin. Invest. 99: 1287–1294.

Bouhnik, Y., B. Flourié, L. D'Agay-Abensour, P. Pochart, G. Gramet, M. Durand and J. Rambaud. 1997. Administration of transgalacto-oligosaccharides increases fecal bifidobacteria and modifies colonic fermentation metabolism in healthy humans. J. Nutr. 127: 444–448.

Bouhnik, Y., L. Raskine, G. Simoneau, E. Vicaut, C. Neut, B. Flourié, F. Brouns and F.R. Bornet. 2004. The capacity of nondigestible carbohydrates to stimulate fecal bifidobacteria in healthy humans: a double-blind, randomized, placebo-controlled, parallel-group, dose-response relation study. Am J. Clin. Nutr. 80: 1658–1664.

Braun, M., C. Palacios, K. Wigertz, L.A. Jackman, R.J. Bryant, L.D. McCabe, B.R. Martin, G.P. McCabe, M. Peacock and C.M. Weaver. 2007. Racial differences in skeletal calclium retention in adolescent girls on a range of controlled calcium intakes. Am. J. Clin. Nutr. 85: 1657–1663.

Brommage, R., C. Binacua, S. Antille and A.L. Carrie. 1993. Intestinal calcium absorption in rats is stimulated by dietary lactulose and other resistant sugars. J. Nutr. 123: 2186–2194.

Bronner, F. 2009. Recent developments in intestinal calcium absorption. Nutr. Rev. 67: 109–113.

Bronner, F., D. Pansu and W.D. Stein. 1986. An anlysis of intestinal calcium transport across the rat intestine. J. Physiol. 250: G561–G569.

Cheung, A.L.T.F., G. Wilcox, K.Z. Walker, N.P. Shah, B. Strauss, J.F. Ashton and L. Stojanovska. 2011. Fermentation of calcium-fortified soya milk does not appear to enhance acute calcium absorption in osteopenic post-menopausal women. Br. J. Nutr. 105: 282–286.

Chonan, O., K. Matsumoto and M. Watanuki. 1995. Effect of galactooligosaccharides on calcium absorption and preventing bone loss in ovariectomized rats. Biosci. Biotech. Biochem. 59: 236–239.

Chonan, O. and M. Watanuki. 1995. Effect of galactooligosaccharides on calcium absorption in rats. J. Nutr. Sci. Vitaminol. 41: 95–104.

Chonan, O. and M. Watanuki. 1996. The effect tof 6'-galactooligosaccharides on bone mineralization of rats adapted to different levels of dietary calcium. Int. J. Vit. Nutr. Res. 66: 244–249.

Chonan, O., R. Takahashi and M. Watanuki. 2001. Role of activity of gastrointestinal microflora in absorption of calcium and magnesium in rats fed β1-4 linked galactooligosaccharides. Biosci. Biotechnol. Biochem. 65: 1872–1875.

Cochet, B., A. Jung, M. Griessen, P. Bartholdi, P. Schaller and A. Donath. 1983. Effects of lactose on intestinal calcium absorption in normal and lactase-deficient subjects. Gastroenterology 84: 935–940.

Collins, M.D. and G.R. Gibson. 1999. Probiotics, prebiotics, and synbiotics: Approaches for modulating the microbial ecology of the gut. Am. J. Clin. Nutr. 69: 1052S–1057S.

Coppa, G.V., O. Gabrielli, P. Pierani, C. Catassi, A. Carlucci and P.L. Giorgi. 1993. Changes in carbohydrate composition in human milk over 4 months of lactation. Pediatrics 91: 637–641.

Coudray, C., J. Bellanger, C. Castiglia-Delavaud, V. Vermorel and Y. Rayssignuier. 1997. Effect of solluble or partly soluble dietary fibres supplementation on absorption and balance of calcium, magnesium, iron and zinc in healthy young men. Eur. J. Clin. Nutr. 51: 375–380.

Coudray, C., J.C. Tressol, E. Gueux and Y. Rayssiguier. 2003. Effects of inulin-type fructans of different chain length and type of branching on intestinal absorption and balance of calcium and magnesium in rats. Eur. J. Nutr. 42: 91–98.

Coudray, C., M. Rambeau, C. Feillet-Coudray, J.C. Tressol, C. Demigne, E. Gueux, A. Mazur and Y. Rayssiguier. 2005a. Dietary inulin intake and age can significantly affect intestinal absorptin of calcium and magnesium in rats: A stable isotope approach. Nutr. J. 4: 29.

Coudray, C., C. Feillet-Coudray, J.C. Tressol, E. Gueux, S. Thien, L. Jaffrelo, A. Mazur and Y. Rayssiguier. 2005b. Stimulatory effect of inulin on intestinal absorption of calcium and magnesium in rats is modulated by dietary calcium intakes short- and long-term balance studies. Eur. J. Nutr. 44: 293–302.

Coudray, C., C. Feillet-Coudray, E. Gueux, A. Mazur and Y. Rayssiguier. 2006. Dietary inulin intake and age can affect intestinal absorption of zinc and copper in rats. J. Nutr. 136: 117–122.

Coxam, V. 2005. Inulin-type fructans and bone health: State of the art and perspectives in the management of osteoporosis. Br. J. Nutr. 93: S111–S123.

Delzenne, N., J. Aertssens, H. Verplaetse, M. Roccaro and M. Roberfroid. 1995. Effect of fermentable fructo-oligosaccharides on mineral, nitrogen and energy digestive balance in the rat. Life Sci. 57: 1579–1587.

Demigné, C., M.A. Levrat and C. Rémésy. 1989. Effects of feeding fermentable carbohydrates on the cecal concentrations of minerals and their fluxes between the cecum and blood plasma in the rat. J. Nutr. 119: 1625–1630.

Demigné, C., H. Jacobs, C. Moundras, M.J. Davicco, M.N. Horcajada, A. Bernalier and V. Coxam. 2008. Comparison of native or reformulated chicory fructans, or non-purified chicory, on rat cecal fermentation and mineral metabolism. Eur. J. Nutr. 47: 366–374.

Devareddy, L., D.A. Khalil, K. Korlagunta, S. Hooshmand, D.D. Bellmer and B.H. Arjmandi. 2006. The effects of fructo-oligosaccharides in combination with soy protein on bone in osteopenic ovariectomized rats. Menopause 13: 692–699.

Diplock, A.T., P.J. Aggett, M. Ashwell, F. Bornet, E.B. Fern and M.B. Roberfroid. 1999. Scientific concepts of functional foods in Europe: consensus document. Br. J. Nutr. 81: S1–S28.

Douglas, L.C. and M.E. Sanders. 2008. Probiotics and prebiotics in dietetics practice. J. Am. Diet. Assoc. 108: 510–521.

Ducros, V., J. Arnaud, M. Tahiri, C. Coudray, F. Bornet, C. Bouteloup-Demange, F. Brouns, Y. Rayssiguier and A.M. Roussel. 2005. Influence of short-chain fructo-oligosaccharides (sc-FOS) on absorption of Cu, Zn, and Se in healthy postmenopausal women. J. Am. Coll. Nutr. 24: 30–37.

Ervin, R.B., C.-Y. Wang, J.D. Wright and J. Kennedy-Stephenson. 2004. Dietary intake of selected minerals for the United States population: 1999–2000. *In:* Advance Data from Vital and Health Statistics. National Center for Health Statistics, Hyattsville, Maryland, pp. 341.

Fanaro, S., G. Boehm, J. Garssen, J. Knol, F. Mosca, B. Stahl and V. Vigi. 2005. Galacto-oligosaccharides and long-chain fructo-oligosaccharides as prebiotics in infant formulas: A review. Acta Paediatr Suppl. 94: 22–26.

Fanaro, S., B. Marten, R. Bagna, V. Vigi, C. Fabris, L. Peña-Quintana, F. Argüelles, K.E. Scholz-Ahrens, G. Sawatzki, R. Zelenka, J. Schrezenmeir, M. de Vrese and E. Bertino. 2008. Galactooligosaccharides are bifidogenic and safe at weaning: A double-blind randomized multicenter study. J. Pediatr. Gastroenterol. Nutr. 48: 82–88.

[FAO-WHO] Food and Agriculture Organization of the United Nations, World Health Organization. 2001. Health and nutritional properties of probiotics in food including powder milk with live lactic acid bacteria. Report of a joint FAO/WHO expert consultation on evaluation of health and nutritional properties of probiotics in food including powder milk with live lactic acid bacteria. Report No. 10-1-2001.

Fleet, J.C. and R.D. Schoch. 2010. Molecular mechanisms for regulation of intestinal calcium absorption by vitamin D and other factors. Crit. Rev. Clin. Lab. Sci. 47: 181–195.

[FNB-IOM] Food and Nutrition Board, Institute of Medicine. 2010. Dietary Reference Intakes for Calcium and Vitamin D. National Academy Press, Washington, DC.

Fujita, H., K. Sugimoto, S. Inatomi, T. Maeda, M. Osanai, Y. Uchiyama, Y. Yamamoto, T. Wada, T. Kojima, H. Yokozaki, T. Yamashita, S. Kato, N. Sawada and H. Chiba. 2008. Tight junction proteins claudin-2 and -12 are critical for vitamin D-dependent Ca2+ absorption between enterocytes. Mol. Biol. Cell. 19: 1912–1921.

Fukushima, A., A. Ohta, K. Sakai and K. Sakuma. 2005. Expression of calbindin-D9k, VDR and Cdx-2 messenger RNA in the process by which fructooligosaccharides increase calcium absorption in rats. J. Nutr. Sci. Vitaminol. 51: 426–432.

German, J.B., S.L. Freeman, C.B. Lebrilla and D.A. Mills. 2008. Human milk oligosaccharides: Evolution, structures and bioselectivity as substrates for intestinal bacteria. Nestle Nutr. Workshop Ser. Pediatr. Program 62: 205–222.

Gibson, G.R., E.B. Beatty, X. Wang and J.H. Cummings. 1995. Selective stimulation of bifidobacteria in the human colon by oligofructose and inulin. Gastroenterol. 108: 975–982.

Gibson, G.R., K.P. Scott, R.A. Rastall, K.M. Tuohy, A. Hotchkiss, A. Dubert-Ferrandon, M. Gareau, E.F. Murphy, D. Saulnier, G. Loh, S. Macfarlane, N. Delzenne, Y. Ringel, G. Kozianowski, R. Dickmann, I. Lenoir-Wijnkoop, C. Walker and R. Buddington. 2010. Dietary prebiotics: Current status and new definition. Food Sci. Tech. Bull. Funct. Foods 7: 1–19.

Goldstein, D.A. 1990. Serum calcium. *In:* H.K. Walker, W.D. Hall and J.W. Hurst [eds.]. Clinical methods: The history, physical, and laboratory examinations, ed. 3. Butterworths, Boston, USA, pp. 677.

Griessen, M., B. Cochet, F. Infante, A. Jung, P. Bartholdi, A. Donath, E. Loizeau and B. Courvoisier. 1989. Calcium absorption from milk in lactase-deficient subjects. Am. J. Clin. Nutr. 49: 377–384.

Griffin, I.J., P.M. Davila and S.A. Abrams. 2002. Non-digestible oligosaccharides and calcium absorption in girls with adequate calcium intakes. Br. J. Nutr. 87: S187–S191.

Griffin, I.J., P.M.D. Hicks, R.P. Heaney and S.A. Abrams. 2003. Enriched chicory inulin increases Ca absorption mainly in girls with lower Ca absorption. Nutr. Res. 23: 901–909.

Haboubi, N.Y., P. Hudson, Q. Rahman, G.S. Lee and A. Ross. 1988. Small-intestinal transit time in the elderly. Lancet. 35: 783–808.

[HHS-ARS] U.S. Department of Health and Human Services and U.S. Department of Agriculture. 2005. Dietary Guidelines for Americans, ed. 6. U.S. Government Printing Office, Washington, DC.

Heijnen, A.M.P., E.J. Brink, A.G. Lemmens and A.C. Beynen. 1993. Ileal pH and apparent absorption of magnesium in rats fed on diets containing either lactose or lactulose. Br. J. Nutr. 70: 747–756.

Hill, T.R., L. Brennan, A.O. Connor, L. Scully, S. Healy, A.O. Sullivan, B. Mion, G. Dawson, S. Kaluskar, M.J. Gibney, F. Shanahan and K.D. Cashman. 2009. Effect of probiotic and vitamin D supplementation on markers of vitamin D status and bone turnover in healthy adults. Proceedings of the Nutrition Society. 68, E114.

Holloway, L., S. Moynihan, S.A. Abrams, K. Kent, A.R. Hsu and A.L. Friedlander. 2007. Effects of oligofructose-enriched inuln on intestinal absorption of calcium and magnesium and bone turnover markers in postmenopausal women. Br. J. Nutr. 97: 365–372.

Hooshmand, S., S. Juma and B.H. Arjmandi. 2010. Combination of genistin and fructooligosaccharides prevents bone loss in ovarian hormone deficiency. J. Med. Food 13: 320–325.

Hudson, M. and P.D. 1995. Marsh. Human colonic bacteria: Role in nutrition physiology and pathology. *In:* G.R. Gibson and G.T. Macfarlane [eds.]. Carbohydrate metabolism in the colon. CRC Press, Boca Raton, Florida, USA.

Igarashi, M., Y. Iiyama, R. Kato, M. Tomita, N. Omi and I Ezawa. 1994. Effect of Bifidobacterium longum and lactulose on the strength of bone in ovariectomized osteoporosis model rats. Bifidus. 7: 139–147.

Karbach, U. 1992. Paracellular Calcium Transport Across the Small Intestine. J. Nutr. 122: 672–677.

Kelleher, S.L., I. Casas, N. Carbajal and B. Lonnerdal. 2002. Supplementation of infant formula with the probiotic lactobacillus reuteri and zinc: impact on enteric infection and nutrition in infant rhesus monkeys. J. Pediatr. Gastroenterol. Nutr. 35: 162–8.

Kim, Y.Y., K.H. Jang, S.A. Kang, W.K. Ha, E.Y. Lee, Y.H. Cho and R.W. Choue. 2004. The effect of chicory fructan fiber on calcium absorption and bone metabolism in Korean postmenopausal women. Nutr. Sci. 7: 151–157.

Kleessen, B., L. Hartmann and M. Blaut. 2003. Fructans in the diet cause alterations of intestinal mucosal architecture, released mucins and mucosa-associated bifidobacteria in gnotobiotic rats. Br. J. Nutr. 89: 597–606.

Kunz, C., S. Rudloff, W. Baier, N. Klein and S. Strobel. 2000. Oligosaccharides in human milk: structural, functional, and metabolic aspects. Annu. Rev. Nutr. 20: 699–722.

Lan, G.Q., N. Abdullah, S. Jalaludin and Y.W. Ho. 2002. Efficacy of supplementation of a phytase-producing bacterial culture on the performance and nutrient use of broiler chickens fed corn-soybean meal diets. Poult. Sci. 81: 1522–1532.

Lee, W.T., S.S. Leung, D.M. Leung and J.C. Cheng. 1996. A follow-up study on the effects of calcium-supplement withdrawal and puberty on bone acquisition of children. J. Clin. Nutr. 64: 71–77.

Leibson, C.L., A.N. Tosteson, S.E. Gabriel, J.E. Ransom and L.J. Melton. 2002. Mortality, disability, and nursing home use for persons with and without hip fracture: A population-based study. J. Am. Geriatr. 50: 1644–1650.

Levrat, M.A., C. Remesy and C. Demigné. 1991. High propionic acid fermentations and mineral accumulation in the cecum of rats adapted to different levels of inulin. J. Nutr. 121: 1730–1737.

Lloyd, T., M.B. Andon, N. Rollings, J.K. Martel, J.R. Landis, L.M. Demers, D.F. Eggli, K. Kieselhorst and H.E. Kulin. 1993. Calcium supplementation and bone mineral density in adolescent girls. JAMA. 270: 841–844.

Lobo, A.R., C. Colli and T.M.C.C. Filisetti. 2006. Fructooligosaccharides improve bone mass and biomechanical properties in rats. Nutr. Res. 26: 413–420.

Lobo, A.R., J.M. Filho, E.P. Alvares, M.L. Cocato and C. Collli. 2009. Effects of dietary lipid composition and inulin-type fructans on mineral bioavailability in growing rats. Nutrition 25: 216–225.

Lopez, H.W., C. Coudray, M.A. Levrat-Verny, C. Feillet-Coudray, C. Demigné and C. Rémésy. 2000. Fructooligosaccharides enhanced mineral apparent absorption and counteract the deleterious effects of phytic acid on mineral homeostasis in rats. J. Nutr. Biochem. 11: 500–508.

López-Huertas, E., B. Teucher, J.J. Boza, A. Martínez-Férez, G. Majsak-Newman, L. Baró, J.J. Carrero, M. González-Santiago, J. Fonollá and S. Fairweather-Tait. 2006. Absorption of calcium from milks enriched with fructooligosaccharides, caseinophosphopeptides, tricalcium phosphate, and milk solids. Am. J. Clin. Nutr. 83: 310–316.

Lutz, T. and E. Scharrer. 1991. Effect of short-chain fatty acids on calcium absorption by the rat colon. Exp. Physiol. 76: 615–618.

Macfarlane, S., G.T. Macfarlane and J.H. Cummings. 2006. Review article: Prebiotics in the gastrointestinal tract. Aliment Pharmacol. Ther. 24: 701–714.

Magee, D.F. and A.F. Dalley. 1986. Digestion and the structure and function of the gut. Switzerland, Karger, pp. 183.

Martin, B.R., M.M. Braun, K. Wigertz, R. Bryant, Y. Zhao, W. Lee, A. Kempa-Steczko and C.M. Weaver. 2010. Fructo-oligosaccharides and calcium absorption and retention in adolescent girls. J. Am. Coll. Nutr. 29: 382–386.

Mathey, J., C. Puel, S. Kati-Coulibaly, C. Bennetau-Pelissero, M.J. Davicco, P. Lebecque, M.N. Horcajada and V. Coxam. 2004. Fructooligosaccharides maximize bone-sparing effects of soy isoflavone-enriched diet in the ovariectomized rat. Calcif Tissue Int. 75: 169–179.

Matkovic, V., K. Kostial, I. Simonovic, R. Buzinz, A. Brodarec and B.E. Nordin. 1979. Bone status and fracture rates in two regions of Yugoslavia. Am. J. Clin. Nutr. 32: 540–549.

Matkovic, V., P.K. Goel, N.E. Badenhop-Stevens, J.D. Landoll, B. Li, J.Z. Ilich, M. Skugor, L.A. Nagode, S.L. Mobley, E-J. Ha, T.N. Hangartner and A. Clairmont. 2005. Calcium supplementation and bone mineral density in females from childhood to young adulthood: A randomized controlled trial. Am. J. Clin. Nutr. 81: 175–188.

Mineo, H., H. Hara and F. Tomita. 2001. Short-chain fatty acids enhance diffusional Ca transport in the epithelium of the rat cecum and colon. Life Sciences 69: 517–526.

Mitamura, R. and H. Hara. 2006. Ingestion of difructose anhydride III partially restores calcium absorption impaired by vitamin D and estrogen deficiency in rats. Eur. J. Nutr. 45: 242–249.

Mogna, L., S. Nicola, M. Pane, P. Lorenzini, G. Strozzi and G. Mogna. 2012. Selenium and Zinc internalized by Lactobacillus buchneri Lb26 (DSM 16341) and Bifidobacterium lactis Bb1 (DSM 17850) improved bioavailability using a new biological approach. J. Clin. Gastroenterol. 46 Supp. 1.

Morohashi, T., T. Sano, A. Ohta and S. Yamada. 1998. True calcium absorption in the intestine is enhanced by fructooligosaccharide feeding in rats. J. Nutr. 128: 1815–1818.

Moshfegh, A., J. Goldman and L. Cleveland. 2005. What we Eat in America, NHANES 2001-2002: Usual Nutrient Intakes from Food Compared to Dietary Reference Intakes. U.S. Department of Agriculture, Agricultural Research Service website. http://www.ars.usda.gov.

Narva, M., R. Nevala, T. Poussa and R. Korpela. 2004a. The effect of Lactobacillus helveticus fermented milk on acute changes in calcium metabolism in postmenopausal women. Eur. J. Nutr. 43: 61–8.

Narva, M., J. Halleen, K. Vaananen and R. Korpela. 2004b. Effects of Lactobacillus helveticus fermented milk on bone cells *in vitro*. Life Sci. 75: 1727–34.

Nemere, I., M.C. Farach-Carson, B. Rohe, T.M. Sterling, A.W. Norman, B.D. Boyan and S.E. Safford. 2004. Ribozyme knockdown functionally links a 1,25(OH)2D3 membrane binding protein (1,25D3-MARRS) and phosphate uptake in intestinal cells. Proc. Natl. Acad. Sci USA 101: 7392–7397.

[NOF] National Osteoporosis Foundation. 2011. Fast Facts. National Osteoporosis Foundation website. http://www.nof.org.

Norman, A.W., J.E. Bishop, C.M. Bula, C.J. Olivera, M.T. Mizwicki, L.P. Zanello, H. Ishida and W.H. Okamura. 2002. Molecular tools for study of genomic and rapid signal transduction responses initiated by 1 alpha,25(OH)(2)-vitamin D(3). Steroids 67: 457–466.

Nowson, C.A., R.M. Green, J.L. Hopper, A.J. Sherwin, D. Young, B. Kaymakci, C.S. Guest, M. Smid, R.G. Larkins and J. Wark. 1997. A co-twin study of the effect of calcium supplementation on bone density during adolescence. Osteoporos Int. 7: 219–225.

Nzeusseu, A., D. Dienst, V. Haufroid, G. Depresseux, J.P. Devogelaer and D.H. Manicourt. 2006. Inulin and fructo-oligosaccharides differ in their ability to enhance the density of cancellous and cortical bone in the axial and peripheral skeleton of growing rats. Bone 38: 394–399.

Ohta, A., M. Ohtuki, T. Takizawa, H. Inaba, T. Adachi and S. Kimura. 1994. Effects of fructooligosaccharides on the absorption of magnesium and calcium by cecectomized rats. J. Vit. Nutr. 64: 316–323.

Ohta, A., M. Ohtsuki, S. Baba, T. Adachi, T. Sakata and E. Sakaquchi. 1995. Calcium and magnesium absorption from the colon and rectum are increased in rats fed fructooligosaccharides. J. Nutr. 125: 2417–2424.

Ohta, A., S. Baba, M. Ohtsuki, A. Taguchi and T. Adachi. 1997. *In vivo* absorption of calcium carbonate and magnesium oxide from the large intestine in rats. J. Nutr. Sci. Vitaminol. 43: 35–46.

Ohta, A., Y. Motohashi, K. Sakai, M. Hirayama, T. Adachi and K. Sakuma. 1998. Dietary fructooligosaccharides increase calcium absorption and levels of mucosal calbindin-D9k in the large intestine of gastrectomized rats. Scand. J. Gastroenterol. 33: 1062–1068.

Ohta, A., M. Uehara, K. Sakai, M. Takasaki, H. Adlercreutz, T. Morohashi and Y. Ishimi. 2002. A combination of dietary fructooligosaccharides and isoflavone conjugates increases femora bone mineral density and equol production in ovariectomized mice. J. Nutr. 132: 2048–2054.

Pansu, D., C. Bellaton and F. Bronner. 1983. Development changes in the mechanisms of duodenal calcium transport in the rat. J. Physiol. 244: G10–G26.

Parracho, H., A.L. McCartney and G.R. Gibson. 2007. Probiotics and prebiotics in infant nutrition. Proc. Nutr. Soc. 66: 405–411.

Pérez-Conesa, D., G. López, P. Abellán and G. Ros. 2006. Bioavailability of calcium, magnesium and phosphorus in rats fed probiotic, prebiotic and symbiotic powder follow-up infant formulas and their effect on physiological and nutritional parameters. J. Sci. Food Agric. 86: 2327–2336.

Pérez-Conesa, D., G. López and G. Ros. 2007. Effects of probiotics, prebiotic and synbiotic follow-up infant formulas on large intestine morphology and bone mineralisation in rats. J. Sci. Food Agric. 87: 1059–1068.

Petry, N., I. Egli, C. Chassard, C. Lacroix and R. Hurrell. 2012. Inulin modifies the bifidobacteria population, fecal lactate concentration, and fecal pH but does not influence iron absorption in women with low iron status. Am. J. Clin. Nutr. 96: 325–331.

Raschka, L. and H. Daniel. 2005. Mechanisms underlying the effects of inulin-type fructans on calcium absorption in the large intestine of rats. Bone 37: 728–735.

Richardson, J., M. Verghese, I.A. Bonsi, L.T. Walker, L. Shackelford and C.B. Chawan. 2002. Effects of prebiotics on bone mineralisation in Fisher 344 male weanling rats. ITF USA. 46F-8.

Roberfroid, M.B. 1998. Prebiotics and synbiotics: Concepts and nutritional properties. Br. J. Nutr. 80: S197–S202.

Roberfroid, M., G.R. Gibson, L. Hoyles, A.L. McCartney, R. Rastall, I. Rowland, D. Wolvers, B. Watzl, H. Szajewska, B. Stahl, F. Guarner, F. Respondek, K. Whelan, V. Coxam, M.J. Davicco, L. Leotoing, Y. Wittrant, N.M. Delzenne, P.D. Cani, A.M. Neyrinck and A. Meheust. 2010. Prebiotic effects: Metabolic and health benefits. Br. J. Nutr. 104: S1–S63.

Rodrigues, F.C., A.S. Castro, V.C. Rodrigues, S.A. Fernandes, E.A. Fontes, T.T. de Oliveira, H.S. Martino and C.L. de Luces Fortes Ferreira. 2012. Yacon flour and Bifidobacterium longum modulate bone health in rats. J. Med. Food 15: 664–70.

Rodriguez-Cabezas, M.E., D. Camuesco, B. Arribas, N. Garrido-Mesa, M. Comalada, E. Bailón, M. Cueto-Sola, P. Utrilla, E. Guerra-Hernández, C. Pérez-Roca, J. Gálvez and A. Zarzuelo. 2010. The combination of fructooligosaccharides and resistant starch shows prebiotic additive effects in rats. Clin. Nutr. 29: 832–839.

Rondon, L.J., Y. Rayssiguier and A. Mazur. 2008. Dietary inulin in mice stimulates Mg^{2+} absorption and modulates TRPM6 and TRPM7 expression in large intestine and kidney. Magnes Res. 21: 224–231.

Rovensky, J., K. Svik, V. Matha, R. Istok, L. Ebringer, M. Ferencik and M. Stancikova. 2004. The effects of Enterococcus faecium and selenium on methotrexate treatment in rat adjuvant-induced arthritis. Clin. Dev. Immunol. 11: 267–73.

Rowland, I.R. and R. Tanaka. 1993. The effects of transgalactosylated oligosaccharides on gut flora metabolism in rats associated with a human faecal microflora. J. Appl. Bacteriol. 74: 667–674.

Sakai, K., A. Ohta, K. Shiga, M. Takasaki, T. Tokunaga and H. Hara. 2000. The cecum and dietary short-chain fructooligosaccharides are involved in preventing postgastrectomy anemia in rats. J. Nutr. 130: 1608–1612.

Schaafsma, G., W.J. Visser, P.R. Dekker and M. van Schaik. 1988. Effect of dietary calcium supplementatioin with lactose on bone in vitamin D deficient rats. Bone 8: 357–362.

Scheppach, W., H.P. Bartram, A. Richter, F. Richter, H. Liepold, G. Dusel, G. Hofstetter, J. Rüthlein and H. Kasper. 1995. Role of short-chain fatty acids on the human colonic mucosa *in vitro*. J. Parenter Enter Nutr. 16: 43–48.

Scholz-Ahrens, K.E., G. Schaafsma, E.G.H.M. van den Heuvel and J. Schrezenmeir. 2001. Effects of prebiotics on mineral metabolism. Am. J. Clin. Nutr. 73: 459S–464S.

Scholz-Ahrens, K.E., Y. Aćil and J. Schrezenmeir. 2002. Effect of oligofructose or dietary calcium on repeated calcium and phosphorus balances, bone mineralization and trabecular structure in ovariectomized rats. Br. J. Nutr. 88: 365–377.

Sheikh, M.S., L.R. Schiller and J.S. Fordtran. 1990. *In vivo* intestinal absorption of calcium in humans. Miner Electrolyte Metab. 16: 130–146.

Tahiri, M., J.C. Tressol, J. Arnaud, R.F. Bornet, C. Bouteloup-Demange, C. Feillet-Coudray, M. Brandolini, V. Ducros, D. Pépin, F. Brouns, A.M. Roussel, Y. Rayssiguier and C. Coudray. 2003. Effect of short chain fructooligosaccharides on intestinal Ca absorption and Ca status in postmenopausal women: A stable isotope study. Am. J. Clin. Nutr. 77: 449–457.

Takahara, S., T. Morohashi, T. Sano, A. Ohta, S. Yamada and R. Sasa. 2000. Fructooligosaccharide consumption enhances femoral bone volume and mineral concentrations in rats. J. Nutr. 130: 1792–1795.

Takasaki, M., H. Inaba, A. Ohta, Y. Motohashi, K. Sakai, H. Morris and K. Sakuma. 2000. Dietary short-chain fructooligosaccharides increase calbindin-D9k levels only in the large intestine in rats independent of dietary calcium deficiency or serum 1,25 dihydroxy vitamin D levels. Int. J. Vitam. Nutr. Res. 70: 206–213.

Teuri, U., M. Karkkainen, C. Lamberg-Allardt and R. Korpela. 1999. Addition of inulin to breakfast does not acutely affect serum ionized calcium and parathyroid hormone concentrations. Ann. Nutr. Metab. 43: 356–364.

Tremaine, W.J., A.D. Newcomer, B.L. Riggs and D.B. McGill. 1986. Calcium absorption from milk in lactase-deficient and lactase-sufficient adults. Dig. Dis. Sci. 31: 376–378.

Tzortzis, G., A.K. Goulas, J.M. Gee and G.R. Gibson. 2005. A novel galactooligosaccharide mixture increases the bifidobacterial population numbers in a continuous *in vitro* fermentation system and in the proximal colonic contents of pigs *in vivo*. J. Nutr. 135: 1726–1731.

[USDA] U.S. Department of Agricultural Research Service. 2010. Fluid Milk Consumption in the United States: What we Eat in America, NHANES 2005-2006. Food Surveys Research Group Dietary Data Brief, Beltsville Human Nutrition Research Center, Food Surveys Research Group. http://ars.usda.gov/Services/docs.htm?docid=19476.

van den Heuvel, E.G.H.M., G. Schaafsma, T. Muys and W. van Dokkum. 1998. Nondigestible oligosaccharides do not interfere with calcium and nonheme-iron absorption in young, healthy men. Am. J. Clin. Nutr. 67: 445–451.

van den Heuvel, E.G.H.M., T. Muys, W. van Dokkum and G. Schaafsma. 1999a. Oligofructose stimulates calcium absorption in adolescents. Am. J. Clin. Nutr. 69: 544–548.

van den Heuvel, E.G.H.M., T. Muijs, W. van Dokkum and G. Schaafsma. 1999b. Lactulose stimulates calcium absorption in postmenopausal women. J. Bone Miner Res. 14: 1211–1216.

van den Heuvel, E.G., M.H. Schoterman and T. Muijs. 2000. Transgalactooligosaccharides stimulate calcium absorption in postmenopausal women. J. Nutr. 130: 2938–2942.

van den Heuvel, E.G., T. Muijs, F. Brouns and H.F. Hendriks. 2009. Short-chain fructo-oligosaccharides improve magnesium absorption in adolescent girls with a low calcium intake. Nutr. Res. 29: 229–237.

Wang, Y., T. Zeng, S.E. Wang, W. Wang, Q. Wang and H.X. Yu. 2010. Fructo-oligosaccharides enhance the mineral absorption and counteract the adverse effects of phytic acid in mice. Nutrition 26: 305–311.

Weaver, C.M., B.R. Martin, C.H. Nakatsu, A.P. Armstrong, A. Clavijo, L.D. McCabe, G.P. McCabe, S. Duignan, M.H. Schoterman and E.G. van den Heuvel. 2011. Galactooligosaccharides improve mineral absorption and bone properties in growing rats through gut fermentation. J. Agric. Food Chem. 59: 6501–6510.

Weber, P. 1999. The role of vitamins in the prevention of osteoporosis–a brief status report. Int. J. Vitam. Nutr. Res. 69: 194–7.

Yap, K.W., S. Mohamed, A.M. Yazid, I. Maznah and D.M. Meyer. 2005. Dose-response effects of inulin on the faecal fatty acids content and mineral absorption of formula-fed infants. Nutr. Food Sci. 35: 208–219.

Yazawa, K., K. Imai and Z. Tamura. 1978. Oligosaccharides and polysaccharides specifically utilisable by bifidobacteria. Chem. Pharm. Bull. 26: 3306–3311.

Yazawa, K. and Z. Tamura. 1982. Search for sugar sources for selective increase of bifidobacteria. Bifidobacteria Microflora 1: 39–44.

Younes, H., C. Coudray, J. Bellanger, C. Demigné, Y. Rayssiguier and C. Rémésy. 2001. Effects of two fermentable carbohydrates (inulin and resistant starch) and their combination on calcium and magnesium balance in rats. Br. J. Nutr. 86: 479–485.

Yuan, J.-P., J.-H. Wang and X. Liu. 2007. Metabolism of dietary soy isoflavones to equol by human intestinal microflora—implications for health. Mol. Nutr. Food Res. 51: 765–781.

Zafar, T., C.M. Weaver, Y. Zhao, B.R. Martin and M.E. Wastney. 2004a. Nondigestible oligosaccharides increase calcium absorption and suppress bone resorption in ovariectomized rats. J. Nutr. 134: 399–402.

Zafar, T., C.M. Weaver, K. Jones, D.R. Moore and S. Barnes. 2004b. Inulin effects on bioavailability of soy isoflavones and their calcium absorption enhancing ability. J. Agric Food Chem. 52: 2827–2831.

11

Probiotics and Prebiotics in Obesity and Energy Metabolism

Ayşe Binnur Erbağcı

The biological systems regulating food intake and energy expenditure in humans are concerned with energy deprivation rather than nutrient overload. These homeostatic systems formerly preventing starvation and sustaining the species have not adapted to the sedentary lifestyle, high-calorie, and high-fat diet prevalent these days, and now lead to a cluster of metabolic disorders at the interface of nutrition and systemic low-grade inflammation such as obesity, insulin resistance, metabolic syndrome, fatty liver disease, and type 2 diabetes mellitus (DM) (Delzenne et al. 2011, Sacks and Path 2006). Excess weight and obesity are important issues for public health as not only are they preventable, but they are also the fifth leading cause for global deaths. Each year, about 2.8 million people die as a result of being overweight or obese. In addition, 44% of DM, 23% of coronary artery disease and between 7% and 41% of certain cancer cases are attributed to overweight and obesity (WHO 2012, Allison et al. 1999). The prevalence of obesity has more than doubled during the last decades and has reached epidemic proportions worldwide. The estimated number of overweight adults has reached 1.4 billion as of 2008. Of these over 200 million men

University of Gaziantep, Medical Faculty, Department of Medical Biochemistry, 27310 Gaziantep, Turkey.
Email: aerbagci@gantep.edu.tr

and nearly 300 million women were obese. In 2010, more than 40 million children under the age of five were overweight (WHO 2012). Although genetic factors can determine predisposition to obesity, the dramatic increase of obesity prevalence reflects implications of lifestyle changes and environmental factors upon acquired disturbances of the energy metabolism (Bleich et al. 2008).

Many of the mechanistic studies have mainly focused on the biology of human organ or tissue cell systems, and the human genome, to unravel the risk factors for metabolic diseases. Nevertheless, there is an increasing body of literature that directs attention to a third culprit: the gut microbial community (DiBaise et al. 2008, Duncan et al. 2008, Tilg et al. 2009). The gut microbiome represents 150-fold more genes than the human genome, therefore the gut microbiota consisting of 10^{14} bacteria, and its metabolic potential are among the most important external factors contributing to the host response towards nutrients (Delzenne et al. 2011, Diamant et al. 2010, Qin et al. 2010). The gut microbiota and its effect on nutrients regulate several signaling networks associated with development of obesity. Observation that germ-free mice are resistant to diet-induced obesity suggests substantial effect of the gut microbiota on the host metabolism (Ding et al. 2010, Fleissner et al. 2010, Rabot et al. 2010). Obesity develops as a consequence of nutrient overload, decreased energy expenditure, impaired metabolism of hormones regulating satiety, altered *de novo* lipogenesis, fatty acid oxidation, defects of insulin release/signal system, and meta-inflammation (Sacks and Path 2006, Rıfai and Warnick 2006). Interestingly, these molecular mechanisms have intriguing associations with the gut microbiota and nutrition.

Prebiotics, Probiotics and Obesity

A direct causal relationship between the gut microbiota and obesity has been conceived based on the observation that germ-free mice are leaner than conventionally raised counterparts, and are resistant to diet-induced obesity, while colonization increases fat mass by about 50% and reduces insulin sensitivity (Backhed et al. 2004, 2007, Ding et al. 2010, Rabot et al. 2010, Turnbaugh et al. 2006). Furthermore, gut microbiota transplantation from obese mice to germ-free recipients transfers the obese phenotype, confirming contribution of gut microbial community to adipogenesis (Turnbaugh et al. 2006, 2008). New technologies allowing detection of non-cultivating species have revolutionized our knowledge of intestinal bacterial flora by identification and classification of new species. Interestingly, hypervariable region analysis of bacterial 16S rRNA have revealed a microbial inter- and intra-individual variability (Hooper et al. 2001, Hsiao and Fraser-Liggett 2009, Tannock 2001, Turnbaugh et al. 2008). Most of the studies concerning the role of gut microbiota in obesity or related metabolic disorders have shown changes of the relative abundance of phyla, gender, or species of bacteria, which correlate with the host phenotype such as fat mass, biomarkers of glucose homeostasis, and inflammation (Ley 2010, Ley et al. 2005). Initial research done on qualitative changes of the gut microflora in human obese individuals found a shift in bacterial phyla with lower *Bacteroidetes versus Firmicutes* compared to lean controls. In this study, following a low-calorie diet and weight loss, the ratio of *Bacteroidetes* to *Firmicutes* approached a lean type profile after one year (Ley et al. 2006). However, later trials evaluating gut microbiota in obese subjects have yielded contradictory results, with some studies finding decreased or increased *Bacteroidetes* community, while others have not identified a significant difference (Armougom et al. 2009, Collado et al. 2008, Duncan et al. 2008, Zhang et al. 2009). In obese mice, alteration of the gut microbial community is also dependent on age, which should be considered in data interpretation (Murphy et al. 2010).

Alternative hypothesis of more specific modulation of the gut microbial community in obesity at the species level indicates obesity-preventing effect of several *Bifidobacterium* species. Gram-positive *Bifidobacterium* spp. has been shown to reduce the levels of intestinal lipopolysaccharides (LPS) in mice and to improve the mucosal barrier function (Cani et al. 2007a, b, 2008, Turnbaugh et al. 2008). Benefits of *Bifidobacterium* spp. in preventing children from becoming overweight at seven years has been described (Kalliomaki et al. 2008). Likewise, *Bifidobacterium* spp. level was higher in normal-weight than in overweight women and in women with lower weight gain during pregnancy (Collado et al. 2008). However, reduced *Bifidobacterium bifidum* and *Bifidobacterium breve* counts and increased *Bifidobacterium catenulatum* have been described following weight loss (Santacruz et al. 2009). Another randomized study in overweight patients has shown that administration of *Lactobacillus gasseri* decreased body weight and fat mass (Kadooka et al. 2010). On the other hand, *Clostridium* and *Staphylococcus* may trigger low-grade inflammation contributing to the promotion of weight gain (Lundell et al. 2007).

Several research on alterations in gut microbial community of patients with DM have been reported. In fecal samples of diabetic patients, the *Bifidobacterium* genus was reduced compared to healthy controls (Wu et al. 2010). *Faecalibacterium prausnitzii* proportion was also decreased in diabetic *versus* non diabetic obese subjects, which was negatively associated with inflammatory markers before and after roux- and Y gastric bypass surgery (Furet et al. 2010).

Further approaches focused on microbial genes involved in energy metabolism and revealed an enrichment of genes contributing energy harvest in obese mice. Consequently, the microbiome of the obese mice had higher fermentation capacity, and short chain fatty acid (SCFA) levels were increased in the distal gut (Turnbaugh et al. 2006, 2008). Similarly, microbiome of obese humans was enriched in genes involved in carbohydrate sensing and catabolism (Greiner and Backhed 2011, Turnbaugh et al. 2009). Additionally, insulin-resistant mice had increased microbial metabolism of phosphatidylcholine required for lipid transport from the liver, which could have an implication on hepatic steatosis (Dumas et al. 2006).

Fructo-oligosaccharides (FOS), galacto-oligosaccharides (GOS), and inulin type fructans (ITF) are the main prebiotics used in research on obesity and obesity related metabolic disorders (Roberfroid et al. 2010). In experimental animal models, 5–10% prebiotics in their feed showed significant effect on body weight and fat mass. Decrease in fat mass has been observed in epididymal, visceral or subcutaneous white adipose tissue, and has not always led to a significant reduction in body weight. This decrease in fat mass was associated with a reduction of food intake, and was abrogated when ITF are substituted by non-fermentable dietary fibre (Cani et al. 2009a,b, Daubioul et al. 2002).

Increased Energy Yield from the Diet

The gut microbiota has at least two ways of promoting delivery of calories back to the host: increased monosaccharide uptake from the gut, and increased processing of undigested dietary polysaccharides to produce SCFA by microbial glycosylhydrolases (Samuel et al. 2008). Comparative metagenomic studies of cecal microbial community DNA of obese mice and from lean controls have proved that the obesity-associated microbiomes have a greater capacity to ferment carbohydrates to SCFA (Turnbaugh et al. 2006, 2008). Besides, colonization of adult germ-free mice, with saccharolytic *Bacteroides thetaiotaomicron*, plus *Methanobrevibacter smithii* that enhances polysaccharide fermentation by removing the H_2 end-product reveals higher levels of SCFA in the colon, and significantly greater host adiposity than colonization of germ-free animals with either organism alone (Samuel and Gordon 2006, Sonnenburg et al. 2005). SCFAs may function as energy sources for the colon mucosa, the liver, and partly muscle and adipose tissue (Wolever et al. 1989). However, the cause-effect relationship between SCFAs and increased adiposity may not solely BE INCREASED capture of energy trapped within the nutrient, as SCFAs may act as signal molecules as well (Le Poul et al. 2003). Mice deficient of SCFA receptors were leaner than their wild-type counterparts, supporting the role of SCFA as signal transducers rather than a source of energy (Bjursell et al. 2011, Samuel et al. 2008).

The second mechanism proposed to take part in improved energy harvest from the diet by the host is augmented absorbtion of monosaccharides. Conventionalization of germ-free mice promotes increased monosaccharide uptake from the gut and its transfer to the portal circulation as well as doubling the density of capillaries that underlie the small intestinal villus epithelium (Backhed et al. 2004, Stappenbeck et al. 2002). Together, these findings are consistent with an increase in absorption of monosaccarides, and processing of dietary polysaccharides to SCFA, which may contribute to increased energy yield from the diet.

Increased Metabolic Rate

Several experimental studies have shown microbiota-mediated increase in body fat content, despite steady calorie consumption (Backhed et al. 2004, Samuel et al. 2008), raising the possibility of decreased energy expenditure. Also, conventionalized mice have decreased expression of fasting-induced adipose factor (Fiaf, synonym for angiopoietin-like protein 4) whose overexpression is known to lead to uncoupling in fat tissue (Mandard et al. 2006). However, gut microbiota also promotes bile acid synthesis, which in turn induces expression of the uncoupling protein in brown adipose tissue through increased 3,5,3′-triiodothyronine formation (Watanabe et al. 2006). Analyzed with open-circuit indirect calorimetry, conventionalized mice were found to have 27% higher metabolic rate than the age- and gender-matched leaner germ-free counterparts (Backhed et al. 2004). Additionally, conventionalized mice had significant increases in the steady-state levels of tricarboxylic acid cycle intermediates without alterations in tissue high-energy phosphate stores. The increased oxygen consumption without high-energy phosphate storage may indicate the presence of futile cycles, thus an inefficient metabolism (Backhed et al. 2004). Additionally, ingredients showing a prebiotic effect are shown to modulate thermogenesis. In obese dogs, treatment with short-chain fructans increased uncoupling protein 2, and carnitine palmitoyltransferase 1 expression in the adipose tissue (Respondek et al. 2008).

Increased Hepatic *de novo* Lipogenesis

Acetyl-CoA carboxylase is the initial and the rate-controlling enzyme of the main pathway for *de novo* synthesis of fatty acids in humans, followed by the fatty acid synthase multienzyme complex. These adaptive lipogenic enzymes increase in total amount in the fed state, leading to accumulation of triacylglycerol (TAG) in the liver, and in the adipose tissue (Rıfai and Warnick 2006). The liver responds to increased carbohydrate flow generated via fermentation of polysaccharides by gut microbiata and enhanced monosaccharide uptake, by increasing inefficient metabolism, *de novo* lipogenesis, and exporting these calories in the form of fat for deposition in peripheral tissues (Backhed et al. 2004). Augmented delivery of calories to the liver enhances *de novo* lipogenesis by stimulating two nuclear transcription factors: carbohydrate response element-binding protein (ChREBP) and sterol response element-binding protein 1 (SREBP-1), which in turn induce acetyl-CoA carboxylase and fatty acid synthase formation in the liver (Backhed et al. 2004, Musso et al. 2011). A 14-d conventionalization of germ free mice produced a 2.3-fold increase in liver TAG content, therewithal increased liver mRNAs of ChREBP, SREBP-1, acetyl-CoA carboxylase, and fatty acid synthase complex, yet no significant changes in total liver free fatty acids or cholesterol (Backhed et al. 2004).

Microbial community-related elements play a role in the regulation of host homeostasis. SCFA (C1-C6) are fermentation products of carbohydrates indigestible by gut microbiota (Delzenne et al. 2011). Increased levels of SCFA are associated with increased lipogenesis and very low density lipoprotein (VLDL) production in the liver (Backhed et al. 2004, Velagapudi et al. 2010). Apart from the contribution to energy harvest, propionate, acetate, and to a lesser extent butyrate and pentanoate are signal molecules in host tissues as ligands for two G protein-coupled free fatty acid receptors, FFAR2 and FFAR3 (alternative symbols GPR43, and GPR41). Both FFARs are broadly expressed in tissues, including the distal small intestine, colon, and adipocytes inhibiting *de novo* lipogenesis by a negative feed-back regulation (Brown et al. 2003, Le Poul et al. 2003, Xiong et al. 2004). A microbiota-dependent FFAR3 regulation of fatty acid synthase complex is suggested. Gnotobiotic (colonized with fermentative bacteria) FFAR3-deficient mice had reduced expression of fatty acid synthase in the liver and decreased fasting serum triglycerides compared to wild type animals. However these differences were not attributable to alterations in hepatic expression of genes involved in long-chain fatty-acid transport, trafficking, or fatty-acid reesterification. Together, these data indicate that gut microbiota may induce hepatic lipogenesis via FFAR3 and fatty acid synthase complex-dependent ways (Samuel et al. 2008).

Concerning nutrients showing a prebiotic effect, changes in either TAG accumulation in the liver or serum lipids of hepatic origin have been observed often linked to a decrease in *de novo* lipogenesis (Delzenne and Cani 2008, Delzenne and Williams 2002). ITF supplementation of high carbohydrate diet fed lean rats and/or hamsters revealed decreased liver and serum triglycerides (Delzenne and Williams 2002). In rats fed a high-fat diet containing fructans, plasma triglycerides decreased without any improvement in hepatic lipogenesis and TAG accumulation, suggesting increased peripheral utilization (Kok et al. 1998). However, in obese rats ITF relieved hepatic lipid accumulation with no effect on plasma triglycerides (Daubioul et al. 2000). Improved glycemia and insulin sensitivity are postulated to take part in control of hepatic lipogenesis, since together they have a potent lipogenetic effect. Secondly, fermentation products of prebiotics participate as signal transducers in hepatic TAG synthesis as aforementioned (Morand et al. 1993, Roberfroid et al. 2010, Sakakibara et al. 2006). A clinical trial on healthy human subjects revealed decreased hepatic *de novo* lipogenesis upon prebiotics supplementation (Diraison et al. 2003).

Increased Lipogenesis in Adipose Tissue

Lipoprotein lipase (LPL) is located on the endothelial surface of capillaries and widely distributed throughout the heart, adipose tissue, spleen, and lung but not adult liver. This enzyme readily acts on chylomicrons and VLDL and plays an important role in the delivery of fatty acids, progressively hydrolyzing their TAG contents to free fatty acids, bulk of which is transported into the tissues (Rıfai and Warnick 2006). Fiaf is an *in vitro* and *in vivo* inhibitor of LPL, produced by the adipose tissue, liver and the intestine that links intestinal microbiata to the adaptation in host energy partitioning (Kersten et al. 2000, Yoon et al. 2000). Colonization of germ-free mice promotes suppression of Fiaf expression in the ileum, leading to increased LPL activity and fat storage in white adipose tissue (Backhed et al. 2004, 2007). Germ-free Fiaf –/– animals having the same amount of total body fat as their conventionalized wild-type littermates establish the importance of Fiaf as a prominent mediator of microbial stimulation of peripheral fat storage (Backhed et al. 2004). By contrast, specific microbiota, namely *Lactobacillus paracasei,* has been shown to increase Fiaf expression and reduce body fat in mice fed a high-fat diet. Induction of Fiaf gene expression by *Lactobacillus* was confirmed in co-culture experiments (Aronsson et al. 2010).

Together these findings suggest that the conventional microbiota stimulate hepatic triglyceride production and promote LPL-directed incorporation of these triglycerides into adipocytes (Greiner and Backhed 2011). However Fleisshner et al. suggested that despite increased intestinal mRNA expression, circulating Fiaf levels showed no major changes in GF mice on both high fat and western diet, and Fiaf may not play a causal role in gut microbiota-mediated effects on fat storage (Fleissner et al. 2010).

Role of FFARs in the adipogenesis promoted by SCFA has been shown (Delzenne et al. 2011). FFAR3 knock-out mice gain less weight when colonized with saccharolytic bacteria as compared to mice expressing the receptor (Samuel et al. 2008). Other data have shown that SFCA may exert a stimulatory effect on lipogenesis in the white adipose tissue through FFAR2 activation (Hong et al. 2005).

Increased Fatty Acid Oxidation

Another mechanism which links microbiota to fatty acid and cholesterol metabolism is the increased muscle and liver activity of AMP-activated protein kinase (AMPK) (Backhed et al. 2007, Vrieze et al. 2010). The AMPK system is widely expressed in tissues, including the liver, adipose tissue, brain, and skeletal muscle. In general, it acts as a sensor of cellular energy status that is activated by increases in the cellular AMP:ATP ratio and responds by switching on catabolic pathways (e.g., glucose uptake, glycolysis, lipolysis) while switching off ATP-consuming anabolic pathways (fatty acid, cholesterol, glycogen, and protein synthesis) (Rifai and Warnick 2006, Mhairi et al. 2007). Elevated NAD : NADH ratio, leptin and adiponectin, also increase AMPK activity (Backhed et al. 2007). Decreased HMG-CoA reductase and acetyl-CoA carboxylase-2 activities, increased GLUT4 translocation and expression, increased carnitine palmitoyl-CoA transferase-1 activity are among the first known events downstream of AMPK activation (Rıfai and Warnick 2006, Mhairi et al. 2007). Backhed and colleagues have demonstrated that the lean phenotype of germ-free mice have increased levels of phosphorylated AMPK in muscle and liver (Backhed et al. 2007) which would stimulate glucose uptake, glycolysis and fatty acid oxidation, while inhibiting triglycerides and cholesterol synthesis, improving insulin resistance. Indeed, the AMPK system is the target for the antidiabetic drug metformin (Zhou et al. 2001).

A second complementary but independent mechanism involves supression of intestinal expression of Fiaf by the intestinal microflora. Fiaf not only regulates LPL but also induces expression of the key enzymes involved in fatty acid oxidation (Backhed et al. 2007, Mandart et al. 2006). Conventionally raised transgenic mice with engineered forced expression of Fiaf in adipose tissue exhibited significant reduction of adiposity by stimulating fatty acid oxidation and uncoupling (Mandart et al. 2006). Additionally, Fiaf deficient germ-free animals fed a Western diet were not protected from diet-induced obesity, and had 24–46% decreased expression of medium-chain acylCoA dehydrogenase, carnitine palmitoyl-CoA transferase in muscle, and a significant reduction in expression of the peroxisome proliferator-activated receptor (PPAR) coactivator 1α which activates genes encoding key enzymes of mitochondrial fatty acid oxidation (Backhed et al. 2007, Vega et al. 2000).

Hormones Regulating Satiety

Specific enteroendocrine cells of the gut communicate with the hypothalamus by neural and endocrine pathways to control energy balance. These enteroendocrine cells scattered along the the entire gastrointestinal tractus are the largest population of hormone-producing cells in human body that sense the biochemical milieu of the gut, dietary nutrients, and key metabolic activities of the microbiota, such as polysaccharide fermentation (Hocker and Wiedenmann 1998). In experimental animal models prebiotics feeding has revealed decreased fat mass associated with a reduction of food intake, which was abrogated when ITF prebiotics are substituted by non-fermentable dietary fiber (Cani et al. 2009b, Daubioul et al. 2002). In this context, colonic fermentation products are suggested to play a role in modulating satiety via gastrointestinal peptides.

Glucagon-like peptide 1 (GLP-1) and gastric inhibitory polypeptide (GIP) are the most important incretin hormones secreted from the intestine (Greiner and Backhed 2011). Incretin hormones promote insulin secretion and β-cell proliferation, control glycogen synthesis in muscle cells and promote satiety (Holst 2004, Knauf et al. 2005). It is suggested that dietary fiber can affect incretin function indirectly through affecting the gut microbiota. Several studies obtained in animals demonstrated that FOS and/or ITF reduce food intake, body weight gain and fat mass development, related with significant increases in the portal levels of anorexigenic peptides GLP-1 and PYY. In addition, a decrease in the serum level of orexigenic ghrelin upon prebiotics feeding has been reported (Cani et al. 2004, 2007b, Delzenne et al. 2005, Reimer and Russell 2008). The decrease in food intake and in fat mass after fructans treatment which is abolished in GLP-1 receptor knock-out mice or in mice treated with a GLP-1 receptor antagonist, could constitute a key point explaining the role of incretine hormones in anti-obesity effect of prebiotics (Cani et al. 2007b). Although at present it has not been shown that increased GLP-1 secretion is promoted by increased *Bifidobacterium* genus, treatment with oligofructose has increased number of *Bifidobacterium*, doubled GLP-1-expressing L cells with the increase in proglucagon content in the proximal colon tissue in rats (Cani et al. 2007c).

Use of probiotics in humans has revealed promising results in healthy human subjects such as ITF promoted satiety and reduced prospective food consumption following meals (Cani et al. 2006a). Gut microbiota fermentation of ITF, added to food as a fat-replacer, reduced energy intake during a test day (Archer et al. 2004) and ITF feeding increased plasma GLP-1 levels which was correlated with the increase in expired H_2 (marker of fermentation) (Piche et al. 2003). In accordance, an association of the prebiotics-induced gut microbiotal fermentation with increased postprandial GLP-1 and PYY has been demonstrated (Cani et al. 2009a). In obese pre-menopausal women intake of 0·14 g FOS/kg per day, over three months increased satiety sensation, decreased waist circumference and body mass index (BMI) (Genta et al. 2009). A clinical trial has shown that ITF feeding decreased food intake and fat mass development in obese subjects with higher plasma PYY and lower ghrelin levels following meals (Pamell and Reimer 2009). Finally, treatment with ITF and/or glucans for two days has not affected satiety, suggesting requirement for intestinal microbiota modulation in this process (Peters et al. 2009).

Leptin, a product of the obese gene represents a signal that stores of fat are adequate, exerts a feedback inhibition of food intake, and stimulates energy expenditure by promoting uptake and oxidation of glucose and fatty acids in skeletal muscle (Muoio et al. 1997, Xiong et al. 2004). Leptin levels increase upon colonization in germ-free animals, proportional to the increase in their body fat (Backhed et al. 2004). SCFA are ligands of the FFAR3 that are asserted to have a pivotal role in microbiota-related leptin production in mouse-cultured adipocytes (Samuel et al. 2008, Xiong et al. 2004). Samuel et al. have shown that co-colonization of mice with fermentative microbial community produced increased adiposity and two-fold reductions in the FFAR3, and FFAR2 mRNAs compared to germ-free animals. Microbiota-related weight gain was blunted in FFAR3 knockout mice, and their serum leptin levels were significantly lower than would be expected based solely on the observed decrease in their adiposity. Additionally, PYY levels were increased after co-colonization of germ-free mice, but this increase was significantly blunted in their FFAR3 knockout littermates. Together, these findings implicate gut microbiata in increased adiposity and increased circulating levels of leptin and PYY in ways comprising FFAR3 repression (Samuel et al. 2008). It is possible that increased levels of anorexigenic hormones associated with intestinal microflora may be a robust compensatory mechanism for increased body fat, rather than preventing obesity. However, supplementation with the probiotic bacteria *Lactobacillus plantarum* for six weeks reduced leptin levels without a significant change in BMI in humans (Naruszewicz et al. 2002).

Intestinal Transit Rate

PYY also regulates gut motility. It produces a dose-related inhibition of transit rate along the gut (Lin et al. 2004). FFAR3-deficient gnotobiotic mice with lower PPY levels exhibitied similar gastric empying rates but had increased intestinal transit rates *versus* wild type littermates. FFAR3 -deficient animals had increased amounts of undigested polysaccharides and SCFA in the distal gut without significant differences in the expression of bacterial genes involved in fermentation. Authors explained this difference in cecal SCFA content with reduced intestinal absorption (Samuel et al. 2008).

Bile Acid Metabolism

Apart from their detergent properties, a novel role of bile acids serving as metabolic integrators in the lipid, glucose, and energy metabolism through the farnesoid X receptor (FXR), and the G protein-coupled receptor TGR5 (Greiner and Backhed 2011, Lefebvre et al. 2009) was conceived, based on the observation that addition of cholic acid to the diet increases energy expenditure in mice, preventing development of obesity and insulin resistance. Activation of FXR by bile acids results in improved glucose tolerance, suppressed hepatic lipogenesis, decreased plasma triglycerides and promoted an anti-inflammatory state. Metabolic effect of bile acids on thermogenesis is dependent on TGR5-regulated induction of type 2 iodothyronine deiodinase that converts tyroxine into more potent 3,5,3′-triiodothyronine in cells, which in turn increases expression of the uncoupling protein in brown adipose tissue (Watanabe et al. 2006). In addition, bile acids promote GLP-1 secretion through the activation of TGR5 (Katsuma et al. 2005).

Profound effect of gut microbiota on bile acid metabolism is among the main molecular mechanisms linking prebiotics and energy metabolism. The gut microflora promotes both the synthesis of bile acids and the production of the secondary bile acids. Consequently, germ-free mice have been shown to exhibit a low bile acid diversity compared with colonized counterparts (Claus et al. 2008, Madsen et al. 1976). In addition, a change in the bile acid profile of germ-free rats affected expression of several genes involved in glucose and lipid metabolism (Swann et al. 2010). Otherwise, supplementation with lactic acid bacteria increased cholesterol absorption, excretion, and formation of secondary bile acids through stimulating LDL receptor, cholesterol 7α-hydroxylase, and bile acid deconjugating/dehydroxylating enzyme expressions respectively (Park et al. 2007). Lithocholic acid which is formed by microbial dehydroxylation of chenodeoxycholic acid has the highest agonist activity for TGR (Maruyama et al. 2002). Treatment with TGR5 agonists improved glucose tolerance in obese mice, and TGR5 knockout mice fed a high-fat diet showed impaired glucose tolerance (Thomas et al. 2009). However concerning for prebiotics, supplementation with GOS or FOS did not support the involvement of prebiotics in the bile salt pool size or kinetics (van Meer et al. 2008).

Meta-inflammation

The metabolic pathways that are involved in the sensing and management of nutrients are integrated with pathogen-sensing mechanisms and immune responses on a functional and molecular basis. Dysfunction of these crucial homeostatic systems underlies many chronic metabolic diseases, including atherosclerosis, type 2 DM and obesity (Erbağcı et al. 2001, 2002, Hotamisligil and Erbay 2008, Vrieze et al. 2010). The instinct to survive would have favoured energy efficiency and storage to prepare for times of food deprivation and to maintain defence systems, as maintenance of immunity is a metabolically sumptuous industry and cannot operate efficiently under conditions of energy deficit (Demas et al. 1997). For example, sepsis increases energy consumption by 30–60% (Maier et al. 1994). By contrast, obesity and metabolic syndrome, both of which are characterised by energy surplus, can also impair immune responses and induce chronic inflammation (Hotamisligil and Erbay 2008). On the cellular basis, macrophages and adipocytes are functionally related; they both secrete cytokines, and can be activated by pathogen-associated components, such as bacterial LPS (Chung et al. 2006). It is suggested that under energy-rich conditions, the ancient inflammatory potential of adipose tissue could be reactivated; adipocytes of obese individuals has been shown to produce higher levels of the pro-inflammatory cytokines, tumour-necrosis factor (TNF) and interleukine 1β and 6 (Hotamisligil and Erbay 2008, Uysal et al. 1997, Wellen and Hotamisligil 2005).

Meta-inflammation and Insulin Resistance

Although inflammation process has been shown to contribute to metabolic dysregulation at several metabolic pathways, modulation of insulin signalling is the most crucial, as it is a dominant metabolic pathway in energy homeostasis. Antigenic components of bacteria such as LPS trigger inflammatory signaling pathways and pro-inflammatory cytokine production that inhibit insulin-receptor signaling and lead to insulin resistant states (Hotamisligil 2006). Other inflammatory pathways are induced by organelle stress due to nutrient overload and processing defects. In both conditions, activation of JUN N-terminal kinase and IκB kinase-β leads to the serine phosphorylation of Insulin Receptor Substrate1 (IRS1) impeding the insulin signaling pathway (Hotamisligil and Erbay 2008, Hotamisligil 2006). TNF-α has been shown to induce insulin resistance in obese rats, and insulin sensitivity can be normalized by neutralization of TNF-α receptors or by deletion of TNF-α in animal models (Hotamisligil and Erbay 2008, Rabot et al. 2010, Uysal et al. 1997). Interestingly, advanced glycation end-products, dyslipidemia, and lipid peroxidation developing as a consequence of insulin resistant states, in turn trigger inflammatory signaling pathways, indicating presence of a bi-directional relationship. The chronic inflammatory response that is triggered by nutrients or other intrinsic cues does not resemble classic inflammation in some ways and has been referred to as meta-inflammation or para-inflammation (Hotamisligil 2006, Medzhitov 2008).

Effect of gut microbiata and high-fat diet on gut inflammation was investigated using germ-free and conventionally raised mice. Only conventionally raised mice fed a high-fat diet exhibited an activation of nuclear factor κB (NF-κB) and increased TNF α mRNA in ileum, supporting contribution of both gut microbes and high fat content of the diet in intestinal inflammation (Ding et al. 2010). Additionally, this inflammatory state was correlated with weight gain, higher adiposity, increased plasma levels of insulin, and glucose (Delzenne et al. 2011, Ding et al. 2010).

Another hypothesis linking gut microbiota to meta-inflammation could be butyrate bioavailability as obese participants are characterised by decreased plasma butyrate levels (Vice et al. 2005). Butyrate is an essential energy source for colon epithelial cells, and possesses anti-inflammatory properties (Saemann et al. 2000). Consumption of non-digestible carbohydrates stimulates growth of a particular butyrate-producing bacteria (*Roseburia/Eubacterium rectale* species and *Faecalibacterium prausnitzii*-cluster of *Firmicutes*) and raises plasma levels of butyrate (Mahowald et al. 2009). Butyrate has anti-diabetic effects improving insulin sensitivity and increasing energy expenditure in animal models of diet-induced obesity (Gao et al. 2009, Vrieze et al. 2010).

Rationale Linking Meta-inflammation and the Gut Microbiata

LPSs, the main component of the gram-negative bacteria wall have antigenic properties and serve as endotoxins. Lipid portions of LPS with lipid A containing only saturated fatty acids cause toxicity, while their polysaccharide portions drive immunogenicity (Todar 2011). Released lipid A through lysis of bacteria, initiates a series of immune responses in the circulation. Intestinal alkaline phosphatase is a part of the defence system by virtue of its LPS-degradating properties (Bates et al. 2009). LPSs have been postulated to be the source of endotoxaemia and inflammation associated with the gut microbiota (Cani and Delzenne 2007, 2011, Nakamura and Omaye 2012). LPS receptor Cd14-knockout mice were resistant to chronic inflammation, excessive weight gain and insulin resistance induced by continuous subcutaneous low-rate infusion of LPS (Cani and Delzenne 2007). Additionally, a high-fat diet decreased *bifidobacterium* genus and increased plasma LPSs, while modulation of gut microbiota, e.g., by antibiotic treatment or dietary intervention with FOS, inhibited inflammation, reduced glucose intolerance, and decreased body weight gain in mice (Cani et al. 2007a,b, Membrez et al. 2008). Higher plasma levels of LPS-binding protein (LBP) found in both obese-prone and high-fat-diet fed mice, compared to the controls fed standard diets support contribution of LPSs (Nakarai et al. 2011). LBP levels were high and positively associated with biomarkers of metabolic syndrome in obese humans, compared to normal-weight controls (Sun et al. 2010).

It is suggested that the inflammatory actions of LPSs are mediated by Toll-like receptor 4 (TLR4), an innate immune receptor localized on the surface of various cells. LPS-induced activation of TLR4 leads to increased production of proinflammatory cytokines and chemokines (Erridge 2011, Nakamura and Omaye 2012). Consequently, patients with metabolic syndrome have higher expression and activity of monocyte TLR4 than the control subjects (Jialal et al. 2012), and a positive correlaton exists between serum LPS activity and biomarkers of metabolic syndrome (e.g., triglyceride levels, insulin resistance) in patients with Type 1 DM (Lassenius et al. 2011). A two-fold increase in plasma levels of LPS in obese, diabetic, or high fat diet fed individuals has been reported (Delzenne et al. 2011).

LPSs are internalized into the enterocytes and transported to the lymphatic circulation by chylomicrons along with dietary fats. Increased lipid content of the diet enhances absorption of LPS, and stimulates TLR4 expression of mononuclear cells in normal humans (Deopurkar et al. 2010). Thus increased chylomicron formation may be an important mechanism linking diet to metabolic low-grade inflammation, eventually leading to development of atherosclerosis, obesity and Type 2 DM (Ghoshal et al. 2009).

Decreased gut barrier integrity and decreased intestinal degradation of LPS due to low alkaline phosphatase activity are also suggested to take part in enhanced transport of LPS. In prebiotic-treated animals, LPS absorption decreases through an improvement of the expression and activity of proteins involved in gut barrier-function, namely, Zonula-occludens 1 and Occludin (Delzenne et al. 2011). Additionally, treatment with prebiotics improves endocannabinoid system responsiveness of the gut, consequently decreasing gut permeability, metabolic endotoxemia of LPS and fat mass development (Muccioli et al. 2010).

It is of particular interest that qualitative changes of the gut microbiota in human obese individuals show a shift in bacterial phyla sustaining predominance of gram-positive, rather than gram-negative bacteria that possess LPL (lower Bacteroidetes and more *Firmicutes*) (Armougom et al. 2009, Ley et al. 2006). Although the methodology used for bacterial analysis could explain certain discrepancies between results published by different study groups, it is still logical to assume that penetrance or transport of LPS may be the prominent determinent of metabolic endotoxaemia rather than quantity of LPS-containing bacteria. However, those effects are not verified in human studies (Cani et al. 2007a, 2006b, 2009a,b, Cani and Delzenne 2007, 2009a,b, Delzenne et al. 2007).

Meta-inflammation and Nutrients

Apart from LPS, dietary macronutrients can act as ligands of TLR4 (Dandona et al. 2010, Wong et al. 2009). Saturated fatty acids in the diet have a structural similarity to lipid A derived from LPS, and could be recognized by pathogen sensing systems, subsequently leading to the same consequence, inflammation. Obesity-associated inflammation is stimulated by overnutrition, particularly by saturated fatty acids through TLR4 activation (Wong et al. 2009). Free saturated fatty acids induced by high glucose concentrations exacerbate expression and activity of the TLR4 and increase superoxide generation, NF-κB activity and pro-inflammatory factors in human monocytes (Dasu and Jialal 2011). In addition, free fatty acids and LPS appear to contribute to the reduced levels of GLUT4 found in type 2 DM through blocking activation of PPARγ (Armoni et al. 2005). PPARγ is a transcription factor which triggers adipocyte differentiation; promotes fatty acid uptake and storage in adipose tissue, and improves insulin resistance (Leonardini et al. 2009).

On the contrary, Ω-3 polyunsaturated fatty acids, eicosapentaenoic acid and decosahexaenoic acid exhibit anti-inflammatory and anti-diabetic properties mainly in animal models (Fedor and Kelley 2009, Kalupahana et al. 2011, Kelley and Adkins 2012). Ω-3 fatty acids control inflammation and adiposity by upregulating adiponection and β-oxidation respectively, through activation of PPARγ (Kalupahana et al. 2011). Conjugated linoleic acid (CLA) is another agonist of redox-sensitive transcription factors PPARγ and NF-κB (Nakamura and Omaye 2009). CLA is a group of geometric and positional isomers of Ω-6 linoleic acid and is found mainly in dairy products. CLA are produced from linoleic acid by *Lactobacillus rhamnosus, Lactobacillus acidophilus, Lactobacillus plantarum, Lactobacillus paracasei,* and *Lactobacillus casei* in mice (Kishino et al. 2002, Ewaschuk et al. 2006, Lawson et al. 2001). CLA has health-promoting properties including anti-oxidant, anti-inflammatory, anti-atherogenic, and anti-obesity effects (Nakamura and Omaye 2009). Anti-obesity effects of CLA in humans are suggested to be associated with decreased appetite, increased energy expenditure, suppression of lipogenesis, and induction of adipocyte apoptosis (Kennedy et al. 2010).

Insulin Resistance and Diabetes Mellitus

Diabetes is a complex, heterogeneous disorder defined as having a fasting blood glucose of 126 mg/dL or greater. Type 1 DM results from selective autoimmune destruction of the pancreatic beta cell, leading to insulin deficiency, while insulin resistance is essential in the pathogenesis of type 2 DM along with impaired beta-cell function. In 2012, the commonly encountered spectrum of diabetes showed that only about 10% of cases have severe insulin deficiency, while majority of patients with diabetes are overweight and have a combination of insulin resistance and impaired insulin secretion. Metabolic syndrome is a clinical phenotype, which includes insulin resistance, greater waist circumference, hypertension, hypertriglyceridemia and low HDL cholesterol levels. It is suggested that the increased body weight in the general population is the most important factor associated with increased prevalence of type 2 DM and metabolic syndrome (Polonsky 2012, Rıfai and Warnick 2006, Scott et al. 2011).

Effect of intestinal microflora on insulin sensitivity has been shown on germ-free mice fed a high-fat diet. Germ-free animals were resistant to diet-induced insulin resistance with improved glucose tolerance, reduced fasting/non-fasting insulin levels, and increased phospho-Akt (also known as Protein Kinase B) in adipose tissue (Rabot et al. 2010). Backhed et al. found statistically significant elevations in fasting glucose and insulin levels, an insulin-resistant state, as defined by glucose and insulin tolerance tests, and increased fat content after conventionalization (Backhed et al. 2004). In addition, several studies have demonstrated that antibiotic administration improves oral glucose tolerance in ob/ob and high-fat diet-induced insulin resistant mice (Cani et al. 2007c, Membrez et al. 2008). In two different diabetes models in rat, supplementation with *Lactobacillius acidophilus* and *Lactobacillius casei* decreased plasma glucose, glycosylated haemoglobin, and insulin levels (Yadav et al. 2007), and decreased incremental peaks and delayed reduction of insulin secretion during oral glucose tolerance test respectively (Yadav et al. 2008). These data suggest that not only is intestinal microflora associated with insulin resistance but modulation or reduction of the gut microbiota can be a candidate strategy in managing insulin resistance as well. Similarly, the effect of prebiotics on glucose homeostasis is coherent in animal studies. It is proposed that prebiotics feeding improves glucose tolerance and increases plasma insulin levels through partial restoration of pancreatic β-cell mass. GLP-1 appears to be the mediator of this effect, since prebiotic treatment promotes GLP-1 production and the beneficial effect is abolished in rats that are characterised by defective production of gut peptides (Cani et al. 2005, 2007b, Perrin et al. 2003). Nevertheless, effects of other metabolic mechanisms, such as a decrease in inflammatory tone, could also contribute to the improvement of glucose homeostasis upon treatment with prebiotics (Roberfroid et al. 2010).

Human studies investigating the role of prebiotics or probiotics are limited and contradictory. Supplementation with *Lactobacillus fermentum* capsules for ten weeks did not caused a significant change over time or between treatments in fasting blood glucose levels in volunteers with elevated cholesterol (Simons et al. 2006). However, the incidence of gestational diabetes has been reduced by *Lactobacillus rhamnosus* and

Bifidobacterium lactis administiration during the first trimester of pregnancy (Luoto et al. 2010). Prebiotic supplementation of healthy subjects with short-chain fructans for four weeks decreased basal hepatic glucose production, but had no detectable effect on insulin-stimulated glucose metabolism while treatment of patients with type 2 DM with the same prebiotics affected neither (Luo et al. 1996, 2000). An interventional study showed that ITF increased GLP-1 production and reduced the postprandial glucose response (Cani et al. 2009a).

Although the role of microbiota in regulating host metabolism is more evident for Type 2 DM and insulin resistant states, some reports suggest a link for type 1 DM (Greiner and Backhed 2011). Analysis of the gut microbiota in patients with Type 1 DM showed less diversity than controls. Furthermore, Type 1 DM progression period was associated with alterations in the gut microbiota; at the species level, both *Bacteroides ovatus* and *Bacteroides fragilis* were non-progressors, indicating a potential protective role (Giongo et al. 2011). A direct involvement of an altered gut microbiota was demonstrated in a Type 1 DM animal model. Specific pathogen-free mice lacking MyD88 protein, an adaptor for multiple innate immune receptors, were protected from the disease. MyD88 deficiency changed composition of the distal gut microbiota, and attenuated Type 1 DM in recipients. Thus, intestinal flora and its interaction with innate immune system may be a critical epigenetic factor modifying predisposition to Type 1 DM (Wen et al. 2008).

References

Allison, D.B., K.R. Fontaine, J.E. Manson, J. Stevens and T.B. VanItallie. 1999. Annual deaths attributable to obesity in the United States. JAMA 282: 1530–1538.

Archer, B.J., S.K. Johnson, H.M. Devereux and A.L. Baxter. 2004. Effect of fat replacement by inulin or lupin-kernel fibre on sausage patty acceptability, post-meal perceptions of satiety and food intake in men. Br. J. Nutr. 91: 591–599.

Armoni, M., C. Harel, F. Bar-Yoseph, S. Milo and E. Karnieli. 2005. Free fatty acids repress the GLUT4 gene expression in cardiac muscle via novel response elements. J. Biol. Chem. 80: 34786–34795.

Armougom, F., M. Henry, B. Vialettes, D. Raccah and D. Raoult. 2009. Monitoring bacterial community of human gut microbiota reveals an increase in Lactobacillus in obese patients and Methanogens in anorexic patients. PLoS. One 4: 7125.

Aronsson, L., Y. Huang, P. Parini, M. Korach-Andre, J. Hakansson, J.A. Gustafsson, S. Pettersson, V. Arulampalam and J. Rafter. 2010. Decreased Fat Storage by Lactobacillus Paracasei Is Associated with Increased Levels of Angiopoietin-Like 4 Protein (ANGPTL4). PLoS ONE 5: 1–7.

Backhed, F., H. Ding, T. Wang, L.V. Hooper, G.Y. Koh, A. Nagy, C.F. Semenkovich and J.I. Gordon. 2004. The gut microbiota as an environmental factor that regulates fat storage. PNAS 104: 15718–15723.

Backhed, F., J.K. Manchester, C.F. Semenkovich and J.I. Gordon. 2007. Mechanisms underlying the resistance to dietinduced obesity in germ-free mice. Proc. Natl. Acad. Sci. USA 104: 979–984.

Bates, J.M., J. Akerlund, E. Mittge and K. Guillemin. 2009. Intestinal alkaline phosphatase detoxifies lipopolysaccharide and prevents inflammation in response to the gut microbiota. Cell. Host. Microbe. 2: 371–382.

Bjursell, M., T. Admyre, M. Göransson, A.E. Marley, D.M Smith, J. Oscarsson and Y.M. Bohlooly. 2011. Improved glucose control and reduced body fat mass in free fatty acid receptor 2 (Ffar2) deficient mice fed a high fat diet. Am. J. Physiol. Endocrinol. Metab. 300: 211–220.

Bleich, S., D. Cutler, C. Murray and A. Adams. 2008. Why is the developed world obese? Annu. Rev. Public Health. 29: 273–295.

Brown, A.J., S.M. Goldsworthy, A.A. Barnes, M.M. Eilert, L. Tcheang, D. Daniels, A.I. Muir, M.J. Wigglesworth, I. Kinghorn, N.J. Fraser, N.B. Pike, J.C. Strum, K.M. Steplewski, P.R. Murdock, J.C. Holder, F.H. Marshall, P.G. Szekeres, S. Wilson, D.M Ignar, S.M. Foord, A. Wise and S.J. Dowell. 2003. The Orphan G protein-coupled receptors GPR41 and GPR43 are activated by propionate and other short chain carboxylic acids. J. Biol. Chem. 278: 11312–11319.

Cani, P.D., C. Dewever and N.M. Delzenne. 2004. Inulin-type fructans modulate gastrointestinal peptides involved in appetite regulation (glucagon-like peptide-1 and ghrelin) in rats. Br. J. Nutr. 92: 521–526.

Cani, P.D., C.A. Daubioul, B. Reusens, C. Remacle, G. Catillon and N.M. Delzenne. 2005. Involvement of endogenous glucagon-like peptide-1(7-36) amide on glycaemia- lowering effect of oligofructose in streptozotocin-treated rats. J. Endocrinol. 185: 457–465.

Cani, P.D., E. Joly, Y. Horsmans and N.M. Delzenne. 2006a. Oligofructose promotes satiety in healthy human: a pilot study. Eur. J. Clin. Nutr. 60: 567–572.

Cani, P.D., C. Knauf, M.A. Iglesias, D.J. Drucker, N.M. Delzenne and R. Burcelin. 2006b. Improvement of glucose tolerance and hepatic insulin sensitivity by oligofructoserequires a functional glucagon-like Peptide 1 receptor. Diabetes 55: 1484–1490.

Cani, P.D., A.M. Neyrinck, F. Fava, C. Knauf, R.G. Burcelin, K.M. Tuohy, G.R. Gibson and N.M. Delzenne. 2007a. Selective increases of bifidobacteria in gut microflora improve high-fat-diet-induced diabetes in mice through a mechanism associated withendotoxaemia. Diabetologia 50: 2374–2383.

Cani, P.D., J. Amar, M.A. Iglesias, M. Poggi, C. Knauf, D. Bastelica, A.M. Neyrinck, F. Fava, K.M. Tuohy, C. Chabo, A. Waget, E. Delmee, B. Cousin, T. Sulpice, B. Chamontin, J. Ferrieres, J.F. Tanti, G.R. Gibson, L. Casteilla, N.M. Delzenne, M.C. Alessi and R. Burcelin. 2007b. Metabolic endotoxemia initiates obesity and insulin resistance. Diabetes 56: 1761–1772.

Cani, P.D., S. Hostel, Y. Guiot and N.M. Delzenne. 2007c. Dietary non-digestible carbohydrates promote L-cell differentiation in the proximal colon of rats. British Journal of Nutrition 98: 32–37.

Cani, P.D., R. Bibiloni, C. Knauf, A. Waget, A.M. Neyrinck, N.M. Delzenne and R. Burcelin. 2008. Changes in gut microbiota control metabolic endotoxemiainduced inflammation in high-fat diet-induced obesity and diabetes in mice. Diabetes 57: 1470–1481.

Cani, P.D., E. Lecourt, E.M. Dewulf, F.M. Sohet, B.D. Pachikian, D. Naslain, B.F. De, A.M. Neyrinck and N.M. Delzenne. 2009a. Gut microbiota fermentation of prebiotics increases satietogenic and incretin gut peptide production with consequences for appetite sensation and glucose response after a meal. Am. J. Clin. Nutr. 90: 1236–43.

Cani, P.D., S. Possemiers, W.T. Van de, Y. Guiot, A. Everard, O. Rottier, L. Geurts, D. Naslain, A. Neyrinck, D.M. Lambert, G.G. Muccioli and N.M. Delzenne. 2009b. Changes in gut microbiota control inflammation in obese mice through a mechanism involving GLP-2-driven improvement of gut permeability. Gut. 58: 1091–1103.

Cani, P.D. and N.M. Delzenne. 2007. Gut microflora as a target for energy and metabolichomeostasis. Curr. Opin. Clin. Nutr. Metab. Care. 10: 729–734.

Cani, P.D. and N.M. Delzenne. 2009a. Interplay between obesity and associated metabolicdisorders: new insights into the gut microbiota. Curr. Opin. Pharmacol. 9: 737–743.

Cani, P.D. and N.M. Delzenne. 2009b. The role of the gut microbiota in energy metabolism and metabolic disease. Curr. Pharm. Des. 15: 1546–1558.
Cani, P.D. and N.M. Delzenne. 2011. The gut microbiome as therapeutic target. Pharmacol. Ther. 130: 202–212.
Chung, S., K. Lapoint, K. Martinez, A. Kennedy, M. Boysen Sandberg and M.K. McIntosh. 2006. Preadipocytes mediate lipopolysaccharide-induced inflammation and insulin resistance in primary cultures of newly differentiated human adipocytes. Endocrinology. 147: 5340–5351.
Claus, S.P., T.M. Tsang, Y. Wang, O. Cloarec, E. Skordi, F.P. Martin, S. Rezzi, A. Ross, S. Kochhar, E. Holmes and J.K. Nicholson. 2008. Systemic multicompartmental effects of the gut microbiome on mouse metabolic phenotypes. Mol. Syst. Biol. 4: 219.
Collado, M.C., E. Isolauri, K. Laitinen and S. Salminen. 2008. Distinct composition of gut microbiota during pregnancy in overweight and normal-weight women. Am. J. Clin. Nutr. 88: 894–899.
Dandona, P., H. Ghanim, A. Chaudhuri, S. Dhindsa and S.S. Kim. 2010. Macronutrient intakeinduces oxidative and inflammatory stress: potential relevance to atherosclerosis andinsulin resistence. Exp. Mol. Med. 42: 245–253.
Dasu, M.R. and I. Jialal. 2011. Free fatty acids in the presence of hugh glucose amplifymonocyte inflammation via Toll-like receptors. Am. J. Physiol.- Endoc. 300: 145–154.
Daubioul, C.A., H.S. Taper, L.D. De Wispelaere and N.M. Delzenne. 2000. Dietary oligofructose lessens hepatic steatosis, but does not prevent hypertriglyceridemia inobese zucker rats. J. Nutr. 130: 1314–1319.
Daubioul, C., N. Rousseau, R. Demeure, B. Gallez, H. Taper, B. Declerck and N. Delzenne. 2002. Dietary fructans, but not cellulose, decrease triglyceride accumulation in the liver of obese Zucker fa/fa rats. J. Nutr. 132: 967–973.
Delzenne, N.M., P.D. Cani, C. Daubioul and A.M. Neyrinck. 2005. Impact of inulin andoligofructose on gastrointestinal peptides. Br. J. Nutr. 93: S157–S161.
Delzenne, N.M., P.D. Cani and A.M. Neyrinck. 2007. Modulation of glucagon-like peptide 1and energy metabolism by inulin and oligofructose: experimental data. J. Nutr. 137: 2547–2551.
Delzenne, N.M., A.M. Neyrinck and P.D. Cani. 2011. Modulation of the gut microbiota by nutrients with prebiotic properties: consequences for host health in the context ofobesity and metabolic syndrome. Microbial Cell Factories. 10(Suppl 1):S10doi:10.1186/1475-2859-10-S1-S10.
Delzenne, N.M. and C.M. Williams. 2002. Prebiotics and lipid metabolism. Curr. Opin. Lipidol. 13: 61–67.
Delzenne, N.M. and P.D. Cani. 2008. Gut microflora is a key player in host energy homeostasis. Med. Sci. 24: 505–510.
Demas, G.E., V. Chefer, M.I. Talan and R.J. Nelson. 1997. Metabolic costs of mounting anantigen-stimulated immune response in adult and aged C57BL/56J mice. Am. J. Physiol. 273: 1631–1637.
Deopurkar, R., H. Ghanim, J. Friedman, S. Abuaysheh, C.L. Sia, P. Mohanty, P. Viswanathan, A. Chaudhuri and P. Dandona. 2010. Differential effects of cream,glucose, and orange juice on inflammation, endotoxin, and the expression of Toll-like receptor-4 and suppressor of cytokine signaling-3. Diabetes Care 33: 991–997.
Diamant, M., E.E. Blaak and W.N. de Vos. 2010. Do nutrient-gut-microbiota interactions playa role in human obesity, insulin resistance and type 2 diabetes? Obes. Rev. 12(4): 272–81.
DiBaise, J.K., H. Zhang, M.D. Crowell, R. Krajmalnik-Brown, G.A. Decker and B.E. Rittmann. 2008. Gut microbiota and its possible relationship with obesity. Mayo. Clin. Proc. 83: 460–469.

Ding, S., M.M. Chi, B.P. Scull, R. Rigby, N.M. Schwerbrock, S. Magness, C. Jobin and P.K. Lund. 2010. High-fat diet: bacteria interactions promote intestinal inflammation which precedes and correlates with obesity and insulin resistance in mouse. PLoS. One 5: 12191. doi: 10.1371/journal.pone.0012191.

Diraison, F., P. Moulin and M. Beylot. 2003. Contribution of hepatic de novo lipogenesis andreesterification of plasma non esterified fatty acids to plasma triglyceride synthesis during non-alcoholic fatty liver disease. Diabet. Metab. 29: 478–485.

Dumas, M.E., R.H. Barton, A. Toye, O. Cloarec, C. Blancher, A. Rothwell, J. Fearnside, R. Tatoud, V. Blanc, J.C. Lindon, S.C. Mitchell, E. Holmes, M.I. McCarthy, J. Scott, D. Gauguier and J.K. Nicholson. 2006. Metabolic profiling reveals a contribution of gut microbiota to fatty liver phenotype in insulin-resistant mice. Proc. Natl. Acad. Sci. 103: 12511–12516.

Duncan, S.H., G.E. Lobley, G. Holtrop, J. Ince, A.M. Johnstone, P. Louis and H.J. Flint. 2008. Human colonic microbiota associated with diet, obesity and weight loss. Int. J. Obes. 32(11): 1720–1724.

Erbağcı, A.B., M. Tarakçıoğlu, M. Aksoy, R. Kocabaş, M. Nacak, Ş. Aynacıoğlu and C. Sivrikoz. 2002. Diagnostic value of CRP and Lp(a) in coronary heart disease. Acta. Cardiol. 57: 197–204.

Erbağcı, A.B., M. Tarakçıoğlu, Y. Coşkun, E. Sivaslı and E.S. Namıduru. 2001 Mediators of inflammation in children with type1 Diabetes Mellitus. Clin. Biochem. 38: 645–650.

Erridge, C. 2011. The capacity of food stuffs to induce innate immune activation of human monocytes in vitro is dependent on food content of stimulants of Toll-like receptors 2 and 4. Br. J. Nutr. 105: 15–23.

Ewaschuk, J.B., J.W. Walker, H. Diaz and K.L. Madsen. 2006. Bioproduction of conjugated linoleic acid by probiotic bacteria occurs *in vitro* and *in vivo* in mice. J. Nutr. 136: 1483–1487.

Fedor, D. and D.S. Kelley. 2009. Prevention of insulin resistance by n-3 polyunsaturarted fatty acids. Curr. Opin. Clin. Nutr. 12: 138–146.

Fleissner, C.K., N. Huebel, M.M. Abd El-Bary, G. Loh, S. Klaus and M. Blaut. 2010. Absence of intestinal microbiota does not protect mice from diet-induced obesity. Br. J. Nutr. 104: 919–929.

Furet, J.P., L.C. Kong, J. Tap, C. Poitou, A. Basdevant, J.L. Bouillot, D. Mariat, G. Corthier, J. Dore, C. Henegar, S. Rizkalla and K. Clement. 2010. Differential adaptation of human gut microbiota to bariatric surgery-induced weight loss: links with metabolic and low-grade inflammation markers. Diabetes 59: 3049–3057.

Gao, Z., J. Yin, J. Zhang, R.E. Ward, R.J. Martin, M. Lefevre, W.T. Cefalu and J. Ye. 2009. Butyrate improves insulin sensitivity and increases energy expenditure in mice.Diabetes 58: 1509–1517.

Genta, S., W. Cabrera, N. Habib, J. Pons, I.M. Carillo, A. Grau and S. Sanchez. 2009. Yaconsyrup: beneficial effects on obesity and insulin resistance in humans. Clin. Nutr. 28: 182–187.

Ghoshal, S., J. Witta, J. Zhong, W. de Williers and E. Eckhardt. 2009. Chylomicrons promoteintestinal absorption of lipopolysaccharides. J. Lipid Res. 50: 90–97.

Giongo, A., K.A. Gano, D.B. Crabb, N. Mukherjee, L.L. Novelo, G. Casella, J.C. Drew, J. Ilonen, M. Knip, H. Hyöty, R. Veijola, T. Simell, O. Simell, J. Neu, C.H. Wasserfall, D. Schatz, M.A. Atkinson and E.W. Triplett. 2011. Toward defining the autoimmune microbiome for type 1 diabetes. ISME J. 5: 82–91.

Greiner T. and F. Backhed. 2011. Effects of the gut microbiota on obesity and glucosehomeostasis. Trends in Endocrinology and Metabolism 22: 117–123.

Hocker, M. and B. Wiedenmann. 1998. Molecular mechanisms of enteroendocrine differentiation. Ann. N. Y. Acad. Sci. 859: 160–174.

Holst, J.J. 2004. On the physiology of GIP and GLP-1. Horm. Metab. Res. 36: 747–754.

Hong, Y.H., Y. Nishimura, D. Hishikawa, H. Tsuzuki, H. Miyahara, C. Gotoh, K.C. Choi, D.D. Feng, C. Chen, H.G. Lee, K. Katoh, S.G. Roh and S. Sasaki. 2005. Acetate and propionate short chain fatty acids stimulate adipogenesis via GPCR43. Endocrinology 146: 5092–5099.

Hotamisligil, G.S. 2006. Inflammation and metabolic disorders. Nature 444: 860–867.

Hotamisligil, G.S. and E. Erbay. 2008. Nutrient sensing and inflammation in metabolic diseases. Nat. Rev. Immunol. 8(12): 923. doi:10.1038/nri2449.

Hooper, L.V., M.H. Wong, A. Thelin, L. Hansson, P.G. Falk and J.I. Gordon. 2001. Molecular analysis of commensal host-microbial relationships in the intestine. Science 291: 881–884.

Hsiao, W.W. and C.M. Fraser-Liggett. 2009. Human Microbiome Project–paving the way to a better understanding of ourselves and our microbes. Drug. Discov. 14: 331–333.

Jialal, I., B.A. Huet, H. Kaur, A. Chien and S. Devaraj. 2012. Increased Toll-like receptoractivity in patients with metabolic syndrome. Diabetes Care 35: 900–904.

Kadooka, Y., M. Sato, K. Imaizumi, A. Ogawa, K. Ikuyama, Y. Akai, M. Okano, M. Kagoshima and T. Tsuchida. 2010. Regulation of abdominal adiposity by probiotics (*Lactobacillus gasseri* SBT2055) in adults with obese tendencies in a randomized controlled trial. Eur. J. Clin. Nutr. 64: 636–643.

Kalliomaki, M., M.C. Collado, S. Salminen and E. Isolauri. 2008. Early differences in fecalmicrobiota composition in children may predict overweight. Am. J. Clin. Nutr. 87: 534–538.

Kalupahana, N.S., K.J. Claycombe and N. Moustaid-Moussa. 2011. (n-3) Fatty acids alleviate adipose tissue inflammation and insulin resistance: mechanistic insights. Adv. Nutr. 2: 304–316.

Katsuma, S., A. Hirasawa and G. Tsujimoto. 2005. Bile acids promote glucagon-like peptide-1 secretion through TGR5 in a murine enteroendocrine cell line STC-1. Biochem.Biophys. Res. Commun. 329: 386–390.

Kelley, D.S. and Y. Adkins. 2012. Similarities and differences between the effects of EPA and DHA on markers of atherosclerosis in human subjects. P. Nutr. Soc. 28: 1–10.

Kennedy, A., K. Martinez, S. Schmidt, S. Mandrup, K. Lapoint and M.K. McIntosh. 2010. Antiobesity mechanisms of action of conjugated linoleic acid. J. Nutr. Biochem. 21: 171–179.

Kersten, S., S. Mandard, N.S. Tan, P. Escher, D. Metzger, P. Chambon, F.J. Gonzalez, B. Desvergne and W. Wahli. 2000. Characterization of the fasting-induced adipose factor FIAF, a novel peroxisome proliferator-activated receptor target gene. J. Biol. Chem. 275: 28488–28493.

Kishino, S., J. Ogawa, Y. Omura, K. Matsumura and S. Shimuzu. 2002. Conjugated linoleic acid production from linoleic acid by lactic acid bacteria. JAOCS 79: 159–163.

Knauf, C., P.D. Cani, C. Perrin, M.A. Iglesias, J.F. Maury, E. Bernard, F. Benhamed, T. Gremeaux, D.J. Drucker, C.R. Kahn, J. Girard, J.F. Tanti, N.M. Delzenne, C. Postic and R. Burcelin. 2005. Brain glucagon-like peptide-1 increases insulin secretion and muscle insulin resistance to favor hepatic glycogen storage. J. Clin. Invest. 115: 3554–3563.

Kok, N.N., H.S. Taper and N.M. Delzenne. 1998. Oligofructose modulates lipid metabolismalterations induced by a fat-rich diet in rats. J. Appl. Toxicol. 18: 47–53.

Lassenius, M.I., K.H. Pietilainen, K. Kaartinen, P.J. Pussinen, J. Syrjanen, C. Forsblom, I. Porsti, A. Rissanen, J. Kaprio, J. Mustonen, P.H. Groop and M. Lehto. 2011. Bacterialendotoxin activity in human serum is associated with dyslipidemia, insulin resistance, obesity, and chronic inflammation. Diabetes Care 34: 1809–1815.

Lawson, R.E., A.R. Moss and D.I. Givens. 2001. The role of dairy products in supplyingconjugated linoleic acid to man's diet. Nutr. Res. Rev. 14: 153–172.

Lefebvre, P., B. Cariou, F. Lien, F. Kuipers and B. Staels. 2009. Role of Bile Acids and Bile Acid Receptors in Metabolic Regulation. Physiol. Rev. 89: 147–191.

Leonardini, A., L. Laviola, S. Perrini, A. Natalicchio and F. Giorgino. 2009. Cross-Talk between PPARγ and Insulin Signaling and Modulation of Insulin Sensitivity. PPAR Research. doi:10.1155/2009/818945.

Le Poul, E., C. Loison, S. Struyf, J.Y. Springael, V. Lannoy, M.E. Decobecq, S. Brezillon, V. Dupriez, G. Vassart, J. Van Damme, M. Parmentierand M. Detheux. 2003. Functionalcharacterization of human receptors for short chain fatty acids and their role inpolymorphonuclear cell activation. J. Biol. Chem. 278: 25481–25489.

Ley, R.E., F. Backhed, P. Turnbaugh, C.A. Lozupone, R.D. Knight and J.I. Gordon. 2005. Obesity alters gut microbial ecology. Proc. Natl. Acad. Sci. 102: 11070–11075.

Ley, R.E., P.J. Turnbaugh, S. Klein and J.I. Gordon. 2006. Microbial ecology: human gutmicrobes associated with obesity. Nature 444: 1022–1023.

Ley, R.E. 2010. Obesity and the human microbiome. Curr. Opin. Gastroenterol. 26: 5–11.

Lin, H.C., C. Neevel and J.H. Chen. 2004. Slowing intestinal transit by PYY depends onserotonergic and opioid pathways. Am. J. Physiol. 286: 558–563.

Lundell, A.C., I. Adlerberth, E. Lindberg, H. Karlsson, S. Ekberg, N. Aberg, R. Saalman, B. Hock, A. Steinkasserer, B. Hesselmar, A.E. Wold and A. Rudin. 2007. Increased levels of circulating soluble CD14 but not CD83 in infants are associated with earlyintestinal colonization with Staphylococcus aureus. Clin. Exp. Allergy. 37: 62–71.

Luo, J., S.W. Rizkalla, C. Alamowitch, A. Boussairi, A. Blayo, J.L. Barry, A. Laffitte, F. Guyon, F.R. Bornet and G. Slama. 1996. Chronic consumption of short-chain fructooligosaccharides by healthy subjects decreased basal hepatic glucose production but had no effect on insulin-stimulated glucose metabolism. Am. J. Clin. Nutr. 63: 939–945.

Luo, J., M. Van Yperselle, S.W. Rizkalla, F. Rossi, F.R. Bornet and G. Slama. 2000. Chronic consumption of short-chain fructooligosaccharides does not affect basal hepaticglucose production or insulin resistance in type 2 diabetics. J. Nutr. 130(6): 1572–1577.

Luoto, R., K. Laitinen, M. Nermes and E. Isolauri. 2010. Impact of maternal probiotic-supplemented dietary counselling on pregnancy outcome and prenatal and postnatal growth: a double-blind, placebo-controlled study. Br. J. Nutr. 103: 1792–1799.

Madsen, D., M. Beaver, L. Chang, E. Bruckner-Kardoss and B. Wostmann. 1976. Analysis ofbile acids in conventional and germfree rats. J. Lipid. Res. 17: 107–111.

Mahowald, M.A., F.E. Rey, H. Seedorf, P.J. Turnbaugh, R.S. Fulton, A. Wollam, N. Shah, C. Wang, V. Magrini, R.K. Wilson, B.L. Cantarel, P.M. Coutinho, B. Henrissat, L.W. Crock, A. Russell, N.C. Verberkmoes, R.L. Hettich and J.I. Gordon. 2009. Characterizing a model human gut microbiota composed of members of its two dominant bacterial phyla. Proc. Natl. Acad. Sci. USA 7: 5859–5864.

Maier, S.F., L.R. Watkins and M. Fleshner. 1994. Psychoneuroimmunology. The interface between behavior, brain, and immunity. Am. Psychol. 49: 1004–1017.

Mandard, S., F. Zandbergen, E. van Straten, W. Wahli, F. Kuipers, M. Muller and S. Kersten. 2006. The Fasting-induced Adipose Factor/Angiopoietin-like Protein 4 Is PhysicallyAssociated with Lipoproteins and Governs Plasma Lipid Levels and Adiposity. The Journal of Biological Chemistry 281: 934–944.

Maruyama, T., Y. Miyamoto, T. Nakamura, Y. Tamai, H. Okada, E. Sugiyama, T. Nakamura, H. Itadani and K. Tanaka. 2002. Identification of membrane-type receptor for bile acids (M-BAR). Biochem. Biophys. Res. Commun. 298: 714–719.

Medzhitov, R. 2008. Origin and physiological roles of inflammation. Nature 454: 428–435.

Membrez, M., F. Blancher, M. Jaquet, R. Bibiloni, P.D. Cani, R.G. Burcelin, I. Corthesy, K. Mace and C.J. Chou. 2008. Gut microbiota modulation with norfloxacin and ampicillin enhances glucose tolerance in mice. FASEB J. 22: 2416–2426.

Mhairi, C., D. Towler and D. Grahame Hardie. 2007. AMP-Activated Protein Kinase inMetabolic Control and Insulin Signaling. Circ. Res. 100: 328–341.

Morand, C., C. Remesy and C. Demigne. 1993. Fatty acids are potent modulators of lactateutilization in isolated hepatocytes from fed rats. Am. J. Physiol. 264: 816–823.

Muccioli, G.G., D. Naslain, F. Backhed, C.S. Reigstad, D.M. Lambert, N.M. Delzenne and P.D. Cani. 2010. The endocannabinoid system links gut microbiota to adipogenesis. Mol. Syst. Biol. 6: 392.

Muoio, D.M., G.L. Dohm, F.T. Fiedorek, E.B. Tapscott and R.A. Coleman. 1997. Leptindirectly alters lipid partitioning in skeletal muscle. Diabetes 46: 1360–1363.

Murphy, E.F., P.D. Cotter, S. Healy, T.M. Marques, O. O'Sullivan, F. Fouhy, S.F. Clarke, P.W. O'Toole, E.M. Quigley, C. Stanton, P.R. Ross, R.M. O'Doherty and F. Shanahan. 2010. Composition and energy harvesting capacity of the gut microbiota: relationship to diet, obesity and time in mouse models. Gut. 59: 1635–1642.

Musso, G., R. Gambino and M. Cassander. 2011. Interactions between gut microbiota and host metabolism predisposing to obesity and diabetes. Annu. Rev. Med. 62: 361–380.

Nakamura, Y.K. and S.T. Omaye. 2009. Conjugated linoleic acid isomers' roles in theregulation of PPARγ and NF-κBDNA binding and subsequent expression ofantioxidant enzymes in human umbilical vein endothelial cells. Nutrition 25: 800–811.

Nakamura, Y.K. and S.T. Omaye. 2012. Metabolic diseases and pro- and prebiotics: Mechanistic insights. Nutrition & Metabolism 9: 60.

Nakarai, H., A. Yamashita, S. Nagayasu, M. Iwashita, S. Kumamoto, H. Ohyama, M. Hata, Y. Soga, A. Kushiyama, T. Asano, Y. Abiko and F. Nishimura. 2011. Adopocytemacrophage interaction may mediate LPS-induced low-grade inflammation: potential link with metabolic complications. Innate. Immun. 18: 164– 170.

Naruszewicz, M., M.L. Johansson, D.Z. Downar and H. Bukowska. 2002. Effect of *Lactobacillus plantarum* 299v on cardiovascular disease risk factors in smokers. Am.J. Clin. Nutr. 76: 1249–1255.

Pamell, J.A. and R.A. Reimer. 2009. Weight loss during oligofructose supplementation isassociated with decreased ghrelin and increased peptide YY in overweight and obeseadults. Am. J. Clin. Nutr. 89: 1751–1759.

Park, Y.H., J.G. Kim, Y.W. Shin, S.H. Kim and K.Y. Whang. 2007. Effect of dietary inclusion of *Lactobacillus acidophilus* ATCC 43121 on cholesterol metabolism in rats. J. Microbiol. Biotechnol. 17: 655–662.

Perrin, I.V., M. Marchesini, F.C. Rochat, E.J. Schiffrin and B. Schilter. 2003. Oligofructose does not affect the development of Type 1 diabetes mellitus induced by dietary proteins in the diabetes-prone BB rat model. Diab. Nutr. Metabol. 16: 94–101.

Peters, H.P., H.M. Boers, E. Haddeman, S.M. Melnikov and F. Qvyjt. 2009. No effect of added beta-glucan or of fructooligosaccharide on appetite or energy intake. Am. J. Clin. Nutr. 89: 58–63.

Piche, T., S.B. des Varannes, S. Sacher-Huvelin, J.J. Holst, J.C. Cuber and J.P. Galmiche. 2003. Colonic fermentation influences lower esophageal sphincter function ingastroesophageal reflux disease. Gastroenterology 124: 894–902.

Polonsky, K.S. 2012. The Past 200 Years in Diabetes. N. Engl. J. Med. 367: 1332–1340.

Qin, J., R. Li, J. Raes, M. Arumugam, K.S. Burgdorf, C. Manichanh, T. Nielsen, N. Pons, F. Levenez, T. Yamada, D.R. Mende, J. Li, J. Xu, S. Li, D. Li, J. Cao, B. Wang, H.Liang, H. Zheng, Y. Xie, J. Tap, P. Lepage, M. Bertalan, J.M. Batto, T. Hansen, P.D. Le, A. Linneberg, H.B. Nielsen, E. Pelletier, P. Renault, T. Sicheritz-Ponten, K. Turner, H. Zhu, C. Yu, S. Li, M. Jian, Y. Zhou, Y. Li, X. Zhang, S. Li, N. Qin, H. Yang, J. Wang, S. Brunak, J. Dore, F. Guarner, K. Kristiansen, O. Pedersen, J. Parkhill, J. Weissenbach, P. Bork, S.D. Ehrlich and J. Wang. 2010. A human gut microbial gene catalogue established by metagenomic sequencing. Nature 464: 59–65.

Rabot, S., M. Membrez, A. Bruneau, P. Gerard, T. Harach, M. Moser, F. Raymond, R. Mansourian and C.J. Chou. 2010. Germ-free C57BL/6J mice are resistant to high-fat-diet-induced insulin resistance and have altered cholesterol metabolism. The FASEB Journal 24: 4948–4959.

Reimer, R.A. and J.C. Russell. 2008. Glucose tolerance, lipids, and GLP-1 secretion inJCR:LA-cp rats fed a high protein fiber diet. Obesity 16: 40–46.

Respondek, F., K.S. Swanson, K.R. Belsito, B.M. Vester, A. Wagner, L. Istasse and M. Diez. 2008. Shortchain fructooligosaccharides influence insulin sensitivity and gene expression of fat tissue in obese dogs. J. Nutr. 138: 1712–1718.

Rıfai N. and R. Warnick. 2006. Lipids, lipoproteins, apolipoproteins, and other cardiovascular risk factors. *In:* C.A. Burtis and D.E. Bruns [eds.]. Tietz Textbook of Clinical Chemistry and Molecular Diagnostics. Elsevier, New York, USA, pp. 903–982.

Roberfroid, M., G.R. Gibson, L. Hoyles, A.L. McCartney, R. Rastall, I. Rowland, D. Wolvers, B. Watzl, H. Szajewska, B. Stahl, F. Guarner, F. Respondek, K. Whelan, V. Coxam, M.J. Davicco, L. Leotoing, Y. Wittrant, N.M. Delzenne, P.D. Cani, A.M. Neyrinck and A. Meheust. 2010. Prebiotic effects: metabolic and health benefits. Br. J. Nutr. 104: 1–63.

Sacks D.B. and F.R.C. Path. 2006. Carbohydrates. *In:* C.A. Burtis and D.E. Bruns [eds.]. Tietz Textbook of Clinical Chemistry and Molecular Diagnostics. Elsevier, New York, USA, pp. 837–901.

Saemann, M.D., G.A. Bohmig, C.H. Osterreicher, H. Burtscher, O. Parolini, C. Diakos, J. Stöckl, W.H. Hörl and G.J. Zlabinger. 2000. Antiinflammatory effects of sodiumbutyrate on human monocytes: potent inhibition of IL-12 and up-regulation of IL-10production. FASEB J. 14: 2380–2382.

Sakakibara, S., T. Yamauchi, Y. Oshima, Y. Tsukamoto and T. Kadowaki. 2006. Acetic acid activates hepatic AMPK and reduces hyperglycemia in diabetic KK-A(y) mice. Biochem. Biophys. Res. Commun. 344: 597–604.

Samuel, B.S., A. Shaito, T. Motoike, F.E. Rey, F. Backhed, J.K. Manchester, R.E. Hammer, S.C. Williams, J. Crowley, M. Yanagisawa and J.I. Gordon. 2008. Effects of the gut microbiota on host adiposity are modulated by the short-chain fatty-acid binding G protein-coupled receptor, Gpr41. PNAS 105: 16767–16772.

Samuel, B.S. and J.I. Gordon. 2006. A humanized gnotobiotic mouse model of host-archaealbacterial mutualism. Proc. Natl. Acad. Sci. 103: 10011–10016.

Santacruz, A., A. Marcos, J. Warnberg, A. Marti, M. Martin-Matillas, C. Campoy, L.A. Moreno, O. Veiga, C. Redondo-Figuero, J.M. Garagorri, C. Azcona, M. Delgado, M. Garcia-Fuentes, M.C. Collado and Y. Sanz. 2009. Interplay between weight loss and gut microbiota composition in overweight adolescents. Obesity 17: 1906–1915.

Scott, R., M. Donoghoe, G.F. Watts, R. O'Brien, C. Pardy, M.R.Taskinen, T.M. Davis, P.G. Colman, P. Manning, G. Fulcher and A.C. Keech. 2011. FIELD Study Investigators. Impact of metabolic syndrome and its components on cardiovascular disease event rates in 4900 patients with type 2 diabetes assigned to placebo in the FIELD randomised trial. Cardiovasc. Diabetol. 21;10: 102.

Simons, L.A., S.G. Amansec and P. Conway. 2006. Effect of *Lactobacillus* fermentum on serum lipids in subjects with elevated serum cholesterol. Nutrition, Metabolism & Cardiovascular Diseases 16: 531–535.

Sonnenburg, J.L., J. Xu, D.D. Leip, C.H. Chen, B.P. Westover, J. Weatherford, J.D. Buhler and J.I. Gordon. 2005. Glycan foraging in vivo by an intestine-adapted bacterial symbiont. Science 307: 1955–1959.

Stappenbeck, T.S., L.V. Hooper and J.I. Gordon. 2002. Developmental regulation of intestinal angiogenesis by indigenous microbes via Paneth cells. Proc. Natl. Acad. Sci. 99: 15451–15455.

Sun, L., Z. Yu, X. Ye, S. Zou, H. Li, D. Yu, H. Wu, Y. Chen, J. Dore, K. Clement, F.B. Hu and X. Lin. 2010. A marker of endotoxemia is associated with obesity and related metabolic disorders in apparently healthy Chinese. Diabetes Care 33: 1925–1932.

Swann, J.R., E.J. Want, F.M. Geier, K. Spagou, I.D. Wilson, J.E. Sidaway, J.K. Nicholson and E. Holmes. 2010. Microbes and Health Sackler Colloquium: systemic gutmicrobial modulation of bile acid metabolism in host tissue compartments. Proc. Natl. Acad. Sci. 108: 4523–4530.

Tannock, G.W. 2001. Molecular assessment of intestinal microflora. Am. J. Clin. Nutr. 73: 410–414.

Thomas, C., A. Gioiello, L. Noriega, A. Strehle, J. Oury, G. Rizzo, A. Macchiarulo, H. Yamamoto, C. Mataki, M. Pruzanski, R. Pellicciari, J. Auwerx and K. Schoonjans.2009. TGR5-mediated bile acid sensing controls glucose homeostasis. Cell Metab. 10:167–177.

Tilg, H., A.R. Moschen and A. Kaser. 2009. Obesity and the microbiota. Gastroenterology 136: 1476–1483.

Todar, K. 2011. Online Textbook of Bacteriology. Bacterial Endotoxin. Textbookofbacteriology.net/endotoxin.html.

Turnbaugh, P.J., R.E. Ley, M.A. Mahowald, V. Magrini, E.R. Mardis and J.I. Gordon. 2006. An obesity-associated gut microbiome with increased capacity for energy harvest. Nature. 444(7122): c1027–1031.

Turnbaugh, P., F. Backhed, L. Fulton and J.I. Gordon. 2008. Diet-induced obesity is linked to marked but reversible alterations in the mouse distal gut microbiome. Cell Host. Microbe. 17: 213–223.

Turnbaugh, P.J., M. Hamady, T. Yatsunenko, B.L. Cantarel, A. Duncan, R.E. Ley, M.L. Sogin, W.J. Jones, B.A. Roe, J.P. Affourtit, M. Egholm, B. Henrissat, A.C. Heath, R. Knight and J.I. Gordon. 2009. A core gut microbiome in obese and lean twins. Nature 457: 480–484.

Uysal, K.T., S.M. Wiesbrock, M.W. Marino and G.S. Hotamisligil. 1997. Protection fromobesity-induced insulin resistance in mice lacking TNF-afunction. Nature 389: 610–614.

van Meer, H., G. Boehm, F. Stellaard, A. Vriesema, J. Knol, R. Havinga, P.J. Sauer and H.J. Verkade. 2008. Prebiotic oligosaccharides and the enterohepatic circulation of bile salts in rats. Am. J. Physiol. Gastrointest. Liver Physiol. 294: G540–G547.

Velagapudi, V.R., R. Hezaveh, C.S. Reigstad, P. Gopalacharyulu, L. Yetukuri, S. Islam, J. Felin, R. Perkins, J. Boren, M. Oresic and F. Backhed. 2010. The gut microbiotamodulates host energy and lipid metabolism in mice. J. Lipid Res. 51: 1101–1112.

Vega, R.B., J.M. Huss and D.P. Kelly. 2000. The coactivator PGC-1 cooperates with peroxisome proliferator-activated receptor alpha in transcriptional control of nuclear genes encoding mitochondrial fatty acid oxidation enzymes. Mol. Cell Biol. 20: 1868–1876.

Vice, E., J.D. Privette, R.C. Hickner and H.A. Barakat. 2005. Ketone body metabolism in lean and obese women. Metabolism 54: 1542–1545.

Vrieze, A., F. Holleman, E.G. Zoetendal, W.M. de Vos, J.B.L. Hoekstra and M. Nieuwdorp. 2010. The environment within: how gut microbiota may influence metabolism and body composition Diabetologia 53: 606–613.

Watanabe, M., S.M. Houten, C. Mataki, M.A. Christoffolete, B.W. Kim, H. Sato, N. Messaddeq, J.W. Harney, O. Ezaki, T. Kodama, K. Schoonjans, A.C. Bianco and J. Auwerx. 2006. Bile acids induce energy expenditure by promoting intracellularthyroid hormone activation. Nature 439: 484–489.

Wellen, K.E. and G.S. Hotamisligil. 2005. Inflammation, stress, and diabetes. J. Clin. Invest. 115: 1111–1119.

Wen, L., R.E. Ley, P.Y. Volchkov, P.B. Stranges, L. Avanesyan, A.C. Stonebraker, C. Hu, F.S. Wong, G.L. Szot, J.A. Bluestone, J.I. Gordon and A.V. Chervonsky. 2008. Innate immunity and intestinal microbiota in the development of Type 1 diabetes. Nature 455: 1109–1113.

WHO (World Health Organization) 2012. Noncommunicable diseases: a major health challenge of the 21st century. World Health Statistics. Geneva, Switzerland, 34–37 pp.

Wolever, T.M., F. Brighenti, D. Royall, A.L. Jenkins and D.J. Jenkins. 1989. Effect of rectal infusion of short chain fatty acids in human subjects. Am. J. Gastroenterol. 84: 1027–1033.

Wong, S.W., M.J. Kwon, A.M.K. Choi, H.P. Kim, K. Nakahara and D.H. Hwang. 2009. Fatty acids modulate Toll-like receptor 4 activation through regulation of receptor dimerization and recruitment into lipid rafts in a reactive oxygen species-dependent manner. J. Biol. Chem. 284: 27384–27392.

Wu, X., C. Ma, L. Han, M. Nawaz, F. Gao, X. Zhang, P. Yu, C. Zhao, L. Li, A. Zhou, J. Wang, J.E. Moore, B.C. Millar and J. Xu. 2010. Molecular characterisation of the faecal microbiota in patients with type II diabetes. Curr. Microbiol. 61: 69–78.

Xiong, Y., N. Miyamoto, K. Shibata, M.A. Valasek, T. Motoike, R.M. Kedzierski and M. Yanagisawa. 2004. Short-chain fatty acids stimulate leptin production in adipocytes through the G protein-coupled receptor GPR41. Proc. Natl. Acad. Sci. 101: 1045–1050.

Yadav, H., S. Jain and P.R. Sinha. 2007. Antidiabetic effect of probiotic dahi containing *Lactobacillus acidophilus* and *Lactobacillus casei* in high fructose fed rats. Nutrition 23: 62–68.

Yadav, H., S. Jain and P.R. Sinha. 2008. Oral administration of dahi containing probiotic*Lactobacillus acidophilus* and *Lactobacillus casei* delayed the progression ofstreptozotocin-induced diabetes in rats. J. Dairy. Res. 75: 189–195.

Yoon, J.C., T.W. Chickering, E.D. Rosen, B. Dussault, Y. Qin, A. Soukas, J.M. Friedman, W.E. Holmes and B.M. Spiegelman. 2000. Peroxisome proliferator-activated receptor gamma target gene encoding a novel angiopoietin-related protein associated with adipose differentiation. Mol. Cell. Biol. 20: 5243–5349.

Zhang, H., J.K. DiBaise, A. Zuccolo, D. Kudrna, M. Braidotti, Y. Yu, P. Parameswaran, M.D. Crowell, R. Wing, B.E. Rittmann and R. Krajmalnik-Brown. 2009. Human gutmicrobiota in obesity and after gastric bypass. Proc. Natl. Acad. Sci. 106: 2365–2370.

Zhou, G., R. Myers, Y. Li, Y. Chen, X. Shen, J. Fenyk-Melody, M. Wu, J. Ventre, T. Doebber, N. Fujii, N. Musi, M.F. Hirshman, L.J. Goodyear and D.E. Moller. 2001.Role of AMP-activated protein kinase in mechanism of metformin action. J. Clin. Invest. 108: 1167–1174.

12

Probiotics and Prebiotics and the Gut Microbiota

Eamonn M.M. Quigley

Introduction

Thanks to a phenomenal rate of evolution in the methods available for the detection and accurate annotation of the individual species and strains that inhabit the gastrointestinal tract (Fraher et al. 2012), coupled with the delineation of their metabolic activity and the development of customized bioinformatics approaches which facilitate inter- and intra-subject analysis, there has been an explosion of interest in the microbiota (a term preferred to flora as the former includes archaea, fungi and viruses as well as bacteria), in health and disease (Guarner and Malagelada 2003, Sekirov et al. 2010). *In vitro* and *in vivo* studies in a variety of model systems facilitated, for example, by novel imaging methodologies (Cronin et al. 2008), have revealed the nature and the complexity of the interactions between the microbiota and the host. These advances in biology have generated the expectation that the microbiota may provide new avenues for the development of new diagnostic and therapeutic approaches to a number of gastrointestinal and non-gastrointestinal disorders (Prakash et al. 2011).

The Gut Microbiota: An Overview

The human microbiota is a complex ecosystem which may contain as many as 1000 to 1150 bacterial species and between 10^{13} to 10^{14} microorganisms,

Alimentary Pharmabiotic Centre, Department of Medicine, Cork University Hospital, Cork, Ireland. Email: e.quigley@ucc.ie with the greatest density and diversity of bacteria being found in the distal small bowel and colon (Guarner and Malagelada 2003, Sekirov 2010, Eckburg et al. 2005). The number of bacteria within the gut is about 10 times that of all cells in the human body and the microbiome contains more than 150 times as many non-redundant genes as the human genome. At birth, the entire intestinal tract is sterile; bacteria enter the gut at birth and with the first feed (O'Toole and Claesson 2010). Following infancy, the composition of the intestinal microflora remains relatively constant until later life (O'Toole and Claesson 2010). When disturbed, the flora has, in general, a considerable capacity to restore itself.

Because of the normal motility of the intestine and the antimicrobial effects of gastric acid, the stomach and proximal small intestine contain relatively small numbers of bacteria in healthy subjects. The microbiology of the terminal ileum represents a transition zone between the jejunum containing predominantly aerobic species and the dense population of anaerobes found in the colon. Bacterial colony counts may be as high as 10^9 colony forming units (CFU)/mL immediately proximal to the ileocecal valve, with a predominance of gram-negative organisms and anaerobes. On crossing into the colon, the bacterial concentration and variety of the enteric flora changes dramatically. Concentrations as high as 10^{12} CFU/mL may be found; comprised mainly of anaerobes such as *Bacteroides, Porphyromonas, Bifidobacterium, Lactobacillus* and *Clostridium*, with anaerobic bacteria outnumbering aerobic bacteria by a factor of 100–1000:1. At any given level of the gut, the composition of the flora also demonstrates variation along its diameter with certain bacteria tending to be adherent to the mucosal surface while others predominate in the lumen.

As modern molecular methods begin to describe in complete detail the human microbiome and expose its true diversity (Fraher et al. 2012), as well as the potential of such factors as age and diet to influence its composition (Claesson et al. 2012), some common patterns (enterotypes) are also beginning to emerge (Arumugam et al. 2011).

The normal enteric bacterial flora influences a variety of intestinal functions and plays a key role in nutrition, maintaining the integrity of the epithelial barrier, the development of mucosal immunity, gut motility and sensation, host metabolism and, even, in influencing such apparently distant phenomena as mood and behavior (Guarner and Malagelada 2003, Sekirov et al. 2010, Prakash et al. 2011, Pennisi 2011, Mazmanian et al. 2005, Lesniewska et al. 2006, Khan and Collins 2005, Collins 1996, Dumas et al. 2006, Rook and Lowry 2008, Bercik et al. 2012). Studies employing individual commensal/probiotic organisms have played a key role in delineating these interactions between the microbiota and the host (Resta-Lenert and Barrett 2003, Zeng et al. 2008, Valeur et al. 2004, Verdu et al. 2004, Marteau et al. 2002, Rousseaux et al. 2007, Desbonnet et al. 2008) and in defining the precise mechanism(s) whereby these effects are achieved (Qin et al. 2009, Klaenhammer et al. 2012, Foligne et al. 2007, Grangette et al. 2005, Heuvelin et al. 2009, Van Baarlen et al. 2009, Yan et al. 2011, Duncker et al. 2008). A key finding in these and many other studies is the specificity of a given strain for effects on any one of these host-microbe interactions ; no two commensal bacteria are the same, no two probiotic strains can be expected to exert the same effects (Shanahan 2011).

Disturbances in the microbiota have been described in relation to a wide range of human and animal diseases and disorders; in some instances this relationship is clearly causal, in others that primacy of the observed changes (dysbiosis) remains to be defined. These issues are addressed in much greater detail elsewhere in this volume. What follows is a brief overview.

The Gut Flora in Disease

It has only been in very recent years that the true extent of the consequences of disturbances in the microbiota, or in the interaction between the flora and the host, to health has been recognized (Sekirov et al. 2010). Some of these are relatively obvious: for example, when components are eliminated or suppressed by broad-spectrum antibiotics the stage is set for other, potentially pathogenic, organisms to step in and cause disease. The classical example of this is antibiotic-associated diarrhea and its deadliest manifestation, *Clostridium difficile* colitis (Neu et al. 2008). Traditionally, it has been believed that the microbiota promptly returns to its pre-treatment state following a course of oral antibiotics; more recent data suggests that subtle antibiotic-induced changes may be more persistent (Dethlefsen and Relman 2011).

An abnormal interaction between host and microbiota, in this case an immature gut being exposed to what at term would be considered a "normal" microbiome, is thought to be involved in a devastating form of intestinal inflammation that may occur in premature, low birth weight, infants: necrotizing enterocolitis (Lawley et al. 2012).

In other situations, bacteria may simply be where they should not be: impaired motility and/or acid secretion from the stomach promote an environment conducive to the proliferation, in the small intestine, of organisms normally confined to the colon, and small bowel bacterial overgrowth (SIBO) ensues (Quigley and Abu-Shanab 2010). In other situations, such as inflammatory bowel disease (IBD), the host-microbiota immune interaction malfunctions, with the host coming to recognize commensals not as friend, but as foe, and mounting an inappropriate inflammatory response (Sartor 2010). Furthermore, there is also some evidence for the presence of an altered microbiota in IBD; as elsewhere, the primacy of these microbial changes to the pathogenesis of the disorder remains unclear (Tannock 2010, Willing et al. 2010). The same applies to irritable bowel syndrome; changes in the microbiota (relative to matched controls) have been described and could be seen to be of pathophysiological significance but this has yet to be proven (Jeffery et al. 2012).

If damage to the intestinal epithelium, from whatever source, renders the gut wall leaky and permits enteric bacteria (in whole or in part) to gain direct access to the submucosal compartments or translocate into the systemic circulation, the stage is set for the development of potentially catastrophic sepsis syndromes; a scenario all too familiar to those who work in intensive care units (Quigley 2011).

Metabolic Functions of the Microbiota

Most recently, qualitative changes in the microbiota have been invoked in the pathogenesis of a global epidemic, obesity and its attendant consequences, the metabolic syndrome and non-alcoholic fatty liver disease (NAFLD) (Tilg and Kaser 2011). These clinical observations, as well as a large body of experimental work, have supported the concept of an important role for the microbiota in human and animal metabolism. As a consequence of such work, it has been postulated, for example, that a shift in the composition of the flora towards a population where bacteria that are more avid extractors of absorbable nutrients deliver more calories to the host and thus contribute to obesity.

Some of the metabolic effects of the microbiota have been recognized for decades. These include the salvage of unabsorbed dietary sugars such as lactose and alcohols by bacterial disaccharidases and their conversion into short-chain fatty acids (SCFA's), the de-conjugation of bile salts, the synthesis of vitamins such as folate and vitamin K, and the metabolism of certain drugs. However, it is only recently that the full metabolic potential of the microbiome has come to be recognized and the potential contributions of the microbiota to the metabolic status of the host, in health, and to obesity and related disorders, in disease, appreciated.

The application of genomics, metabolomics and transcriptomics can now reveal, in immense detail, the metabolic potential of a given organism (Saulnier et al. 2011). That microbe-microbe interaction can also play a critical role in homeostasis and disease in man (Hibbing et al. 2011) and is exemplified by an observation made 20 years ago, namely, the consumption of hydrogen by methanogenic organisms (Strocchi and Levitt 1992).

Recently, the role of the microbiota and its interaction with diet in the pathogenesis of obesity *per se* has been extensively investigated (Backhed et al. 2005, Ley et al. 2005) and pertinent findings include the ability of Gram negative anaerobes, such as *Bacteriodes thetaiotamicron,* to cleave most glycosidic linkages, degrade plant polysaccharides, and thereby supply the host with 10–15% of its calorific requirements (Backhed et al. 2005,

Ley et al. 2005, Turnbaugh et al. 2006, Ley et al. 2006). The microbiota of obese individuals, as well as the caecal microbiota of *ob/ob* mice are more efficient at the extraction of energy from the diet and in the production of short chain fatty acids (SCFAs) (Turnbaugh et al. 2006, Schwiertz et al. 2010). Furthermore, the microbiota has been shown to stimulate hepatic triglyceride production through suppression of the lipoprotein lipase (LPL) inhibitor, fasting-induced adipose factor (Fiaf; also known as angiopoietin-like 4), thereby leading to continued expression of LPL, a key regulator of fatty acid release from triglycerides in liver (Bachked et al. 2004). The gut microbiota can also modulate systemic lipid metabolism through modification of bile acid metabolic patterns, impacting directly on the emulsification and absorption properties of bile acids and thus, indirectly, on the storage of fatty acids in liver. The microbiota has also been implicated in the development of insulin resistance (Bachked et al. 2004), a fundamental abnormality in the metabolic syndrome, by affecting energy balance, glucose metabolism, and the low-grade inflammatory state that has been associated with obesity and related metabolic disorders. Its role in choline metabolism (Dumas et al. 2006, Wang et al. 2011, Rak and Rader 2011) as well as in activation of pro-inflammatory cytokines (e.g., tumour necrosis factor α; TNFα) appear relevant to the development of non-alcoholic fatty liver disease (NAFLD) and progression to non-alcoholic steato-hepatitis (NASH). Most recently, studies in experimental models have shown that defective/deficient inflammasome sensing and related dysbiosis result in an abnormal accumulation of bacterial products in the portal circulation and promote progression of NAFLD/NASH (Henao-Mejia et al. 2012).

Impact of Probiotics on the Gut Microbiota: What Factors Influence Colonization

What is the fate of an administered probiotic? Several factors contribute to the survival of a live organism and to its ability to engage with the microbiota and the host and to carry out any putative biological effect that has been claimed for it.

First and foremost the viability of the product at the time of consumption must be considered. It stands to reason that the viability of the probiotic product should have been tested under the very conditions (i.e. room temperature vs refrigeration, humidity, exposure to light) and for the duration of its alleged shelf life and that, based on these tests, the manufacturer should be able to guarantee the presence of live organisms in the number needed to exert the given effect (if, indeed, this is known). Regrettably, many "probiotic" products have not been subjected to this very rudimentary step in quality control.

Next comes the issue of formulation. The range of products allegedly containing probiotic organisms is vast and seems to be limited only by the imagination and audacity of manufacturers; yet, few have tested the impact of the mode of formulation and presentation on the viability and biological impact of organisms. For example, does the vehicle in which the probiotic is presented impair or enhance its viability? One approach to "protecting" the probiotic is microencapsulation; this strategy has been shown to increase by five-fold the colonization and faecal recovery rates for lactobacilli and bifidobacteria (Del Piano et al. 2011). The administration of foods with prebiotic effects (Spiller 1994) at the same time as a probiotic or a combination of a prebiotic with a probiotic in the one formulation (so-called synbiotic) has obvious potential to impact on the colonization potential of the probiotic (Bouhnik et al. 1997, Macfarlane et al. 2006, Russo et al. 2012). The profound effects of diet, *per se*, on the normal microbiota (Claesson et al. 2012) and, by analogy, its potential to influence probiotic colonization and effects, though little studied, must also be remembered.

The next hurdle is survival as the probiotic transits the gut; here it must confront gastric acid, bile, pancreatic and digestive enzymes, gastric mucus and other mucosal protective factors as well as competition from commensal bacteria. Before considering any one of these factors in any detail it is appropriate to make some mention of the methodological issues that must be confronted in attempting to document probiotic viability and colonization. Several issues must be grappled with.

1. What is the site of action of the probiotic: is it in the lumen or at the mucosal surface and will its optimal interaction with the host occur in the stomach, small intestine or colon? In general effects that are immunologically mediated will likely occur at the epithelial surface (and, perhaps, most effectively in the small intestine which bears that greatest surface area of immune cells) whereas metabolic effects occur in the lumen (and predominantly in the colon). It follows that while faecal sampling may provide a reasonable surrogate for quantifying a probiotic whose primary effect is proposed to be metabolic, direct sampling from the biofilm or the mucosa of the small intestine (no easy task!) may be necessary to assess the colonization of an immunologically active probiotic. Suffice it to say that in most instances and regardless of the putative mode of action or proposed site of action of a probiotic, transit through the gut and viability have been assessed by faecal sampling alone (Marteau et al. 2003, Valeur et al. 2004).
2. How is viability/colonization assessed? If one accepts the commonly used definition of a probiotic as a live organism, then only techniques which quantify live organisms are appropriate and those that assess amounts of bacterial DNA or other components of the organism, be it alive or dead, are not.

3. How is the probiotic organism identified and differentiated from the host's intrinsic microbiome? Many probiotic organisms used in man are derived from the commensal microbiota of normal human beings so the differentiation of the probiotic form what already exists in the gut may prove problematic without the use of a deliberate strategy, such as tagging or the use of rifampicin-resistant strains, which allows the detection of the administered organism.
4. Have dose-responses been studied? Unfortunately, the literature on probiotics in man is virtually bereft of dose-response studies of any form, including studies of the influence of various doses of a given probiotic on its recovery from the intestine or the faeces.
5. How does one assess the impact of the administered probiotic on the host microbiome? While most studies of the impact of a probiotic on the host microbiome have revealed relatively minor changes, here again the sensitivity of the methodology employed will be all-important.

The resistance of probiotic organisms to gastric acid varies tremendously and needs to be defined before embarking on a therapeutic approach (Marteau et al. 2003). Other factors such as the buffering effects of meals or the use of anti-secretory agents will also have an impact. The next challenge is presented by bile; in general, bile is less "toxic" than acid but their combined effects may dramatically reduce the viability of some probiotics. Bacterial resistance to bile may be genetically determined (Fang et al. 2009). In one review (Marteau et al. 2003), faecal recovery rates ranging from as low as 0.01% of the administered dose to as high as 30% were documented. When longitudinal sampling has been performed most studies have found that excretion of the administered organism declines dramatically within two weeks of the cessation of probiotic feeding. This observation suggests that, while the administered organism can survive and even proliferate (as counts higher than those administered are sometimes recorded) in an environmental niche as long as it is being administered, internal homeostatic mechanisms ensure its elimination and the restoration of the status quo once feeding has ceased. Many bacterial properties may be relevant to this phenomenon, including production of anti-bacterial peptides (Carr et al. 2007, Rea et al. 2010), adherence and other intrinsic bacterial properties that facilitate host engagement (Fanning et al. 2012).

Effects of Probiotics & Prebiotics on Xenobiotic Metabolism by Gut Flora

Bacteria in the gut microbiota can metabolise ingested xenobiotics into either less active or more bioactive metabolites (Macfarlane and Macfarlane 2007, Haiser and Turnbaugh 2012). Some of these interactions, such as the metabolism by gut bacteria of polycyclic aromatic hydrocarbons, result in the production of toxic metabolites. There is considerable interest in metabolic products from these interactions that may be carcinogenic and, thereby, play a role in colon cancer. It stands to reason, therefore, that modulation or manipulation of the microbiota through the administration of probiotics could be an attractive and effective strategy to beneficially influence xenobiotic metabolism (Haiser and Turnbaugh 2012). To date, however, evidence from studies in man have been relatively scanty (Rabot et al. 2010); though evidence for impacts of the microbiota on digoxin metabolism, activation of azo bond-containing prodrugs and alteration of L-dopamine pharmacokinetics is beginning to appear (Haiser and Turnbaugh 2010). In an animal model of inflammatory bowel disease, a disorder characterized by a down-regulation of xenobiotic receptors, the administration of the probiotic cocktail, VSL#3, stimulated xenobiotic pathways, a potentially beneficial effect (Reiff et al. 2009).

References

Arumugam, M., J. Raes, E. Pelletier, D. Le Paslier, T. Yamada, D.R. Mende, G.R. Fernandes, J. Tap, T. Bruls, J.M. Batto et al. 2011. Enterotypes of the human gut microbiome. Nature 473: 174–80.

Bäckhed, F., H. Ding, T. Wang, L.V. Hooper, G.Y. Koh, A. Nagy et al. 2004. The gut microbiota as an environmental factor that regulates fat storage. PNAS 101: 15718–23.

Bäckhed, F., R.E. Ley, J.L. Sonnenburg, D.A. Peterson and J.I. Gordon. 2005. Host-bacterial mutualism in the human intestine. Science 307: 1915–20.

Bercik, P., S.M. Collins and E.F. Verdu. 2012. Microbes and the gut-brain axis. Neurogastroenterol. Motil 1–9.

Bouhnik, Y., B. Flourie, L. D'Agay-Abensour, P. Pochart, G. Gramet, M. Durand and J.-C. Rambaud. 1997. Administration of transgalacto-oligosaccharides increases fecal bifidobacteria and modifies colonic fermentation metabolism in healthy humans. J. Nutr. 127: 444–448.

Claesson, M.J., I.B. Jeffery, S. Conde, S.E. Power, E.M. O'Connor, S. Cusack, H.M.B. Harris, M. Coakley, B. Lakshminarayanan, O. O'Sullivan, G.F. Fitzgerald, J. Deane, M. O'Connor, N. Harnedy, K. O'Connor, D. O'Mahony, D. van Sinderen, M. Wallace, L. Brennan, C. Stanton, J. Marchesi, A.P. Fitzgerald, C.Hill, R.P. Ross, F. Shanahan and P.W. O'Toole. 2012. Gut microbiota composition correlates with diet and health in the elderly. Nature 488: 178–184.

Collins, S.M. 1996. The immunomodulation of enteric neuromuscular function: implications for motility and inflammatory disorders. Gastroenterology 111: 1683–1699.

Corr, S.C., Y. Li, C.U. Riedel, P.W. O'Toole, C. Hill and C.G. Gahan 2007. Bacteriocin production as a mechanism for the antiinfective activity of *Lactobacillus salivarius* UCC118. Proc. Natl. Acad. Sci. USA 104: 7617–21.

Cronin, M., R.D. Sleator, C. Hill , G.F. Fitzgerald and D. van Sinderen. 2008. Development of a luciferase-based reporter system to monitor *Bifidobacterium breve* UCC2003 persistence in mice. BMC Microbiol. 8: 161.

Del Piano, M., S. Carmagnola, M. Ballare, M. Sartori, M. Orsello, M. Balzarini, M. Pagliarulo, R. Tari, A. Anderloni, G.P. Strozzi, L. Mogna, F. Sforza and L. Capurso. 2011. Is microencapsulation the future of probiotic preparations? The increased efficacy of gastro-protected probiotics. Gut. Microbes 2: 120–123.

Desbonnet, L., L. Garrett, G. Clarke, J. Bienenstock and T.G. Dinan. 2008. The probiotic *Bifidobacterium infantis*: a assessment of potential antidepressant properties in the rat. J. Psychiatr. Res. 43: 164–174.
Dethlefsen, L. and D.A. Relman. 2011. Incomplete recovery and individualized responses of the human distal gut microbiota to repeated antibiotic perturbation. Proc. Natl. Acad. Sci. USA 108(Suppl. 1): 4554–61.
Dumas, M.-E., R.H. Barton, A. Toye, O. Cloarec, C. Blancher, A. Rothwell, J. Fearnside, R. Tatoud, V. Blanc, J.C. Lindon, S.C. Mitchell, E. Holmes, M.I. McCarthy, J. Scott, D. Gaugier and J.K. Nicholson. 2006. Metabolic profiling reveals a contribution of gut microbiota to fatty liver phenotype in insulin-resistant mice. PNAS 103: 12511–12516.
Dumas, M.E., R.H. Barton, A. Toye, O. Cloarec, C. Blancher, A. Rothwell et al. 2006. Metabolic profiling reveals a contribution of gut microbiota to fatty liver phenotype in insulin-resistant mice. PNAS 103: 12511–6.
Duncker, S.C., L. Wang, P. Hols and J. Bienenstock. 2008. The d-alanine content of lipoteichoic acid is crucial for *Lactobacillus plantarum*-mediated protection from visceral pain perception in a rat colorectal distension model. Neurogastroenterol. Motil. 20: 843–850.
Eckburg, P.B., E.M. Bik, C.N. Bernstein, E. Purdom, L. Dethlefsen, M. Sargent, S.R. Gill, K.E. Nelson and D.A. Relman. 2005. Diversity of the human microbial flora. Science 308: 1635–8.
Fang, F., Y. Li, M. Bumann, E.J. Raftis, P.G. Casey, J.C. Cooney, M.A. Walsh and P.W. O'Toole . 2009. Allelic variation of bile salt hydrolase genes in *Lactobacillus salivarius* does not determine bile resistance levels. J. Bacteriol. 191: 5743–5757.
Fanning, S., L.J. Hall, M. Cronin, A. Zomer, J. MacSharry, D. Goulding, M.O. O'Connell Motherway, F. Shanahan, K. Nally, G. Dougan and D. van Sinderen. 2012. Bifidobacterial surface-exopolysaccharide facilitates commensal-host interaction through immune modulation and pathogen protection. PNAS 109: 2108–2113.
Foligne, B., G. Zoumpopoulou, J. Dewulf, A. Ben Younes, F. Chareyre, J.-C. Sirard, B. Pot and C. Grangette. 2007. A key role of dendritic cells in probiotic functionality. PLoS One2: e313.
Fraher, M.H., P.W. O'Toole and E.M.M. Quigley. 2012. Techniques used to characterise the intestinal microbiota: a guide for the clinician. Nat. Rev. Gastroenterol. 9: 312–22.
Grangette, C., S. Nutten, E. Plaumbo, S. Morath, C. Hermann, J. Dewulf, B. Pot, T. Hartung, P. Hols and A. Mercenier. 2005. Enhanced anti-inflammatory capacity of a *Lactobacillus plantarum* mutant synthesizing modified teichoic acids. PNAS 102: 10321–10326.
Guarner, F. and J.R. Malagelada. 2003. Gut flora in health and disease. Lancet 361: 512–9.
Haiser, H.J. and P.J. Turnbaugh. 2012. Developing a metagenomic view of xenobiotic metabolism. Pharmacol. Res. [epub ahead of print].
Henao-Mejia, J., E. Elinav, C. Jin, L. Hao, W.Z. Mehal, T. Strowig et al. 2012. Inflammasome-mediated dysbiosis regulates progression of NAFLD and obesity. Nature 482: 179–85.
Heuvelin, E., C. Lebretin, C. Grangette, B. Pot, N. Cerf-Bensussan and M. Heyman. 2009. Mechanisms involved in alleviation of intestinal inflammation by Bifidobacterium breve soluble factors. PLoS One 4: e5184.
Hibbing, M.E., C. Fuqua, M.R. Parsek and S.B. Peterson. 2010. Bacterial competition: surviving and thriving in the microbial jungle. Nat. Rev. Microbiol. 8: 15–25.
Jeffery, I.B., E.M.M. Quigley, L. Ohman, M. Simren and P.W. O'Toole. 2012. The microbiota link to irritable bowel syndrome. An emerging story. Gut Microbes 3: 572–576.
Khan, W.I. and S.M. Collins. 2005. Gut motor function: immunological control in enteric infection and inflammation. Clin. Exp. Immunol. 143: 389–397.
Klaenhammer, T.R., M. Kleerebezem, M.V. Kopp and M. Rescigno. 2012. The impact of probiotics and prebiotics on the immune system. Nat. Rev. Immunol. 12: 728–734.
Lawley, T.D., S. Clare, A.W. Walker, M.D. Stares, T.R. Connor, C. Raisen, D. Goulding, R. Rad, F. Schreiber, C. Brandt, L.J. Deakin, D.J. Pickard, S.H. Duncan, H.J. Flint, T.G. Clark, J. Parkhill and G. Dougan. 2012. Targeted restoration of the intestinal microbiota with a simple, defined bacteriotherapy resolves relapsing *Clostridium difficile* disease in mice. PLOS Pathogens 8: e1002995.

Lesniewska, V., I. Rowland, H.N. Laerke, G. Grant and P.J. Naughton. 2006. Relationship between dietary-induced changes in intestinal commensal micro flora and duodenojejunal myoelectric activity monitored by radiotelemetry. Exp. Physiol. 91: 229–237.

Ley, R.E., F. Bäckhed, P. Turnbaugh, C.A. Lozupone, R.D. Knight and J.I. Gordon. 2005. Obesity alters gut microbial ecology. PNAS 102: 11070–5.

Ley, R.E., P.J. Turnbaugh, S. Klein and J.I. Gordon. 2006. Microbial ecology—Human gut microbes associated with obesity. Nature 444: 1022–1023.

Macfarlane, G.T. and S. Macfarlane. 2007. Models for intestinal fermentation: association between food components, delivery systems, bioavailability and functional interactions in the gut. Curr. Opinion. Biotechnol. 18: 156–62.

Macfarlane, S., G.T. Macfarlane and J.H. Cummings. 2006. Review article: prebiotics in the gastrointestinal tract. Aliment Pharmacol. Ther. 24: 701–714.

Marteau, P. and F. Shanahan. 2003. Basic aspects and pharmacology of probiotics: an overview of pharmacokinetics, mechanisms of action and side-effects. Best Pract. Res. Clin. Gastroenterol. 17: 725–740.

Marteau, P., E. Cuillerier, S. Meance, M.F. Gerhardt, A. Myara, M. Bouvier, C. Bouley, F. Tondu, G. Bommelaer and J.C. Grimaud. 2002. *Bifidobacterium animalis* strain DN-173 010 shortens the colonic transit time in healthy women: a double blind, randomized, controlled study. Aliment Pharmacol. Ther. 16: 587–593.

Mazmanian, S.K., C.H. Liu, A.O. Tzianabos and D.L. Kasper. 2005. An immunomodulatory molecule of symbiotic bacteria directs maturation of the host immune system. Cell 122: 107–118.

Neu, J., M. Mshvildadze and V. Mai. 2008. A roadmap for understanding and preventing necrotizing enterocolitis. Curr. Gastroenterol. Rep. 10: 450–7.

O'Toole, P.W. and M.J. Claesson. 2010. Gut microbiota: Changes throughout the lifespan from infancy to elderly. Int Dairy J. 20: 281–291.

Pennisi, E. 2011. Girth and the gut (bacteria). Science 332: 32–33.

Prakash, S., L. Rodes, M. Coussa-Charley and C. Tomaro-Duchesneau. 2011. Gut microbiota: next frontier in understanding human health and development of biotherapeutics. Biologics: Targets and Therapy. 5: 71–86.

Qin, H., Z. Zhang, X. Hang and Y. Jiang. 2009. *L. plantarum* prevents *Enteroinvasive Escherichia coli*-induced tight junction proteins changes in intestinal epithelial cells. BMC Microbiol. 9: 63.

Quigley, E.M. 2011. Passing the bug-Translocation, bacteremia, and sepsis in the intensive care unit patient: Is intestinal decontamination the answer? Crit. Care Med. 39: 1202–3.

Quigley, E.M. and A. Abu-Shanab. 2010. Small intestinal bacterial overgrowth. Infect. Dis. Clin. North Am. 24: 943–59.

Rabot, S., J. Rafter, G.T. Rijkers, B. Watzl and J.-M. Antoine. 2010. Guidance for substantiating the evidence for beneficial effects of probiotics: impact of probiotics on digestive system metabolism. J. Nutr. 140: 677S–689S.

Rak, K. and D.J. Rader. 2011. Cardiovascular disease: the diet-microbe morbid union. Nature 472: 40–1.

Rea, M.C., C.S. Sit, E. Clayton, P.M. O'Connor, R.M. Whittal, J. Zheng, J.C. Vederas, R.P. Ross, C. Hill and C.D. Thuricin. 2010. A posttranslationally modified bacteriocin with a narrow spectrum of activity against Clostridium difficile. Proc. Natl. Acad. Sci. USA 107: 9352–7.

Reiff, C., M. Melday, G. Rucklidge, M. Reid, G. Duncan, S. Wohlgemuth, G. Hormannsperger, G. Loh, M. Blaut, E. Collie-Duguid, D. Haller and D. Kelly. 2009. Balancing inflammatory, lipid and xenobiotic signaling pathways by VSL#3, a biotherapeutic agent, in the treatment of inflammatory bowel disease. Inflamm Bowel Dis. 15: 1721–1736.

Resta-Lenert, S. and K.E. Barrett. 2003. Live probiotics protect intestinal epithelia cells from the effects of infection with enteroinvasive *Escherichia coli* (EIEC). Gut. 52: 988–997.

Rook, G.A.W. and C.A. Lowry. 2008. The hygiene hypothesis and psychiatric disorders. Trends Immunol. 29: 150–158.

Rousseaux, C., X. Thuru, A. Gelot, N. Barnich, C. Neut, L. Dubuquoy, C. Dubuquoy, E. Merour, K. Geboes, M. Chamaillard, A. Ouwehand, G. Leyer, D. Carcano, J.-F. Colombel, D. Ardid and P. Desreumaux. 2007. *Lactobacillus acidophilus* modulates intestinal pain and induces opioid and cannabinoid receptors. Nat. Med. 13: 35–37.

Russo, P., P. Lopez, V. Capozzi, P. Fernandez de Palencia, M.T. Duenas , G. Spano and D. Fiocco. 2012. Beta-glucans improve growth, viability and colonization of probiotic organisms. In J. Mol. Sci. 13: 6026–6039.

Sartor, R.B. 2010. Genetics and environmental interactions shape the intestinal microbiome to promote inflammatory bowel disease versus mucosal homeostasis. Gastroenterology 139: 1816–9.

Saulnier, D.M., F. Santos, S. Roos, T.-A. Mistretta, J.K. Spinler, D. Molenaar, B. Teusink and J. Versalovic. 2011. Exploring metabolic pathway reconstruction and genome-wide expression profiling in *Lactobacillus reuteri* to define functional probiotic features. PLoS ONE6: e18783.

Schwiertz, A., D. Taras, K. Schafer, S. Beijer, N.A. Bos, C. Donus and P.D. Hardt. 2010. Microbiota and SCFA in lean and overweight healthy subject. Obesity 18: 190–195.

Sekirov, I., S.L. Russell, L.C. Antunes and B.B. Finlay. 2010. Gut microbiota in health and disease. Physiol. Rev. 90: 859–904.

Shanahan, F. 2011. Molecular mechanisms of probiotic action: it's all in the strains! Gut.

Spiller, R.C. 1994. Pharmacology of dietary fibre. Pharmacol. Ther. 62: 407–427.

Strocchi, A. and M.D. Levitt. 1992. Factors affecting hydrogen production and consumption by human fecal flora. J. Clin. Invest. 89: 1304–11.

Tannock, G.W. 2010. The bowel microbiota and inflammatory bowel diseases. Int. J. Inflamm. 2010: 954051.

Tilg, H. and A. Kaser. 2011. Gut microbiome, obesity, and metabolic dysfunction. J. Clin. Invest. 121: 2126–32.

Turnbaugh, P.J., R.E. Ley, M.A. Mahowald, V. Magrini, E.R. Mardis and J.I. Gordon. 2006. An obesity-associated gut microbiome with increased capacity for energy harvest. Nature 444: 1027–1031.

Valeur, N., P. Engel, N. Carbajal, E. Connolly and K. Ladefoged. 2004. Colonization and immunomodulation by *Lactobacillus reuteri* ATCC 55730 in the human gastrointestinal tract. Appl. Environ. Microbiol. 70: 1176–1181.

Van Baarlen, P., F.J. Troost, S. van Hemert, C. van der Meer, W.M. de Vos, P.J. de Groot, G.J.E.J. Hooiveld, R.-J.M. Brummer and M. Kleerebezeem. 2009. Differential NF-κB pathways induction by *Lactobacillus plantarum* in the duodenum of healthy humans correlating with immune tolerance. PNAS 106: 2371–2376.

Verdu, E.F., P. Bercik, G.E. Bergonzelliand X.X. Huang, P. Blennerhasset, F. Rochat, M. Fiaux, R. Mansourian, I. Corthesy-Theulaz and S.M. Collins. 2004. *Lactobacillus paracasei* normalizes muscle hypercontractility in a murine model of post-infective gut dysfunction. Gastroenterology 127: 826–837.

Wang, Z., E. Klipfell, B.J. Bennett, R. Koeth, B.S. Levison, B. Dugar et al. 2011. Gut flora metabolism of phosphatidylcholine promotes cardiovascular disease. Nature 472: 57–6.

Willing, B.P., J. Dicksved, J. Halfvarson, A.F. Andersson, M. Lucio, Z. Zheng, G. Järnerot, C. Tysk, J.K. Jansson and L. Engstrand. 2010. A pyrosequencing study in twins shows that gastrointestinal microbial profiles vary with inflammatory bowel disease phenotypes. Gastroenterology 139: 1844–1854.

Yan, F., H. Cao, T.L. Cover, M.K. Washington, Y. Shi, L.S. Liu, R. Chaturverdi, R.M. Peek Jr., K.T. Wilson and D.B. Polk. 2011. Colon-specific delivery of a probiotic-derived soluble protein ameliorates intestinal inflammation in mice through an EGFR-dependent mechanism. J. Clin. Invest. 121: 2242–53.

Zeng, J., Y.-Q. Li, X.-L. Zuo., Y.-B. Zhen, J. Yang and C.-H. Liu. 2008. Clinical trial: effect of active lactic acid bacteria on mucosal barrier function in patients with diarrhoea-predominant irritable bowel syndrome. Aliment Pharmacol. Ther. 28: 994–1002.

13

Probiotics and Prebiotics in Immune System Protection

Esther Nova, Cristina Abad, Sonia Gómez-Martínez, Tamara Pozo-Rubio* and *Ascensión Marcos*

Introduction

Many studies have been published that support the fact that probiotics help regulate as well as enhance various aspects of the innate and adaptive immune response, both in animals and in humans. These studies, however, vary in the grade of evidence that they provide on the potential of prebiotics and probiotics to have an effect on immune system protection and overall health. The consumption of probiotics has been shown to influence various aspects of the innate nonspecific immune system like promotion of mucin production, inhibition of pathogens, decrease in gut permeability, macrophage activation and phagocytic capacity, and Natural Killer (NK) cell activity. Regarding the adaptive immune system, the effects observed are an increase in the production of antibodies (IgA, IgM and IgG), and there is also an influence in the orchestration of both branches of the immune system by the production of cytokines and other regulatory elements. The documented effects, however, may be different depending on the species

Immunonutrition Group. Department of Metabolism and Nutrition. Institute of Food Science, Technology and Nutrition. ICTAN-CSIC. C/Jose Antonio Novais, 10. 28040 Madrid (Spain).
Emails: enova@ictan.csic.es; cris_abad89@hotmail.com; sgomez@ictan.csic.es; tpozo@ictan.csic.es; amarcos@ictan.csic.es
* Corresponding author

of the probiotic used and its specific strain (Paineau et al. 2008), it's state of viability (Gill and Rutherfurd 2001), and it's level of consumption (Donnet-Hughes et al. 1999). This modulation of the immune system can be made through innate cell-surface pattern-recognition receptors expressed in monocytes, macrophages and dendritic cells (DCs) or via direct lymphoid cell activation (Isolauri et al. 2001, Cross 2002, Ruiz et al. 2005). It is important to remember that it is inaccurate to generalize findings observed in a single probiotic species to all probiotics, since in numerous intervention studies made with potentially probiotic strains, no effect was observed. Based on this, the reasonable approach to the study of the immune modulation exerted by potential probiotics in humans, is to have evidence of the functionality of the particular strain *in vitro* and/or in animal studies and after that, to perform a human study. The differences between strains are due to the differences in their cell wall protein profile or the CpG content of their DNA (Akira and Takeda 2004). Microbe sensing by intestinal epithelial cells establishes a crosstalk between the epithelium layer and the adjacent immune cells triggering the expression of a number of immune modulators including defensins, cytokines, and chemokines. These mediators will have an impact on the regulation of the immune function of other cells in the mucosal site and are necessary to maintain intestinal homeostasis (Marques and Boneca 2011). There also exists a tendency on the immunomodulatory activity of the probiotics to be predominantly characterized by interleukin (IL)-12 production with development of a T helper (Th)-1 response and enhancement of the immune system, with potential applications in the prevention and treatment of infectious diseases. However, they may induce predominantly IL-10 production, promote T Foxp3+ regulatory cells' development and control excess immune responses, thereby improving inflammatory and allergic diseases. New research also shows that in addition to this two effector pathways, certain probiotic strains can activate T cell subsets such as Th9 and Th17 (De Roock et al. 2011, López et al. 2010). Therefore, probiotics are currently studied in the context of all those diseases that can benefit from the immunomodulatory activity of microbes passing through or colonizing the gut. However, the immune modulation potential of probiotics will be addressed in this chapter in the context of health maintenance and/or prevention of disease states.

Prebiotics that induce the establishment and/or growth of bacterial communities in the intestine help develop immunologic structures of the intestinal mucosa. It has been observed in new born mice that oral inulin administration promotes mucosal immune system maturation as derived from a higher number of IgA-committed B cells in Peyer patches (Roller et al. 2004). This effect seems to be mediated by the growth promotion of *Bifidobacterium* species in the intestine. Prebiotics included in infant formula for term infants are believed to favor the development of a short chain fatty acid (SCFA) pattern and microbiota composition that resembles those of breastfed infants (Knol et al. 2005). The influence of prebiotics on postnatal development of the immune system is likely to be significant but the study of these interactions is complex. Not only the bifidogenic effect but also direct interaction with immune cells through adhesion molecules and other receptors can occur (Schumacher et al. 2006).

Enhancement of immune protection by prebiotics has been documented in an epidemiological study of 326 healthy infants who, if consuming a mixture of short-chain galacto- long-chain fructo-oligosaccharides (scGOS/lcFOS) supplemented formula, presented a reduced incidence of different infectious symptoms during the first year of life than infants fed a control formula (Bruzzese et al. 2006). Despite this positive finding, a recent epidemiological study of 830 infants randomized to receive supplemented formula (scGOS, lcFOS and pectin-derived acidic oligosaccharides) or non-supplemented formula, found no significant differences in the median number of fever episodes in the first 6 months or 12 months of life (van Stuijvenberg et al. 2011).

Specific elements and aspects of the immune system will be addressed in the following sections of this chapter.

Dendritic Cells

Dendritic cells (DC) sense antigen in tissues before migrating to draining lymph nodes, where they have the unique ability to activate and influence functional differentiation of naïve T cells. Signals from DC can determine whether tolerance or an active immune response occur to a particular antigen (Banchereau and Steinman 1998, Steinman and Nussenzweig 2002). The Th differentiation begins with the activation of the DCs by several pathogen-associated molecular patterns (PAMPs) that can have antagonistic or synergistic effects; this differentiation secondarily determines the polarization of the effector T cell responses (Mazzoni and Segal 2004). The cytokine environment during priming is another key element driving the polarization of naïve CD4+ T cells. The production of IL-10 by DCs can limit mucosal inflammation either by direct anti-inflammatory effects or by enhancing the activity of regulatory T cells (Treg) (Hart et al. 2004).

Probiotics

Different probiotic strains may modulate the production of cytokines to induce one of the possible polarizations, as mentioned above. These strain-dependent effects are probably linked to interactions between specific bacterial surface structures and the pattern-recognition receptors (PRRs), such as the DC-SIGN, a member of the CLR (C-type leptin receptor) family, which interacts with certain strains, but not with others (Smits et al. 2005). *Lactobacillus casei, L. gasseri, L. johnsonii, L. reuteri*, can induce a Th1 response (Mohamadzadeh et al. 2005, Christensen et al. 2002, Chuang et al. 2007), *L. rhamnosus Lcr35* induces a Th1 response and pro-Th17 immune response and therefore a proinflammatory and antiinfectious response (Evrard et al. 2011), while the strain *L. reuteri* DSM12246 induces IL-10 production by mouse bone-marrow derived DCs (Christensen et al. 2002). Therefore, the lactobacilli-induced cytokine response in DCs seems to be strain dependent, although the type and origin of the cell used in the experiments should also be taken into account.

One way in which DCs regulate the differentiation of T cells is through the expression of CD80 and CD86 coestimulatory molecules, which bind to T cell membrane proteins. Some strains of bifidobacteria have been shown to downregulate the expression of CD80 (Hart et al. 2004). *Bifidobacterium breve* and *B. infantis* also reduced the level of CD40 expression on DC. CD40 signal increases IL-12 production by DC and enhances their survival (Bjorck et al. 1997).

There exists a communication between the gut DCs and probiotic bacteria performed by non-digestible peptides as seen in studies with *L. plantarum* species. The STp peptide, derived from the extracellular proteins of this probiotic strain, showed capacity to modulate the phenotype acquired by DC. In addition, the intestinal STp-pulsed DCs induced more skin-homing CLA protein expression on stimulated T-cells than basal intestinal DC. IL-10 production by stimulated T-cells was also increased. An important finding that STp-containing proteins were absent in the intestinal microenvironment from inflammatory bowel disease patients, suggests that it has a potential role as a homeostasis biomarker (Bernardo et al. 2012).

Prebiotics

It has been shown that long-chain inulin increases DCs in the Peyer's patches of growing female rats (Ryz et al. 2009). In addition, a recently developed β-galactomannan (βGM) prebiotic has been shown to induce the activation of DCs when cocultured with *Salmonella*. The expression of TNF (tumour necrosis factor)-α, GM-CSF (granulocyte-macrophage colony-stimulating factor), and the chemokine CXCL8 mRNA was higher on porcine monocyte-derived DCs when βGM was added compared to that of the control with Salmonella (Badia et al. 2012).

T Cell Responses and Cytokine Production

The antigen-presenting activity of cells (APCs), such as monocytes, macrophages and DCs leads to the activation of the adaptive immune system.

The recognition of conserved molecular patterns of bacterial components through Toll-like receptors leads to the activation of a variety of transcription factors, which triggers the production of cytokines (Karlsson et al. 2002). Proinflammatory cytokines, such as TNF-α and IL-6, are among the first cytokines produced in response to bacteria. Cytokines produced by APCs together with certain surface receptors are instrumental in the development of T-cell differentiation to Th1, Th2, or Treg (Cools et al. 2007). IL-12 is a major Th1-promoting factor (Trinchieri 1993), whereas IL-10 downregulates the production of gamma interferon (IFN-γ) and IL-12 (D'Andrea et al. 1993, Latvala et al. 2011). This response of the innate immunity is quick and the production of cytokines modulates the specific response depending on which cytokines are produced and their concentrations. IL-12/IFN-γ promotes Th1 differentiation, IL-4/(IL-2, IL-7, TSLP) promotes Th2, whereas tumour growth factor (TGF)-β/(IL-6, IL-21, IL-23) lead to Th17 cells, and TGFβ/IL-2 towards Treg cells (Zhu and Paul 2010).

Probiotics

Although the therapeutic potential of probiotics has been attributed frequently to the skewing of the Th2/Th1 balance towards a Th1 profile (Chuang et al. 2007), more recently it has been recognized that immune homeostasis is defined also by immune responses other than these, and Th17 and Treg cells are to be considered in a more complex concept of immune balance (de Roock et al. 2010). T cell percentage of the whole lymphocyte population as well as the number of CD4+ and CD8+ positive cells are usually not affected by the intervention with probiotic strains in healthy subjects as reviewed by Lomax and Calder (2009).

Some strains have the capacity of activating different types of Th cells while others seem to just activate one type. *L. salivarius* activates Foxp3+ cells and Th17 cells as seen in studies *in vitro* with human peripheral blood mononuclear cells (PBMC) while *L. lactis* induces Th1 response (de Roock et al. 2011). Also, the magnitude of $CD25^+$ ($Foxp3^+$) cells induction by bacteria in PBMCs has been shown to differ between probiotic strains, with some, such as *L. acidophilus* W55 being potent inducers and others not at all compared with medium alone (de Roock et al. 2010). No differences were observed between Lactic acid bacteria (LAB) in the production of IL-17, IFN-γ or IL-13.

Although *in vitro* research using human PBMC cultures can give a good indication of what the immune modulatory properties of a probiotic can be, the *in vitro* effects could differ from *in vivo* observations (de Roock et al. 2011). Both types of experimental approaches are needed. The induction of Treg cells is thought to be beneficial in combating both autoimmune diseases and allergies (López et al. 2012). Autoimmune diseases will likely benefit from Treg inducing probiotics while Th1 and Th2 inducing probiotics will aggravate it (Delcenserie et al. 2008).

Many studies on the immune modulatory capacity of probiotics, with emphasis on cytokine induction/inhibition and Th1/Th2 balance, have been performed in allergy-models in mice, mainly ovoalbumin (OVA)-sensitized mice. Since a chapter in this book is dedicated to the effects of probiotics on allergy, we are going to present here, only *in vitro* and *in vivo* studies which do not belong to this category, since it has been shown that the magnitude of the probiotic effect differs between healthy and allergic subjects (Ghadimi et al. 2008). Probiotic strains of the *Lactobacillus* and *Bifidobacterium* genera, have shown capacity to modulate cytokine production by intestinal epithelial cells, monocyte-derived DC and PBMCs in *in vitro* experiments (Candela et al. 2008, Latvala et al. 2008, Niers et al. 2005, Pozo-Rubio et al. 2011). Experiments have tested the effect of probiotic LAB on unstimulated cytokine production by immune cells, while others have used them in combination with other stimuli.

Live *L. rhamnosus* GG, *L. gasseri* (PA16/8), *Bifidobacterium bifidum* (MP20/5), and *Bifidobacterium longum* (SP07/3), and also their genomic DNA were tested for their effects on the Th1/Th2 production by PBMCs. The Gram-positive bacteria and their genomic DNA inhibited *Staphylococcus* enterotoxin A (SEA)- and *Dermatophagoides pteronyssinus* (Dpt)-stimulated secretion of Th2 cytokines (IL-4 and IL-5) and enhanced the stimulation of IFN-γ (Ghadimi et al. 2008). Oral administration of *L. paracasei* KW3110 induced IL-12 mRNA expression in mice Peyer's patches and transiently increased blood IL-12 levels (Ichikawa et al. 2009). In a human study with anorexia nervosa patients, yogurt containing only *L. bulgaricus* and *Streptococcus thermophilus* consumed during 10 weeks significantly enhanced PHA-stimulated IFN-γ production, while no effect was observed on other cytokine levels (Nova et al. 2006). *L. casei* Shirota did not change IFN-γ, IL-1β and IL-2 production by stimulated PBMCs in healthy men (Spanhaak et al. 1998) while *B. lactis* HN019 increased stimulated IFN-α production in elderly healthy adults (Arunachalam et al. 2000).

A probiotic yogurt containing the strains *L. gasseri* CECT 5714 and *L. coryniformis* CECT 5711 plus *S. thermophilus* increased serum levels of the anti-inflammatory cytokine IL-10 after 2 weeks of consumption, while having no effect on the production of the proinflammatory cytokines TNF-α or IL-12 (Olivares et al. 2006a). Other probiotics, such as *L. salivarius* CECT5713 can also increase the production of IL 10 in healthy adults (Sierra et al. 2010). These regulatory cytokines are potential key factors in immune response homeostasis, since they are involved in the T-regulatory and Th3 response, which counter balances the Th1 and Th2 responses (Akbari et al. 2003). Another way in which probiotics such as *S. thermophilus* can downregulate the inflammatory effects is by inducing the SOCS3 mRNA expression that controls the expression of proinflammatory cytokine genes (Latvala et al. 2011). The IL-10/IL-12 ratio is considered a good indicator of the anti-inflammatory effect of a certain probiotic strain (Foligne et al. 2007, Latvala et al. 2011).

Prebiotics

Recently published mice studies, using both vaccination and allergy models, have shown that nondigestible carbohydrate supplementation stimulated Tregs. Depletion of these cells abrogated the nondigestible carbohydrate-dependent allergy attenuation and increased vaccination response (Schouten et al. 2010, van't Land et al. 2010). In addition, adoptive transfer experiments showed that the beneficial effects could be transferred by Tregs, suggesting that Tregs are the key players in nondigestible carbohydrate-induced immune improvement (Schouten et al. 2012). In these studies, the nondigestible carbohydrate-enriched diet was administered directly to neonates or pre-pubertal mice. Recently, one study has been published on the immune effects of supplementation of nondigestible oligosaccharides during pregnancy (van Vlies et al. 2012). It showed that supplementation of a mixture of scGOS/lcFOS (ratio 9:1) seems to exert distinct effects in pregnant mice compared to virgin. Compared with virgin mice, supplementation appears to elicit a more tolerogenic immune reaction in pregnant mice as shown by an increase in the percentage of alternatively activated macrophages in placentas of scGOS/lcFOS-fed mice, together with the increased whole blood IL-4 production and IL-10 expression and supplementation does not increase the Th1-dependent delayed-type hypersensitivity response in pregnant mice as it does in virgin mice (van Vlies et al. 2012).

A prebiotic supplementation of 1.3g oligosaccharides daily for 12 weeks in elderly subjects (84±7 y.) significantly decreased TNF-α mRNA and IL-6 mRNA in PBMCs, while no change was observed in the placebo group. However, no differences were found in fecal gut microbiota in the studied subjects (Schiffrin et al. 2007). The effect of a mixture of long-chain FOS, GOS and acidic oligosaccharides on immune system biomarkers was studied in young children (aged 9–24 months) with acute diarrhea. Only serum TNF-α decreased in the supplemented group, while interleukin-1 (IL-1), IL-1RA, IL-6, IL-8, IL-10, TNF-α and sIL-2R remained unmodified compared to the control group (Vaisman et al. 2010).

A double-blind, randomized, placebo-controlled study, aimed to explore the effect of an infant milk formula with 6 g/l scGOS/lcFOS (ratio 9:1) showed that the prebiotic oligosaccharides did not change the serum level of IL-2, IL 4, IL-5, IL-10, TNF-α and IFN-γ in healthy infants with a balanced immune system during the first 6 months of life in comparison to standard infant formula and in comparison to exclusive breastfeeding (Raes et al. 2010).

Antibody Production

Probiotics

The daily consumption of probiotics has been proven to enhance mucosal (secretory IgA, sIgA) and systemic antibody responses (Rinne et al. 2005, Cukrowska et al. 2002). Consumed prior and following vaccination, increases in total or specific IgM, IgG or IgA levels in serum and salivary IgA have been documented in several studies. Both, the supplementation of *B. animalis* ssp. *lactis* (BB-12©) and *L. paracasei* ssp. *paracasei* (*L. casei* 431©) led to the significant increase in specific IgG1 and IgG3 in a randomized double-blind, placebo-controlled study (Rizzardini et al. 2012). *L. fermentum* CECT 5716 increased significantly anti-influenza specific-IgA and total IgM and IgG in plasma (Olivares et al. 2007) and LGG increased anti-poliovirus specific IgA (de Vrese et al. 2005).

It has been shown that the consumption of *B. lactis* Bb12 in formula fed infants increases the sIgA production by the intestinal tract. It has also been demonstrated that negative immune-related effects of not breastfeeding and cesarean delivery can be mitigated by including Bb12 in infant formula, thereby providing infants a safe, dietary, immune-modulating bacterial introduction (Holscher 2012). It has been observed that Th1, Th2 and Th17 functions increase with age. Feeding *L. paracasei* subsp. *paracasei* strain F19 during weaning tended to decrease the Th0 response and increase the Th1 and Th17 response, although it was a modest effect (West et al. 2012). It also increases the specific IgG immune response to vaccination (West et al. 2008), although the effect seems to be greater in infants with a short duration of breast feeding, especially if they are colonized by the probiotic.

On the other hand, the number of IgA and IgM-secreting cells in infants with atopic dermatitis receiving LGG daily during 3 months decreased with the probiotic treatment, suggesting a beneficial modulation in this particular population (Nermes et al. 2011).

Prebiotics

A stimulation of the vaccination response in a dose-dependent manner and modulation towards a predominant Th1 response was documented in mice when a mixture of prebiotics was administered through intervention before the first vaccination (Vos et al. 2007). In humans, Firmansyah et al. (2000)

reported increased post-vaccination IgG antibodies in plasma induced by a mixture of scFOS and lcFOS. Results from other studies, although with different mixtures of prebiotics and different sampling periods after vaccination have not found and enhancement of the antiboby response (van Hoffen et al. 2009, Stam et al. 2011, Bunout et al. 2002).

However, modulation of humoral immunity in the first stages of life through a mixture of scGOS and lcFOS was observed in infants at risk for allergy, since a more anti-allergic immunoglobulin profile was found in supplemented infants compared to the placebo group (van Hoffen et al. 2009). No significant difference in the response to the diphtheria, tetanus and polio vaccine was observed in these infants. More recently, a study in healthy infants showed that the *Haemophilus influenza* type b (Hib) and tetanus specific antibody levels in infants fed a mixture of 3 prebiotics or a control diet were similar during the first year of life. Despite this normal response in all infants, the authors hypothesize that the mixture mainly promotes Th1 and Treg dependent immune responses and induces a down regulation of IgE-mediated allergic responses, while not affecting vaccination responses (Stam et al. 2011).

NK Cells

Probiotics

Probiotics also have the capacity to modulate the activity of NK cells. NK cell activity has been shown to be enhanced by the consumption of LAB in healthy adults (Nanno et al. 2011, Sierra et al. 2010), the elderly (Moro-Garcia et al. 2012) and smokers (Morimoto et al. 2005). NK cells are active in tumor surveillance and in the control of viral infection, as well as functioning as immunoregulatory cells via the secretion of interferons, and are thus important contributors to cell-mediated immunity. The NK cells interact with the DCs matured by LAB (Fink et al. 2007). The consumption of LAB consistently induces activation and promotes proliferation and cytotoxicity in NK cells. However, the effect produced by the intake of the probiotic seems to be lost after the supplementation is ended (Gill et al. 2001a) which reflects the fact that probiotic bacteria hardly ever permanently colonize the host (Rizzello et al. 2011). Several strains of probiotics such as *L. casei* Shirota, *L. rhamnosus GG, L. plantarum NCIMB 8826, L. reuteri NCIMB 11951, bifidobacteria (Bifidobacterium longum SP 07/3* and *B. bifidum MF 20/5),* and *Bacillus coagulans*: GBI-30, have increased the CD69 and CD25 expression percentage in NK cells in *in vitro* assays (Dong et al. 2012, Jensen et al. 2010). These results suggest that probiotics enhance activation of the NK cells, without any obvious strain specificity (Dong et al. 2012).

Monocytes, macrophages and dendritic cells have an important role in the activation of the innate immune response. On one hand, IL-12 production by macrophages seems to be enhanced by gram-positive bacteria (Hessle et al. 2000) and on the other hand it has been reported that early production of IL-12 by macrophages contributes to the maturation of NK cells and leads to a Th1 preferential response (Niers et al. 2005). Thus, *L. casei* Shirota stimulates the secretion of IL-12 by macrophages and monocytes producing an augmentation in the NK cell activity (Kodama et al. 1999). The main action the probiotics have on NK cells is the augmentation of their cytotoxic activity rather than the increase in the NK cell count (Takeda et al. 2006, Takeda and Okumura 2007), although an increase of NK cell percentage has also been observed in some cases (Olivares et al. 2006b). In agreement with this, several human studies have found that probiotic supplementation enhanced NK cell activity, and interestingly this increase seems to be consistently accompanied by the increase in the activity of phagocytic cells (Gill et al. 2001a,c, Chiang et al. 2000).

Finally, it has been documented that the probiotic strain *L. casei* DN114001 prevents the decrease in NK cell numbers observed in stress conditions such as strenuous exercise in recreational athletes (Pujol et al. 2000) and examination induced stress in university students (Marcos et al. 2004).

Prebiotics

A mixture of GOS has been shown to increase the NK cell activity of healthy elderly people in a double-blind, placebo-controlled, crossover study (Vulevic et al. 2008). Another study has assessed a unique prebiotic mixture of scGOS/lcFOS/pectin hydrolysate-derived acidic oligosaccharides (15 or 30 g) in a double-blind, randomized, placebo-controlled study performed in highly active antiretroviral therapy-naïve HIV-infected adults. The results also showed a significant increase in NK cell activity with the most pronounced effect reached at the 15g dose (Gori et al. 2011).

Effects of prebiotics on NK cells have also been observed in animal studies. Smad3-deficient mice were supplemented with GOS before and after infection with *Helicobacter hepaticus*. Percentage of NK cells and NK expression of CCR9 receptor were increased in supplemented mice (Gopalakrishnan el al. 2012). In a study of female C57BL/6 old mice supplemented with isomalto-oligosaccharide, the percentage of NK cells in liver and NK cell activity in the spleen were increased (Mizubuchi et al. 2005). In other study, B6C3F1 mice were fed diets containing cellulose or with cellulose replaced entirely with oligofructose (OF) or inulin (IN). After a period of 6 weeks with OF and IN supplementation, both prebiotics increased NK cell activity compared to the cellulose group (Kelly-Quagliana et al. 2003). In a long-term study consisting of a 33-week intervention period, Roller et al. (2004) focused on the effects of prebiotics (inulin-based enriched with oligofructose) and synbiotics (combination of inulin-based enriched with oligofructose and *L. rhamnosus* GG and *B. lactis* Bb12) on the gut associated lymphoid tissue of rats with azoxymethane (AOM)-induced colon cancer. The decrease in the NK cell-like cytotoxic activity associated with the AOM treatment was prevented in the groups receiving the prebiotic or the symbiotic (Roller et al. 2004).

Finally, an *in vitro* study evaluating the effects of FOS from *Asparagus racemosus* on NK cell activity measured in human PBMCs showed a significant increase of this activity by this 2→1 linked FOS (Thakur et al. 2012).

Phagocytic Activity

Probiotics

The study by Schiffrin et al. (1995) was the first to show that supplementation with *L. acidophilus* La 1 or with *B. bifidum* Bb12 increased the global phagocytic activity of blood phagocytes, especially granulocytes. Another strain, such as *L. johnsonii* La1 has also shown this effect (Donnet-Hughes et al. 1999). The strain *B. lactis* HN019 has been evaluated in several independent studies carried out in healthy elderly adults and results showing increased phagocytosis activity have been consistent (Arunachalam et al. 2000, Chiang et al. 2000, Gill et al. 2001a). The same has been shown for *L. rhamnosus* HN001 in middle aged and elderly subjects (Gill et al. 2001b, Sheih et al. 2001). Both of these strains, *B. lactis* HN019 and *L. rhamnosus* HN001 have shown in these studies an increase in the tumoricidal activity of NK cells (Gill et al. 2001a,c, Chiang et al. 2000).

Parra et al. (2004) studied the effects of *Lactobacillus casei* DN114001 fermented milk consumption on the immune response capacity in middle-age volunteers. After the trial the probiotic-treated group increased oxidative burst capacity of monocytes but not of granulocytes. On the contrary, no differences in global phagocytic activity were observed after the ingestion of this same strain in a study carried out in university students undergoing exams (Marcos et al. 2004).

In a study with elderly subjects consuming cheese fermented with two probiotic strains (10^9 CFU/day *L. rhamnosus* HN001 and *L. acidophilus* NCFM) an increase in both the phagocytic activity and the cytotoxicity activity of NK cells was observed. The phagocytic activity increased also in the control subjects consuming the control cheese with starter strains. A similar finding was observed by Olivares et al. (2006a) who testing a yogurt containing *L. gasseri* CECT 5714 and *L. coryniformis* CECT 5711 plus

S. thermophilus observed an increase in the percentage of phagocytic activity, not only in the probiotic group but also in the control group consuming plain yogurt. This suggests that, as mentioned for NK cytotoxicity activity, the promotion of phagocytosis is a characteristic of a wide number of LAB strains. Thus, something to bear in mind is how results are presented in each study and if the effect of the fermented milk per se has been considered with the appropriate control group. Finally, another useful hint when testing phagocytosis is to take into account batch variations in commercial kits, since systematic changes among subjects in all groups might be accounted for by batch-dependent shifts in measured activity (Christensen et al. 2006).

Prebiotics

A prebiotic supplementation of 1.3g oligosaccharides daily for 12 weeks in elderly subjects (84±7 y.) significantly decreased the serum levels of sCD14, a marker of macrophage activation, suggesting that this prebiotic can improve low-level inflammatory processes in this population (Schiffrin et al. 2007). The effect of 8g/day of β-fructans consumed during 4 weeks on immune parameters and functions was assessed in healthy adults (Lomax et al. 2012). Although the prebiotic showed a bifidogenic effect, it did not change significantly any of the immune parameters measured including phagocytic activity. A cross-over study with β-GOS treatment (5.5 g/day) or placebo (maltodextrine) for 10 weeks each, with a 4-week wash-out period showed an increase in phagocytosis after the intervention with the prebiotic in comparison with the placebo period (Vulevic et al. 2008).

References

Akbari, O., P. Stock, R.H. DeKruyff and D.T. Umetsu. 2003. Role of regulatory T cells in allergy and asthma. Curr. Opin. Immunol. 15: 627–633.

Akira, S. and K. Takeda. 2004. Toll-like receptor signalling. Nat. Rev. Immunol. 4: 499–511.

Arunachalam, K., H.S. Gill and R.K. Chandra. 2000. Enhancement of natural immune function by dietary consumption of *Bifidobacterium lactis* (HN019). Eur. J. Clin. Nutr. 54: 263–267.

Badia, R., M.T. Brufau, A.M. Guerrero-Zamora, R. Lizardo, I. Dobrescu, R. Martin-Venegas, R. Ferrer, H. Salmon, P. Martínez and J. Brufau. 2012. β-Galactomannan and *Saccharomyces cerevisiae* var. *boulardii* modulate the immune response against *Salmonella enterica* serovar *Typhimurium* in porcine intestinal epithelial and dendritic cells. Clin. Vaccine Immunol. 19: 368–76.

Banchereau, J. and R.M. Steinman. 1998. Dendritic cells and the control of immunity. Nature 392: 245–252.

Bernardo, D., B. Sánchez, H.O. Al-Hassi, E.R. Mann, M.C. Urdaci, S.C. Knight and A. Margolles. 2012. Microbiota/host crosstalk biomarkers: regulatory response of human intestinal dendritic cells exposed to Lactobacillus extracellular encrypted peptide. PLoS One 7:e36262.

Bjorck, P., J. Banchereau and L. Flores-Romo. 1997. CD40 ligation counteracts Fas-induced apoptosis of human dendritic cells. Int Immunol. 9: 365–72.

Bruzzese, E., M. Volpicelli, F. Salvini, M. Bisceglia, P. Lionetti, M. Cinquetti, G. Iacono and A. Guarino. 2006. Early administration of GOS/FOS prevents intestinal and respiratory infections in infants. J. Pediatr Gastroenterol. Nutr. 42. E95.

Candela, M., F. Perna, P. Carnevali, B. Vitali, R. Ciati, P. Gionchetti, F. Rizzello, M. Campieri and P. Brigidi. 2008. Interaction of probiotic Lactobacillus and Bifidobacterium strains with human intestinal epithelial cells: adhesion properties, competition against enteropathogens and modulation of IL-8 production. Int. J. Food Microbiol. 125: 286–292.

Chiang, B.L., Y.H. Sheih, L.H. Wang, C.K. Liao and H.S. Gill. 2000. Enhancing immunity by dietary consumption (*Bifidobacterium lactis* HN019): optimization and definition of cellular immune responses. Eur. J. Clin. Nutr. 54: 849–855.

Christensen, H.R., C.N. Larsen, P. Kaestel, L.B. Rosholm, C. Sternberg, K.F. Michaelsen and H. Frøkiaer. 2006. Immunomodulating potential of supplementation with probiotics: a dose-response study in healthy young adults. FEMS Immunol. Med. Microbiol. 47: 380–390.

Christensen, H.R., H. Frøkiaer and J.J. Pestka. 2002. Lactobacilli differentially modulate expression of cytokines and maturation surface markers in murine dendritic cells. J. Immunol. 2002 Jan 1; 168(1): 171–8.

Chuang, L., K.G. Wu, C. Pai, P.S. Hsieh, J.J. Tsai, J.H. Yen and M.Y. Lin. 2007. Heat-killed cells of lactobacilli skew the immune response toward T helper 1 polarization in mouse splenocytes and dendritic cell-treated T cells. J. Agric Food Chem. 55: 11080–11086.

Cools, N., P. Ponsaerts, V.F. Van Tendeloo and Z.N. Berneman. 2007. Balancing between immunity and tolerance: an interplay between dendritic cells, regulatory T cells, and effector T cells. J. Leukoc. Biol. 82: 1365–1374.

Cross, M.L. 2002. Microbes versus microbes: immune signals generated by probiotic lactobacilli and their role in protection against microbial pathogens. FEMS Immunol. Med. Microbiol. 34: 245–253.

Cukrowska, B., R. LodInová-ZádnIková, C. Enders, U. Sonnenborn, J. Schulze and H. Tlaskalová-Hogenová. 2002. Specific proliferative and antibody responses of premature infants to intestinal colonization with nonpathogenic probiotic *E. coli* strain Nissle 1917. Scand J. Immunol. 55: 204–209.

D'Andrea, A., M. Aste-Amezaga, N.M. Valiante, X. Ma, M. Kubin and G. Trinchieri. 1993. Interleukin 10 (IL-10) inhibits human lymphocyte interferon gamma-production by suppressing natural killer cell stimulatory factor/IL-12 synthesis in accessory cells. J. Exp. Med. 178: 1041–1048.

De Roock, S., M. van Elk, M.E. van Dijk, H.M. Timmerman, G.T. Rijkers, B.J. Prakken, M.O. Hoekstra and I.M. de Kleer. 2010. Lactic acid bacteria differ in their ability to induce functional regulatory T cells in humans. Clin. Exp. Allergy 40(1): 103–10.

De Roock, S., M. van Elk, M.O. Hoekstra, B.J. Prakken, G.T. Rijkers and I.M. de Kleer. 2011. Gut derived lactic acid bacteria induce strain specific CD4(+) T cell responses in human PBMC. Clin. Nutr. 30: 845–851.

de Vrese, M., P. Rautenberg, C. Laue, M. Koopmans, T. Herremans and J. Schrezenmeir. 2005. Probiotic bacteria stimulate virus-specific neutralizing antibodies following a booster polio vaccination. Eur. J. Nutr. 44: 406–413.

Delcenserie, V., D. Martel, M. Lamoureux, J. Amiot, Y. Boutin and D. Roy. 2008. Immunomodulatory effects of probiotics in the intestinal tract. Curr. Issues Mol. Biol. 10: 37–54.

Dong, H., I. Rowland and P. Yaqoob. 2012. Comparative effects of six probiotic strains on immune function *in vitro*. Br. J. Nutr. 108: 459–470.

Donnet-Hughes, A., F. Rochat, P. Serrant, J.M. Aeschlimann and E.J. Schiffrin. 1999. Modulation of nonspecific mechanisms of defense by lactic acid bacteria: effective dose. J. Dairy Sci. 82: 863–869.

Evrard, B., S. Coudeyras, A. Dosgilbert, N. Charbonnel, J. Alamé, A. Tridon and C. Forestier. 2011. Dose-dependent immunomodulation of human dendritic cells by the probiotic *Lactobacillus rhamnosus* Lcr35. PLoS One 6:e18735.

Fink, L.N., L.H. Zeuthen, H.R. Christensen, B. Morandi, H. Frøkiaer and G. Ferlazzo. 2007. Distinct gut-derived lactic acid bacteria elicit divergent dendritic cell-mediated NK cell responses. Int. Immunol. 19: 1319–1327.

Firmansyah, A., G.D. Pramita, A.L. Fâssler, F. Haschke and H. Link-Amster. 2000. Improved humoral immune response to measles vaccine in infants receiving infant cereal with fructo-oligosaccharides. J. Ped. Gastroenterol. Nutr. 31: S134.

Foligne, B., G. Zoumpopoulou, J. Dewulf, A. B. Younes, F. Chareyre, J.C. Sirard, B. Pot and C. Grangette. 2007. A Key Role of Dendritic Cells in Probiotic Functionality PLoS ONE 2: e313.

Ghadimi, D., R. Fölster-Holst, M. de Vrese, P. Winkler, K.J. Heller and J. Schrezenmeir. 2008. Effects of probiotic bacteria and their genomic DNA on TH1/TH2-cytokine production by peripheral blood mononuclear cells (PBMCs) of healthy and allergic subjects. Immunobiology 213: 677–692.

Gill, H.S. and K.J. Rutherfurd. 2001. Viability and dose-response studies on the effects of the immunoenhancing lactic acid bacterium *Lactobacillus rhamnosus* in mice. Br. J. Nutr. 86: 285–289.

Gill, H.S., K.J. Rutherfurd, M.L. Cross and P.K. Gopal. 2001a. Enhancement of immunity in the elderly by dietary supplementation with the probiotic *Bifidobactedum lactis* HN019. Am. J. Clin. Nutr. 74: 833–839.

Gill, H.S., K.J. Rutherfurd and M.L. Cross. 2001b. Dietary probiotic supplementation enhances natural killer cell activity in the elderly: An investigation of age-related immunological changes. J. Clin. Immunol. 21: 264–271.

Gill, H.S., K.J. Rutherfurd, M.L. Cross and P.K. Gopal. 2001c. Dietary probiotic supplementation to enhance cellular immunity in the elderly. Br. J. Biomed. Sci. 58: 94–6.

Gopalakrishnan, A., J.F. Clinthorne, E.A. Rondini, S.J. McCaskey, E.A. Gurzell, I.M. Langohr, E.M. Gardner and J.I. Fenton. 2012. Supplementation with galacto-oligosaccharides increases the percentage of NK cells and reduces colitis severity in Smad3-deficient mice. J. Nutr. 142: 1336–42.

Gori, A., G. Rizzardini, B. van't Land, K.B. Amor, J. van Schaik, C. Torti, T. Quirino, C. Tincati, A. Bandera, J. Knol, K. Benlhassan-Chahour, D. Trabattoni, D. Bray, A. Vriesema, G. Welling, J. Garssen and M. Clerici. 2011. Specific prebiotics modulate gut microbiota and immune activation in HAART-naive HIV-infected adults: results of the "COPA" pilot randomized trial. Mucosal Immunol. 4: 554–563.

Hart, A.L., K. Lammers, P. Brigidi, B. Vitali, F. Rizzello, P. Gionchetti, M. Campieri, M.A. Kamm, S.C. Knight and A.J. Stagg. 2004. Modulation of human dendritic cell phenotype and function by probiotic bacteria. Gut. 53: 1602–1609.

Hessle, C., B. Andersson and A.E. Wold. 2000. Gram-positive bacteria are potent inducers of monocytic interleukin-12 (IL-12) while gram-negative bacteria preferentially stimulate IL-10 production. Infect. Immun. 68: 3581–3586.

Holscher, H.D., L.A. Czerkies, P. Cekola, R. Litov, M. Benbow, S. Santema, D.D. Alexander, V. Perez, S. Sun, J.M. Saavedra and K.A. Tappenden. 2012. *Bifidobacterium lactis* Bb12 enhances intestinal antibody response in formula-fed infants: a randomized, double-blind, controlled trial. JPEN J. Parenter Enteral Nutr. 36: 106S–117S.

Ichikawa, S., M. Miyake, R. Fujii and Y. Konishi. 2009. Orally administered *Lactobacillus paracasei* KW3110 induces *in vivo* IL-12 production. Biosci. Biotechnol. Biochem. 73: 1561–1565.

Isolauri, E., Y. Sutas, P. Kankaanpaa, H. Arvilommi and S. Salminen. 2001. Probiotics: effects on immunity. Am. J. Clin. Nutrit. 73: 444S–450S.

Jensen, G.S., K.F. Benson, S.G. Carter and J.R. Endres. 2010. GanedenBC30 cell wall and metabolites: anti-inflammatory and immune modulating effects *in vitro*. BMC Immunol. 11: 15.

Karlsson, H., C. Hessle and A. Tudin. 2002. Innate immune responses of human neonatal cells to bacteria from the normal gastrointestinal flora. Infect. Immun. 70: 6688–6696.

Kelly-Quagliana, K.A., P.D. Nelson and R.K. Buddington. 2003. Dietary oligofructose and inulin modulate immune functions in mice. Nutr. Res. 23: 257–267.

Knol, J., P. Scholtens, C. Kafka, J. Steenbakkers, S. Gro, K. Helm, M. Klarczyk, H. Schöpfer, H.M. Böckler and J. Wells. 2005. Colon microflora in infants fed formula with galacto- and fructo-oligosacharides: More like breast fed infants. J. Pediatr Gastroenterol. Nutr. 40: 36–42.

Kodama, T., K. Takeda, O. Shimozato, Y. Hayakawa, M. Atsuta, K. Kobayash, M. Ito, H. Yagita and K. Okumura. 1999. Perforin-dependent NK cell cytotoxicity is sufficient for anti-metastatic effect of IL-12. Eur. J. Immunol. 29: 1390–1396.

Latvala, S., M. Miettinen, R.A. Kekkonen, R. Korpela and I. Julkunen. 2011. *Lactobacillus rhamnosus* GG and *Streptococcus thermophilus* induce suppressor of cytokine signalling 3 (SOCS3) gene expression directly and indirectly via interleukin-10 in human primary macrophages. Clin. Exp. Immunol. 165: 94–103.

Latvala, S., T.E. Pietila, V. Veckman, R.A. Kekkonen, S. Tynkkynen, R. Korpela and I. Julkunen. 2008. Potentially probiotic bacteria induce efficient maturation but differential cytokine production in human monocyte-derived dendritic cells. World J. Gastroenterol. 14: 5570–5583; discussion 5581–2.

Lomax, A.R., L.V. Cheung, K.M. Tuohy, P.S. Noakes, E.A. Miles and P.C. Calder. 2012. β2-1 Fructans have a bifidogenic effect in healthy middle-aged human subjects but do not alter immune responses examined in the absence of an *in vivo* immune challenge: results from a randomised controlled trial. Br. J. Nutr. 108: 1818–1828.

Lomax, A.R. and P.C. Calder. 2009. Probiotics, immune function, infection and inflammation: a review of the evidence from studies conducted in humans. Curr. Pharm. Des. 15: 1428–1518.

López, P., I. González-Rodríguez, B. Sánchez, M. Gueimonde, A. Margolles and A. Suárez. 2012. Treg-inducing membrane vesicles from *Bifidobacterium bifidum* LMG13195 as potential adjuvants in immunotherapy. Vaccine 30: 825–829.

López, P., M. Gueimonde, A. Margolles and A. Suárez. 2010. Distinct Bifidobacterium strains drive different immune responses *in vitro*. Int. J. Food Microbiol. 138: 157–165.

Marcos, A., J. Wärnberg, E. Nova, S. Gómez, A. Alvarez, R. Alvarez, J.A. Mateos and J.M. Cobo. 2004. The effect of milk fermented by yogurt cultures plus *Lactobacillus casei* DN-114001 on the immune response of subjects under academic examination stress. Eur. J. Nutr. 43: 381–389.

Marques, R. and I.G. Boneca. 2011. Expression and functional importance of innate immune receptors by intestinal epithelial cells. Cell Mol. Life Sci. 68: 3661–3673.

Mazzoni, A. and D.M. Segal. 2004. Controlling the Toll road to dendritic cell polarization. J. Leukoc Biol. 75: 721–730.

Mizubuchi, H., T. Yajima, N. Aoi, T. Tomita and Y. Yoshikai. 2005. Isomalto-oligosaccharides polarize Th1-like responses in intestinal and systemic immunity in mice. J. Nutr. 135: 2857–2861.

Mohamadzadeh, M., S. Olson, W.V. Kalina, G. Ruthel, G.L. Demmin, K.L. Warfield, S. Bavari and T.R. Klaenhammer. 2005. Lactobacilli activate human dendritic cells that skew T cells toward T helper 1 polarization. Proc. Natl. Acad. Sci. USA 102: 2880–2885.

Morimoto, K., T. Takeshita, M. Nanno, S. Tokudome and K. Nakayama. 2005. Modulation of natural killer cell activity by supplementation of fermented milk containing *Lactobacillus casei* in habitual smokers. Prev. Med. 40: 589.

Moro-García, M.A., R. Alonso-Arias, M. Baltadjieva, C. Fernández Benítez, M.A. Fernández Barrial, E. Díaz Ruisánchez, R. Alonso Santos, M. Alvarez Sánchez, J. Saavedra Miján and C. López-Larrea. 2012. Oral supplementation with *Lactobacillus delbrueckii* subsp. *bulgaricus* 8481 enhances systemic immunity in elderly subjects. Age (Dordr). DOI 10.1007/s11357-012-9434-6.

Nanno, M., I. Kato, T. Kobayashi and K. Shida. 2011. Biological effects of probiotics: what impact does *Lactobacillus casei* shirota have on us? Int. J. Immunopathol. Pharmacol. 24: 45S–50S.

Nermes, M., J.M. Kantele, T.J. Atosuo, S. Salminen and E. Isolauri. 2011. Interaction of orally administered *Lactobacillus rhamnosus* GG with skin and gut microbiota and humoral immunity in infants with atopic dermatitis. Clin. Exp. Allergy 41: 370–377.

Niers, L.E., H.M. Timmerman, G.T. Rijkers, G.M. van Bleek, N.O. van Uden, E.F. Knol, M.L. Kapsenberg, J.L. Kimpen and M.O. Hoekstra. 2005. Identification of strong interleukin-10 inducing lactic acid bacteria which down-regulate T helper type 2 cytokines. Clin. Exp. Allergy 35: 1481–1489.

Nova, E., O. Toro, P. Varela, I. López-Vidriero, G. Morandé and A. Marcos. 2006. Effects of a nutritional intervention with yogurt on lymphocyte subsets and cytokine production capacity in anorexia nervosa patients. Eur. J. Nutr. 45: 225–233.

Olivares, M., M.P. Díaz-Ropero, N. Gómez, F. Lara-Villoslada, S. Sierra, J.A. Maldonado, R. Martin, J.M. Rodriguez and J. Xaus. 2006a. The consumption of two new probiotic strains *Lactobacillus gasseri* CECT 5714 and *Lactobacillus coryniformis* CECT 5711, boosts the immune system of healthy humans. Int. Microbiol. 9: 47–52.

Olivares, M., M.P. Díaz-Ropero, N. Gómez, S. Sierra, F. Lara-Villoslada, R. Martín, R.M.J. Rodriguez and J. Xaus. 2006b. Dietary deprivation of fermented foods causes a fall in innate immune response. Lactic acid bacteria can counteract the immunological effect of this deprivation. J. Dairy Res. 73: 492–498.

Olivares, M., M.P. Díaz-Ropero, S. Saleta Sierra, F. Lara-Villoslada, J. Fonollá, M. Navas, J.M. Rodríguez and J. Xaus. 2007. Oral intake of *Lactobacillus fermentum* CECT5716 enhances the effects of influenza vaccination. Nutrition 23: 254–260.

Paineau, D., D. Carcano, D. Leyer, S. Darquy, M.A. Alyanakian, G. Simoneau, J.F. Bergmann, D. Brassart, F. Bornet and A.C. Ouwehand. 2008. Effects of seven potential probiotic strains on specific immune responses in healthy adults: a double-blind, randomized, controlled trial. FEMS Immunol. Med. Microbiol. 53: 107–113.

Parra, D., B.M. De Morentin, J.M. Cobo, A. Mateos and J.A. Martinez. 2004. Monocyte function in healthy middle-aged people receiving fermented milk containing *Lactobacillus casei*. J. Nutr. Health Aging 8: 208–211.

Pozo-Rubio, T., J.R. Mujico, A. Marcos, E. Puertollano, I. Nadal, Y. Sanz and E. Nova. 2011. Immunostimulatory effect of faecal Bifidobacterium species of breast-fed and formula-fed infants in a peripheral blood mononuclear cell/Caco-2 co-culture system. Br. J. Nutr. 106: 1216–1223.

Pujol, P., J. Huguet, F. Drobnic, M. Banquells, O. Ruiz and P. Galilea. 2000. The effect of fermented milk containing *Lactobacillus casei* on the immune response to exercise. Sports Med. Train Rehab. 9: 209–223.

Raes M., P.A. Scholtens, P. Alliet, K. Hensen, H. Jongen, G. Boehm, Y. Vandenplas and J.L. Rummens. 2010. Exploration of basal immune parameters in healthy infants receiving an infant milk formula supplemented with prebiotics. Pediatr Allergy Immunol. 21: e377–385.

Rinne, M., M. Kalliomaki, H. Arvilommi, S. Salminen and E. Isolauri. 2005. Effect of probiotics and breastfeeding on the bifidobacterium and lactobacillus/enterococcus microbiota and humoral immune responses. J. Pediatr. 147: 186–91.

Rizzardini, G., D. Eskesen, P.C. Calder, A. Capetti, L. Jespersen and M. Clerici. 2012. Evaluation of the immune benefits of two probiotic strains *Bifidobacterium animalis* ssp. *lactis*, BB-12® and *Lactobacillus paracasei* ssp. *paracasei*, *L. casei* 431® in an influenza vaccination model: a randomised, double-blind, placebo-controlled study. Br. J. Nutr. 107: 876–884.

Rizzello, V., I. Bonaccorsi, M.L. Dongarrà, L.N. Fink and G. Ferlazzo. 2011. Role of natural killer and dendritic cell crosstalk in immunomodulation by commensal bacteria probiotics. J. Biomed. Biotechnol. 473097.

Roller, M., G. Rechkemmer and B. Watzl. 2004. Prebiotic inulin enriched with oligofructose in combination with the probiotics *Lactobacillus rhamnosus* and *Bifidobacterium lactis* modulates intestinal immune functions in rats. J. Nutr. 134: 153–156.

Ruiz, P.A., M. Hoffmann, S. Szcesny, M. Blaut and D. Haller. 2005. Innate mechanisms for *Bifidobacterium lactis* to activate transient pro-inflammatory host responses in intestinal epithelial cells after the colonization of germ-free rats. Immunology 115: 441–450.

Ryz, N.R., J.B. Meddings and C.G. Taylor. 2009. Long-chain inulin increases dendritic cells in the Peyer's patches and increases *ex vivo* cytokine secretion in the spleen and mesenteric lymph nodes of growing female rats, independent of zinc status. Br. J. Nutr. 101: 1653–1663.

Schiffrin, E.J., F. Rochat, H. Link-Amster, J.M. Aeschlimann and A. Donnet-Hughes. 1995. Immunomodulation of human blood cells following the ingestion of lactic acid bacteria. J. Dairy Sci. 78(3): 491–497.

Schiffrin, E.J., D.R. Thomas, V.B. Kumar, C. Brown, C. Hager, M.A. Van't Hof, J.E. Morley and Y. Guigoz. 2007. Systemic inflammatory markers in older persons: the effect of oral nutritional supplementation with prebiotics. J. Nutr. Health Aging 11: 475–479.

Schouten, B., B.C. van Esch, G.A. Hofman, L. Boon, L.M. Knippels and L.E. Willemsen. 2010. Oligosaccharide-induced whey-specific CD25(+) regulatory T-cells are involved in the suppression of cow milk allergy in mice. J Nutr. 140: 835–841.

Schouten, B., B.C. van Esch, G.A. Hofman, S. de Kivit, L. Boon and L.M. Knippels. 2012. A potential role for CD25+ regulatory T-cells in the protection against casein allergy by dietary non-digestible carbohydrates. Br. J. Nutr. 107: 96–105.

Schumacher, G., G. Bendas, B. Stahl and C. Beermann. 2006. Human milk oligosaccharides affect P-selectin binding capacities: *in vitro* investigation. Nutrition 22: 620–627.

Sheih, Y.H., B.L. Chiang, L.H. Wang, C.K. Liao and H.S. Gill. 2001. Systemic immunity-enhancing effects in healthy subjects following dietary consumption of the lactic acid bacterium *Lactobacillus rhamnosus* HN001. J. Am. Coll. Nutr. 20: 149–156.

Sierra, S., F. Lara-Villoslada, L. Sempere, M. Olivares, J. Boza and J. Xaus. 2010. Intestinal and immunological effects of daily oral administration of *Lactobacillus salivarius* CECT5713 to healthy adults. Anaerobe. 16(3): 195–200.

Smits, H.H., A. Engering, D. van der Kleij, E.C. de Jong, K. Schipper K. T.M. van Capel, B.A. Zaat M. Yazdanbakhsh, E.A. Wierenga, Y. van Kooyk and M.L. Kapsenberg. 2005. Selective probiotic bacteria induce IL-10-producing regulatory T cells *in vitro* by modulating dendritic cell function through dendritic cell-specific intercellular adhesion molecule 3-grabbing nonintegrin. J. Allergy Clin. Immunol. 115: 1260–1267.

Spanhaak, S., R. Havenaar and G. Schaafsma. 1998. The effect of consumption of milk fermented by *Lactobacillus casei* strain Shirota on the intestinal microflora and immune parameters in humans. Eur. J. Clin. Nutr. 52: 899–907.

Stam, J., M. van Stuijvenberg, J. Garssen, K. Knipping and P.J. Sauer. 2011. A mixture of three prebiotics does not affect vaccine specific antibody responses in healthy term infants in the first year of life. Vaccine 29: 7766–7772.

Steinman, R.M. and M.C. Nussenzweig. 2002. Avoiding horror autotoxicus: the importance of dendritic cells in peripheral T cell tolerance. Proc. Natl. Acad. Sci. USA 99: 351–358.

Takeda, K. and K. Okumura. 2007. Effects of a fermented milk drink containing *Lactobacillus casei* strain Shirota on the human NK-cell activity. J. Nutr. 137: 791S–793S.

Takeda, K., T. Suzuki, S.I. Shimada, K. Shida, M. Nanno and K. Okumura. 2006. Interleukin-12 is involved in the enhancement of human natural killer cell activity by *Lactobacillus casei* Shirota. Clin. Exp. Immunol. 146: 109–115.

Thakur, M., P. Connellan, M.A. Deseo, C. Morris, W. Praznik, R. Loeppert and V.K. Dixit. 2012. Characterization and *in vitro* immunomodulatory screening of fructo-oligosaccharides of Asparagus racemosus Willd. Int. J. Biol. Macromol. 50: 77–81.

Trinchieri, G. 1993. Interleukin-12 and its role in the generation of TH1 cells. Immunol. Today14: 335–338.

Vaisman, N., J. Press, E. Leibovitz, G. Boehm and V. Barak. 2010. Short-term effect of prebiotics administration on stool characteristics and serum cytokines dynamics in very young children with acute diarrhea. Nutrients 2: 683–692.

van Hoffen, E., B. Ruiter, J. Faber, L. M'Rabet, E.F. Knol, B. Stahl, S. Arslanoglu, G. Moro, G. Boehm and J. Garssen. 2009. A specific mixture of short-chain galacto-oligosaccharides and long-chain fructo-oligosaccharides induces a beneficial immunoglobulin profile in infants at high risk for allergy. Allergy 64: 484–487.

van Stuijvenberg, M., A.M. Eisses, C. Grüber, F. Mosca, S. Arslanoglu, G. Chirico, C.P. Braegger, J. Riedler, G. Boehm and P.J. Sauer. 2011. Do prebiotics reduce the number of fever episodes in healthy children in their first year of life: a randomized controlled trial? Br. J. Nutr. 106: 1740–1748.

van Vlies N., A. Hogenkamp, S. Thijssen, G.M. Dingjan, K. Knipping, J. Garssen and L.M. Knippels. 2012. Effects of short-chain galacto- and long-chain fructo-oligosaccharides on systemic and local immune status during pregnancy. J. Reprod Immunol. 94: 161–168.

van't Land, B., M. Schijf, B.C. van Esch, J. van Bergenhenegouwen, J. Bastiaans and B. Schouten. 2010. Regulatory T-cells have a prominent role in the immune modulated vaccine response by specific oligosaccharides. Vaccine 28: 5711–5717.

Vos, A.P., M. Haarman, J.W. van Ginkel, J. Knol, J. Garssen, B. Stahl, G. Boehm and L. M'Rabet. 2007. Dietary supplementation of neutral and acidic oligosaccharides enhances Th1-dependent vaccination responses in mice. Pediatr Allergy Immunol. 18: 304–312.

Vulevic, J., A. Drakoularakou, P. Yaqoob, G. Tzortzis and G.R. Gibson. 2008. Modulation of the fecal microflora profile and immune function by a novel trans-galactooligosaccharide mixture (B-GOS) in healthy elderly volunteers. Am. J. Clin. Nutr. 88: 1438–1446.

West, C.E., L. Gothefors, M. Granström, H. Käyhty, M.L. Hammarström and O. Hernell. 2008. Effects of feeding probiotics during weaning on infections and antibody responses to diphtheria, tetanus and Hib vaccines. Pediatr Allergy Immunol. 19: 53–60.

West, C.E., O. Hernell, Y. Andersson, M. Sjöstedt and M.L. Hammarström. 2012. Probiotic effects on T-cell maturation in infants during weaning. Clin. Exp. Allergy 42: 540–549.

Zhu, J. and W.E. Paul. 2010. Heterogeneity and plasticity of T helper cells. Cell Res. 20: 4–12.

14

Probiotics and Prebiotics in Pediatric Diarrheal Disorders

Y. Vandenplas, B. Hauser, E. De Greef, T. Devreker* and *G. Veereman-Wauters*

Introduction

The human body is home to more than 1 trillion microbes, with the gastrointestinal (GI) tract alone harboring a diverse array of commensal microbes that are considered to contribute to host nutrition, developmental regulation of intestinal angiogenesis, protection from pathogens and development of the immune response (Johnson and Versalovic 2012).

Probiotics are non-pathogenic live micro-organisms that resist normal digestion to reach the colon alive, which, when consumed in adequate amounts, have a positive effect on the health of the host. Although probiotics are today a "hot topic", they are not new. More than 2000 years ago, the Roman author Plinius The Old, recommended fermented milk in the treatment of acute gastroenteritis. The word "probiotic" was used for the first time in the 1960s and means "for life" (from the Greek *πρo βίος, pro bios*). The positive effects of certain bacteria have been noted for more than a century (Lilly and Stillwell 1965). As early as in 1906, Tissier noted that significant stool colonization with bifidobacteria was protective against the likelihood of the development of diarrhea in children (Tissier 1906). There

UZ Brussel, Vrije Universiteit Brussel, Brussels, Belgium.
* Corresponding author: yvanvandenplas@uzbrussel.be

are bacterial and yeast probiotics. The best known bacterial probiotics are lactobacilli and bifidobacteria. *S boulardii* is a non-pathogenic yeast isolated from the lychee fruit and introduced in France for the treatment of diarrhea since 1950. The number of commercialized products and the number of publications on probiotics in different conditions has literally exploded during recent years.

Every probiotic would need a set of minimal requirements, including strain designation and shelf life. The lowest category would include yogurts that reduce adverse effects in lactose-intolerant individuals; other 'category 1' products would require only minimally documented studies in humans. For the 'middle category', at least two randomized controlled studies would be needed to show how the probiotic acts, with results published in peer-reviewed journals. The 'third category' would be reserved for products targeting vulnerable people such as infants and elderly. It would include recombinant strains and species not previously used in foods and supplements, for example, bacteria producing neurochemicals that could improve cognitive function or memory. Strict adjudication would be required for products in 'category 3' (Reid 2012).

Prebiotics are non-digestible food ingredients that stimulate the growth and/or activity of bacteria in the digestive system in ways claimed to be beneficial to health. They were first identified and named by Marcel Roberfroid in 1995. A prebiotic is a selectively fermented ingredient that allows specific changes, both in the composition and/or activity in the gastrointestinal microflora that confers benefits upon host well-being and health. Prebiotics must "survive" acidic conditions, must evade digestion in the small intestine and must be selectively fermented in the colon. Researchers now also focus on the distinction between short-chain, long-chain, and full-spectrum prebiotics. "Short-chain" prebiotics, e.g., oligofructose, contain 2–8 links per saccharide molecule and are typically fermented more quickly in the colon ascendens providing nourishment to the bacteria in that area. Long-chain prebiotics, e.g., inulin, contain 9-64 links per saccharide molecule, and tend to be fermented more slowly, nourishing bacteria predominantly in the colon transversum and descendens and sigmoid. Full-spectrum prebiotics provide the full range of molecular link-lengths from 2–64 links per molecule, and nourish bacteria throughout the colon, e.g., oligofructose-enriched inulin. The majority of research done on prebiotics is based on full-spectrum prebiotics.

The Commercialized Products

Quality control of the commercialized products is fundamental, and is more important for probiotics than for prebiotics (Vanhee et al. 2010). Most probiotics are registered as food supplement, and do not have to fulfill the regulations and quality requirements that exist for medication. Yet, during recent years many of these food supplements are used in medical indications. Food industry includes selected micro-organisms in food, primarily in milk-drinks or yoghurts. Some of these probiotic food supplements are commercialized in capsules, increasing the confusion between "food" and "medication". Temperature and humidity is a major determinant for the viability of *S. boulardii* (Tung et al. 2009). In almost half of the food supplements and 40% of the dairy products the label is not correct (Temmerman et al. 2003). Mislabeling of food supplements is a worldwide problem (Elliott and Teversham 2004). Since there is no legal protection for food supplements as there is for medication, companies

Table 1. Most important microorganisms applied in probiotic products for human use.

Lactic acid bacteria	***Bifidobacterium***	**Other bacteria**	**Yeasts**
Lactobacillus	*Bf. adolescentis*	*Bacillus*	*Saccharomyces cerevisiae* var. boulardii
Lb. acidophilus	*Bf. animalis* subsp. *lactis*	*Bc. cereus*	*Saccharomyces* spp.
Lb. casei/paracasei	*Bf. bifidum*	*Bc. coagulans*	
Lb. crispatus	*Bf. breve*	*Bc. clausii*	
Lb. fermentum	*Bf. longum* subsp. *infantis*	*Bc. pumilus*	
Lb. gallinarum	*Bf. longum* subsp. *longum*	*Bc. subtilis*	
Lb. gasseri		*Escherichia coli* Nissle 1917	
Lb. johnsonii		*Propionibacterium*	
Lb. plantarum		*Pr. acidipropionici*	
Lb. reuteri		*Pr. freudenreichii* subsp. *shermanii*	
Lb. rhamnosus		*Pr. jensenii*	
Lb. salivarius			
Enterococcus faecium			
Lactococcus lactis subsp. *lactis*			
Leuconostoc			
Le. citreum			
Le. mesenteroides subsp. *cremoris*			
Oenococcus oeni			
Pediococcus			
Pd. acidilactici			
Pd. pentosaceus			
Sporolactobacillus inulinus			

may refuse to provide information on the exact strains in the product (Vandenplas 2012).

Fundamental research on the mechanisms of action of specific strains and clinical trials with commercialized products are mandatory since *in vitro* effects of a strain may display opposite behavior *in vivo* (Ibnou-Zekri et al. 2003). Effects demonstrated for one strain cannot be extrapolated to other strains, even if they belong to the same species. Bacterial probiotic strains comprise different *lactobacilli* (L.) (*L. casei GG, L. reuteri, L. LA5, ...)* and bifidobacteria (B.) (*B. Bb12,...*), but to a certain extent also non-pathogenic *Escherichia (E.) coli (E. coli Nissle 1917)* and some strains of enterococci (although relevant transfer of plasmid induced resistance was reported with enterococci). The yeast *Saccharomyces* (*S.) boulardii* is the only non-bacterial probiotic strain known.

Since some commercialized products consist of combinations of different strains, laboratory and even more clinical testing of each combination of these strains is mandatory. *L. acidophilus LB* has been shown to have antibacterial activity against *E. coli*. However, if the *E. coli* is present in the GI tract of the host prior to the *L. acidophilus LB*, as occurs in acute gastroenteritis, its antibacterial activity is strongly reduced by non-specific steric hindrance of the receptor sites (Marteau et al. 1992). Adherence of *B. Bb12* improves in the presence of *L. casei GG*, both in healthy infants and during episodes of diarrhea, suggesting that synergism may as well occur (Juntunen et al. 2001). An effect of dose and duration of administration should also be considered. Low doses and short duration are less effective than high doses and long duration (Ritchie and Romanuk 2012).

More and more "synbiotics" are commercialized: combinations of a prebiotic and most of the time several strains of probiotics.

Prevention of Acute Infectious Gastroenteritis (GE)

Up to 70 to 80% of infectious diarrheas are of viral origin. Till date, rotavirus has been the most prevalent pathogen. However, a global rotavaccination may in the near future alter the epidemiology of infectious gastroenteritis.

The longer an infant is breastfed and the longer breastfeeding is exclusive, the better the protection from infectious diseases such as gastroenteritis. Promotion of (exclusive) breastfeeding should be maximally endorsed.

Probiotics

More than ten years ago, Saavedra et al. demonstrated that *Streptococcus (Str.) thermophilus* and *B. bifidum* (later renamed *B. lactis*) prevent nosocomial acquired diarrhea in a small group of children admitted in a chronic care institution (Saavedra et al. 1994). Saran et al. showed that feeding fermented milk over a period of 6 months to Indian infants, resulted in a significantly better weight gain and a 50% reduction of episodes of infectious diarrhea (Saran et al. 2002). *L. casei GG, L. reuteri, B. lactis* were shown to have a very modest effect (statistically significant, but of questionable clinical importance) on the prevention of community-acquired diarrhea (Szajewska et al. 2006). *L. reuteri* protects for the development of diarrhea in Indonesian children with malnutrition (Agustina et al. 2012). Regarding the prevention of diarrhea acquired in day-care centers, *L. caseii GG, B. lactis, Str. thermophilus, L. reuteri, L. rhamnosus* (not GG), and *L. acidophilus* added to infant formula or given as capsules have been tested in various trials either alone or in comparison with each other. The evidence of efficacy is only modest for the prevention of diarrhea, but somewhat better for prevention of upper respiratory infections (Guandalini 2011).

Literature on the efficacy of *L. casei GG* in the prevention of acute GE is contradictory. There are data showing that *L. casei GG* reduces nosocomial infection, especially for Rotavirus gastroenteritis (Szajewska et al. 2001). But, a double-blind randomized study in 220 children did not show a statistically significant protective effect of *L. casei GG* for nosocomial rotaviral infection (Mastretta et al. 2002).

The viable strain *B. lactis BB12* did not reduce diarrhea in 90 healthy infants living in residential nurseries or foster care centers when compared with placebo (28.3% vs 38.7%; Relative Risk (RR) 0.7 (95% CI 0.4–1.3)) (Chouraqui et al. 2004). A formula fermented with *B. breve c50* and *Str. thermophilus* 065 was well accepted and resulted in normal growth of infants (Thibault et al. 2004). However, incidence, duration of diarrhea episodes, and number of hospital admissions did not differ significantly between groups. Episodes were less severe in the fermented formula group with fewer cases of dehydration, fewer medical consultations and fewer prescriptions of oral rehydration solutions (Thibault et al. 2004).

Seven children would need to be treated with a probiotic to prevent one patient from developing nosocomial rota-gastroenteritis (Szajewska and Mrukowicz 2005). The protective effect on prevention of diarrhea becomes even less significant if the incidence of diarrhea (episodes per patient-month) rather than the percentage of patients with diarrhea would be used as efficacy parameter (Szajewska and Mrukowicz 2005). The preventive beneficial action of probiotic-enriched formula is less obvious in the developed world.

The European Society of Pediatric Gastroenterology, Hepatology and Nutrition (ESPGHAN) concluded that there was insufficient evidence to recommend the use of infant formula enriched with probiotics mainly because the efficacy shown was insufficiently convincing (Braegger et al. 2011). The conclusions of the American Academy of Pediatrics are quite similar: "available data do not support routine use of probiotics to prevent nosocomial rotavirus diarrhea in child care centers. But, there may be special circumstances in which probiotic use in children in long-term health care facilities or in child care centers is beneficial" (Thomas and Greer 2010). On the other hand, most studies do show some benefit although for different endpoints and not always statistically significant, and some studies are negative (Agostoni et al. 2004). Serious adverse events of probiotics were not reported in the trials with infant formula. In summary: although the evidence is limited, the majority of the studies show a trend of some positive effect and no adverse events.

Prebiotics

The majority of the studies with prebiotics have been performed with a specific galacto-oligosaccharides (GOS)/fructoligosaccahrides (FOS) mixture added to infant formula. Short chain GOS comprises the most natural "lactose-derived" oligosaccharide showing the lowest incidence of side effects (gas production, bloating) compared to similar short chain (sc) oligosaccharides. Long chain FOS comprises a more slowly fermentable substrate to allow fermentation all over the full length of the large intestine. Prebiotic supplementation with a GOS/FOS mixture results in more fecal bifidobacteria and lactobacilli compared with a placebo group (Salvini 2011). These differences between the groups were maintained during the second half of the first year of life without any prebiotic supplementation (Salvini 2011). GOS and FOS alone also stimulate the growth of bifidobacteria (Ben 2008). The effect of the GOS/FOS mixture on the prevention of diarrhea is not convincing (Arslanoglu et al. 2007).

Synbiotics

There was no effect of prebiotics and the probiotic *B. lactis HN019* on diarrhea (6% reduction, 95% Confidence Interval [CI]: –1 to 12%; $p = 0.08$). Incidence of dysentery episodes was reduced by 21% (95% CI: 0 to 38%; $p = 0.05$) (Sazawal et al. 2010). *B. lactis HN019* and GOS fortified milk did not reduce diarrhea overall (10% reduction) but reduced significantly bloody diarrhea (Sazawal et al. 2004). However, iron status was better in the synbiotic group (Sazawal et al. 2004). This effect could be either due to better absorption because of the effects on gut flora or secondary to morbidity prevention effects (Sazawal et al. 2004).

Treatment of Acute Infectious GE

The cornerstone of the treatment of acute GE (oral rehydration solution (ORS) and rapid realimentation) will not be discussed. Treatment of acute infectious GE should focus on the pathophysiological consequences of the condition: loss of water and electrolytes and a disturbance of the gastrointestinal ecosystem.

Probiotics

Probiotics address the second pathophysiological aspect of acute GE, abnormal gastro-intestinal flora. There are almost no studies performed in chronic diarrhea of infectious origin.

Compared to a placebo, *L. casei GG* ($2x10^{11}$ b.d. for 5 days) significantly reduced the duration of hospitalization of children treated with ORS because of rotavirus diarrhea (1.4 *versus (vs)* 2.4 days) (Isolauri et al. 1991). *L. casei GG* ($1x10^{11}$ cfu/g b.d. for 2 days) reduced the number of children with diarrhea after 48 hours in a study in 40 Pakistani children admitted for severe diarrhea and malnutrition (31% (*L. GG*) *vs* 75% (placebo)) (Raza et al. 1995). Another clinical trial with *L. casei GG* ($5x10^{9}$ cfu/g b.d. for 5 days and ORS) in 123 hospitalized children (33% with rotavirus) showed a reduction of the duration of viral diarrhea (2.7 vs 3.7 days) (Shornikova et al. 1997). *L. casei GG* ($3x10^{9}$ cfu/g b.d. for a maximum of 6 days) decreased the duration of diarrhea by half in out-patient children and significantly reduced rotavirus shedding (Guarino et al. 1997). A multi-center European prospective, randomized controlled trial (RCT) with *L. casei GG* (10^{10}cfu/250ml) as add-on to ORS in 287 children with acute diarrhea, showed a significant decrease of the duration of diarrhea by about 10% (a mean duration of diarrhea of 123 hours in the placebo group versus 110 hours in the intervention group) (Guandalini et al. 2000). A more detailed analysis showed that the difference was greatest in the rota-positive group (115 vs 136 hours) and that there was no difference in the subgroup with invasive pathogens (about 1/5th of all inclusions) (124 vs 121 hours duration of diarrhea) (Guandalini et al. 2000). Comparable results have been recently obtained in a double-blind RCT with a mixture of three *L. rhamnosus* strains (573L/1; 573L/2; 573L/3 at a dose 1.2×10^{10} CFU, twice daily, for 5 days) in 87 children with infectious diarrhea. *L. rhamnosus* strains significantly shortened the duration of rotavirus diarrhea (76 + 35 h vs. 115 + 67 h (P = 0.03)) but not of diarrhea of other etiology (Szymanski et al. 2006). Gut colonization with administered strains was 80% and 41% at 5 and 14 days, respectively. Intervention also shortened the time of intravenous rehydration (15 + 14 h vs. 38 + 33 h (P = 0.006)) although factors such as physician variability may have influenced the outcome (Szymanski et al. 2006).

The efficacy of *L. reuteri* was evaluated in 66 hospitalized children with rotavirus diarrhea. Randomisation was done in 3 groups: placebo in one group and two groups with different doses of *L. reuteri* (10^7 and 10^{10} cfu/g once a day for 5 days). The probiotic reduced the duration of diarrhea with a dose-dependent effect (2.5 days in the placebo group vs. 1.9 and 1.5 in the *L. reuteri* groups, respectively) (Shornikova et al. 1997). *L. acidophilus LB* (Lacteol Fort®, a product containing heat-killed lactobacilli, 10^{10} for 5 doses) was tested in 73 children with acute diarrhea (50% rotavirus positive) and did result in a similar reduction of the duration of diarrhea (43 vs 57 hours) (Simakachorn et al. 2000).

However, there are also many studies in which lactobacilli failed to shorten the duration of diarrhea. Several RCTs in developing countries negated the beneficial effect of *L. casei GG* and *L. acidophilus* in acute diarrhea, likely related to the distinct etiological profile (Costa-Ribeiro et al. 2003, Khanna et al. 2005, Salazar-Lindo et al. 2004, Kowalska-Duplaga et al. 2004, Sarker et al. 2005). In children with more severe diarrhea, there was no demonstrable benefit of *L. casei GG* (Khanna et al. 2005, Kowalska-Duplaga et al. 2004). Absence of shortening of the duration of diarrhea was also reported for a mixture of *L acidophilus, B. bifidum* (later renamed *B. lactis*) and *L. bulgaricus* (Kowalska-Duplaga et al. 2004). *L. paracasei* strain *ST11* did not reduce the volume of stool output in rotavirus infection but improved the outcome of non-rotavirus diarrhea in children in Bangladesh (Sarker et al. 2005).

The first double-blind, prospective RCT with the yeast *S. boulardii* was performed more than 15 years ago: diarrhea persisted for more than 7 days in 12% of the placebo group and in only 3% of the *S. boulardii* group (Höchter et al. 1990). Since then, several double-blind, prospective RCTs were performed with *S. boulardii* in children with acute GEs and showed, a significantly better outcome in comparison to the placebo. Kurugol treated 200 children, with acute diarrhea, with 250 mg *S. boulardii* or placebo for 5 days: duration of both diarrhea and hospital stay decreased by approximately 24 hours (Kurugol and Koturoglu 2005). Villaruel and co-workers showed in an ambulatory care in Argentina that in the group treated with *S. boulardii,* diarrhea persisted for more than 7 days in 7% compared to 27% in the placebo group, with a greater effect if treatment was started within the first two days of the disease (Villarruel et al. 2007).

S. boulardii improved tolerance of feeding in children with chronic Giardia Lamblia infection (Castañeda et al. 1995). *S. boulardii* is also effective in amebiasis and HIV-diarrhea (Mansour-Ghanaei et al. 2003, Saint-Marc et al. 1991). An open-label RCT in Pakistani children with acute infectious GE showed that administration of 500 mg *S. boulardii* for 5 days significantly reduced the frequency of stools and duration of diarrhea (3.5 days versus 4.8 days, p=0.001) and resulted, two months later, in a 50% decrease in re-infection rates and 30% better weight gain (Billoo et al. 2006). We showed in 150 Indian and 50 Indonesian children a trend in reduction of the duration of diarrhea and a significant reduction in symptomatic children on day three (unpublished data).

Three meta-analyses concluded that efficacy was demonstrated for *L. rhamnosus GG*, *acidophilus* and *bulgaricus* (Huang et al. 2002, Szajewska and Mrukowicz 2001, Van Niel et al. 2002). In particular, the duration of (viral) diarrhea was significantly reduced (about 17 hours or 0.7 days) (RR 0.40) (Szajewska and Mrukowicz 2001). The efficacy of *L. casei GG* appeared related to the logarithm of the dose (>10^{11} as the most efficient dose) (Huang et al. 2002). In acute diarrhea of diverse causes, probiotics reduce the duration of the diarrhea by about 50% (35–71%) (Sazawal et al. 2006). Probiotics may also reduce the relapse rate of diarrhea (Francavilla et al. 2012). A Cochrane review included 63 trials, of which 56 were pediatric, including 8014 patients (Allen et al. 2010). The trials varied in the definition used for acute diarrhea and the end of the diarrheal illness, as well as in the risk of bias. The trials were undertaken in a wide range of different settings and also varied greatly in organisms tested, dosage, and participants' characteristics. No adverse events were attributed to the probiotic intervention. Probiotics reduced the duration of diarrhea, although the size of the effect varied considerably between studies. The average of the effect was significant for mean duration of diarrhea (mean difference 24.76 hours; 95% CI 15.9 to 33.6 hours), diarrhea lasting ≥4 days (risk ratio 0.41; 0.32 to 0.53), and stool frequency on day 2 (mean difference 0.80; 0.45 to 1.14) (Allen et al. 2010). The differences in effect size between studies was not explained by study quality, probiotic strain, the number of different strains, the viability of the organisms, dosage of organisms, the causes of diarrhea, or the severity of the diarrhea, or whether the studies were done in developed or developing countries (Allen et al. 2010). A major problem of meta-analyses is the different definitions used to describe diarrhea (Johnston et al. 2010). In 138 RCTs reporting on pediatric acute diarrhea/diseases, there were 64 unique definitions of diarrhea, 69 unique definitions of diarrhea resolution and 46 unique primary outcomes (Johnston et al. 2010). A shortening of the duration of diarrhea, as well as a reduced hospital stay suggests a relevant social and economic benefit of biotherapeutic treatment in adjunction to ORS in acute infectious GE in children (Guarino et al. 2012). Benefit related to cost should get more attention (Guarino et al. 2012).

Numerous clinical trials have been published evaluating different probiotics in the treatment of acute GE. But the trials vary in relation to strains tested, dosage, methodological quality, diarrhea definitions and outcomes. Most studies show a statistically significant effect that is of only moderate clinical benefit, with a greatest effect in viral and watery diarrhea (Szajewska et al. 2006). In general, meta-analyses of published trials conclude in a reduction of diarrheal duration of approximately 24 hours (17 to 30 hours) for selected strains of lactobacilli (such as *L. casei GG, L. acidophilus, L. Bulgaricus* and *L. reuteri*) and *S. boulardii*. Greater efficacy has been shown if the probiotic is administered early in the disease. However, authors also conclude that mainly business pressures force usage of probiotics (and antisecretory drugs such as racecadotril) as important in the management of acute diarrhea while their relevance yet has to be established (Alam and Bhatnagar 2006). However, the following conclusions seem to win the dispute: many probiotic strains decrease the duration of diarrhea with about 24 hours; they decrease the duration of hospitalisation by a similar duration. The administration of probiotics leads to a relevant cost-benefit (Vandenplas and De Hert 2012).

There is a need for trials comparing the efficacy of different strains, as has been recently done comparing *S boulardii* to *B. Lactis*, suggesting efficacy for the latter and not the first (Erdoğan et al. 2012). *L. reuteri* was reported more effective than *L. casei CRL431* (Agustina et al. 2012).

Prebiotics

There are almost no studies with prebiotics in the treatment of acute infectious gastroenteritis. The few studies performed are negative (Hoekstra et al. 2004).

Synbiotics

Recently, studies have been published with synbiotics in the treatment of infectious GE showing a comparable effect to probiotics alone (Passariello et al. 2012, Vandenplas and De Hert 2011). Shamir et al. showed a reduction in duration of acute GE from 1.96 + 1.24 to 1.43 + 0.71 days ($p = 0.017$) with addition of 10^9 CFU *Str. Thermophilus, B. lactis, L. acidophilus,* 10 mg zinc and 0.3 gram FOS per day (Shamir et al. 2005). Also, prebiotics and zinc added to ORS reduced the duration of diarrhea (Passariello et al. 2011).

Antibiotic Associated Diarrhea (AAD)

Antibiotic treatment is known to disturb the GI microflora, which results in a range of clinical symptoms, especially diarrhea. The incidence of AAD in children in first line health care is about 10%, independent of the reason for antibiotic administration (Turck et al. 2003). The risk for AAD is increased in young children (18% in children younger than 2 years), and when specific antibiotics such as amoxicillin-clavulanate are administered (23% with the latter) (Turck et al. 2003). However, the vast majority of AADs are mild to moderate and hospitalization is seldom required. AAD is only clinically relevant in a minority of cases.

According to a recent meta-analysis, probiotics reduce the risk of AAD in children (Johnston et al. 2011). Preplanned subgroup analysis showed that reduction of the risk of AAD was associated with the use of *L. caseii GG* (95% CI 0.15 to 0.6), *S. boulardii* (95% CI 0.07–0.6), or *B. lactis* and *Str. thermophilus* (95% CI 0.3 to 0.95) (Szajewska et al. 2006, Johnston et al. 2011). The number needed to treat is seven: 7 patients need to be given probiotics to have one patient less with AAD (Johnston et al. 2011). Only *S. boulardii* was reported to be effective in the treatment of AAD caused by Clostridium Difficile (C. dif.) (Johnston et al. 2011, McFarland 2006). *S. boulardii* may be effective for secondary prevention in specific patient populations with particular concurrent antibiotic treatment (Tung et al. 2009). However, there is no evidence to support the use of any probiotic to prevent the recurrence of C. Dif. infection or to treat existing C. Dif. diarrhea (Szajewska et al. 2006).

The American Academy of Pediatrics concluded that RCTs showed a beneficial effect for probiotics in the prevention of AAD in children (Thomas and Greer 2010). This was also the result of an analysis of AAD independent of age (Ritchie and Romanuk 2012). According to a Cochrane review published in 2011, including 16 pediatric studies (3432 participants), *Bacillus* spp., *Bifidobacterium* spp., *Lactobacilli* spp., Lactococcus spp., Leuconostoc cremoris, *Saccharomyces* spp. or *Streptococcus* spp. alone or in combination have all been evaluated in the prevention of AAD (Johnston et al. 2011). Nine studies used a single strain, four combined two and one combined three probiotic strains, one study was done with a probiotic food supplement containing ten strains, and one study had two probiotic arms that used three and two strains respectively. Overall, the incidence of AAD in the probiotic group was 9% compared to 18% in the control group (2874 participants; RR 0.52; 95% CI 0.38 to 0.72) (Johnston et al. 2011). However, this benefit was not statistically significant in an extreme plausible (60% of children lost to follow-up in probiotic group and 20% lost to follow-up in the control group had diarrhea) intention to treat (ITT) sensitivity analysis. If the data are analyzed that way, the incidence of AAD in the probiotic group was 16% compared to 18% in the control group (3392 participants; RR

0.81; 95% CI 0.63 to 1.04). However, ITT subgroup analysis was marginally significant for high dose probiotics: AAD in the probiotic group was 17% compared to 22% in the control group (1776 participants; RR 0.72; 95% CI 0.53 to 0.99). None of the 11 trials (n = 1583) that reported on adverse events documented any serious adverse events (Johnston et al. 2011). *L. GG* was reported equally effective as *S. boulardii* (D'Souza et al. 2002). Age may, as well, be important. According to one meta-analysis, L. does reduce AAD in adults, but not in children (Kale-Pradhan et al. 2010). However, a recent meta-analysis did not find an age-related difference (Ritchie and Romanuk 2012).

In most studies, the probiotic is started together with antibiotic treatment (Saavedra et al. 1994). Despite heterogeneity in probiotic strain, dose, and duration, as well as in study quality, the overall evidence suggests a protective effect of probiotics in preventing AAD. A GRADE analysis indicated that the overall quality of the evidence for the primary endpoint (incidence of diarrhea) was low due to issues with risk of bias (due to high loss to follow-up) and imprecision (sparse data) (Johnston et al. 2011). Another shortcoming is that the severity of the AAD is not considered: most of the time AAD is very mild and does not need any intervention.

Prebiotics and synbiotics

To the best of our knowledge, there are no data suggesting efficacy of prebiotics and synbiotics in the prevention of AAD (Szajewska et al. 2012).

Traveler's Diarrhea

Traveler's diarrhea is a frequent condition of great socio-economic impact. However, original data on this topic are very scarce, and there are no data specific for children. Different randomized trials have been performed evaluating the efficacy of probiotics in the prevention of traveler's diarrhea. One trial with *L. Acidophilus* and two with *L. caseii GG* showed negative results (Katelaris et al. 1995, Hilton et al. 1997, Oksanen et al. 1990). One trial with *S. boulardii* reported a small but significant preventive effect in a subgroup, suggesting geographical differences in efficacy (Kollaritsch 1989). In a review, McFarland concluded that there is comparable evidence for efficacy for *L. rhamnosus GG, L. casei DN-114001* and *S. boulardii*, and no efficacy for *L. acidophilus* (McFarland 2010). However, a recent meta-analysis concluded that there is no efficacy of probiotics in the prevention of traveler's diarrhea (Ritchie and Romanuk 2012).

There are no data on prebiotics or synbiotics on the prevention of traveler's diarrhea.

Inflammatory Bowel Disease (IBD)

The concept of dysbiosis, a breakdown of balance between "protective" and "harmful" intestinal bacteria, has been generally accepted as one of the pathophysiologic abnormalities in IBD patients. Many animal and in vitro studies favor this hypothesis. Literature suggests a theoretical role for probiotics in the maintenance of remission in IBD in adults, especially in pouchitis (Sartor 2004). Individuals with pouchitis have a reduced number of lactobacilli and bifidobacteria and probiotics have been shown to prevent relapse of pouchitis (Ritchie and Romanuk 2012). A 2 year follow-up study with *L. GG* in children with Crohn's disease in remission, resulted in a relapse rate of 31% in the probiotic group versus 17% in the placebo group (non significant difference) (Bousvaros 2005). There was also no difference in the time frame before the relapse (Bousvaros 2005). In children with active distal ulcerative colitis (UC), rectal infusion of *L. reuteri* decreases mucosal inflammation and changes the mucosal expression levels of some cytokines involved in the mechanisms of IBD (Oliva et al. 2012). VSL#3 decreases the relapse rate of ulcerative colitis (Miele et al. 2009). Given present data, adjunct VSL#3 use for pediatric UC induction and maintenance of remission is not cost-effective (Park et al. 2011). However, although there is a lack of evidence of benefit, almost 80% of the children with IBD take probiotics regularly (Day 2004).

To the best of our knowledge, there are no data on prebiotics and synbiotics in IBD.

Irritable Bowel Syndrome (IBS)

Probiotics

An RCT comparing *L. casei GG* with placebo for 6 weeks showed overall negative results in 50 children and young adults (6–20 years), although there was a lower incidence of perceived abdominal distension in the *L.GG* group (Bausserman and Michail 2005). In another cross-over trial with a two-weeks wash-out period in between, patients (n:59) with IBS were randomized to receive either VSL#3 or placebo for 12 weeks (Guandalini et al. 2010). Although placebo was effective in as many as half of the patients for some parameters, VSL#3 was significantly superior for the primary endpoint, the subjective assessment of relief of symptoms and in 3 of 4 secondary endpoints (abdominal pain/discomfort, abdominal bloating/

gassiness, and family assessment of life disruption) (Guandalini et al. 2010). No significant difference was found in the stool pattern (Guandalini et al. 2010). Another trial showed that *L. casei GG*, but not the placebo, caused a significant reduction of both frequency and severity of abdominal pain compared to baseline (Francavilla et al. 2010). At week 12, treatment success was achieved in 48 children in the *L. casei GG* group compared with 37 children in the placebo group (P <.03) (Francavilla et al. 2010). At entry, 59% of the children had an abnormal intestinal permeability test: *L. GG*, but not placebo, induced a significant decrease in the number of patients with abnormal permeability (Francavilla et al. 2010).

A meta-analysis of pediatric studies showed that, compared with placebo, *L. casei GG* administration is associated with a significantly higher rate of treatment responders (defined as no pain or a decrease in pain intensity) (3 RCTs, n = 167; RR 1.70, 95% CI 1.27-2.27, NNT 4, 95% CI 3-8) (Horvath et al. 2011).

Including adult data, Lactobacilli compared to placebo showed a pooled odds ratio for improvement of symptoms of 1.17 (95% CI 0.62, 2.21) (Fedorak and Madsen 2004). A significant reduction of a composite score of IBS symptoms (the sum of scores for abdominal pain/discomfort, bloating/distension and bowel movement difficulty) occurred in adults with *B. infantis* but not with *L. Salivarius* (O'Mahony et al. 2005). The significant differences in cytokine-profiles measured with both probiotics illustrate strain specificity.

Prebiotics

A Cochrane review from 2009 included 2 trials (83 participants) and showed that fiber supplements were not better than placebo (Huertas-Ceballos et al. 2009).

Safety and Side Effects

Probiotics are "Generally Regarded As Safe" and side-effects in ambulatory care are almost never reported. Large scale epidemiological studies in countries where probiotic use is endemic demonstrate (in adults) low rates of systemic infection, between 0.05 and 0.40% (Borriello et al. 2003). Documented invasive infections have been primarily noted to occur in immuno-compromised adults. Lactobacilli have been reported (in adults) to cause cases with sepsis, meningitis and infections localized in different organs (Mackay et al. 1999, Rautio et al. 1999). Invasive infections in infants and children are exceedingly rare (Cabana et al. 2006). Two cases of bacteremia related to Lactobacillus intake were reported in an infant and a child without underlying gastrointestinal disease or known immune deficiency (Land et al. 2005). Sepsis with probiotic lactobacilli has been reported in children with short gut. Probiotic enterococci may be of higher risk given possible plasmid transfer in immuno-compromised patients. Fungemia with *S. boulardii.* has been reported in about 50 patients (Enache-Angoulvant 2005). A central venous catheter is the main risk factor (Enache-Angoulvant 2005). Fungemia has even been reported in patients with deep central venous lines hospitalized next to a patient treated with the yeast (Cassone et al. 2003). Translocation from the gastro-intestinal tract in the systemic circulation has not been reported for the yeast. These case reports emphasize that probiotic yeast supplementation should be used with caution in children with indwelling central venous catheters. The potential benefits of supplementation should be weighed against the risk of development of an invasive infection resulting from probiotic therapy.

In order to minimize the risk for side-effects such as fungemia and bacteremia, more research should be done with inactivated or non-viable preparations. These modified probiotic preparations may be the preferred product in at-risk situations. It may not always be necessary to administer the intact probiotic organisms to achieve benefits. At the basic research level, products of probiotics such as secreted proteins or DNA can block inflammation and stop the death of epithelial cells (Jijon et al. 2004).

Conclusion

Probiotics and prebiotics added to infant formula have an influence on GI flora composition. However, the evidence for a clinically relevant health care benefit of pro- and prebiotics added to infant formula remains limited. Although it has to be acknowledged that i) negative effects have not been shown and ii) the vast majority of the trials show some benefit although not always significant. Nevertheless, the addition of these ingredients to infant formula brings the second choice infant feeding closer to the golde standard, the breastfed baby. The species *L. acidophilus, L. Plantarum* and *B. infantis* showed no efficacy (Ritchie and Romanuk 2012). Although probiotics can be helpful for specific disorders, they have been broadly prescribed for disorders without clear evidence to support their use (Michail et al. 2006). Prebiotics are not helpful in therapeutic indications. There is evidence that probiotics shorten the duration of acute GE with about 24 hours and decrease the risk of developing AAD. Both for lactobacilli and *S. boulardii* greater efficacy has been shown if treatment is started early in the course of the disease. Because of strain-specificity, only those organisms that have been clinically tested can be recommended. During recent years, some promising results have been published on the use of probiotics in the maintenance of remission of ulcerative colitis and in the treatment of irritable bowel syndrome. However, more data are needed in these areas before recommendations can be made.

Prebiotics have not been shown to be of any benefit except when added to infant formula. Prebiotics added to infant formula promote the development of a bifidobacteria. There is a trend that this bifidobacteria dominated flora may induce a trend to decreased episodes of infectious GE.

References

Agostoni, C., I. Axelsson, C. Braegger, O. Goulet, B. Koletzko, K.F. Michaelsen, J. Rigo, R. Shamir, H. Szajewska, D. Turck and L.T. Weaver. 2004. ESPGHAN Committee on Nutrition. Probiotic bacteria in dietetic products for infants: a commentary by the ESPGHAN Committee on Nutrition. J. Pediatr Gastroenterol. Nutr. 38: 365–74.

Agustina, R., E.J. Kok, O. van de Rest, U. Fahmida, A. Firmansyah, W. Lukito, E.J. Feskens, E.G. van den Heuvel, R. Albers and I.M. Bovee-Oudenhoven. 2012. Randomized trial of probiotics and calcium on diarrhea and respiratory tract infections in Indonesian children. Pediatrics 129: e1155–64.

Alam, S. and S. Bhatnagar. 2006. Current status of anti-diarrheal and anti-secretory drugs in the management of acute childhood diarrhea. Indian J. Pediatr 73: 693–6.

Allen, S.J., E.G. Martinez, G.V. Gregorio and L.F. Dans. 2010. Probiotics for treating acute infectious diarrhoea. Cochrane Database Syst. Rev. (11): CD003048.

Arslanoglu, S., G.E. Moro and G. Boehm. 2007. Early supplementation of prebiotic oligosaccharides protects formula-fed infants against infections during the first 6 months of life. J. Nutr. 137: 2420–4.

Bausserman, M. and S. Michail. 2005. The use of Lactobacillus GG in irritable bowel syndrome in children: a double-blind randomized control trial. J. Pediatr 147: 197–201.

Ben, X.M. 2008. Low level of GOS in infant formula stimulates growth of intestinal Bifidobacteria and Lactobacilli. World J. Gastroenterol. 14: 6564–8.

Billoo, A.G., M.A. Memon, S.A. Khaskheli, G. Murtaza, K. Iqbal, M. Saeed Shekhani and A. Siddiqi. 2006. Role of a probiotics (*Saccharomyces boulardii*) in the management and prevention of diarrhea. World J. Gastroenterol. 12: 4557–60.

Borriello, S.P., W.P. Hammes, W. Holzapfel, P. Marteau, J. Schrezenmeir, M. Vaara and V. Valtonen. 2003. Safety of probiotics that contain *Lactobacilli* or *Bifidobacteria*. Clin. Infect. Dis. 36: 775–80.

Bousvaros, A. 2005. A randomized, double-blind trial of Lactobacillus GG versus placebo in addition to standard maintenance therapy for children with Crohn's disease. Inflamm. Bowel Dis. 11: 833–9.

Braegger, C., A. Chmielewska, T. Decsi, S. Kolacek, W. Mihatsch, L. Moreno, M. Pieścik, J. Puntis, R. Shamir, H. Szajewska, D. Turck and J. van Goudoever. 2011. ESPGHAN Committee on Nutrition. Supplementation of infant formula with probiotics and/or prebiotics: a systematic review and comment by the ESPGHAN committee on nutrition. J. Pediatr Gastroenterol. Nutr. 52: 238–50.

Cabana, M.D., A.L. Shane, C. Chao and M. Oliva-Henker. 2006. Probiotics in primary care pediatrics. Clin. Pediatr. 45: 405–10.

Cassone, M., P. Serra, F. Mondello, A. Girolamo, S. Scafetti, E. Pistella and M. Venstaditti. 2003. Outbreak of Saccharomyces cerevisiae subtype boulardii fungemia in patients neighboring those treated with a probiotic preparation of the organism. Clin. Microbiol. 41: 5340–3.

Castañeda, C., E. Garcia, M. Santa Cruz, M. Fernandez and P. Monterrey. 1995. Effects of *Saccharomyces boulardii* in children with chronic diarrhea, especially cases due to giardiasis. Revista Mexicana de Puericultura y Pediatria 2: 12–6.

Chouraqui, J.P., L.D. Van Egroo and M.C. Fichot. 2004. Acidified milk formula supplemented with Bifidobacterium lactis: impact on infant diarrhea in residential care settings. J. Pediatr Gastroenterol. Nutr. 38: 288–92.

Costa-Ribeiro, H., T.C. Ribeiro, A.P. Mattos, S.S. Valois, D.A. Neri, P. Almeida, C.M. Cerqueira, E. Ramos, R.J. Young and J.A. Vanderhoof. 2003. Limitations of probiotic therapy in acute, severe dehydrating diarrhea. J. Pediatr Gastroenterol. Nutr. 36: 112–5.

D'Souza, A.L., C. Rajkumar, J. Cooke and C.J. Bulpitt. 2002. Probiotics in prevention of antibiotic associated diarrhoea: meta-aanlysis. BMJ 324: 1361.

Day, A.S. 2004. Use of complementary and alternative medicines by children and adolescents with inflammatory bowel disease. Paediatr Child Health 40: 681–4.

Elliott, E. and K. Teversham. 2004. An evaluation of nine probiotics available in South Africa, August 2003. SAMJ 94: 121–4.

Enache-Angoulvant, A. 2005. Invasive Saccharomyces infection: a comprehensive review. Clin. Infect. Dis. 41: 1559–68.

Erdoğan, O., B. Tanyeri, E. Torun, E. Gönüllü, H. Arslan, U. Erenberk and F. Oktem. 2012. The comparison of the efficacy of two different probiotics in Rotavirus gastroenteritis in children. J. Trop. Med. 787240.

Fedorak, R.N. and k.I. Madsen. 2004. Probiotics and prebiotics in gastrointestinal disorders. Curr. Opin. Gastroenterol. 20: 146–55.

Francavilla, R., E. Lionetti, S. Castellaneta, F. Ciruzzi, F. Indrio, A. Masciale, C. Fontana, M.M. La Rosa, L. Cavallo and A. Francavilla. 2012. Randomised clinical trial: *Lactobacillus reuteri* DSM 17938 vs. placebo in children with acute diarrhoea—a double-blind study. Aliment Pharmacol. Ther. 36: 363–9.

Francavilla, R., V. Miniello, A.M. Magistà, A. De Canio, N. Bucci, F. Gagliardi, E. Lionetti, S. Castellaneta, L. Polimeno, L. Peccarisi, F. Indrio and L. Cavallo. 2010. A randomized controlled trial of Lactobacillus GG in children with functional abdominal pain. Pediatrics 126: e1445–52.

Guandalini, S. 2011. Probiotics for prevention and treatment of diarrhea. J. Clin. Gastroenterol. 45 Suppl. S149–53.

Guandalini, S., G. Magazzù, A. Chiaro, V. La Balestra, G. Di Nardo, S. Gopalan, A. Sibal, C. Romano, R.B. Canani, P. Lionetti and M. Setty. 2010. VSL#3 improves symptoms in children with irritable bowel syndrome: a multicenter, randomized, placebo-controlled, double-blind, crossover study. J. Pediatr Gastroenterol. Nutr. 51: 24–30.

Guandalini, S., L. Pensabene, M.A. Zikri, J.A. Dias, L.G. Casali, H. Hoekstra, S. Kolacek, K. Massar, D. Micetic-Turk, A. Papadopoulou, J.S. de Sousa, B. Sandhu, H. Szajewska and Z. Weizman. 2000. Lactobacillus GG administered in oral rehydration solution in children with acute diarrhea: a multicenter European trial. J. Pediatr Gastroenterol. Nutr. 30: 214–6.

Guarino, A., C. Dupont, A.V. Gorelov, F. Gottrand, J.K. Lee, Z. Lin, A. Lo Vecchio, T.D. Nguyen and E. Salazar-Lindo. 2012. The management of acute diarrhea in children in developed and developing areas: from evidence base to clinical practice. Expert Opin. Pharmacother. 13: 17–26.

Guarino, A., R. Berni Canani, M.I. Spagnuolo, M. Bisceglia, M.C. Boccia and A. Rubino. 1997. Oral bacterial therapy reduces the duration of symptoms and viral excretion in children with mild diarrhea. J. Pediatr Gastroenterol. Nutr. 25: 516–9.

Hilton, E., P. Kolakowski, C. Singer and M. Smith. 1997. Efficacy of *Lactobacillus* GG as a diarrheal preventive in travelers. J. Travel Med. 4: 41–3.

Höchter, W., D. Chase and G. Hagenhoff. 1990. *Saccharomyces boulardii* bei akuter Erwachsenendiarrhoe. Münch Med. Wschr. 132: 188–92.

Hoekstra, J.H., H. Szajewska, M.A. Zikri, D. Micetic-Turk, Z. Weizman, A. Papadopoulou, A. Guarino, J.A. Dias and B. Oostvogels. 2004. Oral rehydration solution containing a mixture of non-digestible carbohydrates in the treatment of acute diarrhea: a multicenter randomized placebo controlled study on behalf of the ESPGHAN working group on intestinal infections. J. Pediatr Gastroenterol. Nutr. 39: 239–45.

Horvath, A., P. Dziechciarz and H. Szajewska. 2011. Meta-analysis: *Lactobacillus rhamnosus* GG for abdominal pain-related functional gastrointestinal disorders in childhood. Aliment Pharmacol. Ther. 33: 1302–10.

Huang, J.S., A. Bousvaros, J.W. Lee, A. Diaz and E.J. Davidson. 2002. Efficacy of probiotic use in acute diarrhea in children: a meta-analysis. Dig. Dis. Sci. 47: 2625–34.

Huertas-Ceballos, A.A., S. Logan, C. Bennett and C. Macarthur. 2009. Dietary interventions for recurrent abdominal pain (RAP) and irritable bowel syndrome (IBS) in childhood. Cochrane Database Syst. Rev. (1): CD003019.

Ibnou-Zekri, N., S. Blum, E.J. Schiffrin and T. von der Weid. 2003. Divergent patterns of colonization and immune response elicited from two intestinal Lactobacillus strains that display similar properties *in vitro*. Infect. Immun. 71: 428–36.

Isolauri, E., M. Juntunen, T. Rautanen, P. Sillanaukee and T. Koivula. 1991. A human Lactobacillus strain (*Lactobacillus casei* sp. strain GG) promotes recovery from acute diarrhea in children. Pediatrics 88: 90–7.

Jijon, H., J. Backer, H. Diaz, H. Yeung, D. Thiel, C. McKaigney, C. De Simone and K. Madsen. 2004. DNA from probiotic bacteria modulates murine and human epithelial and immune function. Gastroenterology 126: 1358–73.

Johnson, C.L. and J. Versalovic. 2012. The human microbiome and its potential importance to pediatrics. Pediatrics 129: 950–60.

Johnston, B.C., J.Z. Goldenberg, P.O. Vandvik, X. Sun and G.H. Guyatt. 2011. Probiotics for the prevention of pediatric antibiotic-associated diarrhea. Cochrane Database Syst. Rev. 11: CD004827.

Johnston, B.C., L. Shamseer, B.R. da Costa, R.T. Tsuyuki and S. Vohra. 2010. Measurement issues in trials of pediatric acute diarrheal diseases: a systematic review. Pediatrics 126: e222–31.

Juntunen, M., P.V. Kirjavainen, A.C. Ouwehand, S.J. Salminen and E. Isolauri. 2001. Adherence of probiotic bacteria to human intestinal mucus in healthy infants and during ROTAvirus infection Clin. Diagn. Lab Immunol. 8: 293–6.

Kale-Pradhan, P.B., H.K. Jassal and S.M. Wilhelm. 2010. Role of Lactobacillus in the prevention of antibiotic-associated diarrhea: a meta-analysis. Pharmacotherapy 30: 119–26.

Katelaris, P.H., I. Salam and M.J. Farthing. 1995. Lactobacilli to prevent traveler's diarrhea? N. Eng. J. Med. 333: 1360–1.

Khanna, V., S. Alam and A. Malik. 2005. Efficacy of tyndalized *Lactobacillus acidophilus* in acute diarrhea. Indian J. Pediatr 72: 935–8.

Kollaritsch, H. 1989. Traveller's diarrhea among Austrian tourists to warm climate countries: II. Clinical features. Eur. J. Epidemiol. 5: 355–62.

Kowalska-Duplaga, K., K. Fyderek, H. Szajewska and R. Janiak. 2004. Efficacy of Trilac in the treatment of acute diarrhoea in infants and young children—a multicentre, randomized, double-blind placebo-controlled study. Pediatria Współczesna 3: 295–9.

Kurugol, Z. and G. Koturoglu. 2005. Effects of *Saccharomyces boulardii* in children with acute diarrhoea. Acta Paediatr 94: 44–7.

Land, M.H., K. Rouster-Stevens, C.R. Woods, M.L. Cannon, J. Cnota and A.K. Shetty. 2005. Lactobacillus sepsis associated with probiotic therapy. Pediatrics 115: 178–81.

Lilly, D.M. and R.H. Stillwell. 1965. Probiotic growth promoting factors produced by microorganisms. Science 147: 747–8.

Mackay, A.D., M.B. Taylor, C.C. Kibbler, J.M. Hamilton-Miller. 1999. *Lactobacillus endocarditis* caused by a probiotics organism. Clin. Microbiol. Infect.5: 290–2.

Mansour-Ghanaei, F., N. Dehbashi, K. Yazdanparast and A. Shafaghi. 2003. Efficacy of *Saccharomyces boulardii* with antibiotics in acute amoebiasis. World J. Gastroenterol. 9: 1832–3.

Marteau, P., P. Pochart, Y. Bouhnik, S. Zidi, I. Goderel and J.C. Rambaud. 1992. Survival of *Lactobacillus acidophilus* and *Bifidobacterium* sp. in the small intestine following ingestion in fermented milk. A rational basis for the use of probiotics in man. Gastroenterol. Clin. Biol. 16: 25–8.

Mastretta, E., P. Longo, A. Laccisaglia, L. Balbo, R. Russo, A. Mazzaccara and P. Gianino. 2002. Effect of a *Lactobacillus* GG and breast-feeding in the prevention of rotavirus nosocomial infection. J. Pediatr Gastroenterol. Nutr. 35: 527–31.

McFarland, L.V. 2006. Meta-analysis of probiotics for prevention of antibiotic associated diarrhea and the treatment of Clostridium difficile disease. Am. J. Gastroenterol. 101: 812–22.

McFarland, L.V. 2010. Probiotics and diarrhea. Ann. Nutr. Metab. 57 Suppl: 10–1.

Michail, S., F. Sylvester, G. Fuchs and R. Issenman. 2006. (NASPGHAN Nutrition Report Committee) Clinical efficacy of probiotics: review of the evidence with focus on children. J. Pediatr Gastroenterol. Nutr. 43: 550–7.

Miele, E., F. Pascarella, E. Giannetti, L. Quaglietta, R.N. Baldassano and A. Staiano. 2009. Effect of a probiotic preparation (VSL#3) on induction and maintenance of remission in children with ulcerative colitis. Am. J. Gastroenterol. 104: 437–43.

O'Mahony, L., J. McCarthy, P. Kelly, G. Hurley, F. Luo, K. Chen, G.C. O'Sullivan, B. Kiely, J.K. Collins, F. Shanahan and E.M. Quigley. 2005. Lactobacillus and bifidobacterium in irritable bowel syndrome: symptom responses and relationship to cytokine profiles. Gastroenterology 128: 541–51.

Oksanen, P.J., S. Salminen, M. Saxelin, P. Hamalainen, A. Ihantola-Vormisto, L. Muurasniemi-Isoviita, S. Nikkari, T. Oksanen, I. Porsti and E. Salminen. 1990. Prevention of traveller's diarrhoea by Lactobacillus GG. Ann. Med. 22: 53–6.

Oliva, S., G. Di Nardo, F. Ferrari, S. Mallardo, P. Rossi, G. Patrizi, S. Cucchiara and L. Stronati. 2012. Randomised clinical trial: the effectiveness of *Lactobacillus reuteri* ATCC 55730 rectal enema in children with active distal ulcerative colitis. Aliment Pharmacol. Ther. 35: 327–34.

Park, K.T., F. Perez, R. Tsai, A. Honkanen, D. Bass and A. Garber. 2011. Cost-effectiveness analysis of adjunct VSL#3 therapy versus standard medical therapy in pediatric ulcerative colitis. J. Pediatr Gastroenterol. Nutr. 53: 489–96.

Passariello, A., G. Terrin, G. Cecere, M. Micillo, G. De Marco, M. Di Costanzo, L. Cosenza, L. Leone, R. Nocerino and R.B. Canani. 2012. Randomised clinical trial: efficacy of a new synbiotic formulation containing *Lactobacillus paracasei* B21060 plus arabinogalactan and xilooligosaccharides in children with acute diarrhoea. Aliment Pharmacol. Ther. 35: 782–8.

Passariello, A., G. Terrin, G. De Marco, G. Cecere, S. Ruotolo, A. Marino, L. Cosenza, M. Tardi, R. Nocerino and R. Berni Canani. 2011. Efficacy of a new hypotonic oral rehydration solution containing zinc and prebiotics in the treatment of childhood acute diarrhea: a randomized controlled trial. J. Pediatr 158: 288–92.e1.

Rautio, M., H. Jousimies-Somer, H. Kauma, I. Pietarinen, M. Saxelin, S. Tynkkynen and M. Koskela. 1999. Liver abcess due to a *Lactobacillus rhamnosus* strain indistinguishable from a L. rhamnosus strain GG. Clin. Infect. Dis. 28: 1159–60.

Raza, S., S.M. Graham and S.J. Allen. 1995. Lactobacillus GG promotes recovery from acute nonbloody diarrhea in Pakistan. Pediatr Infect. Dis. J.14: 107–11.

Reid, G. 2012. Microbiology: Categorize probiotics to speed research Nature 485: 446.

Ritchie, M.L. and T.N. Romanuk. 2012. A meta-analysis of probiotic efficacy for gastrointestinal diseases. Plos One 7: e34938.

Saavedra, J.M., N.A. Bauman, I. Oung, J.A. Perman and R.H. Yolken. 1994. Feeding of Bifidobacterium bifidum and Streptococcus thermophilus to infants in hospital for prevention of diarrhoea and shedding of rotavirus. Lancet 344: 1046–9.

Saint-Marc, T., L. Rossello-Prats and J.L. Touraine. 1991. Efficacy of *Saccharomyces boulardii* in the treatment of diarrhea in AIDS. Ann. Med. Interne (Paris) 142: 64–5.

Salazar-Lindo, E., P. Miranda-Langschwager, M. Campos-Sanchez, E. Chea-Woo and R.B. Sack. 2004. Lactobacillus casei strain GG in the treatment of infants with acute watery diarrhea: A randomized, double-blind, placebo controlled clinical trial [ISRCTN67363048] BMC Pediatrics 4: 18.

Salvini, F.J. 2011. A specific prebiotic mixture added to starting infant formula has long-lasting bifidogenic effects. Nutr. 141: 1335–9.

Saran, S., S. Gopalan and T.P. Krishna. 2002. Use of fermented foods to combat stunting and failure to thrive. Nutrition 18: 393–6.

Sarker, S.A., S. Sultana, G.J. Fuchs, N.H. Alam, T. Azim, H. Brussow and L. Hammarstrom. 2005. *Lactobacillus paracasei* strain ST11 has no effect on rotavirus but ameliorates the outcome of nonrotavirus diarrhea in children from Bangladesh. Pediatrics 116: e221–8.

Sartor, R.B. 2004. Therapeutic manipulation of the enteric microflora in inflammatory bowel diseases: antibiotics, probiotics, and prebiotics. Gastroenterology 126: 1620–33.

Sazawal, S., G. Hiremath, U. Dhingra, P. Malik, S. Deb and R.K. Black. 2006. Efficacy of probiotics in prevention of acute diarrhoea: a meta-analysis of masked, randomised, placebo-controlled trials. Lancet Infect. Dis. 6: 374–82.

Sazawal, S., U. Dhingra, A. Sarkar, P. Dhingra, S. Deb, D. Marwah, V.P. Menon, J. Kumar and R.E. Black. 2004. Efficacy of milk fortified with a probiotic *Bifidobacterium lactis* (DR-10TM) and prebiotic galacto-oligosaccharides in prevention of morbidity and on nutritional status. Asia Pac. J. Clin. Nutr. 13(Suppl):S28.

Sazawal, S., U. Dhingra, G. Hiremath, A. Sarkar, P. Dhingra, A. Dutta, P. Verma, V.P. Menon and R.E. Black. 2010. Prebiotic and probiotic fortified milk in prevention of morbidities among children: community-based, randomized, double-blind, controlled trial. PLoS One 5:e12164.

Shamir, R., I.R. Makhoul, A. Etzioni and N. Shehadeh. 2005. Evaluation of a diet containing probiotics and zinc for the treatment of mild diarrheal illness in children younger than one year of age. J. Am. Coll Nutr. 24: 370–5.

Shornikova, A.V., E. Isolauri, L. Burkanova, S. Lukovnikova and T. Vesikari. 1997. A trial in the Karelian Republic of oral rehydration and Lactobacillus GG for treatment of acute diarrhoea. Acta Paediatr 86: 460–5.

Shornikova, A.V., I.A. Casas, H. Mykkanen, E. Salo and T. Vesikari. 1997. Bacteriotherapy with *Lactobacillus reuteri* in Rotavirus gastroenteritis. Pediatr Infect. Dis. J. 16: 1103–7.

Simakachorn, N., V. Pichaipat, P. Rithipornpaisarn, C. Kongkaew, P. Tongpradit and W. Varavithya. 2000. Clinical evaluation of the addition of lyophilized, heat-killed *Lactobacillus acidophilus* LB to oral rehydration therapy in the treatment of acute diarrhea in children. J. Pediatr Gastroenterol. Nutr. 30: 68–72.

Szajewska, H. and J.Z. Mrukowicz. 2001. Probiotics in the treatment and prevention of acute infectious diarrhea in infants and children: a systematic review of published randomized, double-blind, placebo-controlled trials. J. Pediatr Gastroenterol. Nutr. 33: S17–25.

Szajewska, H. and J.Z. Mrukowicz. 2005. Use of probiotics in children with acute diarrhea. Paediatr Drugs 7: 111–22.

Szajewska, H., M. Kotowska, J.Z. Mrukowicz, M. Armanska and W. Mikolajczyk. 2001. Efficacy of Lactobacillus GG in prevention of nosocomial diarrhea in infants. J. Pediatr 138: 361–5.

Szajewska, H., M. Setty, J. Mrukowicz and S. Guandalini. 2006. Probiotics in gastrointestinal disease in children: hard and not-so-hard evidence of efficacy. J. Pediatr Gastroenterol. Nutr. 42: 454–75.

Szajewska, H., Z. Weizman, M. Abu-Zekry, A.J. Kekez, C.P. Braegger, S. Kolacek, D. Micetic-Turk, M. Ruszczyński and T. Vukavic. 2012. (ESPGHAN Working Group on Probiotics and Prebiotics) Inulin and Fructo-oligosaccharides for the prevention of antibiotic-associated diarrhea in children: report by the ESPGHAN Working Group on Probiotics and Prebiotics. J. Pediatr Gastroenterol. Nutr. 54: 828–9.

Szymanski, H., J. Pejcz, M. Jawien, A. Chmielarczyk, M. Strus and P.B. Heczko. 2006. Treatment of acute infectious diarrhoea in infants and children with a mixture of three *Lactobacillus rhamnosus* strains—a randomized, double-blind, placebo-controlled trial. Aliment Pharmacol. Ther. 23: 247–53.

Temmerman, R., B. Pot, G. Huys and J. Swings. 2003. Identification and antibiotic susceptibility of bacterial isolates from probiotic products. Int. J. Food Microbiol. 81: 1–10.

Thibault, H., C. Aubert-Jacquin and O. Goulet. 2004. Effects of long-term consumption of a fermented infant formula (with Bifidoacterium Breve c50 and Streptococcus thermophilus 065) on acute diarrhea in healthy infants. J. Pediatr Gastroenterol. Nutr. 39: 147–52.

Thomas, D.W. and F.R. Greer. 2010. Clinical report: probiotics and prebiotics in pediatrics. Pediatrics126: 1217–31.

Tissier, H. 1906. Traitement des infections intestinales par la méthode de la transformation de la flore bactérienne de l'intestin. CR Soc. Biol. 60: 359–61.

Tung, J.M., L.R. Dolovich and C.H. Lee. 2009. Prevention of Clostridium difficile infection with *Saccharomyces boulardii*: a systematic review. Can J. Gastroenterol. 23: 817–21.

Turck, D., J.P. Bernet, J. Marx, H. Kempf, P. Giard, O. Walbaum, A. Lacombe, F. Rembert, F. Toursel, P. Bernasconi, F. Gottrand, L.V. McFarland and K. Bloch. 2003. Incidence and risk factors of oral antibiotic associated diarrhea in an outpatient pediatric population. J. Pediatr Gastroenterol. Nutr. 37: 22–6.

Van Niel, C.W., C. Feudtner, M.M. Garrison and D.A. Christakis. 2002. Lactobacillus therapy for acute infectious diarrhea in children: a meta-analysis. Pediatrics 109: 678–84.

Vandenplas, Y. 2012. Identification of probiotics by specific strain name. Aliment Pharmacol. Ther. 35: 860.

Vandenplas, Y. and S. De Hert. 2012. Cost/benefit of synbiotics in acute infectious gastroenteritis: spend to save. Benef. Microbes 3: 189–94.

Vandenplas, Y. and S.G. De Hert. 2011. Probiotical-study group. Randomised clinical trial: the synbiotic food supplement Probiotical vs. placebo for acute gastroenteritis in children. Aliment Pharmacol. Ther. 34: 862–7.

Vanhee, L.M.E., F. Goeme, H.J. Nelis and T. Coenye. 2010. Quality control of fifteen probiotic products containing *Saccharomyces boulardii*. J. Appl. Microbio. 109: 1745–52.

Villarruel, G., D.M. Rubio, F. Lopez, J. Cintioni, R. Gurevech, G. Romero and Y. Vandenplas. 2007. *Saccharomyces boulardii* in acute childhood diarrhea: a randomized, placebo-controlled study. Acta Paediatr 96: 538–41.

15

Probiotics for Gastrointestinal Diseases

Hale Akpınar

Introduction

For more than a century it has been known that certain microorganisms may impart health benefits to the host when administered in adequate amounts. These microorganisms, termed probiotics, have recently become an important topic in basic and clinical investigations. Probiotics are commonly used by patients with gastrointestinal (GI) complaints or diseases and are also being recommended by the clinicians in the daily practice of gastroenterology (Williams et al. 2010). The goal of this chapter is to provide an overview of the rationale and data for the role of probiotics for treating commonly encountered GI disorders.

The Human Microbiome and Probiotic Mechanisms

The human GI tract is host to over 500 bacterial species which facilitate digestion, nutrient provision, and shape our immune system (Kau et al. 2011). Our intestinal bacteria weigh up to 1 kg and bacterial cells outnumber human cells by 10:1. The bacterial genome may outnumber the human genome by 100:1. Several vitamin Bs, vitamin K, folate, and

Dokuz Eylül University School of Medicine, Gastroenterology Department, 35340 Inciralti Izmir Turkey.
Email: hale.akpinar@deu.edu.tr

short-chain fatty acids are produced by these bacteria. Up to 10% of an individual's daily energy needs can be derived from the by-products of bacterial fermentation. GI microbiota are also critical for normal immune system development (Macpherson and Harris 2004). Probiotics have several putative mechanisms such as modulation of immune or sensory-motor function, enhancement of mucosal barrier function and antipathogen effects (Fig. 1) (Ng et al. 2009, Madsen 2011). Soluble products secreted by probiotics also mediate important physiologic benefits (Yan et al. 2011). The mechanisms by which probiotics exert benefits varies by specific probiotic strain and likely depends on the clinical indication (Shanahan 2010). In the future, greater understanding of probiotic-specific mechanisms could allow for precise selection of a particular probiotic strain to target a patient's specific pathogenic defect and clinical problem (Ciorba 2012).

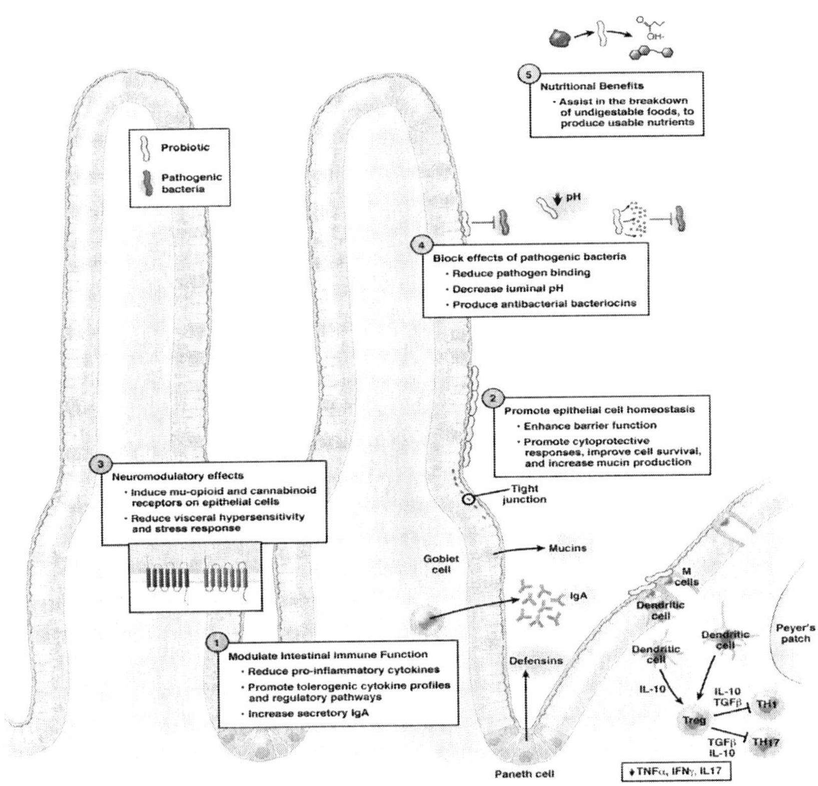

Fig. 1. Mechanisms of action of probiotics in the gastrointestinal tract (Ciorba 2012).

Color image of this figure appears in the color plate section at the end of the book.

Lactobacillus and *Bifidobacterium* species are the most commonly used probiotics. However, one of the first probiotics, which is still in use, is the non-pathogenic *Escherichia coli* Nissle 1917 (*ECN*). *Saccharomyces boulardii* is a probiotic yeast strain with the potential advantage of having resistance to most antibiotics. Some of the more commonly available probiotics include (LaMont and Grover 2013):

- VSL#3® (*Bifidobacterium breve, B. longum, B. infantis, Lactobacillus acidophilus, L. plantarum, L. paracasei, L. bulgaricus, Streptococcus thermophilus*)
- Align® (*B. infantis*)
- Culturelle® (*L. rhamnosus GG*)
- DanActive® (*L. casei*)
- Mutaflor® (*ECN1917*)
- Florastor® (*S. boulardii*)

Lactic acid–producing bacteria have been used for centuries in food fermentation. Many yogurts contain live-active *lactobacillus* cultures and are considered functional food products. Most are not considered as probiotics. This term is only used for products with an adequate number of microorganisms at time of consumption specifically shown to confer health benefits in controlled human trials. Yogurts fortified with an adequate number of viable bacteria shown to exert benefit in controlled trials are classified as probiotics (Ciorba 2012). If sustained benefit from a probiotic is desired, continued consumption is usually required.

In this chapter, data for probiotic use in several GI disorders will be reviewed. Probiotics with strong supportive data are among the treatment modalities for antibiotic-associated diarrhea (AAD) and viral gastroenteritis. However, the duration of symptoms in these conditions is typically short regardless of probiotic use. In ulcerative colitis (UC), pouchitis and irritable bowel syndrome (IBS), adequate data exist to recommend for clinicians a therapeutic trial of specific probiotic strains or preparations in selected patients. In these conditions probiotics are usually administered as adjunctive therapy, rather than first-line therapy. In hepatic encephalopathy (HE), Crohn's disease (CD) and *Clostridium difficile*–associated diarrhea (CDAD), conventional medical therapies remain the standard. Practical relevant probiotic concepts are summarized in Table 1 (Ciorba 2012).

Table 1. Practical Considerations Relevant to Probiotics in Clinical Practice.

• Common side effects are transient but include gas and bloating
• Different probiotic strains possess unique properties for benefiting host physiology
• One probiotic does not fit all GI illnesses; probiotic selection should be based on the clinical indication and take into consideration the strain and dosage used in clinical trials
• Probiotic therapies may be used to supplement rather than replace conventional therapies
• Continued consumption is likely required, if sustained benefit from a probiotic is desired
• Avoid probiotics in the critically ill and those with severe immune compromise

Probiotic Therapy for Gastrointestinal Conditions

Acute Infectious Diarrhea (AID)

Acute diarrhea, where the most common cause is infection, continues to be a leading cause of morbidity, hospitalization and mortality worldwide. Non-antimicrobial approaches to therapy have become increasingly important with the consideration of antimicrobial resistance. Probiotics may be considered in patients with presumed infectious diarrheal illness (Gadewar and Fasano 2005). Multiple systematic reviews have demonstrated a modest reduction in the duration of infectious diarrhea with the use of probiotics, although there was heterogeneity among studies (Van Niel et al. 2002, MacFarland 2007, Szajewska et al. 2007, Britton and Versalovic 2008, Bernaola Aponte et al. 2010, Szejewska and Skórka 2009, Allen et al. 2010).

The data of both prevention and treatment of AID with probiotics come mostly from pediatric studies. The benefit of probiotics in preventing AID is moderate (Thomas and Greer 2010, Guandalini 2011). *L. rhamnosus* GG (LGG), *Lactobacillus* and *L. casei* all have shown benefit, with an approximate NNT of 7 in the child care center setting (Weizman et al. 2005, Pedone et al. 2000, Szajewska et al. 2001). According to the American Academy of Pediatrics, probiotics for preventing AID are not universally endorsed, but they may have a role in special circumstances (Thomas and Greer 2010). The US Center for Disease Control also states that data are not sufficient to support the use of probiotics such as LGG to prevent traveler's diarrhea of bacterial origin.

The data supporting treatment of AID with probiotics are stronger. Probiotics reduces both severity and duration of diarrhea by approximately 1 day or 36 hours (Allen et al. 2010, Szajewska and Mrukowicz 2001, Szymański et al. 2006, Hom 2011, Dinleyici et al. 2012a,b). A 2010 meta-analysis that included 63 randomised controlled trials (RCT) (using several different probiotic preparations) in patients found that probiotics reduced the overall risk of diarrhea lasting four or more days by 59% (RR 0.41, 95% CI 0.32–0.53) (Allen et al. 2010). The two most commonly studied probiotics were LGG and *S. boulardii*. The American Academy of Pediatrics supports the recommendation of LGG early in the course of AID to reduce symptom duration (Ciorba 2012). However, it is unclear whether probiotics reduce important complications of diarrheal illness such as dehydration and malnutrition.

Antibiotic Associated Diarrhea (AAD)

The use of antibiotics that disturb the gastrointestinal flora is associated with diarrhea in as many as 30% of patients (Barbut and Meynard 2002). Symptoms range from mild and self-limiting to severe. AAD is an important reason for nonadherence with antibiotic treatment. There is an increasing evidence for the effectiveness of probiotics in preventing or treating AAD. Potentially, probiotics maintain or restore gut microecology during or after antibiotic treatment through receptor competition, competition for nutrients, inhibition of epithelial and mucosal adherence of pathogens, introduction of lower colonic pH favoring the growth of nonpathogenic species, stimulation of immunity, or production of antimicrobial substances (Cremonini et al. 2002, Hempel et al. 2012).

Many studies have evaluated a variety of probiotics in the management of AAD. However, many of them were small, had different end-points, and had important methodological problems which inhibited their comparisons. Systematic reviews suggest that probiotics are effective in reducing the incidence of diarrhea in patients who are taking antibiotics (Hempel et al. 2012, Johnston et al. 2011, Szajewska et al. 2006, MacFarland 2006, Videlock and Cremonini 2012). A 2011 Cochrane review evaluating more than 3400 pediatric patients from 16 studies concluded that probiotics had a protective effect in preventing AAD (NNT:7) (Jonhston et al. 2011). Studies using LGG and *S. boulardii* produced the most convincing results (Szajewska et al. 2006). The American Academy of Pediatrics also recommends probiotic use for prevention of, but not treatment of AAD (Thomas and Greer 2010). In the adult population, probiotics also appear effective in preventing AAD. A meta-analysis evaluating studies on various probiotics and antibiotic regimens published between 1977 and 2005 found that both LGG and *S. boulardii* offered a reduction (69% and 63%, respectively) in risk of AAD development (combined RR 0.31 and 0.37, respectively) (MacFarland 2006). One of the largest systematic reviews from 2012 identified 82 randomized trials of probiotics for the prevention of AAD (Hempel et al. 2012). *Lactobacillus,* either alone or in combination with other organisms, were used in the majority of trials (69%), while 16 trials (20%) used *S. boulardii* or *Hansen* CBS 5926. A meta-analysis of 63 trials (11,811 participants) indicated that participants assigned to probiotics had a 42% lower risk of developing AAD than participants in the control groups (RR 0.58; 95% CI 0.50–0.68, NNT: 13). Among the 17 trials that used *Lactobacillus* alone, there was a 36% decreased risk of AAD (RR 0.64; 95% CI 0.47–0.86), whereas the risk was decreased by 52% (RR 0.48; 95% CI 0.35–0.65) among the 15 trials that used *S. boulardii*. In addition, there was no difference between children, adults (18 to 65 years), and older adults (>65 years) for the risk reduction.

Clostridium difficile-Associated Diarrhea (CDAD)

C. difficile–associated diarrhea (CDAD) is a common nosocomial and community-based medical condition which has increased incidence, morbidity, and mortality in the last few years. Many patients receive combinations of antibiotics or multiple antibiotics, which results in the risk of developing CDAD or its recurrence. It is typically linked to antibiotic induced disturbance of the intestinal microbiota. CDAD is now increasingly identified in patients without recent antibiotic exposure (Kelly and LaMont 2008). Metronidazole and vancomycin are the mainstay of the treatment of CDAD (Lo Vecchio and Zacur 2012). New treatment methods that include fidaxomicin, monoclonal antibodies and fecal microbiota transplantation are emerging, and show promise for the treatment of *C. difficile* infection (Khanna and Pardi 2012). Approximately 15–30% of patients experience a symptomatic recurrence after discontinuation of antibiotics. It still remains an important clinical problem. In 1994, a trial reported that *S. boulardii* (500 mg twice a day) given for 4 weeks after antibiotic therapy reduced overall CDAD recurrence rates (MacFarland et al. 1994). However, a follow-up study which was designed to be confirmatory, did not find *S. boulardii* to significantly reduce CDAD recurrence after standard therapy (Surawicz et al. 2000). *Lactobacillus* probiotics as single species or combination probiotic products have been tested for preventing CDAD recurrence. While some results have been promising, most studies are underpowered, have methodological flaws, or have not been reproduced (Na and Kelly 2011). Probiotic-based primary prevention may be an approach to the management of CDAD. A meta-analysis including 3818 participants in 20 trials demonstrated that probiotics reduced the incidence of CDAD by 66% (pooled RR, 0.34 [95% CI, 0.24 to 0.49]; I(2) = 0%). In a population with a 5 % incidence of antibiotic-induced CDAD (median control group risk), probiotic prophylaxis would prevent 33 episodes (CI, 25 to 38 episodes) per 1000 persons (Johnston et al. 2012). The moderate quality evidence of this meta-analysis suggests that probiotic prophylaxis results in a large reduction in CDAD. However, current society guidelines and expert opinion panels state that existing data are not sufficient to justify recommending probiotics for preventing primary or recurrent CDAD (Floch et al. 2011, Cohen et al. 2010).

Collagenous Colitis

The microscopic colitis is diagnosed by a triad of watery diarrhea, normal endoscopic and characteristic histologic findings (Chetty and Govender 2012). Collagenous colitis, a subtype of microscopic colitis, is a diarrheal illness characterized by the presence of a thickened subepithelial collagenous plate and lymphocytic infiltrate in the colonic mucosa. A possible benefit of ECN was suggested in an open-label study of 14 patients (Tromm et al. 2004). The authors hypothesized that the benefit may have been due to an antagonistic effect of the probiotic against strains of Yersinia species. In a second placebo-controlled trial, a combination of *L. acidophilus* and *B. animalis* strains had no significant effect on primary end points but were associated with some improvement in symptoms (Wildt et al. 2006).

Inflammatory Bowel Disease (IBD)

Ulcerative colitis (UC)

Several published RCTs have shown benefit of probiotics in the management of ulcerative colitis (UC). ECN at 200 mg/day was found to be effective as 1500 mg meselamine for maintaining remission in UC (Kruis et al. 2004). High dose VSL#3 has shown therapeutic efficacy in 2 RCTs evaluating patients with mild to moderately active UC (Tursi et al. 2010). Another study including 144 adults with relapsing UC, showed that VSL#3 group had significantly higher remission rates (42.9% vs. 15.9%) and endoscopic healing (32% vs. 14.7%) (Sood et al. 2009). Recent Cochrane review concluded that there were insufficient data to demonstrate the efficacy of probiotics in maintaning remission in UC (Naidoo et al. 2011). *Lactobacillus* and *B. infantis* 35624 also have been ineffective for maintaining remission (Fujimori et al. 2009, Ciorba 2012). As a result, existing evidence suggests that ECN and VSL#3 have modest efficacy, similar to mesalamine, in inducing and maintaining remission for mild-to-moderately active UC (Ciorba 2012).

Crohn's disease (CD)

Probiotic use in the management of CD is not supported by currently available RCT data (Ciorba 2012).

Pouchitis

Chronic or recurrent pouchitis is an important complication occurring in approximately 10–20% of UC patients after ileal anal pouch formation surgery. VSL#3 was shown to be beneficial in prophylaxis against pouchitis onset after surgery (Gionchetti et al. 2003) and in maintaining clinical remission after antibiotic induction (Mimura et al. 2004). These trials were conducted in Europe and included approximately 20 patients per group. A practice-based report from the United States (US) found only 19% of patients who started taking VSL#3 after treatment with antibiotics, to still be taking the probiotic at 8 months (Shen et al. 2005). A single study from the Netherlands found that compared with a historical cohort, patients taking LGG had a delayed onset of pouchitis at 3 years (7% vs. 29%) (Gosselink et al. 2004). Clinical expert-generated guidelines concur that probiotics (VSL#3) can be effective for preventing recurrence of pouchitis (Floch et al. 2011, Pardi et al. 2009, Holubar et al. 2010).

Diverticular colitis

Patients with diverticular disease can develop a segmental colitis, most commonly in the sigmoid colon. Combination therapy with VSL#3 and oral beclomethasone dipropionate was found to be beneficial in a case series (Tursi et al. 2005).

Radiation Enteritis

A meta-analysis of 4 RCTs of probiotics in radiation induced diarrhea did not detect an overall benefit despite significant effects in some of the individual studies (Fuccio et al. 2009). Three studies of 632 subjects explored prophylactic administration and one study examined treatment. Basic research indicates that intestinal bacteria contribute to radiation-induced injury and repair, so this therapeutic approach is open to investigation (LaMont and Grover 2013).

Constipation

Constipation is a common functional gastrointestinal disorder that affects patients of all ages. Treatment modalities include diet and lifestyle changes, bulking agents, stool softeners, osmotic and stimulant laxatives, prucalopride and probiotics in the management of chronic constipation (Liu 2011). A systematic review of 5 RCTs with a total of 377 subjects (194 in the experimental group and 183 in the control group, 266 adults in 3 and 111 children in 2 RCTs) with constipation suggested a favourable effect of treatment with *B. lactis DN-173 010, L. casei Shirota* and ECN on defecation frequency and stool consistency in adults. In children, *L. casei rhamnosus Lcr35,* but not LGG, showed a beneficial effect (Tabbers et al. 2011). The authors concluded that the use of probiotics for the treatment of constipation should be considered investigational (Chmielewska and Szajewska 2010). Recent RCTs of probiotics including *L. paracasei* and *B. longum* have shown some beneficial results in patients with chronic constipation (Riezzo et al. 2012, Guerra et al. 2011). However, in another RCT, the fermented dairy product containing *B. lactis strain DN-173*

010 did increase stool frequency in constipated children but this increase was comparable in the control group (Tabbers et al. 2011).

Irritable Bowel Syndrome (IBS)

Irritable bowel syndrome (IBS) is characterized by recurrent abdominal pain or discomfort, accompanied by abnormal bowel habits which occur over at least 3 months without any underlying organic abnormality (Mertz 2003). Multiple pathological mechanisms are likely to be involved in the development of IBS (Bolino and Bercik 2010). Some studies have shown that there is a link between IBS sypmtoms and the intestinal microbiota. IBS patients may have differences in their intestinal microbiota compared with controls. A relative reduction in *lactobacilli* and *bifidobacteria,* combined with increased numbers of *enterobacteria, coliforms, bacteroides,* and *firmicutes* species, have been noted in groups of patients with IBS (Bolino and Bercik 2010, Rajilic-Stojanovic et al. 2011, Jeffery et al. 2012). Although controversy exists, bacteria likely contribute to some symptoms of IBS including abdominal pain, bloating, and flatulence. They may be related to excessive gas production by bacterial fermentation in the colon. However gas volumes have been reported to be normal in IBS patients (Azpiroz and Malagelada 2005). This mechanism is unlikely to explain symptoms in all patients. Small intestinal bacterial overgrowth (SIBO) has also been controversially implicated in the pathophysiology of IBS (Vanner 2008). New-onset IBS symptoms can develop in up to a third of individuals after an episode of infectious gastroenteritis. It has been thought that gut microbiota changes may occur during this infectious episode and contribute to the IBS symptoms (Quigley 2011a,b). Mucosal adaptive and innate immune response can be activated by abnormal microbiota. It can lead to the increase of intestinal permeability. Therefore nociceptive sensory pathways can be activated and enteric nervous system can be dysregulated (Simren 2013). Epithelial defense mechanisms such as mucus and defensins (host defense peptides) are effected by the bidirectional signalization between microbiota and altered epithelium in IBS (Langhorst 2009). Increased colonic mucosal expression of Toll-like receptors (Brint et al. 2011) and increased circulating antibodies against components of indigenous microbiota have also been detected in IBS patients. These abnormal microbial–host interactions can alter gut permeability, increase microbial antigenic load, and contribute to the sensory-motor dysfunction often observed in IBS.

There is also evidence which supports a role of microbiota-host interactions in IBS coming from clinical trials which demonstrated beneficial impacts for probiotics in IBS (Quigley 2010, 2011a). Recent meta-analysis of RCTs of probiotics in patients with IBS found that these preparations were better than placebo at improving global IBS symptoms (Moayyedi et al. 2010) (Table 1). General recommendations from the American College of Gastroenterology as well as expert consensus panels from both US and Europe are similar (Floch 2011, Brandt et al. 2009, Guarner et al. 2012). There are at least some positive controlled studies showing that probiotic supplementation reduces IBS symptoms in some patients. However, variations in trial design, poor quality of many of the studies, and a paucity of information on the potential mechanisms of actions of probiotics limit that interpretation of available data. Furthermore, benefits for probiotics over placebo in IBS have generally been modest and it is not yet known whether specific probiotics help to reduce specific symptoms and whether products with a single strain are better than those with multiple strains (Aziz et al. 2013). With probiotics, patients might experience a global improvement in symptomatology rather than specific improvement in bowel function (Ciorba 2012). Further, well controlled studies are required.

Table 1. Meta-analyses and systematic reviews of probiotic therapy in irritable bowel syndrome (Aziz et al. 2013).

Authors	**Number of studies (number of subjects)**	**Outcome**
Huertas-Ceballos et al. 2008	3 (168) children only	No benefit Pooled OR for improvement 1.17 (0.62–2.21)
McFarland and Dublin 2008	20 (1414)	Less global IBS Symptoms: RR 0.77 (0.6–0.94) Less abdominal Pain: RR 0.78 (0.69–0.88)
Nikfar et al. 2008	8 (922)	Clinical improvement: RR 1.22 (1.07–1.4)
Hoveyda et al. 2008	14	Outcomes as a dichotomous variable: 7 RCT's (n = 895) OR for overall improvement = 1.6 (1.2–2.2) Continuous data: 6 RCT's (n = 657) SMD for overall improvement = 0.23 (0.07–0.38)
Brenner et al. 2009	16	Only Bifidobacterium infantis 35624 showed significant improvement over placebo in an appropriately designed study
Moayyedi et al. 2010	20 (1628)	Outcomes as a dichotomous variable: 11 RCT's (n = 936) RR of IBS not improving = 0.71; 95% CI = 0.57 to 0.87, NNT = 4 IBS score as a continuous outcome: 15 RCT's (n = 1351) SMD = (0.34; 95% CI=–0.60 to –00.7)
Horvath et al. 2010	3 (167) children only Lactobacillus rhamnosus GG (LGG) only	LGG associated with a significantly higher rate of treatment responders in children with IBS (RR 1.70, 95% CI 1.27–2.27, NNT 4)

RCT, randomized controlled trial; RR, relative risk; OR, odds ratio; CI, confidence interval; NNT, number needed to treat; SMD, standardized mean difference

Lactose Intolerance

Lactose intolerance resulting in gastrointestinal symptoms is a common health concern. Ingestion of lactase-containing probiotics has the potential to aid lactose digestion in patients with lactose intolerance. A systematic review of 10 controlled trials found inconsistent results and suggested further studies on specific strains in which a benefit was suggested (Levri et al. 2005).

Hepatic Encephalopathy (HE)

Hepatic encephalopathy (HE) represents a continuum of transient and reversible neurologic and psychiatric dysfunction. It is a reversible state of impaired cognitive function or altered consciousness in patients with liver disease or portosystemic shunting (Khungar and Poordad 2012). Alterations of intestinal microbiota seem to play an important role in induction and promotion of liver damage progression. Probiotics are able to decrease the permeability of the intestinal wall, and decrease bacterial translocation and endotoxemia in animal models as well as in clinical studies, which is extremely important in the prevention of complications of liver cirrhosis (Lata et al. 2011). According to meta-analyses, probiotics appear to reduce plasma ammonia concentration when compared with placebo or no intervention in patients with HE. Before probiotics can be endorsed as effective therapy for HE, rigorous evaluation in standardized RCT with clinically relevant outcomes is still needed (McGee et al. 2011, Holte et al. 2012).

Pancreatitis

Modulation of the intestinal flora through the administration of probiotics has a rationale as a possible treatment option. A multicenter, double-blind, placebo-controlled randomized trial of multispecies probiotic preparation and placebo demonstrated that probiotics did not reduce the risk of infectious complications and actually increased mortality from mesenteric ischemia in patients with pancreatitis (Besselink et al. 2008, Capurso et al. 2008). As a result, probiotics are not recommended in severe acute pancreatitis. More data on efficacy and safety from a larger and stringently designed study are eagerly waited.

Small Intestinal Bacterial Overgrowth (SIBO)

Small bowel bacterial overgrowth, in which colon-derived bacteria colonize the upper small bowel, is found in a wide variety of adult diseases associated with intestinal failure and dysfunction. Its treatment is based on antibiotics. The role of probiotics in SIBO is unproven (LaMont and Grover 2013). A recent study showed a good outcome with sequential antibiotic-probiotic/prebiotic administration in patients with SIBO (Rosania et al. 2012).

Helicobacter pylori

Helicobacter pylori (Hp) is one of the most widespread infections worldwide. It is known as a cause of gastritis and peptic ulcer. It has also been classified as a group A carcinogen for gastric cancer by World Health Organization. The triple treatment including proton pump inhibitor (PPI)-clarithromycin and amoxicillin or metronidazole with the duration of 1 or 2 weeks has been the first line strategy for Hp eradication. The aim of giving probiotics during this therapy is to reduce the adverse effects, to improve tolerability and compliance of multiple antibiotics regimens (Sabbi 2011). Meta-analyses on the studies where *Lactobacilli* were used are heterogeneous (Zou et al. 2009, Sachdeva and Nagpal 2009). A meta-analysis on the use of *S. boulardii* as adjuvant to triple therapy showed promising results (OR¼ 0.46 (95% CI 0.3 to 0.7)) (Szajewska et al. 2010). Therefore the statement regarding the use of probiotics as an adjuvant treatment in reducing side effects in the eradication therapy of Hp has been placed in the Maastrict IV Consensus Report of Europe (Malfertheiner et al. 2012). However more studies need to be performed.

References

Allen, S.J, E.G. Martinez, G.V. Gregorio and L.F. Dans. 2010. Probiotics for treating acute infectious diarrhoea. Cochrane Database Syst. Rev. CD003048.

Aziz, Q., J. Dore, A. Emmanuel, F. Guarner and E.M.M. Quigley. 2013. Gut microbiota and gastrointestinal health: current concepts and future directions. Neurogastroenterol. Motil. 25: 4–15.

Azpiroz, F. and J.R. Malagelada. 2005. The pathogenesis of bloating and visible distension in irritable bowel syndrome Gastroenterol. Clin. North. Am. 34: 257–269.

Barbut, F. and J.L. Meynard. 2002. Managing antibiotic associated diarrhoea. BMJ 324(7350): 1345–1346.

Bernaola Aponte, G., C.A. Bada Mancilla, N.Y. Carreazo Pariasca and R.A. Rojas Galarza. 2010. Probiotics for treating persistent diarrhoea in children. Cochrane. Database. Syst. Rev.: CD007401.

Besselink, M.G., H.C. van Santvoort, E. Buskens, M.A. Boermeester, H. van Goor, H.M. Timmerman, V.B. Nieuwenhuijs, T.L. Bollen, B. van Ramshorst, B.J. Witteman, C. Rosman, R.J. Ploeg, M.A. Brink, A.F. Schaapherder, C.H. Dejong, P.J. Wahab, C.J. van Laarhoven, E. van der Harst, C.H. van Eijck, M.A. Cuesta, L.M. Akkermans and H.G. Gooszen. 2008.

Dutch Acute Pancreatitis Study Group. Probiotic prophylaxis in predicted severe acute pancreatitis: a randomised, double-blind, placebo-controlled trial. Lancet. 371: 651–9.
Bolino, C.M. and P. Bercik. 2010. Pathogenic factors involved in the development of irritable bowel syndrome: focus on a microbial role. Infect. Dis. Clin. North. Am. 24: 961–975.
Brandt, L.J., W.D. Chey, A.E. Foxx-Orenstein, L.R. Schiller, P.S. Schoenfield, B.M. Spiegel, N.J. Talley and E.M. Quigley. 2009. American College of Gastroenterology Task Force on Irritable Bowel Syndrome. An evidence-based position statement on the management of irritable bowel syndrome. Am. J. Gastroenterol. 104(Suppl. 1): S1–35.
Brenner, D.M., M.J. Moeller, W.D. Chey and P.S. Schoenfeld. 2009. The utility of probiotics in the treatment of irritable bowel syndrome: a systematic review. Am. J. Gastroenterol. 104: 1033–1049.
Brint, E.K., J. MacSharry, A. Fannig, F. Shanahan and E.M. Quigley. 2011. Differential expression of toll like receptors in patients with Irritable Bowel. Am. J. Gastroenterol. 106: 329–336.
Britton, R.A. and J. Versalovic. 2008. Probiotics and gastrointestinal infections. Interdiscip. Perspect. Infect. Dis. 290769.
Capurso, G., M. Marignani, M. Piciucchi, E. Merola and G. Delle Fave. 2008. Probiotics and severe acute pancreatitis. J. Clin. Gastroenterol. 42 Suppl. 3 Pt 1: S148.
Chetty, R. and D. Govender. 2012. Lymphocytic and collagenous colitis: an overview of so-called microscopic colitis. Nat. Rev. Gastroenterol. Hepatol. 9(4): 209–218.
Chmielewska, A. and H. Szajewska. 2010. Systematic review of randomised controlled trials: probiotics for functional constipation. World. J. Gastroenterol. 16(1): 69–75.
Ciorba, M.A. 2012. A gastroenterologist's guide to probiotics. Clin. Gastroenterol. Hepatol. 10(9): 960–968.
Cohen, S.H., D.N. Gerding, S. Johnson, C.P. Kelly, V.G. Loo, L.C. McDonald, J. Pepin and M.H. Wilcox. 2010. Society for Healthcare Epidemiology of America; Infectious Diseases Society of America. Clinical practice guidelines for *Clostridium difficile* infection in adults: 2010 update by the society for healthcare epidemiology of America (SHEA) and the infectious diseases society of America (IDSA). Infect. Control. Hosp. Epidemiol. 31: 431–455.
Cremonini, F., S. Di Caro, L. Santarelli, M. Gabrielli, M. Candelli, E.C. Nista, A. Lupascu, G. Gasbarrini and A. Gasbarrini. 2002. Probiotics in antibiotic-associated diarrhoea. Dig. Liver. Dis. 34(suppl. 2): S78–S80.
Dinleyici, E.C., N. Dalgic, S. Guven, M. Ozen, A. Kara, V. Arica, O. Metin-Timur, M. Sancar, Z. Kurugol, G. Tanir, D. Ozturk, S. Aydogdu, M. Tutanc, M. Eren and Y. Vandenplas. 2012a. The effect of a multispecies synbiotic mixture on the duration of diarrhea and length of hospital stay in children with acute diarrhea in Turkey: Single blinded randomized study. Eur. J. Pediatr. [Epub ahead of print]
Dinleyici, E.C., M. Eren, M. Ozen, Z.A. Yargic and Y. Vandenplas. 2012 b. Effectiveness and safety of Saccharomyces boulardii for acute infectious diarrhea. Expert. Opin. Biol. Ther. 12(4): 395–410.
Floch, M.H., W.A. Walker, K. Madsen, M.E. Sanders, G.T. Macfarlane, H.J. Flint, L.A. Dieleman, Y. Ringel, S. Guandalini, C.P. Kelly and L.J. Brandt. 2011. Recommendations for probiotic use-2011 update. J. Clin. Gastroenterol. 45(Suppl.): S168–S171.
Fuccio, L., A. Guido, L.H. Eusebi, L. Laterza, D. Grilli, V. Cennamo, L. Ceroni, E. Barbieri and F. Bazzoli. 2009. Effects of probiotics for the prevention and treatment of radiation-induced diarrhea. J. Clin. Gastroenterol. 43: 506–513.
Fujimori, S., K. Gudis, K. Mitsui, T. Seo, M. Yonezawa, S. Tanaka, A. Tatsuguchi and C. Sakamoto. 2009. A randomized controlled trial on the efficacy of synbiotic versus probiotic or prebiotic treatment to improve the quality of life in patients with ulcerative colitis. Nutrition 25: 520–525.
Gadewar, S. and A. Fasano. 2005. Current concepts in the evaluation, diagnosis and management of acute infectious diarrhea. Curr. Opin. Pharmacol. 5(6): 559–565.

Gionchetti, P., F. Rizzello, U. Helwig, A. Venturi, K.M. Lammers, P. Brigidi, B. Vitali, G. Poggioli, M. Miglioli and M. Campieri. 2003. Prophylaxis of pouchitis onset with probiotic therapy: a double-blind, placebo-controlled trial. Gastroenterology 124: 1202–1209.

Gosselink, M.P., W.R. Schouten, L.M. van Lieshout, W.C. Hop, L.D. Laman and J.G. Ruseler-van Embden. 2004. Delay of the first onset of pouchitis by oral intake of the probiotic strain *Lactobacillus rhamnosus GG*. Dis. Colon. Rectum. 47: 876–884.

Guandalini, S. 2011. Probiotics for prevention and treatment of diarrhea. J. Clin. Gastroenterol. 45(Suppl): S149–S153.

Guarner, F., A.G. Khan, J. Garisch, R. Eliakim, A. Gangl, A. Thomson, J. Krabshuis, T. Lemair, P. Kaufmann, J.A. de Paula, R. Fedorak, F. Shanahan, M.E. Sanders, H. Szajewska, B.S. Ramakrishna, T. Karakan and N. Kim. 2012. World Gastroenterology Organization. World Gastroenterology Organisation Global Guidelines: probiotics and prebiotics October 2011. J. Clin. Gastroenterol. 46(6): 468–481.

Guerra, P.V., L.N. Lima, T.C. Souza, V. Mazochi, F.J. Penna, A.M. Silva, J.R. Nicoli and E.V. Guimarães. 2011. Pediatric functional constipation treatment with Bifidobacterium-containing yogurt: a crossover, double-blind, controlled trial. World. J. Gastroenterol. 17(34): 3916–3921.

Hempel, S., S.J. Newberry, A.R. Maher, Z. Wang, J.N. Miles, R. Shanman, B. Johnsen and P.G. Shekelle. 2012. Probiotics for the prevention and treatment of antibiotic-associated diarrhea: a systematic review and meta-analysis. JAMA 307(18): 1959–1969.

Holte, K., A. Krag and L.L. Gluud. 2012. Systematic review and meta-analysis of randomized trials on probiotics for hepatic encephalopathy. Hepatol. Res. 42(10): 1008–1015.

Holubar, S.D., R.R. Cima, W.J. Sandborn and D.S. Pardi. 2010. Treatment and prevention of pouchitis after ileal pouch-anal anastomosis for chronic ulcerative colitis. Cochrane. Database. Syst. Rev. CD001176.

Hom, J. 2011. Do probiotics reduce the duration and symptoms of acute infectious diarrhea? Ann. Emerg. Med. 58(5): 445–446.

Horvath, A., P. Dziechciarz and H. Szajewska. 2011. Meta-analysis: Lactobacillus rhamnosus GG for abdominal pain-related functional gastrointestinal disorders in childhood. Aliment. Pharmacol. Ther. 33: 1302–1310.

Hoveyda, N., C. Heneghan, K.R. Mahtani, R. Perera, N. Roberts and P. Glasziou. 2009. A systematic review and meta-analysis: probiotics in the treatment of irritable bowel syndrome. BMC. Gastroenterol. 9: 15.

Huertas-Ceballos, A.A., S. Logan, C. Bennett and C. MacArthur. 2009. Dietary interventions for recurrent abdominal pain (RAP) and irritable bowel syndrome in childhood. Cochrane. Database. Syst. Rev. 23: CD003019.

Jeffery, I.B., P.W. O'Toole, L. Öhman, M.J. Claesson, J. Deane, E.M. Quigley and M. Simrén. 2012. An irritable bowel syndrome subtype defined by species-specific alterations in faecal microbiota. Gut. 61: 997–1006.

Johnston, B.C., J.Z. Goldenberg, P.O. Vandvik, X. Sun and G.H. Guyatt. 2011. Probiotics for the prevention of pediatric antibiotic-associated diarrhea. Cochrane. Database. Syst. Rev. 11: CD004827.

Johnston, B.C., S.S. Ma, J.Z. Goldenberg, K. Thorlund, P.O. Vandvik, M. Loeb and G.H. Guyatt. 2012. Probiotics for the prevention of Clostridium difficile-associated diarrhea: a systematic review and meta-analysis. Ann. Intern. Med. 157(12): 878–888.

Kau, A.L., P.P. Ahern, N.W. Griffin, A.L. Goodman and J.I. Gordon. 2011. Human nutrition, the gut microbiome and the immune system. Nature 474: 327–336.

Kelly, C.P. and J.T. LaMont. 2008. *Clostridium difficile*–more difficult than ever. N. Engl. J. Med. 359: 1932 1940.

Khanna, S. and D.S. Pardi. 2012. *Clostridium difficile* infection: new insights into management. Mayo. Clin. Proc. 87(11): 1106–17.

Khungar, V. and F. Poordad. 2012. Hepatic encephalopathy. Clin. Liver. Dis. 16(2): 301–320.

Kruis, W., P. Fric, J. Pokrotnieks, M. Lukás, B. Fixa, M. Kascák, M.A. Kamm, J. Weismueller, C. Beglinger, M. Stolte, C. Wolff and J. Schulze. 2004. Maintaining remission of ulcerative colitis with the probiotic *Escherichia coli Nissle* 1917 is as effective as with standard mesalazine. Gut. 53: 1617–1623.

Langhorst, J., A. Junge, A. Rueffer, J. Wehkamp, D. Foell, A. Michalsen, F. Musial and G.J. Dobos. 2009. Elevated human beta defensin-2 levels indicate an a ctivation of the innate immune system in patients with irritable bowel syndrome. Am. J. Gastroenterol. 104: 404–410.

LaMont, J.T. and S. Grover. 2013. Probiotics for gastrointestinal disease. Topic 2603 Version 16.0 ©2013 UpToDate®.

Lata, J., J. Jurankova, M. Kopacova and P. Vitek. 2011. Probiotics in hepatology. World. J. Gastroenterol. 28; 17(24): 2890–2896.

Levri, K.M., K. Ketvertis, M. Deramo, J.H. Merenstein and F. D'Amico. 2005. Do probiotics reduce adult lactose intolerance? A systematic review. J. Fam. Pract. 54(7): 613–20.

Liu, L.W. 2011. Chronic constipation: current treatment options. Can. J. Gastroenterol. 25 Suppl. B: 22B–28B.

Lo Vecchio, A. and G.M. Zacur. 2012. *Clostridium difficile* infection: an update on epidemiology, risk factors, and therapeutic options. Curr. Opin. Gastroenterol. 28: 1–9.

Macpherson, A.J. and N.L. Harris. 2004. Interactions between commensal intestinal bacteria and the immune system. Nat. Rev. Immunol. 4: 478–485.

Madsen, K.L. 2011. Interactions between microbes and the gut epithelium. J. Clin. Gastroenterol. 45(Suppl): S111–S114.

Malfertheiner, P., F. Megraud, C.A. O'Morain, J. Atherton, A.T. Axon, F. Bazzoli, G.F. Gensini, J.P. Gisbert, D.Y. Graham, T. Rokkas, E.M. El-Omar and E.J. Kuipers. 2012. European Helicobacter Study Group. Management of Helicobacter pylori infection- the Maastricht IV/ Florence Consensus Report. Gut. 61(5): 646–664.

McFarland, L.V., C.M. Surawicz, R.N. Greenberg, G.W. Elmer, K.A. Moyer, S.A. Melcher, K.E. Bowen, J.L. Cox, Z. Noorani et al. 1994. A randomized placebo-controlled trial of *Saccharomyces boulardii* in combination with standard antibiotics for *Clostridium difficile* disease. JAMA 271: 1913–1918.

McFarland, L.V. 2006. Meta-analysis of probiotics for the prevention of antibiotic associated diarrhea and the treatment of *Clostridium difficile* disease. Am. J. Gastroenterol. 101: 812–822.

McFarland, L.V. 2007. Meta-analysis of probiotics for the prevention of traveler's diarrhea. Travel. Med. Infect. Dis. 5: 97–105.

McFarland, L.V. and S. Dublin. 2008. Meta-analysis of probiotics for the treatment of irritable bowel syndrome. World. J. Gastroenterol. 14: 2650–2661.

McGee, R.G., A. Bakens, K. Wiley, S.M. Riordan and A.C. Webster. 2011. Probiotics for patients with hepatic encephalopathy. Cochrane. Database. Syst. Rev. 9(11): CD008716.

Mertz, R. 2003. Irritable bowel syndrome. N. Engl. J. Med. 349: 2136–2146.

Mimura, T., F. Rizzello, U. Helwig, G. Poggioli, S. Schreiber, I.C. Talbot, R.J. Nicholls, P. Gionchetti, M. Campieri and M.A. Kamm. 2004. Once daily high dose probiotic therapy (VSL#3) for maintaining remission in recurrent or refractory pouchitis. Gut. 53: 108–114.

Moayyedi, P., A.C. Ford, N.J. Talley, F. Cremonini, A.E. Foxx-Orenstein, L.J. Brandt and E.M. Quigley. 2010. The efficacy of probiotics in the treatment of irritable bowel syndrome: a systematic review. Gut. 59: 325–332.

Na, X. and C. Kelly. 2011. Probiotics in *Clostridium difficile* infection. J. Clin. Gastroenterol. 45(Suppl): S154–S158.

Naidoo, K., M. Gordon, A.O. Fagbemi, A.G. Thomas and A.K. Akobeng. 2011. Probiotics for maintenance of remission in ulcerative colitis. Cochrane. Database. Syst. Rev. CD007443.

Ng, S.C., A.L. Hart, M.A. Kamm, A.J. Stagg and S.C. Knight. 2009. Mechanisms of action of probiotics: recent advances. Inflamm. Bowel. Dis. 15: 300–310.

Nikfar, S., R. Rahimi, F. Rahimi, S. Derakhshani and M. Abdollahi. 2008. Efficacy of probiotics in irritable bowel syndrome: a meta-analysis of randomized, controlled trials. Dis. Colon. Rectu. m. 51: 1775–1780.

Pardi, D.S., G. D'Haens, B. Shen, S. Campbell and P. Gionchetti. 2009. Clinical guidelines for the management of pouchitis. Inflamm. Bowel. Dis. 15: 1424–1431.

Pedone, C.A., C.C. Arnaud, E.R. Postaire, C.F. Bouley and P. Reinert. 2000. Multicentric study of the effect of milk fermented by *Lactobacillus casei* on the incidence of diarrhoea. Int. J. Clin. Pract. 54(9): 568–571.

Quigley, E.M. 2010. Prebiotics and probiotics; modifying and mining the microbiota. Pharmacol. Res. 61: 213–218.

Quigley, E.M. 2011a. Gut microbiota and the role of probiotics in therapy. Curr. Opin. Pharmacol. 11: 593–603.

Quigley, E.M. 2011b. Therapies aimed at the gut microbiota and inflammation: antibiotics, prebiotics, probiotics, synbiotics, anti-inflammatory therapies. Gastroenterol. Clin. North. Am. 40: 207–222.

Rajilić-Stojanović, M., E. Biagi, H.G. Heilig, K. Kajander, R.A. Kekkonen, S. Tims and W.M. de Vos. 2011. Global and deep molecular analysis of microbiota signatures in fecal samples from patients with irritable bowel syndrome. Gastroenterology 141: 1792–1801.

Riezzo, G., A. Orlando, B. D'Attoma, V. Guerra, F. Valerio, P. Lavermicocca, S. De Candia and F. Russo. 2012. Randomised clinical trial: efficacy of Lactobacillus paracasei-enriched artichokes in the treatment of patients with functional constipation—a double-blind, controlled, crossover study. Aliment. Pharmacol. Ther. 35(4): 441–450.

Rosania, R., F. Giorgio, M. Principi, A. Amoruso, R. Monno, A. Di Leo and E. Ierardi. 2012. Effect of probiotic or prebiotic supplementation to antibiotic therapy in the small intestinal bacterial overgrowth: A comparative evaluation. Curr. Clin. harmacol. Dec. 10.

Sabbi, T. 2011. Short review about Helicobacter pylori infection in pediatric age: epidemiological and clinical findings, diagnosis, therapy and role of probiotics. Pediatr. Med. Chir. 33(5-6): 221–6.

Sachdeva, A. and J. Nagpal. 2009. Effect of fermented milk-based probiotic preparations on H.pylori eradication: a systematic review and meta-analysis of randomized-controlled trials. Eur. J. Gastroenterol. Hepatol. 1: 45–53.

Shanahan, F. 2010. Probiotics in perspective. Gastroenterology 139: 1808–1812.

Shen, B., A. Brzezinski, V.W. Fazio, F.H. Remzi, J.P. Achkar, A.E. Bennett, K. Sherman and B.A. Lashner. 2005. Maintenance therapy with a probiotic in antibiotic-dependent pouchitis: experience in clinical practice. Aliment. Pharmacol. Ther. 22: 721–728.

Simrén, M., G. Barbara, H.J. Flint, B.M. Spiegel, R.C. Spiller, S. Vanner, E.F. Verdu, P.J. Whorwell and E.G. Zoetendal. 2013. Intestinal microbiota in functional bowel disorders: a Rome foundation report. Gut. 62(1): 159–176.

Sood, A., V. Midha, G.K. Makharia, V. Ahuja, D. Singal, P. Goswami and R.K. Tandon. 2009. The probiotic preparation, VSL#3 induces remission in patients with mild-to-moderately active ulcerative colitis. Clin. Gastroenterol. Hepatol. 7(11): 1202–1209.

Surawicz, C.M., L.V. McFarland, R.N. Greenberg, M. Rubin, R. Fekety, M.E. Mulligan, R.J. Garcia, S. Brandmarker, K. Bowen, D. Borjal and G.W. Elmer. 2000. The search for a better treatment for recurrent *Clostridium difficile* disease: use of high-dose vancomycin combined with *Saccharomyces boulardii*. Clin. Infect. Dis. 31(4): 1012–1017.

Szajewska, H., M. Kotowska, J.Z. Mrukowicz, M. Armańska and W. Mikołajczyk. 2001. Efficacy of *Lactobacillus GG* in prevention of nosocomial diarrhea in infants. J. Pediatr. 138(3): 361–365.

Szajewska, H. and J.Z. Mrukowicz . 2001. Probiotics in the treatment and prevention of acute infectious diarrhea in infants and children: a systematic review of published randomized, double-blind, placebo-controlled trials. J. Pediatr. Gastroenterol. Nutr. 33 Suppl. 2: S17–25.

Szajewska, H., M. Ruszczyński and A. Radzikowski A. 2006. Probiotics in the prevention of antibiotic-associated diarrhea in children: a meta-analysis of randomized controlled trials. J. Pediatr. 149: 367–372.

Szajewska, H., A. Skórka, M. Ruszczyński and D. Gieruszczak-Białek. 2007. Meta-analysis: *Lactobacillus GG* for treating acute diarrhoea in children. Aliment. Pharmacol. Ther. 25: 871–881.

Szajewska, H. and A. Skórka. 2009. *Saccharomyces boulardii* for treating acute gastroenteritis in children: updated meta-analysis of randomized controlled trials. Aliment. Pharmacol. Ther. 30: 960–961.

Szajewska, H., A. Horvath and A. Piwowarczyk. 2010. Meta-analysis: the effects of Saccharomyces boulardii supplementation on Helicobacter pylori eradication rates and side effects during treatment. Aliment. Pharmacol. Ther. 32: 1069–1079.

Szymański, H., J. Pejcz, M. Jawień, A. Chmielarczyk, M. Strus and P.B. Heczko. 2006. Treatment of acute infectious diarrhoea in infants and children with a mixture of three *Lactobacillus rhamnosus* strains–a randomized, double-blind, placebo-controlled trial. Aliment. Pharmacol. Ther. 23(2): 247–53.

Tabbers, M.M., A. Chmielewska, M.G. Roseboom, N. Crastes, C. Perrin, J.B. Reitsma, O. Norbruis, D.W. Thomas and F.R. Greer. 2010. American Academy of Pediatrics Committee on Nutrition; American Academy of Pediatrics Section on Gastroenterology, Hepatology, and Nutrition. Probiotics and prebiotics in pediatrics. Pediatrics 126: 1217–1231.

Tabbers, M.M., Chmielewska, M.G. Roseboom, N. Crastes, C. Perrin, J.B. Reitsma, O. Norbruis, H. Szajewska and M.A. Benninga. 2011. Fermented milk containing *Bifidobacterium lactis DN-173 010* in childhood constipation: a randomized, double-blind, controlled trial. Pediatrics 127(6): e1392–1399.

Tromm, A., U. Niewerth, M. Khoury, E. Baestlein, G. Wilhelms, J. Schulze and M. Stolte. 2004. The probiotic E. coli strain Nissle 1917 for the treatment of collagenous colitis: first results of an open-label trial. Z. Gastroenterol. 42(5): 365–369.

Tursi, A., G. Brandimarte, G.M. Giorgetti and W. Elisei. 2005. Beclomethasone dipropionate plus VSL#3 for the treatment of mild to moderate diverticular colitis: an open, pilot study. J. Clin. Gastroenterol. 39(7): 644–645.

Tursi, A., G. Brandimarte, A. Papa, A. Giglio, W. Elisei, G.M. Giorgetti, G. Forti, S. Morini, C. Hassan, M.A. Pistoia, M.E. Modeo, S. Rodino', T. D'Amico, L. Sebkova, N. Sacca', E. Di Giulio, F. Luzza, M. Imeneo, T. Larussa, S. Di Rosa, V. Annese, S. Danese and A. Gasbarrini. 2010. Treatment of relapsing mild-to-moderate ulcerative colitis with the probiotic VSL#3 as adjunctive to a standard pharmaceutical treatment: a doubleblind, randomized, placebo-controlled study. Am. J. Gastroenterol. 105: 2218–2227.

Vanner, S. 2008. The small intestinal bacterial overgrowth. Irritable bowel syndrome hypothesis: implications for treatment. Gut. 57: 1315–1321.

Van Niel, C.W., C. Feudtner, M.M. Garrison and D.A. Christakis. 2002. Lactobacillus therapy for acute infectious diarrhea in children: a meta-analysis. Pediatrics. 109: 678–684.

Videlock, E.J. and F. Cremonini. 2012. Meta-analysis: probiotics in antibiotic-associated diarrhoea. Aliment. Pharmacol. Ther. 35: 1355–1369.

Weizman, Z., G. Asli and A. Alsheikh. 2005. Effect of a probiotic infant Formula on infections in child care centers: comparison of two probiotic agents. Pediatrics 115: 5–9.

Wildt, S., L.K. Munck, L. Vinter-Jensen, B.F. Hanse, I. Nordgaard-Lassen, S. Christensen, S. Avnstroem, S.N. Rasmussen and J.J. Rumessen. 2006. Probiotic treatment of collagenous colitis: a randomized, double-blind, placebo-controlled trial with Lactobacillus acidophilus and Bifidobacterium animalis subsp. Lactis. Inflamm. Bowel. Dis. 12(5): 395–401.

Williams, M.D., C.Y. Ha and M.A. Ciorba. 2010. Probiotics as therapy in gastroenterology: a study of physician opinions and recommendations. J. Clin. Gastroenterol. 44: 631–636.

Yan, F., H. Cao, T.L. Cover, M.K. Washington, Y. Shi, L. Liu, R. Chaturvedi, R.M. Peek Jr., K.T. Wilson and D.B. Polk. 2011. Colon-specific delivery of a probiotic derived soluble protein ameliorates intestinal inflammation in mice through an EGFR-dependent mechanism. J. Clin. Invest. 121: 2242–2253.

Zou, J., J. Dong and X. Yu. 2009. Meta-analysis: Lactobacillus containing quadruple therapy versus standard triple first-line therapy for Helicobacter pylori eradication. Helicobacter. 14: 97–107.

16

Probiotics and *Helicobacter pylori*

Ruggiero Francavilla,[1,*] *Mari Elena Tripaldi,*[2] *Massimiliano Praitano,*[2] *Elena Lionetti*[3] and *Vito Leonardo Miniello*[4]

Helicobacter pylori

Despite the fact that *Helicobacter pylori* (*H. pylori*) was discovered more than 30 years ago and that the Nobel Prize in Medicine or Physiology was awarded to Marshall and Warren few years ago, *H. pylori* infection is still a challenging subject for many researchers and physicians especially when it deals with treatment.

H. pylori is a spiral-shaped, flagellated, microaerophilic Gram-negative bacillus that colonizes the gastric mucosa of about half of the human population, with the highest prevalence in developing countries. The infection is transmitted within the family mainly in childhood (Malaty et al. 2002), likely by fecal–oral transmission; there is also some evidence of *H. pylori* presence in the oral cavity.

[1] Consultant in Paediatric Gastroenterology, Interdisciplinary Department of Medicine, University of Bari–Italy.
Email: rfrancavilla@me.com
[2] Interdisciplinary Department of Medicine, University of Bari–Italy.
[3] Lecturer in Pediatrics, Department of Pediatrics, University of Catania, Italy.
[4] Consultant in Paediatric Nutrition, Interdisciplinary Department of Medicine, University of Bari–Italy.
* Corresponding author

The clinical course of *H. pylori* infection is highly variable and is influenced by both microbial and host factors. The pattern and distribution of gastritis correlate strongly with the risk of clinical sequelae, such as duodenal or gastric ulcers, mucosal atrophy, gastric carcinoma, or gastric lymphoma (Dixon 2001). Patients with antral-predominant gastritis (the most common form of *H. pylori* gastritis), are predisposed to duodenal ulcers, whereas patients with corpus-predominant gastritis and multifocal atrophy are more likely to have gastric ulcers, gastric atrophy, intestinal metaplasia, and ultimately gastric carcinoma. Fortunately, the majority of the *H. pylori*-infected population remains asymptomatic. The pathological outcome of *H. pylori* infection results both from direct bacterial action and from host response and susceptibility (Suerbaum and Michetti 2002).

The role of *H. pylori* infection in dyspepsia remains controversial: an increased prevalence of *H. pylori* has been reported in this condition, but inconsistent long-term symptom relief has been observed with bacterial eradication in large, randomized trials (McColl et al. 1998). A Cochrane review suggests that eradication of *H. pylori* improves symptoms in less then 9% of patients with dyspepsia without ulcers (Moayyedi et al. 2001) and a systematic review and economic evaluation of *H. pylori* eradication treatment for non-ulcer dyspepsia shown that 13 patients need to be treated to cure one for dyspepsia (Moayyedi et al. 2000).

H. pylori infection plays a crucial role in the pathogenesis of chronic active gastritis and peptic ulcer disease in both adults and children (McColl et al. 1998). An increasing amount of evidence also supports the hypothesis that *H. pylori* is an important co-factor in the development of gastric cancer. (Uemura et al. 2001). Indeed, this bacterium is able to influence gastric cell proliferation and apoptosis, thus modulating the levels of some growth factors and inflammatory cytokines such as tumour necrosis factor alpha (TNF-α) (Ierardi et al. 2003).

The goal of *H. pylori* treatment is the complete elimination of the organism. Reinfection rates are low and therefore, the benefit of treatment is durable. Clinically relevant *H. pylori*–eradication regimens must have cure rates of at least 80% without major side effects and with minimal induction of bacterial resistance. Because luminal acidity influences the effectiveness of some antimicrobial agents that are active against *H. pylori*, antibiotics are combined with proton-pump inhibitors or ranitidine bismuth citrate.

For first-line empirical treatment, the Maastricht IV Consensus Report recommends to use clarithromycin-containing treatments in areas of low clarithromycin resistance, whereas bismuth-containing quadruple treatments are recommended in areas of high clarithromycin resistance: If this regimen is not available sequential treatment or a non-bismuth quadruple treatment is recommended (Malfertheiner et al. 2012). These regimens have the disadvantages of being expensive, risking poor compliance, causing side effects, and in particular encouraging resistance emergence, both in *H. pylori* and in commensal organisms exposed gratuitously (Deltenre et al. 1998).

Recent review studies report eradication rates of standard triple therapy of <75% (Oderda et al. 2007, Suerbaum and Michetti 2002). It has been shown that a novel 10-day sequential regimen, characterised by the sequential administration of three antibiotics, is highly efficacious in eradicating *H. pylori* in both adults and children (Vaira et al. 2007, Francavilla et al. 2005). This regimen consists of omeprazole plus amoxicillin for five days followed by omeprazole plus chlarithromycin and tinidazole for the next five days. One group has reported in children a significantly higher eradication rate than that achieved by the standard triple therapy (97.3 versus 75.7%; $p<0.02$) (Francavilla et al. 2008) even in *H. pylori* clarithromycin- resistant strains. (Francavilla et al. 2010).

Nowadays, there is considerable interest in alternative therapies (e.g., targeting urease, a known virulence factor) or adjunctive treatment against *H. pylori* (Go 2002) to reduce some of the drawbacks associated with the antibiotic consumption. To these aims, probiotics have been included as "possible" tools for management of the infection (Hamilton-Miller 2003) and extensive work has currently been carried out on their possible role in the treatment and prophylaxis of *H. pylori* infections.

In the present chapter we will:

a) provide the available evidence of the effect of probiotics on *H. pylori* infection
b) discuss the possible mechanisms of action of probiotics on *H. pylori* infection and
c) discuss the effect of the addition of probiotics to *H. pylori* eradication therapy.

Effect of Probiotics on *H. pylori* Infection

In vitro studies

Several *in vitro* studies have shown that various Lactobacilli can inhibit *H. pylori* growth. Most of the studies involve lactobacilli or their metabolic products, because of their ability to adhere to the gastric mucosa and even transiently reside in the stomach where they may reach a concentration of up to 10^3 CFU/mL of fluid (Isolauri 2001).

Strains with this ability include *Lactobacillus acidophilus*: *L. acidophilus* strain CRL 639 (Lorca et al. 2001), *L. acidophilus* in a lyophilized culture (Lactisyn) (Bhatia et al. 1989), *L. acidophilus* LB (Coconnier et al. 1998), *L. acidophilus* strain NAS and DDS-1 (Rasic et al. 1995); *L. casei* rhamnosus dairy starter (Midolo et al. 1995); *L. johnsonii* La1 (Michetti et al. 1999) and *L. salivarius* WB 1004 (Aiba et al. 1998). Lactobacilli are known to produce by catabolism, relatively large amounts of lactate, and this has been considered as the inhibitory and/or the bactericidal factor by some authors (Midolo et al. 1995, Borruel et al. 2003). Indeed, lactic acid could inhibit the *H. pylori* urease (Lesbros-Pantoflickova et al. 2007) and in addition could exert its antimicrobial effect by lowering of the pH, although in opposition with this hypothesis it has been recently shown that lactic acid released by gastric mucosa enhances the growth of *H. pylori* (Takahashi et al. 2007).

Michetti et al. showed the ability of *Lactobacillus acidophilus* (johnsonii) La1 culture supernatant to down-regulate *H. pylori* infection and treatment of *H. pylori* infected subjects with a drinkable La1 culture supernatant interferes with *H. pylori* infection assessed by a significant reduction in urea breath test delta over baseline values (Michetti et al. 1999). Similar results on the ability to modulate *Helicobacter pylori* colonization were reported in children in 2003 by Gotteland et al. (2003). Coconnier et al. showed the effects of a culture supernatant of *L. acidophilus* strain LB, against *H. pylori*, both *in vitro* and *in vivo* (Coconnier et al. 1998). More recently, Yang and Sheu have prospectively followed 38 children with *H. pylori* infection and 38 age- and sex-matched non-infected controls; all followed a 4-week ingestion of probiotics-containing yogurt (supplement of *Lactobacillus acidophilus*, *Bifidobacterium lactis*, *Lactobacillus bulgaricus*, and *Streptococcus thermophiles*) and bacterial load monitored by urea breath test. The authors were able to show that in the *H. pylori*-infected children four weeks, yogurt ingestion significantly reduced the values of 13C-UBT (19.2 ± 1.7 vs. 22.4 ± 2.1‰, $p = .047$) (Yang and Sheu 2012).

Other authors have clearly shown that for some strains a substance other than lactate also contributes to the antibacterial effects (Lorca et al. 2001, Coconnier et al. 1998, Michetti et al. 1999, Yang and Sheu 2012, Johnson et al. 2003). Lorca et al. tested the effect of 17 different *lactobacillus* strains on *H. pylori* activity and confirmed that the general bactericidal effect shown by lactobacilli is the result of acid production although some strains, such as *L. acidophilus* CRL 639, which show other specific anti-*H. pylori* activities, such as the release of a proteinaceous compound. In detail, Lorca et al. showed that *L. acidophilus* CRL 639 may exert its anti-*H. pylori* action through the secretion of an autolysin, a proteinaceous compound released after cell lysis (Lorca et al. 2001).

In vitro studies have demonstrated that *L. reuteri* ATCC 55730 exert a significant inhibitory effect on *H. pylori* growth (Johnson et al. 2003). A substance named reuterina is responsible for this effect. The probiotic strain *Bacillus subitilis* 3 has also been shown to inhibit the growth of *H. pylori* by the secretion of bacteriocins similar to anticoumacins, belonging to isocoumarin group of antibiotics (Pinchuk et al. 2001). Other probiotic bacteria, such as *L. acidophilus* LB (Coconnier et al. 1998), *L. casei* strain Shirota (Cats et al. 2003), and *L. johnsonii* La1 (Michetti et al. 1999) were shown to exert an inhibitory effect on *H. pylori* by a lactic acid- and pH-independent mechanism. *Bacillus subtilis* has been shown to produce, *in vitro*, at least two antibiotics (one was named amicoumacin), which are able to inhibit *H. pylori* growth, independently of pH or organic acid concentration (Pinchuk et al. 2001). However, the exact nature of antimicrobial substances secreted by most of these strains remains to be determined.

These findings may lead to the conclusions that although the most relevant antibacterial mechanism of action seems to be lactic acid production, some strains of lactobacilli may also exert different antimicrobial effects, and more importantly, that the anti-*H. pylori* activity is extremely strain specific.

Probiotics and preventive effect on gastric mucosal lesions and H. pylori Adhesion to Gastric Cells

Characteristics of gastric mucosal injury include reductions in gastric mucus, motility, prostaglandin levels, and increases in free radical generation, acid back diffusion, histamine levels, leukotriene production, and gastric vascular permeability (Glavin and Szabo 1992).

Many reports suggest that Lactobacilli and their products may have protective effects against mucosal injury in the stomach. Yogurt-containing *L. gasseri* OLL2716, can compete with *H. pylori* in the gastric mucus layer, resulting in decreased colonisation of the organism (Fujimura et al. 2006). *L. gasseri* OLL2716 has been shown to be effective in improving *H. pylori*-induced gastric mucosal inflammation in humans (Sakamoto et al. 2001). In oral administration of *L. casei* strain Shirota in *H. pylori* SS1-infected mice, levels of *H. pylori* colonisation were significantly reduced in the antrum and body mucosa. Uchida et al. (2004) reported that the ingestion of *L. gasseri* OLL2716 significantly inhibits the formation of acute gastric lesions caused by HCl in rats and increases the rate of prostaglandin E2 generation in the gastric mucosa preventing gastric ulcers. Lactobacillus rhamnosus GG has also been reported to enhance gastric healing (Lam et al. 2007a), to stimulate mucus secretion, and to increase transmucosal resistance in the gastric mucosa (Lam et al. 2007b). Therefore, lactobacilli strains may possess the defence against the injuries caused by *H. pylori* infection or acid mediated gastric lesion.

H. pylori can bind tightly to epithelial cells via multiple bacterial surface components and its adhesion to epithelial cells is important in determining the outcome in *H. pylori*–associated diseases since in the gastric mucosa, *H. pylori* interacts with epithelial cells through secretory components or as a result of adherence (Guruge et al. 1998). Several possible mechanisms are implicated in the inhibition of *H. pylori* adhesion: exertion of anti-adhesion activity by secreting antimicrobial substances or competing with adhesion sites.

Animal studies demonstrated that previous colonization by probiotics prevented or reduced *H. pylori* infection in germ-free mice (Johnson-Henry et al. 2004). Thus, regardless of the mechanisms involved in the inhibition of the adherence of *H. pylori* to epithelial cells, probiotics could prevent *H. pylori* colonization of the gastric mucosa by inhibiting its adhesion to epithelial cells. In this context, a study from Mukai et al. is particularly interesting (Mukai et al. 2002); he showed that two of nine *L. reuteri* strains, JCM 1081 and TM 105, were able to bind to asialo-GM1 and sulphatide and to inhibit binding of *H. pylori* to both glycolipids; a probiotic that shares glycolipid-binding specificity with *H. pylori* may compete with pathogens for the receptor site making it possible to hypothesize a future application as anti-adhesion drugs (Felley and Michetti 2003). Also, W. confusa strain PL9001 was shown to inhibit the binding of *H. pylori* to the human gastric cell line MKN-45 (Nam et al. 2002). Kabir et al. demonstrated that *Lactobacillus salivarius* WB 1004 may inhibit the attachment of *H. pylori* to both murine and human gastric epithelial cells and reduces IL-8 release *in vitro*. In a gnotobiotic murine model, *L. salivarius* was able to compete for the colonization of the gastric mucosa sustained by *H. pylori* (Kabir et al. 1997).

In a recent study, Chen et al. studied the antagonistic activities of *Lactobacillus gasseri* and *L. plantarum* by agar plate diffusion assay and test to determine the growth and urease activity of *Helicobacter pylori* cocultured with lactobacilli and the adherence of *H. pylori* to human gastric epithelial cells in the presence of lactobacilli. The authors showed that the 2 Lactobacillus strains had significant anti-*H. pylori* activity, and that this activity may be contributed by the cell-free supernatants of lactobacilli and live Lactobacillus strains *in vitro* indicating that the two Lactobacillus strains could inhibit *H. pylori* adherence to human gastric epithelial cells (Chen et al. 2012).

These results suggest that selected probiotics strains could be of help in preventing the infection in an early stage of colonization of the gastric mucosa by *H. pylori* (Guruge et al. 1998). It has recently been shown that, two years after *H. pylori* eradication, 30% of children became re-infected (Magistà et al. 2005), therefore the possibility to reduce this phenomenon by the simple administration of a probiotic is fascinating.

Probiotics and Mucosal barrier

An increase of gastric mucosal permeability and the subsequent tissue reaction to luminal aggressive factors such as gastric acid is considered to be one of the leading mechanism of mucosal injury (i.e., by aspirin)

(Sigthorsson et al. 1998). The fact that the reversible increase of the permeability precedes the non-reversible histo-pathological changes in the stomach is one of the strongest points of evidence supporting this theory, and a method that prevents these changes may thus prevent the incidence of aspirin-induced gastropathy.

A human study performed by Gotteland et al. (2001) showed that the regular ingestion of probiotics may protect the integrity of the gastric mucosal barrier against indomethacin, a potent non-steroidal anti-inflammatory drug. In a recent study, Akama et al. investigated whether *Lactobacillus gasseri* OLL2716 can protect the gastric mucosal integrity from aspirin using the urinary sucrose excretion (USE) test. The effects of probiotics were examined in both healthy volunteers administered a single high-dose of aspirin and patients undergoing low-dose aspirin therapy. The authors concluded that the probiotic treatment was able to decrease the elevation in the USE value significantly after either the single dose ($p = 0.018$) and the long term use of aspirin ($p = 0.033$), while no significant difference was found in the period without LG21 ($p = 0.113$) suggesting that the regular ingestion of *Lactobacillus gasseri* OLL2716 may protect the integrity of the gastric mucosal permeability against aspirin (Akama et al. 2011).

H. pylori is known to suppress MUCI and MUC5A gene expression in the human gastric cell line (Byrd et al. 2000). *In vitro* studies have shown that *L. plantarum* strain 299v and *L. rhamnosus* GG increase the expression of MUC2 and MUC3 genes (Mack et al. 1999) and the subsequent extracellular secretion of mucin by colon cell cultures (Mack et al. 2003). This property can mediate the ability of these strains to restore the mucosal permeability of gastric mucosa or inhibit the adherence of pathogenic bacteria, including *H. pylori* (Lesbros-Pantoflickova et al. 2007). Pantoflikova et al. have shown a significant increase of mucus thickness after long-term probiotic intake (*L. jonhsonii* Lj1) both in antrum and corpus (Pantoflickova et al. 2003).

Probiotics and the Immunomodulation

Probiotics are thought to influence immune responses and hence they have also been used to treat a number of human conditions in which immune deregulation is considered the underlying cause. Many rigorous studies have suggested that probiotics may be able to reduce host-related immune diseases and modulate the intestinal microbiota. The main effects of probiotics on the immune system in different life stages in humans have been extensively studied and targeted for specific age groups (Romeo et al. 2010). Probiotic bacteria induce multiple effects on the host by changing the intestinal luminal environment, epithelial and mucosal barrier function and the mucosal immune system. In murine models, several strains of lactobacilli enhance both the innate and adaptive immune response through induction of dendritic cell maturation and further stimulate the lymphocytes to release pro-inflammatory cytokines, including tumour necrosis factor-alpha (TNF-α), interferon-gamma (IFN-γ) and interleukin-12 (IL-12) (Perdigon et al. 1999). *H. pylori* infection or lipopolysaccharide stimulation led to significantly increased expression of inflammatory mediators including TNF-α, interleukin-8 (IL-8), inducible nitric oxide synthase and cyclo-oxygenase-2 in gastric epithelial cells, leading to gastric inflammation. (Lee et al. 2010). Probiotics could modify the immune response of the host. *L. salivarius* WB 1004 has shown, *in vitro,* to reduce IL-8 secretion by gastric epithelial cells (Vitini et al. 2000) and in animal studies to increase the number of IgA producing cells in the small intestine (Borruel et al. 2003).

The reduction of inflammation has been demonstrated directly on gastric biopsies by Pantoflikova et al. (2003) and indirectly by the decrease of serum gastrin-17 in *H. pylori*-infected patients after probiotic dietetic supplementation (*L. jonhsonii* Lj1 and *L. rhamnosus* GG, *L. rhamnosus* LC705, Propionibacterium freudenreichii JS, *B. lactis* Bb12) (Myllyluoma et al. 2007). In a recent study, pre-treatment of gastric epithelial cells with *L. plantarum* MG208, *L. rhamnosus* MG316 and *L. acidophilus* MG501, significantly attenuated the expression of these inflammatory mediators in accordance with the blocking action of nuclear factor-kappa B (NF-κβ) nuclear translocation. *L. salivarius* WB 1004 has been shown, *in vitro,* to reduce IL-8 secretion by gastric epithelial cells (Borruel et al. 2003).

Recent studies have defined potentially new probiotic strains of *L. reuteri,* a small minority of which showed strong anti-inflammatory combined with antipathogenic effects. *L. reuteri* ATCC PTA 6475 produces and exports substances that can interfere with TNF-α production in human macrophages (Lin et al. 2008), and suppress NF-κB activation, affecting apoptosis (Iyer et al. 2008), while still retaining its basic antipathogen activity during both planktonic and biofilm growth (Jones and Versalovic 2009). Initial human studies on this strain in our clinic show good safety and tolerability (personal data). Clinical studies on a combination of the anti-inflammatory effects of this strain with the earlier known anti-*H. pylori* effect of *L. reuteri* DSM 17938 is under investigation in our Unit and preliminary data confirm the anti *H. pylori* effect of the strains. Finally, in animal studies certain lactic acid bacteria (*L. casei, L. acidophilus, L. rhamnosus, L. delbrueckii* ssp. bulgaricus, *S. thermophilus, L. plantarum* and *Lactococcus lactis*) were able to increase the number of immunoglobulin A (IgA)-producing cells associated with the lamina propria of the small intestine (Vitini et al. 2000).

The fact remains that the specific interaction of probiotics with the immune system and the exact mechanism by which they can exert a beneficial effect are still unclear; moreover, the immunoadjuvant capacity observed would be a property of the strain assayed and cannot be generalised to genus or species.

Clinical Studies

Probiotics and Helicobacter pylori-induced Gastritis in Man

On the basis of the above-mentioned results, three studies directly assessed the effect of the administration of probiotics on *H. pylori* gastritis by the histological examination of gastric biopsies (see Pantoflickova et al. 2003, Felley et al. 2001, Wang et al. 2004).

Certain lactobacilli are resistant to the low pH of the stomach and may adhere to and transiently reside in the human stomach. It has been postulated, on the basis of the results of *in vitro* and animal studies, that probiotics could possibly compete with and down-regulate *H. pylori* infection in humans. *Lactobacillus acidophilus* and the related species *L. johnsonii* are among these strains of human origin which can tolerate very low acidic pH and can survive passage through the gastrointestinal tract: as a result, they might transiently reside in or alter the niche occupied by *H. pylori* and hence modify *H. pylori* infection and its possible outcome. Furthermore, a drinkable, whey-based, *L. johnsonii* strain La1 culture supernatant has been shown to have a partial, acid-independent long-term suppressive effect on *H. pylori* in humans. Felley et al. were the firsts to investigate the effect of *L. johnsonii* La1-acidified milk (LC-1) on *H. pylori* infection. Fifty-three volunteers infected with *H. pylori* were randomized to receive either LC-1 or a placebo (180 ml twice a day) for 3 weeks. Oesophagogastroduodenoscopy and biopsies were performed at inclusion and repeated 4–8 weeks after the end of the treatment. The authors showed that LC-1 ingestion induced: a) decrease in *H. pylori* density in the antrum (P= 0.02) and the corpus (P= 0.04);

Table1. Mechanisms of inhibition of *H. pylori* by probiotics *in vitro*.

Author (ref.)	Probiotic	Mechanism of inhibition
Aiba et al. 1998	*L. acidophilus* 4356	Lactic acid
	L. casei 393	Lactic acid
	L. salivarius WB1040	Lactic acid
Cats et al. 2003	*L. casei* strain Shirota	Heat-labile substance
Coconnier et al. 1998	*L. acidophilus* LB	Heat-stable protein
Kim et al. 2008	*L. lactis* BH5	Bacteriocin
Lorca et al. 2001	*L. acidophillus*	CRL639 autolysins
Nam et al. 2002	*W. confusa* PL9001	Class II bacteriocin
Michetti et al. 1999	*L. johnsonii* La1	Heat-stable substance
Midolo et al. 1995	*L. acidophilus*	Lactic acid
	L. casei subsp. Rhamnosus	Lactic acid
Mukai et al. 2002	*L. reuteri* TM 105	Glycolipid-binding proteins
Pinchuk et al. 2001	*B.* subtilis 3	Anticoumacin A, B, C
Sgouras et al. (2004)	*L. casei* strain Shirota	Lactic acid

b) reduced inflammation and gastritis activity in the antrum (P= 0.02 and P= 0.01, respectively) and of activity in the corpus (P= 0.02) providing the first evidence that *H. pylori* infection and gastritis can be down-regulated by LC-1 (Felley et al. 2001).

Pantoflikova et al. confirmed the *in vitro* and *in vivo* inhibitory effect of *Lactobacillus johnsonii* (Lj1), contained in fermented milk (LC1), on *H. pylori* gastritis without the co-administration of antibiotics. Fifty *H. pylori* positive healthy volunteers were randomized in a double-blind study to LC1 or placebo. Gastric biopsies from the antrum and corpus were obtained before and after 3 and 16 weeks of treatment, for histology and quantitative cultures, and the analysis of the data showed that the severity and activity of antral gastritis was reduced (P = 0.04), *H. pylori* density decreased in the antrum (P = 0.04), and mucus thickness increased after 16 weeks of LC1 consumption as compared to placebo (P = 0.03). The authors concluded that LC1 intake had a favorable, albeit weak effect on *H. pylori* associated gastritis, particularly in the antrum, and propose that a regular ingestion of fermented milk containing L. johnsonii may reduce the risk of developing disorders associated with high degrees of gastric inflammation and mucus depletion (Pantoflickova et al. 2003).

Wang et al., based on the evidence that ingesting lactic acid bacteria exerts a suppressive effect on *Helicobacter pylori* infection, tested if the administration of AB-yogurt (107 colony-forming units of *Lactobacillus acidophilus* La5 or Bifidobacterium lactis Bb12 /mL) to subjects with asymptomatic *H. pylori* could inhibit *H. pylori* growth. In an intervention study, 59 adult volunteers infected with *H. pylori* were given AB-yogurt twice daily after a meal for 6 weeks. Eleven subjects positive for *H. pylori* infection were treated with milk placebo as control subjects. *H. pylori* bacterial loads were determined with use of the 13C-urea breath test, which was performed before and 4 and 8 weeks after the start of AB-yogurt supplementation. Administration of AB-yogurt decreased the urease activity of *H. pylori* after 6 weeks of therapy ($P < 0.0001$) and the examination of antral biopsies showed reduced *H. pylori* density and gastritis activity (P = 0.006 and P = 0.015, respectively) from 14 subjects while no significant change was observed in the gastric body. The authors concluded that a regular intake of yogurt containing Bb12 and La5 effectively suppressed *H. pylori* infection in humans (Wang et al. 2004).

Probiotics and Helicobacter pylori Interactions—13C–Urea Breath-test Values

In most studies, the effect of probiotic treatment on the level of *H. pylori* infection has been estimated indirectly by the 13C–urea breath test (13C-UBT) delta over baseline (Table 2) (Michetti et al. 1999, Gotteland and

Table 2. Summary of clinical trials of probiotics in *H. pylori* infection: effects on breath test values.

Reference and type of study	**Type of patient**	**Number**	**Probiotic regimen**	**Results**
Michetti et al. 1999 (25) DB, R, P	Asymptomatic	20 adults	*L. johnsonii* La1 supernatant 4 times a day for 2 wks	Breath test values reduced *
Sakamoto et al. 2001 (71) O	Asymptomatic	31 adults	*L. gasseri* OLL2716 (yogurt) once a day for 8 wks	Breath test values reduced
Wendakoon CN et al. (68) DB, R, P	Asymptomatic	27 adults	*Three Lactobacillus* spp. *(L. acidophilus* and *L. casei) and one commercial starter culture (L. acidophilus, L. bulgaricus* and *Streptococcus thermophilus)*	Breath test values not reduced
Gotteland et al. 2003 (30) O	Asymptomatic	11 adults	*L. johnsonii* La1 (yogurt) 8 times a day for 2 wks	Breath test values reduced
Cruchet et al. 2003 (70) DB, R, P	Asymptomatic	252 children	*L. johnsonii* La1 (yogurt) once a day for 4 wks	Breath test values reduced*
Cats et al. 2003 (33) O, C	Asymptomatic	20 adults	*L. casei Shirota* (fermented milk) 3 times a day for 3 wks	Breath test values not reduced
Linsalata et al. 2004 (72) DB, R, P	Dyspeptic	22 adults	*L. brevis* CD2 (tablets) 9 times a day for 3 wks	Breath test values reduced* Ornithine decarboxylase activity reduced*
Wang et al. 2004 (67) DB, R, P	Asymptomatic	70 adults	*B. lactis* + *L. acidophilus* La5 (yogurt) twice a day for 6 wks	Breath test values reduced*
Gotteland et al. 2005 (75) O, R, C	Asymptomatic	182 children	*S. boulardii* plus inulin for 8 wks	Breath test values reduced *

Table 2. contd....

Table 2. contd.

Reference and type of study	Type of patient	Number	Probiotic regimen	Results
Miki et al. 2007 (73) DB, R, P	Asymptomatic	79 adults	*B. bifidum* (fermented milk) once a day for 12 wks	Breath test values reduced* Serum pepsinogen reduced*
Myllyluoma et al. 2007 (62) O	Dyspeptic	12 adults	*L.* GG, *L. rhamnosus LC705, Propionibacterium freudenreichii* JS, *B. lactis* Bb12 (fermented milk) Once a day for 8 wks	Breath test values reduced Serum gastrin 17 reduced
Francavilla et al. 2008 (68) DB, R, P	Dyspeptic	40 adults	*L. reuteri* (tablets) once a day for 4 wks	Breath test values reduced* HPsA values reduced*

DB: double-blind; R: randomised; P: placebo controlled; O: open.
*statistically significant ($p<0.05$) vs. controls

Cruchet 2003, Cats et al. 2003, Mack et al. 2003, Myllyluoma et al. 2007, Wang et al. 2004, Francavilla et al. 2008, Cruchet et al. 2003, Sakamoto et al. 2001, Linsalata et al. 2004) Our group has recently studied the interaction between *L. reuteri* ATCC 55730 (SD2112) and *H. pylori* by both the 13C-UBT and the *H. pylori* stool antigen (HpSA) test before and after probiotic administration (Francavilla et al. 2008); it is well-known that both tests are a semiquantitative measurement of the bacterial load (Chang et al. 2002). In a double-blind placebo-controlled study, 40 *H. pylori*-positive subjects received *L. reuteri* ATCC 55730 twice daily for four weeks or placebo. At entry, all underwent upper endoscopy, 13C-UBT and HpSA, while after four weeks of treatment (probiotic or placebo) only 13C-UBT and HpSA were repeated. Afterwards, a standard sequential treatment was administered in all patients. We have demonstrated that *L. reuteri*, but not placebo, was able to reduce the intragastric bacterial load as demonstrated by the simultaneous significant decrease of both 13C-UBT (from 33.8±15 to 27.3±12.1‰; $p<0.05$) (Fig. 1) and HpSA (from 18.1±6.4 to 14.4±5.2 net optical density value/control value [net/co]; $p<0.05$) (Fig. 2).

Michetti et al. evaluated the effect of a drink made of whey-based *L. johnsonii* La1 supernatant in 20 *H. pylori*-infected volunteers (Michetti et al. 1999). They were treated for 14 days with 50ml of La1 supernatant combined with either omeprazole or placebo. Four weeks after the end of treatment, the 13C-UBT values were still significantly below the pre-treatment values regardless of treatment group. In another four studies performed in this field with subjects treated with L. johnsonii La1 yoghurt (Cruchet et al. 2003), yoghurts containing *L. acidophilus* La5 and *B. lactis* Bb12 (Wang et al. 2004), *L. gasseri* OLL 2716 (Sakamato et al. 2001), a milk containing B. bifidum BF-158 or a drink consisting of equal doses of L. GG, *L. rhamnosus* LC705,

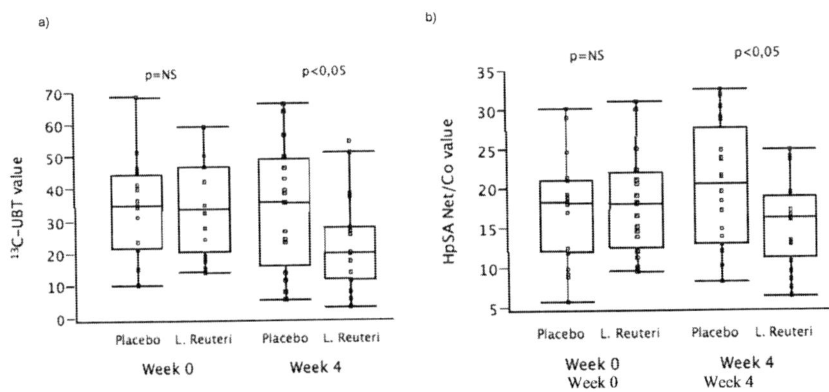

Fig. 1. *H. pylori* bacterial load assessed by 13C-UBT (a) and HpSA (b) before and after placebo or *L. retuteri* ATCC 55730 supplementation.

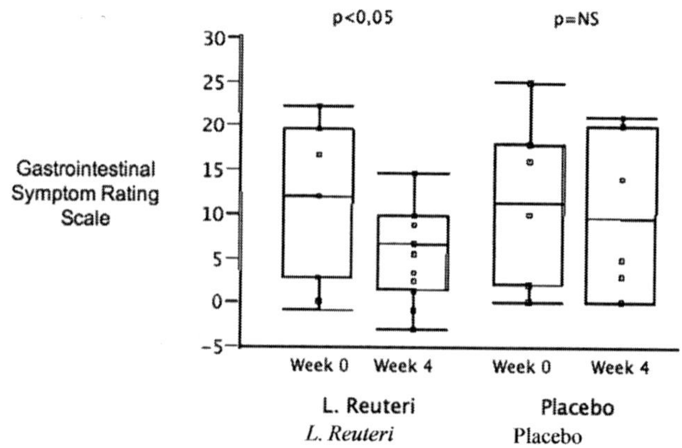

Fig. 2. Gastrointestinal Symptom Rating Scale (GSRS) assessed before and after placebo or *L. reuteri* supplementation.

P. freudenreichii JS and *B. lactis* Bb12 (Myllyluoma et al. 2007), a decrease in 13C-UBT values reflected a decrease in the *H. pylori* bacterial load. In addition, markers of gastric inflammation such as ornithine decarboxylase activity (Linsalata et al. 2004), serum pepsinogen levels (Sakamoto et al. 2001, Miki et al. 2007) or serum gastrin 17 (Myllyluoma et al. 2007) also decreased in the active treatment group compared with the control group. Two studies reported no effect of probiotic treatment on *H. pylori* infection (Cats et al. 2003, Wendakoon et al. 2002) (Table 2).

Finally, two studies have been performed in children to explore this issue. In the first, Cruchet et al. investigated whether regular ingestion of a dietary product containing *Lactobacillus johnsonii* La1 or *L. paracasei* ST11 would interfere with *H. pylori* colonization in children. In a double blind, randomized, controlled clinical trial carried out in school children from a low socioeconomic area of Santiago, 252 children colonized by *H. pylori* were randomized into five groups to receive a product containing live *Lactobacillus johnsonii* Lal and L. helveticus or Living L. paracasei STI1 and L. helveticus (groups 1 and 3), heat-killed La1 or ST11 (Heat-killed *L. johnsonii* Lal and *L. helveticus* and Heat-killed *L. paracasei* STI1 and *L. helveticus*), or vehicle (group 5) everyday for 4 weeks. A (Vaira et al. 2007) C-urea breath test was carried out at entry and at the end of this period. The authors report a 12% decrease in 13C-UBT values in children receiving live Living *Lactobacillus johnsonii* Lal and L. helveticus whereas no differences were observed in the other groups concluding that regular ingestion of a product containing Lactobacillus La1 may represent an interesting alternative to modulate *H. pylori* colonization in children infected by this pathogen (Cruchet et al. 2003).

In the second study, the authors evaluated the effect of a probiotic, *Lactobacillus acidophilus* LB (LB), or a synbiotic, *Saccharomyces boulardii* plus inulin (SbI), on *Helicobacter pylori* (Hp) colonization in children. 141 Hp-positive children were randomly distributed into three groups to receive either antibiotic treatment (lanzoprazole, clarythromycin and amoxicillin) for 8 days, or SbI or LB daily for 8 weeks. A second C-UBT was carried out at this time. Spontaneous clearance was evaluated in the same way in 81 infected, untreated children. The differences in the UBT delta over baseline values before and after treatments were evaluated. Hp was eradicated in 66%, 12% and 6.5% of the children from the Ab, SbI and LB groups, respectively, while no spontaneous clearance was observed in the children without treatment. A moderate but significant difference in UBT delta over baseline was detected in children receiving living SbI, but not in those receiving LB. Therefore the authors conclude that *S. boulardii* seems promising as an agent that interferes with Hp in colonized individuals (Gotteland et al. 2005).

Probiotics and Helicobacter pylori Eradication Rate

The administration of probiotics alone does not lead to the eradication of *H. pylori*. In children, three studies recorded a slight efficacy of some probiotic strains (i.e., *S. boulardii* plus inulin, *L. acidophilus* LB, La1 alone or in association with cranberry juice and *L. gasseri* OLL2716) on the eradication of the infection (Gotteland et al. 2005, Gotteland et al. 2008, Boonyaritichaikij et al. 2009), although there was only a temporary inhibition of *H. pylori* that disappeared in most of the cases once the administration of the inhibiting factors was interrupted (Gotteland et al. 2008, Boonyaritichaikij et al. 2009).

It has been suggested that the use of probiotics as an adjuvant to PPI–antibiotic treatment could improve the success of *H. pylori* eradication. It has been hypothesized that, in addition to the mechanisms mentioned above, lactic acid or other potentially antimicrobial substances secreted by probiotic bacteria can increase the potential of antibiotic therapy to have an antimicrobial effect. In addition, better compliance, as a result of reduced side effects, may play a role.

Several clinical trials have been carried out in both adults and children, providing conflicting results (Table 3) (Canducci et al. 2000, Armuzzi et al. 2001, Armuzzi et al. 2001, Cremonini et al. 2002, Sheu et al. 2002, Guo et al. 2004, Nista et al. 2004, Tursi et al. 2004, Myllyluoma et al. 2005, Cao et al. 2005, Sykora et al. 2005, Goldman et al. 2006, Lionetti et al. 2006, Cindoruk et al. 2007, Park et al. 2007, Kim et al. 2008, Hurduc et al. 2009, Szajewska et al. 2009). In 2007, Tong et al. published a meta-analysis on the effect of supplementation with probiotics on eradication rates and adverse events

Table 3. Summary of clinical trails of probiotics in *H. pylori* infection: effects on eradication rates.

Reference and type of study	Eradication therapy	Probiotic regimen	Eradication ratein probiotics group	Eradication ratein control group	Odds Ratio
Canducci et al. 2000 (78) O, R, P	Rabeprazole + clarithromycin + amoxicillin for 1 wk	*L. acidophilus* for 10 days	52/60	42/60	2.8
Armuzzi et al. 2001 (79) DB, R, P	Rabeprazole + clarithromycin + tinidazole for 1 wk	*L. GG* for 2 wk	25/30	24/30	1.2
Armuzzi et al. 2001 (80) DB, R, P	Pantoprazole + clarithromycin + tinidazole for 1 wk	*L. GG* for 2 wk	48/60	46/60	1.2
Cremonini et al. 2002 (81) DB, P, R	Rabeprazole + clarithromycin + tinidazole for 1 wk	*L. GG, S. boulardii, L. acidophilus +B. lactis* for 3 wk	51/64	16/21	1.2
Sheu et al. 2002 (82) R, P	Lansoprazole + chlarithromycin + amoxicillin for 1 wk	*Lactobacillus* + *Bifidobacterium* for 5 wk	73/80	63/80	2.8
Guo et al. 2004 (83) R,P	Omeprazole + amoxicillin + furanzolidone for 1 wk	*Clostridium butyricum* for 1 wk	44/47	44/50	2
Nista et al. 2004 (84) DB, R, P	Rabeprazole + clarithromycin + amoxicillin for 1 wk	*Bacillus clausi*i for 2 wk	39/54	37/52	1
Tursi et al. 2004 (85) O, C	Ranitidine bismuth citrate + esomeprazole + amoxixillin + tinidazole for 10 days	*L. casei* subsp. *casei* for 10 days	33/35	30/35	2.7
Myllyluoma et al. 2005 (86) DB, R, P	Lansoprazole + clarithromycin + amoxicillin for 1 wk	L. GG + *L. rhamnosus* LC + *Propionibacterium freudenreichii* JS, *B. lactis* Bb12 for 4 wk	21/23	19/24	2.6
Cao et al. 2005 (87) DB, R, P	Omeprazole + amoxicillin + metronidazole for 1 wk	*B. longum* + *L . acidophilus* + *Faecal streptococci*	62/64	59/64	2.7
Sikora et al. 2005 (88) DB, R, P	Omeprazole + amoxicillin + clarithromycin for 1 wk	*L. casei* for 1 wk	33/39	27/47	4

Goldman et al. 2006 (89) DB, R, P	Omeprazole + amoxicillin + clarithromycin for 1 wk	*B. animalis* + *L. casei* for 3 months	15/33	12/32	1.4
Lionetti et al. 2006 (90) DB, R, P	Omeprazole, amoxicillin, clarithromycin, tinidazole (sequential therapy) for 10 days	*L. reuteri* for 20 days	17/20	16/20	1.4
Cindoruk et al. 2007 (91) DB, R, P	Lansoprazole + amoxicillin + clarithromycin for 2 wk	*S. boulardii* for 2 wk	44/62	37/62	1.6
Park et al. 2007 (92) DB, R, C	Omeprazole + amoxicillin + clarithromycin for 1 wk	*Bacillus subtilis* + *Streptococcus faecium* for 9 wk	147/176	129/176	1.8
Kim et al. 2008 (93) DB, R, C	Omeprazole + amoxicillin + clarithromycin for 1 wk	*L. acidophilus* + *L. casei* +*B. longum* + *Streptococcus thermophilus* for 3 wk	133/168	129/179	1.5
Hurduc et al. 2009 (94) O, R, P	Omeprazole, amoxicillin, clarithromycin for 1 wk	*S. boulardii* for 4 wks	45/48	34/42	3.5
Szajewska et al. 2009 (95) DB, R, P	Omeprazole, amoxicillin, clarithromycin for 1 wk	*L. GG* for 1 wk	23/34	22/32	1
Yaşar et al. 2010 (98) DB, R, P	Pantoprazole (40 mg, b.i.d.), amoxicillin (1000 mg b.i.d.), clarithromycin (500 mg b.i.d.)	*25 ml of probiotic-containing yogurt (Bifidobacterium DN-173 010-1010 cfu/g) for 14 days*	25/38	28/45	1,1
Ojetti et al. 2012 (97) SB, R, P	Second-line therapy: esomeprazole, levofloxacin, and amoxicillin for 7 days	*L. reuteri* for 14 days	36/45	20/45	3,0
Manfredi et al. 2012 (99) DB, R, P	Omeprazole, amoxicillin, clarithromycin, tinidazole (sequential therapy) for 10 days	*(Lactobacillus acidophilus; Bifidobacterium bifidum; Streptococcus thermophilus; Lactobacillus bulgaricus)* or *(Lactobacillus acidophilus; Bifidobacterium bifidum; Streptococcus thermophilus; Lactobacillus bulgaricus) for 10 days*	65/73	67/76	1,09

DB: double-blind; SB: single-blind; R: randomised; P: placebo controlled; PR: probiotic group, O: open, C: controlled; PR: probiotic.

during *Helicobacter pylori* eradication therapy. Eleven studies describing *H. pylori* eradication rates were selected for the meta-analysis. Of these selected studies, three reported significantly improved eradication rates, the remaining eight had similar efficacy for *H. pylori* eradication. Pooled eradication rates were achieved in 463 of 554 patients with probiotics supplementation (83.6%; 95% CI = 80.5–86.7%) and in 389 of 520 patients without probiotics (74.8%; 95% CI 71.1–78.5%) by intention-to-treat analysis, the OR was 1.84 (95% CI = 1.34–2.54). Overall, per-protocol eradication rates were 85.4% (95% CI = 82.5–88.4%) and 77.6% (95% CI = 75.8–79.5%) for probiotics supplementation and without probiotics, respectively (OR 1.82; 95% CI = 1.30–2.56). For *H. pylori* eradication failures, the effect of probiotics supplementation on eradication rates was also evaluated and two randomised clinical trials were identified related to probiotics supplementation during *H. pylori* eradication for patients with eradication failure. Pooled eradication rates were achieved in 92 of 104 patients with probiotics supplementation (88.5%; 95% CI = 82.3–94.6%) and in 79 of 104 patients without probiotics (76.0%; 95% CI = 84.2–67.7%): the OR was 2.47 (95% CI = 1.16–5.29). The data led to the conclusion that eradication rates of combining probiotics with standard triple therapy are slightly higher in both the intention-to-treat and per-protocol analysis in naive and in patients in which eradication fail. The sub-analysis of probiotic preparations showed that not all probiotic are the same. Of four trials administrating Lactobacillus, two had reported improved *H. pylori* eradication rate, whole bacillus clausii group and Clostridium butyrium group demonstrated similar eradication rates (Tong et al. 2007).

Since 2007, eight additional studies have been published. Out of these, five reported an increase of eradication rates secondary to the concomitant use of a probiotic (Cindoruk et al. 2007, Park et al. 2007, Kim et al. 2008, Hurduc et al. 2009, Szajewska et al. 2009, Ojetti et al. 2012) with an odd ratio ranging from 1,5 to 3,5 while three studies showed no additional effect (Szajewska et al. 2009, Yaşar et al. 2010, Manfredi et al. 2012).

The major limit to establish whether a probiotic is able to significantly increase the eradication rate is represented by the power of the study. Indeed, due to the high eradication rates that we achieve with current antibiotic treatments, to detect a 10% increase in eradication rate (attributable to probiotic strain) 150 patients need to be enrolled in each arm to have a sufficient power of the esteem.

Probiotics and Helicobacter pylori-related Dyspeptic Symptoms

In our own experience with 40 adults, we were able to demonstrate a favourable effect of *L. reuteri* ATCC 55730 (SD2112) on dyspeptic symptoms induced by *H. pylori*. In this study, *L. reuteri* administration was followed by a significant decrease in the Gastrointestinal Symptom Rating Scale (GSRS) compared with pre-treatment value (7.9±4.1 versus 11.8±8.5; $p<0.05$) that was not present in patients receiving placebo (9.7±8.7 versus 11.4±9.7; p<NS) (Fig. 2) (Francavilla et al. 2008). The effect is strain-specific since in a randomised study the administration of *L. brevis* did not show any effect on symptoms (Linsalata et al. 2004). However, we cannot be sure that the effect we observed reflects the improvement of *H. pylori* status since the symptoms of dyspepsia have a multifactorial origin and an overlap of manifestations of different conditions, such as irritable bowel syndrome (IBS), may strongly be involved. IBS is a symptomatic motility and sensory disorder of the lower gastrointestinal tract and it is a widespread condition in the adult population with a prevalence in Europe and North America of about 15–30%. The main symptoms of IBS such as abdominal discomfort, bloating, and altered bowel activity are those that were significantly decreased in the *L. reuteri* group, and it is known that such symptoms can be controlled by the administration of probiotics, and therefore our finding may reflect a control of IBS rather than of *H. pylori*-related dyspepsia. More studies on larger samples are needed before definite conclusions can be drawn on this issue.

Probiotics and Antibiotic-associated Gastrointestinal Side Effects During Helicobacter pylori Eradication Therapy

H. pylori eradication fails in about 25–30% of cases, mainly because of the occurrence of resistance to antibiotics and/or antibiotic associated side effects (Deltenre et al. 1998). Several studies evaluated whether probiotic supplementation may help to prevent or reduce drug-related side effects during standard *H. pylori* eradication therapy in adults (Table 4) (Armuzzi et al. 2001, Cremonini et al. 2002, Sheu et al. 2002, Nista et al. 2004, De Bortoli et al. 2007, Park et al. 2007, Cindoruk et al. 2007, Imase et al. 2008, Plummer et al. 2005, Lionetti et al. 2006, Tursi et al. 2004). The first study used L. GG, (100) the second used different probiotic preparations (L. GG or *S. boulardii* or a combination of *L. acidophilus* and *B. lactis*) (Cremonini et al. 2002), the third used a Lactobacillus- and Bifidobacterium-containing yoghurt (Sheu et al. 2002) and finally Nista-administered B. clausii (Nista et al. 2004). All found that probiotics were superior to placebo for the prevention of side effects such as diarrhoea, nausea and unpleasant taste. Furthermore, in a double-blind, randomised, placebo-controlled study performed in 338 volunteers, Myllyluoma found that probiotic supplementation significantly alleviates *H. pylori*-treatment-associated symptoms (Myllyluoma et al. 2005) in fact, the probiotic therapy, consisting of four different strains (L. GG, *L. rhamnosus* LC, *P. freudenreichii* spp. shermanii Js, *B. breve* Bb99), reduced the total symptom score, which took into account both the frequency and the severity

Table 4. Summary of clinical trails of probiotics in *H. pylori* infection: effects on antibiotic-associated gastrointestinal side-effects.

Reference and type of study	**Type of patient**	**Number**	**Eradication therapy**	**Probiotic regimen**	**Results**
Armuzzi et al. 2001 (71) DB, R, P	Asymptomatic	60 adults	Rabeprazole + clarithromycin + tinidazole for 1 wk	*L. GG* for 2 wk	Diarrhea, nausea, unpleasant taste, significantly less in PR
Cremonini et al. 2002 (81) DB, P, R	Asymptomatic	85 adults	Rabeprazole + clarithromycin + tinidazole for 1 wk	*L. GG*, *S. boulardii*, *L. acidophilus* + *B. lactis* for 3 wk	Diarrhea, unpleasant taste significantly less in all PR groups
Sheu et al. 2002 (82) R, P	Dyspeptic	160 adults	Lansoprazole + clarithromycin + amoxicillin for 1 wk	*Lactobacillus* + *Bifidobacterium* for 5 wk	Vomiting, constipation, diarrhea, unpleasant taste significantly less in PR
Nista et al. 2004 (84) DB, R, P	Asymptomatic	120 adults	Rabeprazole + clarithromycin + amoxicillin for 1 wk	*Bacillus clausii* for 2 wk	Diarrhea, nausea, epigastric pain, significantly less in PR
Tursi et al. 2004 (85) O, C	Dyspeptic	70 adults	Ranitidine bismuth citrate + esomeprazole + amoxixillin + tinidazole for 10 days	*L. casei* subsp. *casei* for 10 days	Diarrhea, epigastric pain less significantly in PR
Myllyluoma et al. 2005 (86) DB, R, P	Asymptomatic	47 adults	Lansoprazole + clarithromycin + amoxicillin for 1 wk	*L.* GG + *L. rhamnosus* LC + *Propionibacterium freudenreichii* JS, *B. lactis* Bb12 for 4 wk	Total symptom score significantly decreased in PR
Duman DG et al. 2005 (115) DB, R, C	Dyspeptic	376 adults	Omeprazole + amoxicillin + clarithromycin for 2 wk	*S. boulardii* for 2 wk	Diarrhea significantly less in PR

Lionetti et al. 2006 (90) DB, R, P	Dyspepsia	40 children	Omeprazole, amoxicillin, clarithromycin, tinidazole (sequential therapy) for 10 days	*L. reuteri* for 20 days	Epigastric pain, abdominal distension, belching, halitosis significantly less in PR
Cindoruk et al. 2007 (91) DB, R, P	Dyspeptic	124 adults	Lansoprazole + amoxicillin + clarithromycin for 2 wk	*S. boulardii* for 2 wk	Diarrhea, epigastric pain significantly less in PR
Park et al. 2007 (92) DB, R, C	Dyspeptic	352 adults	Omeprazole + amoxicillin + clarithromycin for 1 wk	*Bacillus subtilis* + *Streptococcus faecium* for 9 wk	Diarrhea significantly less in PR
De Bortoli et al. 2007 (104) R	Dyspeptic	206 adults	Esomeprazole + clarithromycin + amoxicillin for 1 wk	*L. plantarum*, *L. reuteri*, *L. casei* subsp. *rhamnosus*, *B. infantis*, *B. longum*, *L. salivarius*, *L. acidophilus*, *S. thermophilus*, *L. sporogenes* for 1 wk	Diarrhea, nausea, metallic taste, abdomianl pain, glossitis significantly less in PR
Imase et al. 2008 (107) R	Peptic ulcer	19 adults	Omeprazole + amoxicillin + clarithromycin for 1 wk	*Clostridium butyricum for 1 wk*	Diarrhea significantly less in PR
Kim et al. 2008 (93) DB, R, C	Dyspeptic	347 adults	Omeprazole + amoxicillin + clarithromycin for 1 wk	*L. acidophilus* + *L. casei* + *B. longum* + *Streptococcus thermophilus* for 3 wk	Unpleasant taste significantly less in controls
Hurduc et al. 2009 (94) O, R, P	Dyspeptic	90 children	Omeprazole + amoxicillin + clarithromycin for 1 wk	*S. boulardii* for 4 wks	Overall Incidence of side effects significantly reduced in PR

Table 4. contd....

Table 4. contd.

Reference and type of study	Type of patient	Number	Eradication therapy	Probiotic regimen	Results
Szajewska et al. 2009 (95) DB, R, P	Dyspeptic	66 children	Omeprazole + amoxicillin + clarithromycin for 1 wk	*L. GG* for 1 wk	No significant difference between PR e placebo
Ojetti et al. 2012 (97) DB, R, P	Second-line therapy	90 adults	Esomeprazole, levofloxacin, and amoxicillin for 7 days	*L. reuteri for 14 days*	Increases the eradication rate
Yaşar et al. 2012 (98) DB, R, P		76 adults	Pantoprazole (40 mg, b.i.d.), amoxicillin (1000 mg b.i.d.), clarithromycin (500 mg b.i.d.)	*25 ml of probiotic-containing yogurt (Bifidobacterium DN-173 010-1010 cfu/g) for 14 days*	No significant difference between PR e placebo
Manfredi M. et al., 2012 (") DB, R, P	Asymptomatic	227 adults	Omeprazole, amoxicillin, clarithromycin, tinidazole (sequential therapy) for 10 days	*(Lactobacillus acidophilus; Bifidobacterium bifidum; Streptococcus thermophilus; Lactobacillus bulgaricus) or (Lactobacillus acidophilus; Bifidobacterium bifidum; Streptococcus thermophilus; Lactobacillus bulgaricus) for 10 days*	No significant difference between PR e placebo

DB: double-blind; R: randomised; P: placebo controlled; PR: probiotic group, O: open, C: controlled; PR: probiotic.

of gastrointestinal symptoms. In a recent study, a standard triple therapy was supplemented with a wide range of probiotics (*L. plantarum*, *L. reuteri*, *L. casei* ssp. rhamnosus, *B. infantis* and *B. longum*, *L. salivarius*, *L. acidophilus*, *S. thermophilus* and *L. sporogenes*) and with bovine lactoferrin in 206 *H. pylori*-positive patients (De Bortoli et al. 2007). The authors obtained a significant difference in side-effect occurrence, such as nausea, diarrhoea, metallic taste, abdominal pain and glossitis, in patients receiving probiotics compared with the placebo group and, interestingly, they observed an increased eradication rate, probably due to the concomitant use of lactoferrin. Recently, similar results were also reported for *L. casei*, *B. subtilis* and *S. faecium* (Park et al. 2007) *S. boulardii* (Cindoruk et al. 2007) and Clostridium butyricum (Imase et al. 2008). The rationale for coupling a probiotic to any antibiotic treatment stems from the results of a recent study showing that daily supplementation with viable probiotic bacteria during and post-antibiotic therapy reduces the extent of disruption to the intestinal microbiota as well as the incidence and total numbers of antibiotic-resistant strains in the re-growth population, suggesting that a probiotic should always be associated with an antibiotic (Plummer et al. 2005).

Our group has recently performed the first trial in children to determine whether adding the probiotic *L. reuteri* to an anti-*H. pylori* regimen could help to prevent or minimise the gastrointestinal side effect burden in children (Lionetti et al. 2006). Forty *H. pylori*-positive children were consecutively treated with 10-day sequential therapy, and at the same time they were blindly randomised to receive either *L. reuteri* ATCC 55730 (SD2112) or placebo (maltodextrin) for 20 days starting from the first day of the anti-*H. pylori* regimen. In order to determine the type and severity of side effects, all children completed the GSRS at entry and on day five and 10 of treatment and at follow-up after 20 days. *H. pylori* status was assessed after eight weeks by 13C-UBT.

Overall, in all probiotic-supplemented children, compared with those receiving placebo, there was a significant reduction in the GSRS score during eradication therapy (4.1±2, 95% CI 2.9–5.9 versus 6.2±3, 95% CI5.2–8.3; $p<0.01$), which became markedly evident at the end of follow-up (3.2±2, 95% CI 2.4–4 versus 5.8±3.4, 95% CI 4.8–6.9; $p<0.009$). In summary, children receiving *L. reuteri* complained of epigastric pain less frequently during eradicating treatment (15 versus 45%; $p<0.04$) as well as abdominal distension (0 versus 25%; $p<0.02$), belching (5 versus 35%; $p<0.04$), disorders of defecation (15 versus 45%; $p<0.04$) and halitosis (5 versus 35%; $p<0.04$) thereafter.

Subsequently, further trials have been conducted in children showing that the occurrence of antibiotic-associated side effects were significantly reduced by the addition of *S. boulardii* (8.3 versus 30.9%; $p=0.047$) (Cremonini et al. 2002) while the supplementation of standard triple therapy with *L. rhamnosus* GG did not significantly alter the incidence of antibiotic associated side effects (52.9 versus 40.6%; p=NS) (Szajewaka et al. 2009). Thus, results showed a positive probiotic impact on overall *H. pylori* therapy tolerability, although in the majority of the studies performed with probiotic administration, it did not significantly affect the drop-out rate due to side effects. It is clear that not all probiotics are equal, that the beneficial effects are strain-specific, and that each strain must be evaluated individually.

In the aforementioned meta-analysis by Tong et al. on the effect of supplementation with probiotics on eradication rates and adverse events during *Helicobacter pylori* eradication therapy, it appears clear that there are positive effects of probiotics for this particular indication (Tong et al. 2007).

Conclusions and Perspectives

The use of probiotics in *H. pylori*-colonized subjects with gastric inflammation is supported by many observations. Specific strains of probitics exert *in vitro* bactericidal effects against *H. pylori* through the release of bacteriocins or production of organic acids, and/or inhibit its adhesion to epithelial cells. On the other hand, the antioxidant and anti-inflammatory properties exerted by probiotics may stabilize the gastric barrier function and decrease mucosal inflammation. Some clinical trials have evaluated the effect of probiotics in colonized adults and children. Their results indicate that probiotics generally do not eradicate *H. pylori* but decrease the density of colonization, thereby maintaining lower levels of this bacterium in the stomach; in association with antibiotic treatments, some probiotics increased eradication rates and/or decreased adverse effects due to the antibiotics. These findings confirm that, as suggested by the 2012-Maastricht Consensus Conference on *H. pylori*, certain probiotics show promising results as an adjuvant treatment in reducing side effects related to *H. pylori* treatment. Results so far are encouraging and further clinical trials are called for. The design of such studies should be to clarify which probiotic strains are suitable, in what form, in what dose and for how long. Moreover, the type of patient should be clearly defined and the method of defining outcome should also be standardised as far as possible and cost-effective analysis should be taken into account.

While more and more probiotic compounds become available on the market or are seeking approval, the demonstration of the efficacy of a given probiotic for a specific therapeutic target will help clinicians choose which probiotic to use when dealing with a specific disease (Preidis et al. 2009). We are entering the era of targeted probiotic use.

References

Aiba, Y., N. Suzuki, A.M. Kabir, A. Takagi and Y. Koga. 1998. Lactic acid-mediated suppression of *Helicobacter pylori* by the oral administration of *Lactobacillus salivarius* as a probiotic in a gnotobiotic murine model. Am. J. Gastroenterol. 93: 2097–101.

Akama, F., R. Nishino, S. Makino, K. Kobayashi, K. Kamikaseda, J. Nagano and Y. Koga. 2011. The effect of probiotics on gastric mucosal permeability in humans administered with aspirin. Scand J. Gastroenterol. Jul. 46(7-8): 831–6.

Armuzzi, A., F. Cremonini, F. Bartolozzi et al. 2001. The effect of oral administration of Lactobacillus GG on antibiotic-associated gastrointestinal side-effects during *Helicobacter pylori* eradication therapy. Aliment Pharmacol. Ther. 15: 163–9.

Armuzzi, A., F. Cremonini, V. Ojetti et al. 2001. Effect of Lactobacillus GG supplementation on antibiotic-associated gastrointestinal side effects during *Helicobacter pylori* eradication therapy: a pilot study. Digestion 63: 1–7.

Bhatia, S.J., N. Kochar, P. Abraham, N.G. Nair and A.P. Mehta. 1989. *Lactobacillus acidophilus* inhibits growth of Campylobacter pylori *in vitro*. J. Clin. Microbiol. 27: 2328–30.

Boonyaritichaikij, S., K. Kuwabara, J. Nagano et al. 2009. Long-term administration of probiotics to asymptomatic preschool children for either the eradication or the prevention of *Helicobacter pylori* infection. Helicobacter 14(3): 202–7.

Borruel, N., F. Casellas, M. Antolín et al. 2003. Effects of nonpathogenic bacteria on cytokine secretion by human intestinal mucosa. Am. J. Gastroenterol. 98: 865–70.

Byrd, J.C., C.K. Yunker, Q.S. Xu, L.R. Sternberg and R.S. Bresalier. 2000. Inhibition of gastric mucin synthesis by *Helicobacter pylori*. Gastroenterology 118: 1072–9.

Canducci, F., A. Armuzzi, F. Cremonini et al. 2000. A lyophilized and inactivated culture of *Lactobacillus acidophilus* increases *Helicobacter pylori* eradication rates. Aliment Pharmacol. Ther. 14: 1625–9.

Cats, A., E.J. Kuipers, M.A. Bosschaert, R.G. Pot, C.M. Vandenbroucke-Grauls and J.G. Kusters. 2003. Effect of frequent consumption of a *Lactobacillus casei*-containing milk drink in *Helicobacter pylori*-colonized subjects. Aliment Pharmacol. Ther. 17: 429–35.

Chang, M.C., Y.T. Chang, C.T. Sun et al. 2002. Quantitative correlation of *Helicobacter pylori* stool antigen (HpSA) test with 13C-urea breath test (13C-UBT) by the updated Sydney grading system of gastritis. Hepatogastroenterology 49: 576–9.

Chen, X., X.M. Liu, F. Tian, Q. Zhang, H.P. Zhang, H. Zhang and W. Chen. 2012. Antagonistic activities of lactobacilli against *Helicobacter pylori* growth and infection in human gastric epithelial cells. J. Food Sci. Jan. 77(1): M9–14.

Cindoruk, M., G. Erkan, T. Karakan et al. 2007. Efficacy and safety of *Saccharomyces boulardii* in the 14-day triple anti-*Helicobacter pylori* therapy: a prospective randomized placebo-controlled double-blind study. Helicobacter 12: 309–16.

Coconnier, M.H., V. Lievin, E. Hemery and A.L. Servin. 1998. Antagonistic activity against Helicobacter infection *in vitro* and *in vivo* by the human *Lactobacillus acidophilus* strain LB. Appl. Environ. Microbiol. 64: 4573–80.

Cremonini, F., S. Di Caro, M. Covino et al. 2002. Effect of different probiotic preparations on anti-*Helicobacter pylori* therapy related side effects: a parallel group, triple blind, placebo-controlled study. Am. J. Gastroenterol. 97: 2744–9.

Cruchet, S., M.C. Obregon, G. Salazar et al. 2003. Effect of the ingestion of a dietary product containing *Lactobacillus johnsonii* La1 on *Helicobacter pylori* colonization in children. Nutrition 19: 716–21.

De Bortoli, N., G. Leonardi, E. Ciancia et al. 2007. *Helicobacter pylori* eradication: a randomized prospective study of triple therapy versus triple therapy plus lactoferrin and probiotics, Am. J. Gastroenterol. 102: 951–6.

Deltenre, M., R. Ntounda, C. Jonas and E. De Koster. 1998. Eradication of *Helicobacter pylori*: why does it fail? Ital. J. Gastroenterol. Hepatol. 30(Suppl. 3): 326–8.

Dixon, M.F. 2001. Pathology of gastritis and peptic ulceration. *In*: H.L.T. Mobley, G.L. Mendz and S.L. Hazell (eds.). *Helicobacter pylori*: physiology and genetics. Washington, D.C.: ASM Press 459–69.

Felley, C. and P. Michetti. 2003. Probiotics and *Helicobacter pylori*. Best Pract. Res. Clin. Gastroenterol. 17: 785–91.

Felley, C.P., I. Corthésy-Theulaz, J.L. Rivero et al. 2001. Favourable effect of an acidified milk (LC-1.) on *Helicobacter pylori* gastritis in man. Eur. J. Gastroenterol. Hepatol. 13: 25–9.

Francavilla, R., E. Lionetti and L. Cavallo. 2008. Sequential treatment for *Helicobacter pylori* eradication in children. Gut. 57: 1178.

Francavilla, R., E. Lionetti, S. Castellaneta et al. 2010. Clarithromycin-resistant genotypes and eradication of *Helicobacter pylori*. J. Pediatr 157(2): 228–32.

Francavilla, R., E. Lionetti, S.P. Castellaneta et al. 2005. Improved efficacy of 10-Day sequential treatment for *Helicobacter pylori* eradication in children: a randomized trial. Gastroenterology 129: 1414–19.

Francavilla, R., E. Lionetti, S.P. Castellaneta et al. 2008. Inhibition of *Helicobacter pylori* infection in humans by *Lactobacillus reuteri* ATCC 55730 and effect on eradication therapy: a pilot study. Helicobacter 13: 127–34. 62.

Fujimura, S., S. Kato, M. Oda, M. Miyahara, Y. Ito, K. Kimura, T. Kawamura, M. Ohnuma, H. Tateno and A. Watanabe. 2006. Detection of *Lactobacillus gasseri* OLL2716 strain administered with yogurt drink in gastric mucus layer in humans. Lett. Appl. Microbiol. 43: 578–581.

Glavin, G.B. and S. Szabo. 1992. Experimental gastric mucosal injury: laboratory models reveal mechanisms of pathogenesis and new therapeutic strategies. FASEB J. 6: 825–831.

Go, M.F. 2002. Natural history and epidemiology of *Helicobacter pylori* infection. Aliment Pharmacol. Ther. 16(Suppl. 1): 3–15.

Goldman, C.G., D.A. Barrado, N. Balcarce et al. 2006. Effect of a probiotic food as an adjuvant to triple therapy for eradication of *Helicobacter pylori* infection in children. Nutrition 22: 984–8.

Gotteland, M. and S. Cruchet. 2003. Suppressive effect of frequent ingestion of *Lactobacillus johnsonii* La1 on *Helicobacter pylori* colonization in asymptomatic volunteers. J. Antimicrob. Chemother. 51: 1317–19.

Gotteland, M., L. Poliak, S. Cruchet and O. Brunser. 2005. Effect of regular ingestion of *Saccharomyces boulardii* plus inulin or *Lactobacillus acidophilus* LB in children colonized by *Helicobacter pylori*. Acta Paediatr 94: 1747.

Gotteland, M., M. Andrews, M. Toledo et al. 2008. Modulation of *Helicobacter pylori* colonization with cranberry juice and *Lactobacillus johnsonii* La1 in children. Nutrition 24: 421–6.

Gotteland, M., S. Cruchet and S. Verbeke. 2001. Effect of lactobacillus ingestion on the gastrointestinal mucosal barrier alterations induced by indometacin in humans. Aliment Pharmacol. Ther. 15: 11–7.

Guo, J.B., P.F. Yang and M.T. Wang. 2004. The application of clostridium to the eradication of *Helicobacter pylori*. Chin. J. Celiopathy 4: 163–5.

Guruge, J.L., P.G. Falk, R.G. Lorenz, M. Dans, H.P. Wirth, M.J. Blaser, D.E. Berg and J.I. Gordon. 1998. Epithelial attachment alters the outcome of *Helicobacter pylori* infection. Proc. Natl. Acad. Sci. USA 95: 3925–30.

Hamilton-Miller, J.M. 2003. The role of probiotics in the treatment and prevention of *Helicobacter pylori* infection. Int. J. Antimicrob. Agents 22: 360–6.

Hurduc, V., D. Plesca, D. Dragomir et al. 2009. A randomized, open trial evaluating the effect of *Saccharomyces boulardii* on the eradication rate of *Helicobacter pylori* infection in children. Acta Paediatr 98: 127–31.

Ierardi, E., A. Di Leo, M. Barone et al. 2003. Tumor necrosis factor alpha and apoptosis in *Helicobacter pylori* related progressive gastric damage: a possible mechanism of immune system involvement in epithelial turnover regulation. Immunopharmacol. Immunotoxicol. 25: 203–11.

Imase, K., M. Takahashi, S. Tanaka et al. 2008. Efficacy of Clostridium butyricum preparation concomitantly with *Helicobacter pylori* eradication therapy in relation to changes in the intestinal microbiota. Microbiol. Immunol. 52: 156–61.

Isolauri, E. 2001. Probiotics in human disease. Am. J. Clin. Nutr. 73: 1142–6S.

Iyer, C., A. Kosters, G. Sethi et al. 2008. Probiotic *Lactobacillus reuteri* promotes TNF-induced apoptosis in human myeloid leukemia-derived cells by modulation of NF-kappaB and MAPK signalling. Cell Microbiol. 10: 1442–52.

Johnson, C., J. Dicksved, H. Jonsson and S. Roos. 2003. Anti *Helicobacter pylori* activity among lactic acid bacteria isolated from gastric biopsies and strains of *Lactobacillus reuteri* (abstract). Helicobacter 8: 473.

Johnson-Henry, K.C., D.J. Mitchell, Y. Avitzur, E. Galindo-Mata, N.L. Jones and P.M. Sherman. 2004. Probiotics reduce bacterial colonization and gastric inflammation in *H. pylori*-infected mice. Dig. Dis. Sci.49: 1095–102.

Jones, S.E. and J. Versalovic. 2009. Probiotic *Lactobacillus reuteri* biofilms produce antimicrobial and anti-inflammatory factors. BMC Microbiol. 9: 35.

Kabir, A.M., Y. Aiba, A. Takagi, S. Kamiya, T. Miwa and Y. Koga. 1997. Prevention of *Helicobacter pylori* infection by lactobacilli in a gnotobiotic murine model. Gut. 41: 49–55.

Lam, E.K., E.K. Tai, M.W. Koo, H.P. Wong, W.K. Wu, L. Yu, W.H. So, P.C. Woo and C.H. Cho. 2007a. Enhancement of gastric mucosal integrity by Lactobacillus rhamnosus GG. Life Sci. 80: 2128–2136.

Lam, E.K., L. Yu, H.P. Wong, W.K. Wu, V.Y. Shin, E.K. Tai, W.H. So, P.C. Woo and C.H. Cho. 2007b. Probiotic Lactobacillus rhamnosus GG enhances gastric ulcer healing in rats. Eur. J. Pharmacol. 565: 171–179.

Lee, J.S., N.S. Paek, O.S. Kwon and K.B. Hahm. 2010. Anti-inflammatory actions of probiotics through activating suppressor of cytokine signaling (SOCS) expression and signaling in *Helicobacter pylori* infection: a novel mechanism. J. Gastroenterol. Hepatol. 25(1): 194–202.

Lesbros-Pantoflickova, D., I. Corthésy-Theulaz and A.L. Blum. 2007. *Helicobacter pylori* and probiotics. J. Nutr. 137(Suppl 2): 812–8.

Lin, Y.P., C.H. Thibodeaux, J.A. Penã, G.D. Ferry and J. Versalovic. 2008. Probiotic *Lactobacillus reuteri* suppress proinflammatory cytokines via c-Jun. Inflamm. Bowel. Dis. 14: 1068–83.

Linsalata, M., F. Russo, P. Berloco et al. 2004. The influence of Lactobacillus brevis on ornithine decarboxylase activity and polyamine profiles in *Helicobacter pylori*-infected gastric mucosa. Helicobacter 9: 165–172.

Lionetti, E., V.L. Miniello, S.P. Castellaneta et al. 2006. *Lactobacillus reuteri* therapy to reduce side-effects during anti-*Helicobacter pylori* treatment in children: a randomized placebo controlled trial. Aliment Pharmacol. Ther. 24: 1461–8.

Lorca, G.L., T. Wadstrom, G.F. Valdez and A. Ljungh. 2001. *Lactobacillus acidophilus* autolysins inhibit. *Helicobacter Pylori in vitro* Curr. Microbiol. 42: 39–44.

Mack, D.R., S. Ahrne, L. Hyde, S. Wei and M.A. Hollingsworth. 2003. Extracellular MUC3 mucin secretion follows adherence of *Lactobacillus* strains to intestinal epithelial cells *in vitro*. Gut. 52: 827–33.

Mack, D.R., S. Michail, S. Wei, L. McDougall and M.A. Hollingsworth. 1999. Probiotics inhibit enteropathogenic *E. coli* adherence *in vitro* by inducing intestinal mucin gene expression. Am. J. Physiol. 276: 941–50.

Magistà, A.M., E. Ierardi, S. Castellaneta et al. 2005. *Helicobacter pylori* status and symptom assessment two years after eradication in pediatric patients from a high prevalence area. J. Pediatr Gastroenterol. Nutr. 40: 312–8.

Malaty, H.M., A. El-Kasabany, D.Y. Graham et al. 2002. Age at acquisition of *Helicobacter pylori* infection: a follow-up study from infancy to adulthood. Lancet 359: 931–935.

Malfertheiner, P., F. Megraud, C.A. O'Morain, J. Atherton, A.T. Axon, F. Bazzoli, G.F. Gensini, J.P. Gisbert, D.Y. Graham, T. Rokkas, E.M. El-Omar and E.J. Kuipers. 2012. European Helicobacter Study Group. Management of *Helicobacter pylori* infection—the Maastricht IV / Florence Consensus Report. Gut. May 61(5): 646–64.

Manfredi, M., B. Bizzarri, R.I. Sacchero, S. Maccari, L. Calabrese, F. Fabbian and G.L. De'Angelis. 2012. *Helicobacter pylori* infection in clinical practice: probiotics and a combination of probiotics + lactoferrin improve compliance, but not eradication, in sequential therapy. Helicobacter Aug. 17(4): 254–63.

McColl, K., L. Murray, E. El-Omar et al. 1998. Symptomatic benefit from eradicating *Helicobacter pylori* infection in patients with nonulcer dyspepsia. N. Engl. J. Med. 339: 1869–1874.

Michetti, P., G. Dorta, P.H. Wiesel et al. 1999. Effect of whey-based culture supernatant of *Lactobacillus acidophilus* (johnsonii) La1 on *Helicobacter pylori* infection in humans. Digestion 60: 203–9.

Midolo, P.D., J.R. Lambert, R. Hull, F. Luo and M.L. Grayson. 1995. *In vitro* inhibition of *Helicobacter pylori* NCTC 11637 by organic acids and lactic acid bacteria. J. Appl. Bacteriol. 79: 475–9.

Miki, K., Y. Urita, F. Ishikawa et al. 2007. Effect of Bifidobacterium bifidum fermented milk on *Helicobacter pylori* and serum pepsinogen levels in humans. J. Dairy Sci. 90: 2630–40.

Moayyedi, P., S. Soo, J. Deeks et al. 2001. Eradication of *Helicobacter pylori* for non-ulcer dyspepsia.Cochrane Database Syst. Rev. 1:CD002096-CD002096.

Moayyedi, P., S. Soo, J. Deeks, D. Forman, J. Mason, M. Innes and B. Delaney. 2000. Systematic review and economic evaluation of *Helicobacter pylori* eradication treatment for non-ulcer dyspepsia. Dyspepsia Review Group. BMJ. Sep. 16; 321(7262): 659–64.

Mukai, T., T. Asasaka, E. Sato, K. Mori, M. Matsumoto and H. Ohori. 2002. Inhibition of binding of *Helicobacter pylori* to the glycolipid receptors by probiotic *Lactobacillus reuteri*. FEMS Immunol. Med. Microbiol. 32: 105–10.

Myllyluoma, E., K. Kajander, H. Mikkola et al. 2007. Probiotic intervention decreases serum gastrin-17 in *Helicobacter pylori* infection. Dig. Liver Dis. 39: 516–23.

Myllyluoma, E., L. Veijola, T. Ahlroos et al. 2005. Probiotic supplementation improves tolerance to *Helicobacter pylori* eradication therapy-a placebo-controlled, double-blind, randomized pilot study. Aliment Pharmacol. Ther. 21: 1263–72.

Nam, H., M. Ha, O. Bae and Y. Lee. 2002. Effect of Weissella confusa strain PL9001 on the adherence and growth of *Helicobacter pylori*. Appl. Environ. Microbiol. 68: 4642–5.

Nista, E.C., M. Candelli, F. Cremonini et al. 2004. Bacillus clausii therapy to reduce side-effects of anti-*Helicobacter pylori* treatment: randomized, double-blind, placebo controlled trial. Aliment Pharmacol. Ther. 20: 1181–8.

Oderda, G., P. Shcherbakov, P. Bontems et al. 2007. European Pediatric Task Force on *Helicobacter pylori*. Results from the pediatric European register for treatment of *Helicobacter pylori* (PERTH), Helicobacter. 12: 150–6.

Ojetti, V., G. Bruno, M.E. Ainora, G. Gigante, G. Rizzo, D. Roccarina and A. Gasbarrini. 2012 . Impact of *Lactobacillus reuteri* Supplementation on Anti-*Helicobacter pylori* Levofloxacin-Based Second-Line Therapy. Gastroenterol. Res. Pract. 740381.

Pantoflickova, D., I. Corthesy-Theulaz, G. Dorta, M. Stolte, P. Isler, F. Rochat, M. Enslen and A.L. Blum. 2003. Favorable effect of long-term intake of fermented milk containing *Lactobacillus johnsonii* on *Helicobacter pylori* associated gastritis. Aliment Pharmacol. Ther. 18: 805–13.

Park, S.K., D.I. Park, J.S. Choi et al. 2007. The effect of probiotics on *Helicobacter pylori* eradication. Hepatogastroenterology 54: 2032–6.

Perdigon, G., E. Vintini, S. Alvarez, M. Medina and M. Medici. 1999. Study of the possible mechanisms involved in the mucosal immune system activation by lactic acid bacteria. J. Dairy Sci. 82: 1108–1114.

Pinchuk, I.V., P. Bressollier, B. Verneuil et al. 2001. *In vitro* anti-*Helicobacter pylori* activity of the probiotic strain Bacillus subtilis 3 is due to secretion of antibiotics. Antimicrob. Agents Chemother. 45: 3156–61.

Plummer, S.F., I. Garaiova, T. Sarvotham et al. 2005. Effects of probiotics on the composition of the intestinal microbiota following antibiotic therapy. Int. J. Antimicrob. Agents 26: 69–74.

Preidis, G.A. and J. Versalovic. 2009. Targeting the human microbiome with antibiotics, probiotics, and prebiotics: gastroenterology enters the metagenomics era. Gastroenterology 136: 2015–31.

Rasic, J., I. Klem, D. Jovanic and M. Ac. 1995. Antimicrobial effect of *Lactobacillus acidophilus* and Lactobacillus delbrueckii subsp. bulgaricus against *Helicobacter pylori in vitro*. Arch. Gastroenterohepatol. 14: 158–60.

Romeo, J., E. Nova, J. Warnberg, S. Gomez-Martinez, L.E. Diaz Ligia and A. Marcos. 2010. Immunomodulatory effect of fibres, probiotics and synbiotics in different life-stages. Nutr. Hosp. 25: 341–349.

Sakamoto, I., M. Igarashi, K. Kimura et al. 2001. Suppressive effect of *Lactobacillus gasseri* OLL 2716 (LG21) on *Helicobacter pylori* infection in humans. J. Antimicrob. Chemother. 47: 709–10.

Sakamoto, I., M. Igarashi, K. Kimura, A. Takagi, T. Miwa and Y. Koga. 2001. Suppressive effect of *Lactobacillus gasseri* OLL 2716 (LG21) on *Helicobacter pylori* infection in humans. J. Antimicrob. Chemother. 47: 709–710.

Sheu, B., J. Wu, C. Lo et al. 2002. Impact of supplement with *Lactobacillus*- and *Bifidobacterium*-containing yoghurt on triple therapy for *Helicobacter pylori* eradication. Aliment Pharmacol. Ther. 16: 1669–75.

Sgouras, D., P. Maragkoudakis, K. Petraki, B. Martinez-Gonzalez, E. Eriotou, S. Michopoulos, G. Kalantzopoulos, E. Tsakalidou and A. Mentis. 2004. *In vitro* and *in vivo* inhibition of Helicobacter pylori by Lactobacillus casei strain Shirota. Appl. Environ. Microbiol. 70: 518–26.

Sigthorsson, G., J. Tibble, J. Hayllar, I. Menzies, A. Macpherson and R. Moots. 1998. Intestinal permeability and inflammation in patients on NSAIDs. Gut. 43: 506–11.

Suerbaum, S. and P. Michetti. 2002. *Helicobacter pylori* infection. N. Engl. J. Med. Oct 10; 347(15): 1175–86.

Sykora, J., K. Valeckova, J. Amlerova et al. 2005. Effects of a specially designed fermented milk product containing probiotic *Lactobacillus casei* DN-114 001 and the eradication of *H. pylori* in children: a prospective randomized, double-blind study. J. Clin. Gastroenterol. 39: 692–8.

Szajewska, H., P. Albrecht and A. Topczewska-Cabanek. 2009. Randomized, double-blind, placebo-controlled trial: effect of lactobacillus GG supplementation on *Helicobacter pylori* eradication rates and side effects during treatment in children. J. Pediatr Gastroenterol. Nutr. 48: 431–6.

Takahashi, T., T. Matsumoto, M. Nakamura, H. Matsui, K. Tsuchimoto and H. Yamada. 2007. L-lactic acid secreted from gastric mucosal cells enhances growth of *Helicobacter pylori*. Helicobacter 12: 532–40.

Tong, J.L., Z.H. Ran, J. Shen, C.X. Zhang and S.D. Xiao. 2007. Meta-analysis: the effect of supplementation with probiotics on eradication rates and adverse events during *Helicobacter pylori* eradication therapy. Aliment Pharmacol. Ther. Jan. 15; 25(2): 155–68.

Tursi, A., G. Brandimarte, G.M. Giorgetti and M.E. Modeo. 2004. Effect of *Lactobacillus casei* supplementation on the effectiveness and tolerability of a new second-line 10-day quadruple therapy after failure of a first attempt to cure *Helicobacter pylori* infection. Med. Sci. Monit. 10: 662–6.

Uchida, M. and K. Kurakazu. 2004. Yogurt containing *Lactobacillus gasseri* OLL2716 exerts gastroprotective action against acute gastric lesion and antral ulcer in rats. J. Pharmacol. Sci. 96: 84–90.

Uemura, N., S. Okamoto, S. Yamamoto et al. 2001. *Helicobacter pylori* infection and the development of gastric cancer. N. Engl. J. Med. 345: 784–9.

Vaira, D., A. Zullo, N. Vakil et al. 2007. Sequential therapy versus standard triple-drug therapy for *Helicobacter pylori* eradication: a randomized trial. Ann. Intern. Med. 146: 556–63.

Vitini, E., S. Alvarez, M. Medina et al. 2000. Gut mucosal immunostimulation by lactic acid bacteria. Biocell 24: 223–32.

Wang, K.Y., S.N. Li, C.S. Liu et al. 2004. Effects of ingesting Lactobacillusand Bifidobacterium-containing yoghurt in subjects with colonized *Helicobacter pylori*. Am. J. Clin. Nutr. 80: 737–41.

Wendakoon, C.N., A.B. Thomson and L. Ozimek. 2002. Lack of therapeutic effect of a specially designed yogurt for the eradication of *Helicobacter pylori* infection. Digestion. 65: 16–20.

Yang, Y.J. and B.S. Sheu. 2012. Probiotics-containing yogurts suppress *Helicobacter pylori* load and modify immune response and intestinal microbiota in the *Helicobacter pylori*-infected children. Helicobacter. Aug; 17(4): 297–304.

Yaşar, B., E. Abut, H. Kayadıbı, B. Toros, M. Sezıklı, Z. Akkan, Ö. Keskın and O. Övünç Kurdaş. 2010. Efficacy of probiotics in *Helicobacter pylori* eradication therapy. Turk. J. Gastroenterol. Sep. 21(3): 212–7. 1.

17

Probiotics and Prebiotics in Infections

Shigeru Kamiya

Introduction

Probiotics are known to have beneficial effects on their hosts once consumed, and many investigations have examined the application of probiotics in a medical context. The diseases targeted by probiotic therapy were mainly infectious diseases such as *Clostridium difficile* infection (antibiotic-associated diarrhea:AAD, pseudomembranous colitis:PMC), rotavirus infection, traveler's diarrhea, etc. In addition to intestinal infections, other studies have also reported positive clinical effects in a broad range of infectious diseases including opportunistic infection, postoperative infection, urinary tract infection and respiratory tract infection (Gerritsen et al. 2011, Quigley 2010).

Effects of Probiotics on Intestinal Infections

Two major intestinal infectious diseases on which probiotics were shown to be effective by clinical trials are *C. difficile*-associated diarrhea/disease (CDAD) and rotavirus diarrhea (Culligan et al. 2009).

Department of Infectious Diseases, Kyorin University School of Medicine, Mitaka, Tokyo 181-8611, Japan.
Email: skamiya@ks.kyorin-u.ac.jp

C. difficile-associated diarrhea/disease (CDAD)

The sphere of CDAD includes both AAD and PMC. One of the main causative agents of AAD is *C. difficile* which is a Gram positive obligate anaerobic bacterium. AAD is detected in approximately 20% of patients treated with antibiotics and approximately 15–39% of AAD is caused by disturbance of the intestinal microbiota by antibiotic administration and consequent overgrowth of *C. difficile* with production of large clostridial toxins A and B (Viswanathan et al. 2010). Since 2002, there have been large outbreaks of CDAD in hospitals in Canada, the USA and Europe. The cause of these epidemics is now known to be the BI (restriction endonuclease type BI)/NAP1 (North America PFGE type 1)/027 (PCR-ribotyping) hypervirulent strain of *C. difficile* (Viswanathan et al. 2010, Warmy et al. 2005). It was later shown that this strain carried the binary toxin gene (*cdtB*) and an 18-bp deletion in the *tcdC* ORF in the pathogenicity locus (*PaLoc*), resulting in overproduction of toxins A and B (Warmy et al. 2005). Interestingly, Akerlund et al. (2008) reported that BI/NAP1/027 strains of *C. difficile* have higher efficiencies of sporulation than comparable non-hypervirulent strains.

In an experiment using germfree mice, it was shown that the mortality rate (44%) of gnotobiotic mice associated with *Saccharomyces boulardii* and *C. difficile* was significantly lower than that (84%) of mice monoassociated with *C. difficile* (Corthier et al. 1986). The cytotoxin titer in fecal samples of the *S. boulardii* associated gnotobiotes was decreased to less than 1/1000 that of the monoassociated mice. In addition, it was reported that *S. boulardii* produces a 54 kDa protein with serine-protease activity which deactivates toxins A and B and inhibits binding of toxin A to its receptor on intestinal epithelial cells (Castagliuolo et al. 1999, Czerucha and Rampal 2002).

We examined the effect of the probiotic strain *Clostridiium butyricum* M588 on lethal colitis caused by *C. difficile* in germfree mice (Table 1) (Kamiya et al. 1997). Hypertoxigenic *C. difficile* VPI10463 strain induced hemorrhagic colitis in monoassociated gnotobiotic mice with an 85.7% mortality rate. In contrast, administration of *C. difficile* VPI10463 strain did not induce any pathogenic effects in conventional mice with normal intestinal microbiota (0% mortality rate). However, association of the germfree mice with *C. butyricum* and *C. difficile* reduced the mortality rate to 20%. It was also shown that the *C. difficile* cell number and cytotoxin titer in cecal contents of the di-associated mice were significantly reduced compared with those of monoassociated mice.

There have been many reports that probiotics containing lactic acid bacteria such as *Lactobacillus acidophilus, Lactobacillus bulgaricus, Lactobacillus rhamnosus* and *Enterococcus faecium* are effective in the prevention of AAD in humans (Table 2) (Gotz et al. 1979, Clements et al. 1983, Wunderlich et

Table 1. Effect of pretreatment with *C. butyricum* on cecitis by *C. difficile* in germ free and conventional mice.

Mouse type	*C. butyricum* pretreatment	Deaths/Number of mice tested(%)		
		1 day	2 days	7 days
Germfree mice	–	5/7(71.4)	6/7(85.7)	6/7(85.7)
	+[a)]	1/10(10.0)	2/10(20.0)	2/10(20.0)
Conventional mice	–	0/10(0.0)	0/10(0.0)	0/10(0.0)
	+	0/10(0.0)	0/10(0.0)	0/10(0.0)

a)*C. butyricum* MIYAIRI588 strain was inoculated 5 days before infection with *C. difficile* VPI10463 strain (cited from Kamiya et al. 1997)

Table 2. Effect of probiotics on antibiotic-associated diarrhea.

Antibiotics used	Probiotics used[a)]	Therapeutic effect[b)] (No. of cases)	Ref.
Ampicillin	*L. acidophilus* + *L. bulgaricus*	8% vs. 21% (n=98)	Gotz et al. 1979
Neomycin	*L. acidophilus* + *L. bulgaricus*	20% vs. 42% (n=39)	Clements et al. 1983
Various	*E. faecium* SF68	9% vs. 27% (n=45)	Wunderlich et al. 1989
Beta-lactams	*S. boulardii*	7% vs. 15% (n=193)	McFarland et al. 1995
Clarithromycin + tinidasol	*L. rhamnosus* GG	3% vs. 27% (n=60)	Armuzzi et al. 2001
Clarithromycin + tinidasol	*L. rhamnosus* GG	5% vs. 30% (n=85)	Cremonini et al. 2002

a) L: *Lactobacillus*; E: *Enterococcus*; S: *Saccharomyces*
b) Percentage of patients with diarrhea in probiotics group vs. control

al. 1989, McFarland et al. 1995, Armuzzi et al. 2001, Cremonini et al. 2002). Probiotics containing *L. rhamnosus, E. faecium, S. boulardii* or *Lactobacillus reuteri* were shown to shorten the periods of diarrhea and hospital stay of AAD patients and to increase weight-gain. *S. boulardii* also decreased the overall incidence of AAD.

Seki et al. (2003) showed that *C. butyricum* MIYAIRI 588 strain reduced the incidence of AAD in children who were treated with antibiotics (59% vs. 5–9%). Szajewska et al. (2006) published the result of a meta-analysis investigating the risk of AAD in children with or without probiotic treatment. Six clinical trials were selected using *L. rhamnosus* GG, *L. acidophilus/Bifidobacterium infantis, L. acidophilus/ L. bulgaricus, Bifidobacterium lactis/S. thermophilus* and *S. boulardii*. The risk of AAD was different depending on the kind of probiotic used, being lower in the studies using *S. boulardii* and *L. rhamnosus* GG but higher in those using *L. acidophilus/L. bulgaricus*. These studies concluded that probiotic treatment was effective for prevention of AAD in children, with a relative risk ratio of AAD of 0.44.

A meta-analysis of probiotic efficacy for gastrointestinal diseases has been recently reported (Ritche and Romanuk 2012). The primary outcome for AAD was defined as diarrhea within 2 months of antibiotic exposure. The primary outcome for *C. difficile* disease (CDD/CDAD) was defined as a new episode of diarrhea associated with a positive culture or toxin (A or B) assay within 1 month of exposure to antibiotics, and that of prevention of CDD was a new episode of *C. difficile* diarrhea positive diarrhea within 1 month of a previous CDD episode. In this study, it was shown that AAD (n=27, relative risk=0.43, 95% CI 0.32–0.56) and CDD (n=6, relative risk=0.60, 95% CI 0.41–0.86) yielded significant effect sizes, indicating that probiotics are beneficial in treatment and prevention of AAD and CDD/CDAD.

Fecal transplantation (FT) from healthy individuals to the cases of severe CDAD or recurrent CDAD was first reported by Eiseman et al. in 1958 (Eiseman et al. 1958). Since then, numerous case reports and retrospective case series have demonstrated benefit of FT in patients with the above diseases with a higher cure rate than 90%. Recently, Kelly et al. (2012) reported the efficacy of FT through colonoscopy for relapsing CDAD in 26 patients. Twenty-four patients remained free of significant diarrhea or CDAD, and remaining 2 patients had loose stool and diarrhea 11 and 2 months after FT, respectively, indicating that FT was simple, safe and 92% effective in preventing relapse of CDAD. Similarly, Mattila et al. (2012) reported that FT through colonoscopy was effective therapy for recurrent *C. difficile* infection. The records from 70 patients with recurrent *C. difficile* infection who had undergone FT were reviewed. During the first 12 weeks after FT, symptoms resolved in all patients who did not have BI/NAP1/027 *C. difficile* infections. Of 36 patients with BI/NAP1/027 *C. difficile* infection, 32 (89%) had a favorable response. The difference in the cure rate by FT between historical and new virulent (BI/NAP1/027d) strains was interestingly detected, and further analysis needs to be done.

Rotavirus infection

Huang et al. (Huang et al. 2002) published a meta-analysis to examine the therapeutic effect of probiotics on rotavirus associated diarrhea in children (Table 3) (Isolauri et al. 1991, Kaila et al. 1992, Isolauri et al. 1994, Shornikova et al. 1997a, Shornikova et al. 1997b, Lee et al. 2001, Rosenfeldt et al. 2002). Eighteen clinical studies were selected and the age of patients in these studies was 1 month to 60 months. Diarrhea by rotavirus infection or other unknown agent was detected in all patient groups. Probiotic microorganisms used in these studies included *L. rhamnosus* GG, *B. infantis*, *Bifidobacterium bifidum*, *Enterococcus*, *L. acidophilus*, *L. bulgaricus*, *Lactobacillus delbruckii*, *Lactobacillus reuteri*, *S. thermophilus*, *Bacillus subtilis* and *S. boulardii*. Final analysis indicated that the duration of diarrhea in the probiotics group was significantly shortened by 0.8 days compared with that in the placebo control group, and that administration of lactobacilli in particular shortened the duration of diarrhea by 1.1 days. From these results, it was concluded that probiotic treatment can be effective in the treatment of non-bacterial diarrhea mainly caused by rotavirus infection in children younger than 5 years old.

Table 3. Effect of probiotics on rotavirus infection.

Ref.	Probiotics used	N	Rotavirus infection (%)	Duration of diarrhea	P value
Isolauri et al. 1991	*L. rhamnosus* GG[a)]	24	92[c)]	1.4	<0.001
	L. rhamnosus GG[b)]	23	74	1.4	<0.001
	Control (yogurt)	24	79	2.4	
Kaila et al. 1992	*L. rhamnosus* GG	22	100	1.1	0.001
	Control (yogurt)	17		2.5	
Isolauri et al. 1994	*L. rhamnosus* GG	21	100	1.5	0.002
	Control	21		2.3	
Huang et al. 2002	*L. reuteri*	19	63	1.7	0.07
	Control (placebo)	21	86	2.9	
Shornikova et al.1997b	*L. reuteri* [d)]	21	100	1.5	0.01
	L. reuteri [e)]	20		1.9	>0.05
	Control (placebo)	25		2.5	
Lee et al. 2001	*L. acidophilus, B. infantis*	50	100	3.1	<0.01
	Control	50		3.6	
Rosenfeldt et al. 2002	*L. rhamnosus, L. reuteri*	24	54	3.2	0.05
	Control (placebo)	19	74	4.8	

a) Fermented milk containing probiotics was administered
b) Freeze dried powder containing probiotics was administered.
c) Rotavirus infection rate
d) High doses ($1 \times 10^{10-11}$) of probiotics were administered.
e) Low doses (1×10^{7}) of probiotics were administered.
(Research papers with rotavirus detection rates higher than 50% from the studies in reference 24 are listed)

Szajewska and Skorka (2009) reported another meta-analysis of 7 randomized controlled trials to evaluate the efficacy of *S. boulardii* for treating acute gastroenteritis in children. Although the cause of acute gastroenteritis was not specified as rotavirus in these studies, they showed that administration of *S. boulardii* shortened the duration of diarrhea by 1.1 days overall.

Traveler's diarrhea

Traveler's diarrhea is a common health complaint among returning travelers. Almost all cases of traveler's diarrhea are mild and self-limited, but some cases can worsen with serious systemic symptoms. Causative agents of traveler's diarrhea include enterotoxigenic *E. coli* (ETEC), *Campylobacter, Salmonella, Shigella,* rotavirus, norovirus and *Giardia lamblia*. The incidence of traveler's diarrhea amongst travelers from industrialized nations travelling to tropical and subtropical regions varies between 10 and 60% for a two-week stay (von Sonnenburg et al. 2000).

Effective prevention of traveler's diarrhea by probiotics has been reported by many investigators (Table 4) (Pozo-Olano et al. 1978, Katelaris et al. 1995, Black et al. 1989, Oksanen et al. 1990, Hilton et al. 1997). In some studies, probiotic treatment with either lactobacilli + bifidobacteria + *S. thermophilus* or *L. rhamnosus* GG was shown to significantly inhibit

Table 4. Effect of probiotics on prevention of traveler's diarrhea.

Probiotics used [a)]	Therapeutic effect [b)] (No. of cases)	Ref.
L. acidophilus + *L. bulgaricus*	35% vs. 29% (n=50)	Pozo-Olano et al. 1978
L. fermentum strain KLD	24% vs. 24% (n=282)	Katelaris et al. 1995
lactobacilli + bifidobacteria + *S. thermophilu*	43% vs. 71%* (n=81)	Black et al. 1989
L. rhamnosus strain GG	41% vs. 47% (n=756)	Oksanen et al. 1990
L. rhamnosus strain GG	4% vs. 7%* (n=245)	Hilton et al. 1997

a) *L.*: *Lactobacillus*; *S.*: *Streptococcus*
b) Percentage of symptomatic patients in probiotic vs. control groups
*statistically significant ($p<0.05$)

the incidence of traveler's diarrhea (Black et al. 1989, Hilton et al. 1997). However, in the clinical studies using *L. acidophilus* + *L. bulgaricus*, *Lactobacillus fermentum* or *L. rhamnosus* GG, probiotic treatment did not prevent the incidence of traveler's diarrhea (Pozo-Olano et al. 1978, Katelaris et al. 1995, Oksanen et al. 1990). McFarland reported a meta-analysis of probiotics for the prevention of traveler's diarrhea (McFarland 2007), with 12 of 940 screened studies meeting the inclusion criteria. It was concluded that several probiotics including *S. boulardii* and a mixture of *L. acidophilus* and *B. bifidum* had significant efficacy for traveler's diarrhea prevention and no serious adverse reactions were reported. Recent meta-analysis showed that significant effect sizes were not observed for probiotics for traveler's disease (n=6, relative risk=0.92, 95% CI 0.79–1.05) (Ritche and Romanuk 2012). Therefore, it is clear that more scientific studies need to be performed in this area.

Other intestinal infectious diseases

Cholera

A 120 kDa protein produced by *S. boulardii* was reported to decrease the concentration of cyclic AMP induced by cholera toxin (Czerucha and Rampal 2002). Although this protein does not directly deactivate cholera toxin, it was speculated that it binds to a receptor for cholera toxin on intestinal epithelial cells and negatively regulates adenylate cyclase.

Shigellosis and salmonellosis

In animal experiments using germfree mice, the effects of probiotics containing *L. acidophilus*, *S. boulardii* and *E. coli* on *Shigella flexneri* (streptomycin-sensitive and resistant strains) and *Salmonella typhimurium* infections were investigated (Filho-Lima et al. 2000). The streptomycin-sensitive *S. flexneri* strain was eliminated by administration of probiotics 11 days after the start of infection, but both streptomycin-resistant *S. flexneri* and *S.* Typhimurium were not affected by the probiotics. It was also shown that culture supernatants of *L. rhamnosus* inhibited the growth of pathogenic bacteria such as *S.* Typhimurium, *S. flexneri, Klebsiella, Enterobacter, E. coli* and *Pseudomonas aeruginosa* (Forestier et al. 2001). Asahara et al. (2010) reported a protective effect of *Lactobacillus casei* strain Shirota against lethal infection with multi-drug resistant *Salmonella enterica* serovar Typhimurium DT104 in mice. Explosive intestinal growth and subsequent lethal extraintestinal translocation of *S.* Typhimurium during fosfomycin administration was significantly inhibited by continuous oral administration of *L. casei* Shirota strain, suggesting that probiotic treatment may be useful for prophylaxis against opportunistic intestinal infection by multi-drug resistant pathogens including *S.* Typhimurium DT104.

O157:H7 EHEC (enterohaemorrhagic E. coli)

Asahara et al. (2004) examined the effect of *Bifidobacterium breve* Yakult strain on O157:H7 EHEC. Infection of streptomycin-treated mice with O157:H7 EHEC led to lethality in 90% of mice within 10 days of infection. However, continuous administration of *B. breve* from 6 days prior to EHEC infection gave a 100% survival rate. Additionally, the EHEC cell number in fecal samples of the *B. breve*-treated mice was reduced to less than 1/10 of that in the positive control mice (EHEC infected mice) and concentrations of Shiga toxin-1 (Stx-1) and Shiga toxin-2 (Stx-2) in the cecum were decreased to 1/43 and 1/454 of the positive control, respectively. Takahashi et al. (2004) evaluated the effect of *C. butyricum* MIYAIRI 588 strain on O157:H7 EHEC infection in a germfree mouse model and found that monoassociation with EHEC led to 100% lethality in mice within 7 days after infection. In contrast, pretreatment with *C. butyricum* gave a 100% survival rate after EHEC infection, and both the EHEC cell number and Shiga toxin concentration were significantly reduced.

Campylobacter infection

It has been reported that administration of *B. breve* reduced the *Campylobacter* cell number in patients with *Campylobacter* enterocolitis compared to patients treated with antidiarrheal drugs only (Tojo et al. 1987). In an animal experiment using SPF mice, administration of probiotics containing *Bacillus* spp. was also reported to reduce the mortality rate caused by *Campylobacter* infection (Sorokulova et al. 1997).

Effect of Probiotics on Superinfection Diseases, Acute Pancreatitis and Opportunistic Infections

Superinfection disease

The patients with bacterial infections are treated with antimicrobial agents which have inhibitory effects on not only pathogenic bacteria but normal microbiota. Consequently, by the use of the antimicrobial agents, various indigenous microorganisms of normal microbiota are replaced by different bacteria or fungi, resulting in the occurrence of superinfection. *C. difficile* and *Candida albicans* are two major pathogens causing superinfection diseases.

The effects of probiotics on *C. difficile*-associated diarrhea/diseases (CDAD) have been reviewed in the previous section. Busscher et al. (1997) reported the effects of *S. thermophilus* on the adhesion activity of *C. albicans* and *Candida tropicalis*. *S. thermophilus* able to produce biosurfactant inhibited the adhesion of *C. albicans* and *C. tropicalis* to the surface of silicon rubber to the level of 15% and 51% of the control culture, respectively. No such activity was observed in *S. thermophilus* without production of biosurfactant, suggesting that glycolipid in biosurfactant of *S. thermophilus* might be associated with the inhibition of adhesion of two *Candida* species. In addition, pretreatment of germfree mice with heat-killed *L. acidophilus* or *L. casei* was reported to inhibit the growth of *C. albicans*. Similarly, in continuous flow culture system, it was shown that *Lactobacillus plantarum* inhibited the growth of *C. albicans* (Payne et al. 2003).

Acute pancreatitis

Infected pancreatic necrosis following acute pancreatitis is observed in 24% of the cases within 1 week after the onset of pancreatitis, and is one of risk factors that increase the mortality caused by acute pancreatitis. Olah et al. (2002) examined the effect of synbiotic therapy using heat killed *L. plantarum* and oat fiber on the prevention of complications in the patients with acute pancreatitis. Infected pancreatic necrosis and abscesses occurred in 1 of 22 patients in the synbiotics treated group, compared with 7 of 23 in the control group treated with heat-killed *L. plantarum* and oat fiber. The mean length of hospital stay was 13.7 days in the synbiotics group versus 21.4 days in the control group, indicating the effectiveness of the synbiotic therapy in reducing pancreatic sepsis. Basselink et al. (2008) investigated a randomized, double-blind, placebo-controlled trial to examine the effect of probiotic Enteric 641, consisting of 6 different strains of freeze-dried viable bacteria (*L. acidophilus, L. casei, Lactobacillus salivarius, Lactococcus lactis, B. bifidum* and *B. lactis*), on the patients with severe acute pancreatitis. The patients with severe acute pancreatitis were randomly assigned, within 72 hrs, to receive a multistep probiotic preparation or placebo for 28 days. The primary endopoint was the composite of infectious complications during admission and 90-day follow-up. Although there was no significant difference in the occurrence of infectious complication between probiotics (30%, 46 cases/152 cases) and placebo (28%, 41 cases/144 cases), 24 patients (16%) in the probiotics group died, compared with 9 (6%) in the placebo group (relative risk=2.53, 95% CI 1.22–5.25). It was also shown that 9 patients in the probiotics group developed bowel ischemia, compared with none in the placebo group (p=0.004), claiming that probiotics should not be administered in this category of patients. With respect to their surprising report, Sand and Nordback (2008) have questioned several points about the correlation between probiotics administered and ischemia, and they also commented on the need of extensive preclinical studies of the particular probiotic regimen in the setting of severe systemic disease before further studies in human beings. Zhang et al. (2010) have recently reported a meta-analysis for 7 studies selected to evaluate the use of pre, pro-and synbiotics on acute pancreatitis. It was shown that pre-, pro- or synbiotics treatment had no significant influence on patients with acute panreatitis. It seems that there is a lack of evidence to support the use of probiotics/synbiotics in this area.

Opportunistic infection

Opportunistic infection is defined as the infection with less pathogenic microorganisms in the immunocompromised hosts which include the patients with congenital/acquired immune deficiency diseases, malignancies, treated with long-term antimicrobial agents, organ transplantation, steroid therapy, diabetes mellitus, surgical operation, etc. The immunocompromised hosts also include healthy neonates, pregnant women and elderly persons. Opportunistic pathogens such as *P. aeruginosa, Staphylococcus, Escherichia coli* and *Klebsiella* are important as causative bacteria for hospital infections.

P. aeruginosa infection

P. aerginosa is a typical opportunistic pathogen and causes acute pneumonia in immunocompromised hosts and chronic pneumonia in the patients with cystic fibrosis. Alvarez et al. (2001) examined the effect of probiotics on pulmonary infection of newborn mice (3 weeds old) with *P. aeruginosa*. In the mice pretreated with yogurt containing *L. casei*, an increased number of *P. aeruginosa* and the enhancement of phagocytic activity in pulmonary macrophages were observed in the mice treated with the probiotics. In addition, the concentrations of IgA and IgM in bronchoalveolar lavage fluid were increased in the mice treated with probiotics, suggesting that activation of host immune response caused by probiotics may inhibit the pulmonary infection with *P. aeruginosa. Lactobacillus crispatus* was reported to inhibit the adhesion of *P. aeruginosa* to urethral epithelial cells and its growth (Osset et al. 2001), indicating a possibility of the use of *L. crispatus* for therapy of urinary tract infection.

Postoperative infection

It was reported that approximately 30% of the patients who had abdominal operations exhibited postoperative infections even if they were treated with antibiotics before and after the surgical operation. There have been many reports to evaluate the effect of probiotics on postoperative infections. McNaught et al. (2002) investigated the effect of *L. plantarum* 299V strain on bacterial translocation and the occurrence of postoperative infections before and after abdominal operations. Bacterial translocation (BT) was evaluated by bacterial culture of mesenteric lymph nodes and mesenteric serosa. The patients (n=64) were orally administered with *L. plantarum* for at least 1 week before operation and for several days after operation (median of days treated with probiotics; 9 days). Positive rates of BT in probiotics group (n=64) and control group (n=64) were 8.8% and 8.8%, respectively. Rate of incidence of sepsis in the patients who were treated with antibiotics for prevention of postoperative infections was 13.2% (7 cased/53 cases tested); that in control group was 15.4% (10 cases/65 cases). Recently, Jeppson et al. (2011) summarized the efficacy of the use of probiotics as prophylaxis for postoperative infections for 14 randomised clinical trials. It seemed that in patients undergoing liver transplantation or elective surgery in the upper gastrointestinal tract prophylactic administration of different probiotic strains (mainly lactobacilli strains) in combination with different fibers resulted in a 3-fold reduction in postoperative infections. In addition, a reduction in postoperative inflammation was observed. However, the use of probiotics with fibers in colorectal surgery was not successful in reducing postoperative infections. It was suggested that higher doses of probiotics with longer duration are needed to influence microbiota in the lower gastrointestinal tract or that immune function in colorectal patients may not be so important.

Enteral tube feeding (ETF) diarrhea

Enteral tube feeding is a useful nutritional treatment for the patients with abdominal operation or acute pancreatitis. ETF associated diarrhea is often observed in the patients having ETF, with an incidence of 2.3–68% (Boge et al. 2009). As ETF diarrhea results in loss of electrolytes and increased risk of endogenous infections, prevention of ETF associated diarrhea is clinically important. Bleichner et al. (1997) examined the effect of *S. boulardii* on ETF associated diarrhea. *S. boulardii* was added into solution of ETF. The occurrence of diarrhea in probiotics group (n=64) and placebo group (n=64) was 14.2% and 18.9%, respectively, indicating that administration of *S. boulardii* decreased the occurrence of ETF associated diarrhea significantly. In addition, it was shown that the number of days when diarrhea was detected was 91 days/648 days observed (14.2%) and 134 days/683 days observed (19.6%) in probiotics and placebo groups, respectively. Although the causative agents for diarrhea were not analyzed in the study, diarrhea was significantly correlated with serious infections and positivity of blood bacterial culture. In contrast, Heimburger et al. (Heimburger et al. 1994) reported that the occurrence of ETF associated diarrhea by addition of *L. acidophilus* and *L. bulgaricus* for 5 days was 17%, which is not significantly different from that in placebo group.

Cryptosporidiosis

Cryptosporidiosis is caused by *Cryptosporidium parvum*, and severe and prolonged diarrhea is observed in infant/preschool children and immunocompromised hosts. Alak et al. (1997) investigated the effect of probiotics on *C. parvum* infection in immunocompromised C57BL/6 mice infected with murine leukemia virus. From 4 months after leukemia virus infection, *L. reuteri* was orally administered to the mice for 10 days (PBS was used for control group). Next, the mice were challenged with *C. parvum* oocysts and the number of the oocysts in ileum and feces was quantified. Although there was no significant difference in the number of oocysts at 7 days after the infection between probiotics and control groups, the number of oocysts in the probiotics group was significantly lower at 14 days after the infection than that in control group. In the probiotics group, *C. parvum* oocysts were eliminated from ileum. As the numbers of *L. reuteri* and *C. parvum* oocysts detected in ileum were inversely correlated, it was speculated that the growth of *L. reuteri* in intestine induced the clearance of *C. parvum*.

Pickerd and Tuthill (2004) reported the effect of probiotics on the patients with Cryptosporidiosis. A twelve-year-old female patient who was diagnosed as Celiac disease at the age of 9 suffered from abdominal pain, soft stool, nausea and somnolence for about 4 months. *C. parvum* oocyts were detected in the diarrheal stool of the patient, but no other diarrheagenic pathogens were detected. Oral administration of *L. rhamnosus* GG (10^9 units/day) and *L. casei* Shirota (6.5 x 10^9 units/day) was done for 4 weeks. Nausea and diarrhea were cured and abdominal pain was weakened in 10 days after probiotics treatment, and no *C. parvum* was detected in fecal specimens 4 weeks after the probiotics treatment.

Intestinal entamoebiasis

Intestinal entamoebiasis is caused by infection with *Entamoeba histolytica*. The patients with intestinal entamoebiasis suffer from mucous and bloody stool, lower abdominal pain and colon ulcer, and the severity of the disease is remarkable in the immunocompromised hosts. Mansour-Ghanaei et al. (2003) examined the effect of probiotics of *S. boulardii* on entamoebiasis by *E. histolytica*. The patients with acute intestinal entamoebiasis were treated for 10 days with 2 different regimens (regimen 1, metronidazole 750 mg + iodoquinol 630 mg/day; regimen 2, metronidazole 750 mg + iodoquinol 630 mg + *S. boulardii* powder 250 mg/day) (Table 5). The time required for curing of diarrhea by regimens 1 and 2 were 48.0 hr and 12.0 hr, respectively, and the time required for disappearing of abdominal pain by regimens 1 and 2 were 24.0 hr and 12.0 hr, respectively. These results indicate that administration of *S. boulardii* was effective for curing of diarrhea and abdominal pain. In addition, it was reported that no cysts of *E. histolytica* were detected in the fecal specimens of the patients at 4 weeks after the treatment by regimen 2.

Table 5. Effect of *S. bourlardii* on intestinal entamoebiasis.

	Regimen 1* **(n=27)**	**Regimen 2**** **(n=27)**	
Duration of diarrhea (hr)	48.0±18.5	12.0±3.7	P<0.0001
Duration of abdominal pain (hr)	24.0±7.3	12.0±3.2	P<0.001
Number of patients positive for amebic cysts***	5 (18.5%)	0 (0%)	

*Regimen 1: Metronidazole 750 mg + iodoquinol 650 mg, 10 days
**Regimen 1: Metronidazole 750 mg + iodoquinol 650 mg + *S. boulardii*, 10 days
***Observed at 4 weeks after the treatment (cited from Mansour-Ghanaei et al. 2003)

Effect of probiotics on urinary tract infection

Urogenital microbial flora of a healthy premenopausal woman is generally dominated by the *Lactobacillus* species, the most common of which are *Lactobacillus iners, L. crispatus, Lactobacillus gasseri* and *Lactobacillus jensennii*. All the factors such as hormonal changes, vaginal pH, and glycogen content can affect the colonization of the *Lactobacillus* in the vagina (Waigankar and Patel 2011). It has been reported that low lactobacilli counts in the vagina and urethra are found in women suffering from recurrent urinary tract infections (UTIs) (Bruce et al. 1973). There have been many reports on the inhibitory effects of lactobacilli on pathogens in urogenital tracts. *Lactobacillus helveticus* KS300 strain and *L. rhamnosus* GG strain were reported to inhibit the adhesion of *E. coli* and *Gardnerella vaginalis* to the surface of HeLa cells (Atassi et al. 2006). It was shown that *L. acidophilus* CRL1259 and *Lactobacillus paracasei* CR1289 strains inhibited the adhesion of *S. aureus* to the surface of vaginal epithelial cells, but not that of *E. coli* (Zarate and Nader-Macias 2006). *L. fermentum, L. rhamnosus, L. plantarum* and *L. acidophilus*, originated from vaginal microbiota were reported to have anti-*Candida* action (Strus et al. 2005).

There have been several clinical studies to examine the effect of probiotics on UTI. Reid et al. (2001) reported the effect of oral administration of *L. rhamnosus* GR-1 and *L. fermentum* RC-14 strains (twice/day, 2 weeks) in 10 female patients (9 recurrent *Candida* vaginitis, 2 bacterial vaginosis, 3 UTI): It was shown that probiotic lactobacilli were detected in vagina for 1–2 months after oral administration and that no urogenital symptoms were observed during the period of probiotic treatment. Uehara et al. (2006) reported a pilot study to evaluate the safety and effectiveness of probiotics (*L. crispatus* GAI98322 strain) vaginal suppositories in 9 patients with recurrent UTI. Probiotics suppositories were used every other day before going to bed for 4–12 months (mean 10.1 months). Mean number of episode of UTI after the probiotic treatment was 1.6 +/–1.4, significantly lower than that (5.0 +/–1.6) before the probiotic treatmet. In contrast, Kontiokari et al. (2001) reported that oral administration of *L. rhamnosus* GG (5 days/week for 1 year) did not affect the recurrence rate of UTI. Waigankar and Patel (2011) stated that probiotics must not be considered a panacea in treating UTI and that further available data promise that probiotics will be a strong option for improving and maintaining urogenital health.

Effect of Probiotics on Respiratory Infections

In vivo study

There have been many reports to examine the in vivo effect of probiotics on influenza virus. Yasui et al. (1999) showed that *B. breve* YIT4064 strain protected the mice from influenza virus infection. Similarly, Hori et al. (2001) reported that intranasal inoculation of *L. casei* Shirota strain (200 mg/ml) decreased the titer of influenza virus to the level of less than one tenth and that mouse lethality of probiotics group was decreased to 15% (control group, 69% lethality). Activation of cellular immunity by probiotics is important for prevention of influenza virus infection. Harata et al. (2010) showed that nasal infection by influenza virus was protected by the stimulated NK cells in alveoli induced by intranasal inoculation of *L. rhamnosus* GG.

Clinical evaluation of the probiotics

Double blind, randomized trial study is needed for clinical evaluation of probiotics. The clinical effects of probiotics on host immune response at vaccination by influenza vaccine was reported by Olivares et al. (2007). The effect of *L. fermentum* CECT5715 strain on induction of anitibody to influenza virus was evaluated. *L. fermentum* CECT5715 strain (probiotics group, n=25) or placebo (methylcellulose; control group, n=25) were administered to healthy volunteers for 2 weeks prior to influenza vaccination. Then the volunteers were inoculated with influenza vaccine, followed by treatment with either probiotics or placebo for 2 weeks. In the probiotics group, the concentration of serum TNFalfa, influenza virus-specific IgA/IgM was significantly more increased in the probiotics group than that in placebo group. However, there was no significant difference in the concentration of virus specific IgM and total concentration of IgA between proiotics and control groups. Interestingly, the number of episodes of influenza-like symptoms during 5 months after vaccination in the probiotics group was significantly lower than that in control group, showing the effectiveness of probiotics treatment. Boge et al. (2009) evaluated the effect of probiotics on immune response in the elderly persons over the age of 70, vaccinated with influenza vaccine (H1N1, H3N2, B). Probiotics of *L. casei* DN114001 strain was administered for 4 and 9 weeks before and after vaccination, respectively. Significant increase of antibody to influenzavirus type B was detected in the probiotics group (n=113) compared to control group (n=109), but not to types H1N1 and H3N2. In addition, the positive conversion rate of antibody to types H1N1 and H3N2 at 5 months after vaccination was significantly higher in the probiotics group, suggesting the usefulness of probiotics in influenza vaccination. Hatakka et al. (2001) compared the incidences of pulmonary infection, enteric infection and otolaryngological infection in the children (1–6 years old) treated with or without fermented milk containing *L. casei* GG for 30 weeks. The number of the patients with otitis media and pulmonary infection was significantly lower in the probiotics group than that in the control group. Leyer et al. (2009) evaluated the effect of probiotics treatement for 6 months in children (3–5 years old) on pulmonary infections. Children in groups 1, 2 and 3 were treated with *L. acidophilus* MCFM strain+*Bifidobacterium animalis* subsp. *lactis* Bi-07 strain, *L. acidophilus* MCFM strain and placebo, respectively. Significant decrease in the numbers of the children with fever, cough or nasal discharge, patients administered with antibiotics, and the days of absence at school were found in the groups 1 (n=112) and 2 (n=110) than that in group 3 (n=104). Although the mean number of the patients with the above symptoms in group 1 was lower than that in group 2, there was no statistical significance in the difference. Hojsak et al. (2010) reported that administration of fermented milk containing *L. casei* GG to children (13–83 months old) for 3 months decreased the incidence of pulmonary infections. On the other hand, there was no significant difference in the difference of the incidence of enteric infections.

Effect of Synbiotics on Infectious Diseases

Synbiotics is defined as a combination of probiotics and prebiotics. There have been many reports to examine the effect of synbiotics on intestinal infections, intestinal microbiota and postoperative infections. Most of the evidence regarding the potential health benefits of prebiotics/synbiotics is derived from experimental animal studies and human trials in small number of subjects (Quigley 2012).

Intestinal infections

Asahara et al. (2001) showed that combined treatment of the *S.* Typhimurium infected mice with *B. bifidum/Bifidobacterium pseudocatenulatum* and oligosaccharide decreased significantly the number of *S.* Typhimurium colonized in cecum, suggesting the effectiveness of synbiotics in salmonellosis. Schultz et al. (Schultz et al. 2004) reported the effect of synbiotics for 2 months on severity of inflammatory changes and composition of intestinal microbiota using HLA-B27 beta2-microglobulin transgenic mice (2 months old). *L. acidophilus* La-5, *B. lactis* Bb-12 strains and inulin were used as synbiotics. In addition to synbiotics, metronidazole (MNZ; 50 mg/kg) was administered to the mice for 2 months to disturb normal microbiota (synbiotics group). The mice treated with only MNZ was considered as MNZ group (without symbiotic treatment). Inflammation score in the synbiotics group was 2.2 +/–0.2 which was significantly lower than that (2.9 +/–0.1) in MNZ group. PCR analysis of cecal normal microbiota indicated that the number of *B. animalis* was significantly more increased in the synbiotics group than that in MNZ group. Interestingly, inulin was detected in the cecal contents at 4 months after symbiotic treatment, but probiotic bacteria were not detected, suggesting that inhibitory action for the occurrence of inflammatory changes by synbiotics might be due to prebiotics.

Postoperative infections

Rayes et al. (2002a) examined the effect of synbiotics on the incidence of postoperative infections in 3 groups (n=30/group) of patients after major abdominal surgery (Table 6). The incidence of postoperative infections in groups 1 (parenteral nutrition or oat fiber-free enteral nutrition), 2(enteral oat fiber-containing nutrition with live *L. plantarum*) and 3 (treated with heat-killed *L. plantarum*) was 30%, 10% and 10%, respectively, showing that probiotic treatment decreased the incidence of postoperative infections. Although heat-killed *L. plantarum* also decreased the incidence of postoperative infections, the incidence of postoperative infections in the patients with gastric and pancreatic resections was significantly lower in group 2 than that in group 3. As there was no significant difference in the number of total lymphocytes, CD4 positive lymphocytes, CD8 positive

Table 6. Effect of probiotics on the occurrence of postoperative infections.

	Group 1* (n=30)	Group 2* (n=30)	Group 3* (n=30)	p
Number of the patients with infectious diseases	9	3	3	0.01
Infectious diseases				
pneumonia	6	2	1	
sepsis	1	0	1	
peritonsilitis	1	0	0	
UTI**	0	1	0	
wound infection	1	0	0	
ear infection	0	0	1	

*Group 1: parenteral nutrition or fiber-free enteral nutrition
Group 2: enteric fiber-containing nutrition with living *Lactobacillus*
Group 3: enteric fiber-containing nutrition with heat-killed *Lactobacillus*
** urinary tract infection (cited from Rayes et al. 2002a)

lymphocytes and NK cells among the 3 groups, it was clarified that probiotics used did not stimulate immune activity.

Anderson et al. (2005) evaluated the prevention effect for postoperative infections of synbiotics using *L. acidophilus* La5, *B. lactis* Bb-12, *S. thermophilus, L. bulgaricus* and oligofructose. Seventy two patients were randomised to the synbiotic group and 65 to the placebo group. There were no significant differences between the synbiotic and control groups in bacterial translocation (12.1% versus 10.7%), gastric colonization (41% versus 44%) or septic complications (32% versus 31%).

Rayes et al. (2002b) investigated the effectiveness of synbiotics on the prevention of postoperative infections in the patients with liver transplantation (Table 7). All the patients were supplied with early enteral nutrition. Patients in groups 1, 2 and 3 were treated with standard formula plus selective bowel decontamination (SBD), oat fiber-containing formula plus living *L. plantarum* 299 and oat fiber-containing formula plus heat-killed *L. plantarum* 299, respectively. The incidences of postoperative infections were 48%, 13% and 34% in groups 1, 2 and 3, respectively, indicating a significant difference between groups 1 and 2. In the group 2, the incidence of cholecystitis and pneumonia was lower, and the mean number of enterococci and staphylococci isolated was also lower. Although

Table 7. Effect of synbiotics on the incidence of infections after liver transplantation.

	Group 1* (n=32)	**Group 2*** (n=31)	**Group 3*** (n=32)
Number of the cases with infections (%)	15(48%)	4(13%)	11(34%)
Number of infections	23	4	8
Infections			
cholecystitis	10	2	8
pneumonia	6	1	4
sepsis	3	0	0
UTI**	0	0	3
wound infection	1	0	0
others	3	1	2
Isolates			
enterococci	8	1	8
E. coli	2	0	1
staphylococci	6	1	3
Klebsiella	0	0	1
None	7	2	5

*Group 1: Selective bowel decontamination group
Group b2: (*L. plantarum* + fiber) group
Group 3: (heat-killed *L. plantarum* + fiber) group
**urinary tract infection (cited from Rayes et al. 2002b)

the incidence of postoperative infections in group 3 was lower than that in group1, there was no statistical significance in the difference, showing non-effectiveness of prebiotics of oat fiber.

Conclusion

In this article, the effects and application of probiotics in various infectious diseases were reviewed. Goldin and Gorbach (2008) categorized 4 groups of clinical application for probiotics by level of evidence of efficacy. The first group is the application to cases of acute/antibiotic-associated gastroenteritis in which benefits of probiotics are well proven. The second includes allergic reactions, specifically atopic dermatitis in which there is substantial evidence of efficacy. The third includes applications that have shown promise, for example in childhood respiratory infections, dental caries, inflammatory bowel disease, combating nasal pathogens and the prevention of relapsing *C. difficile*-induced gastroenteritis. The fourth group covers potential future applications for rheumatoid arthritis, irritable bowel syndrome, cancer, alcohol-induced liver disease, diabetes and graft-versus-host disease. The use of probiotics in medical practice is rapidly increasing, and probiotics will soon be a part of the physician's armament for the prevention and treatment of various kinds of diseases including, but not limited to intestinal infectious diseases. However, more evidence-based research on probiotics is required to establish their efficacy in each area of application before they can be safely used.

References

Akerlund, T., I. Persson, M. Unemo, T. Noren, B. Svenungsson, M. Wullt et al. 2008. Increased sporulation rate of epidemic *Clostridium difficile* type 027/NAP1. J. Clin. Microbiol. 46: 1530–1533.

Alak, J.I.B., B.W. Wolf, E.G. Mduruwa, G.E. Pimentel-Smith and O. Adeyemo. 1997. Effect of *Lactobacillus reuteri* on intestinal resistance to *Cryptosporisium parvum* infection in a murine model of acquired immunodeficiency syndrome. J. Infect. Dis. 175: 218–221.

Alvarez, S., C. Herrero, E. Bru and G. Perdigon. 2001. Effect of *Lactobacillus casei* and yogurt administration on prevention of *Pseudomonas aeruginosa* infection in young mice. J. Food Prot. 64: 1768–1774.

Anderson, A.D., C.E. McNaught, P.K. Jain and J. MacFie. 2005. Randomised clinical trial of symbiotic therapy in elective surgical patients. Gut. 53: 241–245.

Armuzzi, A., F. Cremonini, F. Bartolozzi, F. Canducci, M. Candelli, V. Ojetti, G. Cammarota, M. Anti, A. De Lorenzo, P. Pola, G. Gasbarrini and A. Gasbarrini. 2001. The effect of oral administration of *Lactobacillus* GG on antibiotic-associated gastrointestinal side-effects during *Helicobacter pylori* eradication therapy. Aliment Pharmacol. Ther. 15: 163–169.

Asahara, T., K. Nomoto, K. Shimizu, M. Watanuki and R. Tanaka. 2001. Incresased resistance of mice to *Salmonella enterica* serovar *Typhimurium* infection by symbiotic administration of Bifidobacteria and transgalactosylated oligosaccharides. J. Appl. Microbiol. 91: 985–996.

Asahara, T., K. Shimizu, K. Nomoto, T. Hamabata,A. Ozawa and Y. Takeda. 2004. Probiotic bifidobacteria protect mice from lethal infection with Shiga toxin-producing *Escherichia coli* O157:H7. Infect. Immun. 72: 2240–2247.

Asahara, T., K. Shimizu, T. Takada, S. Kado, N. Yuki, M. Morotomi, R. Tanaka and K. Nomoto. 2010. Protective effect of *L. casei* strain Shirota against lethal infection with multi-drug resistant *Salmonella enterica* serovar Typhimurium DT104 in mice. J. Appl. Microbiol. 110: 163–173.

Atassi, F., D. Brassart, P. Grob, F. Graf and A.L. Servin. 2006. *In vitro* antibacterial activity of *Lactobacillus helveticus* strain KS300 against diarrhoeagenic, uropathogenic and vaginosis-associated bacteria. J. Appl. Micribiol. 101: 647–654.

Basselink, M.G., H.C. van Santvoort, E. Buskens, M.A. Boermeester, H. van Goor, H.M. Timmerman, V.B. Nieuwenhuijs, T.L. Bollen, B. van Ramshorst, B.J. Witteman, C. Rosman, R.J. Ploeg, M.A. Brink, A.F. Schaapherder, C.H. Dejong, P.J. Wahab, C.J. van Laarhoven, E. van der Harst, C.H. van Eijck, M.A. Cuesta, L.M. Akkermans and H.G. Gooszen. 2008. Dutch Acute Pancreatitis Study Group: Probiotic prophylaxis in predicted severe acute pancreatitis: a randomized, double-blind, placebo-controlled trial. Lancet 371: 651–659.

Black, F.T., P.L. Andersen and J. Orskov. 1989. Prophylactic efficacy of lactobacilli on traveler's diarrhea. Travel Med. 7: 333–335.

Bleichner, G., H. Blehaut, H. Mentec and D. Moyse. 1997. *Saccharomyces boulardii* prevents diarrhea in critically ill tube-fed patients. Intensive Care Med. 23: 517–523.

Boge, T., M. Rémigy, S. Vaudaine, J. Tanguy, R. Bourdet-Sicard and S. van der Werf. 2009. A probiotic fermented dairy drink improves antibody response to influenza vaccination in the elderly in two randomized controlled trials. Vaccine 27: 5677–5684.

Bruce, A.W., P. Chadwick, A. Hassan and G.F. van Cott. 1973. Recurrent urethritis in women. Can. Med. Assoc. J. 108: 973–976.

Busscher, H.J., C.G. van Hoogmoed, G.I. Geertsema-Doornbusch, M. van der Kuijl-Booij and H.C. van der Mei. 1997. *Streptococcus thermophilus* and its biosurfactants inhibit adhesion by *Candida* spp. on silicone rubber. Appl. Envir. Microbiol. 63: 3810–3817.

Castagliuolo, I., M.F. Riegler, L. Valenick, J.T. LaMont and C. Pothoulakis. 1999. *Saccharomyces boulardii* protease inhibits the effect of *Clostridium difficile* toxins A and B in human colonic mucosa. Infect. Immun. 67: 302–307.

Clements, M.L., M.M. Levine, P.A. Ristaino, V.E. Daya and T.P. Hughes. 1983. Exogenous lactobacilli fed to man. Their fate and ability to prevent diarrheal disease. Prog. Food Nutr. Sci. 7: 29–37.

Corthier, G., F. Dubos and R. Ducluzeau. 1986. Prevention of *Clostridium difficile* induced mortality in gnotobiotic mice by *Saccharomyces boulardii*. Can J. Microbiol. 32: 894–896.

Cremonini, F., S. Di Caro, M. Covino, A. Armuzzi, M. Gabrielli, L. Santarelli, E.C. Nista, G. Cammarota, G. Gasbarrini and A. Gasbarrini. 2002. Effect of different probiotic preparations on anti-Helicobacter pylori therapy-related side effects: a parallel group, triple blind, placebo-controlled study. Am. J. Gastroenterol. 97: 2744–2749.

Culligan, E.P., C. Hill and R.D. Sleator. 2009. Probiotics and gastrointestinal disease: Successes, problems and future prospects. Gut. Pathog. 1: 19.

Czerucha, D. and P. Rampal. 2002. Experimental effects of *Saccharomyces boulardii* on diarrheal pathogens. Microbe. Infect. 4: 733–739.

Eiseman, B., W. Silen, G.S. Bascom and A.J. Kauvar. 1958. Fecal enema as an adjunct in the treatment of pseudomembranous enterocolitis. Surgery 44: 854.

Filho-Lima, J.V., E.C. Vieira and J.R. Nocoli. 2000. Antagonistic effect of *Lactobacillus acidophilus, Saccharomyces boulardii* and *Escherichia coli* combinations against experimental infections with *Shigella flexneri* and *Salmonella enteritidis* subsp. *typhimurium* in gnotobiotic mice. J. Appl. Microbiol. 88: 365–370.

Forestier, C., C. De Champs, C. Vatoux and B. Joly. 2001. Probiotic activities of *Lactobacillus casei rhamnosus*: in vitro adherence to intestinal cells and antimicrobial properties. Res. Microibiol. 152: 167–173.

Gerritsen, J., H. Smidt, G.T. Rijkers and W.M. de Vos. 2011. Intestinal microbiota in human disease: the impact of probiotics. Genes Nutr. 6: 209–240.

Goldin, B.R. and S.L. Gorbach. 2008. Clinical indications for probiotics: an overview. Clin. Infect. Dis. 46(Suppl. 2): S96–100.

Gotz, V., J.A. Romankiewicz, J. Moss and H.W. Murray. 1979. Prophylaxis against ampicillin-associated diarrhea with a lactobacillus preparation. Am. J. Hosp. Pharm. 36: 754–757.

Harata, G., F. He, N. Hiruta, M. Kawase, A. Kubota, M. Hiramatsu and H. Yausi. 2010. Intranasal administration of *Lactobacillus rhamnosus* GG prortects mice from H1N1 influenza virus infection by regulating respiratory immune response. Lett. Appl. Microbiol. 50: 597–602.

Hatakka, K., E. Savilahti, A. Pönkä, J.H. Meurman, T. Poussa, L. Näse, M. Saxelin and R. Korpela. 2001. Effect of long term consumption of probiotic milk on infections in children attending day care centres: double blind, randomised trial. Br. Med. J. 322: 1327–1329.

Heimburger, D.C, D.G. Sockwell and W.J. Geels. 1994. Diarrhoea with enteral feeding: Prospective reappraisal of putative causes. Nutrition 10: 392–396.

Hilton, E., P. Kolakowski, C. Singer and M. Smith. 1997. Efficacy of *Lactobacillus* GG as a diarrheal preventive in travelers. J. Travel Med. 4: 41–43.

Hojsak, I., N. Snovak, S. Abdović, H. Szajewska, Z. Misak and S. Kolacek. 2010. *Lactobacillus* GG in the prevention of gastrointestinal and respiratory tract infections in children who attend day care centers: a randomized, double-blind, placebo-controlled trial. Clin. Nutrition 29: 312–316.

Hori, T., J. Kiyoshima, K. Shida and H. Yasui. 2001. Effect of intranasal administration of *Lactobacillus casei* Shirota on influenza virus infection of upper respiratory tract in mice. Clin. Diag. Lab Immunol. 8: 593–597

Huang, J.S., A. Bousvaros, J.W. Lee, A. Diazn and E.J. Davidson. 2002. Efficacy of probiotic use in acute diarrhea in children. Dig. Dis. Sci. 47: 2625–2634.

Isolauri, E., M. Juntunen, T. Rautanen, P. Silanaukee and T. Koivula. 1991. A human *Lactobacillus* strain (*Lactobacillus* GG) promotes recovery from acute diarrhea in children. Pediatrics 88: 90–97.

Isolauri, E., M. Kaila, H. Mykkanen, W. Ling and S. Salminen. 1994. Oral bacteriotherapy for viral gastroenteritis. Dig. Dis. Sci. 39: 2595–2600.

Jeppson, B., P. Mangell and H. Thorlacius. 2011. Use of probiotics as prophylaxis for postoperative infections. Nutrients 3: 604–612.

Kaila, M., E. Isolauri, E. Soppi, E. Virtanen, S. Laine and H. Arvilommi. 1992. Enhancement of the circulating antibody secreting cell response in human diarrhea by a human *Lactobacillus* strain. Pediatr Res. 32: 141–144.

Kamiya, S., H. Taguchi, H. Yamaguchi, T. Osaki, M. Takahashi and S. Nakamura. 1997. Bacterioprophylaxis using *Clostridium butyricum* for lethal caecitis by *Clostridium difficile* in gnotobiotic mice. Rev. Med. Microbiol. 8(Suppl 1): S57–S59.

Katelaris, P.H., I. Salam and M.J. Farthing. 1995. Lactobacilli to prevent traveler's diarrhea? N. Engl. J. Med. 333: 1360–1361.

Kelly, C.R., L. de Leon and N. Jasutkar. 2012. Fecal microbiota transplantation for relapsing *Clostridium difficile* infection in 26 patients. Methodology and results. J. Clin. Gastroenterol. 46: 145–149.

Kontiokari, T., K. Sundqvist, M. Nuutinen, T. Pokka, M. Koskela and M. Uhari. 2001. Randomised trial of cranberry-lingonberry juice and *Lactobacillus* GG drink for the prevention of urinary tract infections in women. Br. Med. J. 322: 1–5.

Lee, M.C., L.H. Lin, K.L. Huang and H.Y. Wu. 2001. Oral bacterial therapy promotes recovery from acute diarrhea in children. Acta Paediatr Tw. 42: 301–305.

Leyer, G.J., S. Li, M.E. Mubasher, C. Reifer and A.C. Ouwehand. 2009. Probiotic effects on cold and influenza-like symptom incidence and duration in children. Pediatrics 124: e172–179.

Mansour-Ghanaei, F., N. Dehbashi, K. Yazdanparast and A. Shafaghi. 2003. Efficacy of *Saccharomyces boulardii* with antibiotics in acute amoebiasis. World J. Gastroenterol. 9: 1832–1833.

Mattila, E., R. Uusitalo-Seppala, M. Wuorela, L. Lehtola, H. Nurmi, M. Ristikankare, V. Moilanen, K. Salminen, M. Seppala, P.S. Mattila, V.-J. Anttila and R. Arkkila. 2012. Fecal transplantation, through colonoscopy, is effective therapy for recurrent *Clostridium difficile* infection. Gastroenterology 142: 490–496.

McFarland, L.V. 2007. Meta-analysis of probiotics for the prevention of traveler's diarrhea. Travel Med. Infect. Dis. 5: 97–105.

McFarland, L.V., C.M. Surawicz, R.N. Greenberg, G.W. Elmer, K.A. Moyer, S.A. Melcher, K.E. Bowen and J.L. Cox. 1995. Prevention of beta-lactam-associated diarrhea by *Saccharomyces boulardii* compared with placebo. Am. J. Gastroenterol. 90: 439–448.

McNaught, C.E., N.P. Woodcock, J. MacFie and C.J. Mitchell. 2002. A prospective randomized study of the probiotic *Lactobacillus plantarum* 299V on indices of gut barrier function in elective surgical patients. Gut. 51: 827–831.

Oksanen, P.J., S. Salminen, M. Saxelin, P. Hamalainen, A. Ihantola-Vormisto, L. Muurasniemi-Isoviita, S. Nikkari, T. Oksanen, I. Porsti and E. Salminen. 1990. Prevention of traveller's diarrhoea by *Lactobacillus* GG. Ann. Med. 22: 53–56.

Olah, A., T. Belagyi, A. Issekutz, M.E. Gamal and S. Bengmark. 2002. Randomized clinical trial of specific lactobacillus and fibre supplement to early enteral nutrition in patients with acute pancreatitis. Br. J. Surg. 89: 1103–1107.

Olivares, M., M.R. Diaz-Ropero, S. Sierra, F. Lara-Villoslada, J. Fonolla, M. Navas, J.M. Rodriguez and J. Xaus. 2007. Oral intake of *Lactobacillus fermentum* CECT5716 enhances the effects of influenza vaccination. Nutrition 23: 254–60.

Osset, J., R.M. Bartolome, E. Garcia and A. Andreu. 2001. Assessment of the capacity of *Lactobacillus* to inhibit the growth of uropathogens and block their adhesion to vaginal epithelial cells. J. Infect. Dis. 183: 485–491.

Payne, S., G. Gibson, A. Wynne, B. Hudspith, J. Brostoff and K. Tuohy. 2003. *In vitro* studies on colonization resistance of the human gut microbiota to *Candida albicans* and the effects of tetracycline and *Lactobacillus plantarum* LPK. Curr. Issues Intest. Microbiol. 4: 1–8.

Pickerd, N. and D. Tuthill. 2004. Resolution of cryptosporidiosis with probiotic treatment. Postgrad Med. J. 80: 112–113.

Pozo-Olano, J.D., J.H. Warram, R.G. Gomez and M.G. Cavazos. 1978. Effect of a lactobacilli preparation on traveler's diarrhea. A randomized, double blind clinical trial. Gastroenterology 74: 829–830.

Quigley, E.M. 2010. Prebiotics and probiotics: modifying and mining the microbiota. Pharmacol. Res. 61: 213–218.

Quigley, E.M.M. 2012. Prebiotics and probiotics: Their role in the management of gastrointestinal disorders in adults. Nutr. Clin. Pract. 27: 195–200.

Rayes, N., S. Hansen, D. Seehofer, A.R. Muller, S. Serke, S. Bengmark and P. Neuhaus. 2002a. Early enteral supply of fiber and *Lactobacilli* versus conventional nutrition: A controlled trial in patients with major abdominal surgery. Nutrition 18: 609–615.

Rayes, N., D. Seehofer, S. Hansen, K. Boucsein, A.R. Muller, S. Serke, S. Bengmark and P. Neuhaus. 2002b. Early enteral supply of *Lactobacillus* and fiber versus selective bowel decontamination: A controlled trial in liver transplant recipients. Transplantation 74: 123–128.

Reid, G., A.W. Bruce, N. Fraser, C. Heinemann, J. Owen and B. Henning. 2001. Oral probiotics can resolve urogenital infections. FEMS Immunol. Med. Microbiol. 30: 49–52.

Ritche, M.L. and T.N. Romanuk. 2012. A meta-analysis of probiotic efficacy for gastrointestinal diseases. PLos One 7: 1–11, e34938.

Rosenfeldt, V., K.F. Michaelsen, M. Jakobsen, C.N. Larsen, P.L. Moller, M. Tvede, H. Weyrehter, N.H. Valerius and A. Paerregaard. 2002. Effect of probiotic *Lactobacillus* strains on acute diarrhea in a cohort of nonhospitalized children attending day-care centers. Pediatr Infect. Dis. J. 21: 417–419.

Sand, J. and I. Nordback. 2008. Probiotics in severe acute pancreatitis. Lancet 371: 634–635.

Schultz, M., K. Munro, G.W. Tannock, I. Melchner, C. Gottl, H. Schwietz, J. Scholmerich and H.C. Rath. 2004. Effects of feeding a probiotic preparation (SIM) containing inulin on the severity of colitis and on the composition of the intestinal microflora in HLA-B27 transgenic rats. Clin. Diag. Labor Immunol. 11: 581–587.

Seki, H., M. Shiohara, T. Matsumura, N. Miyagawa, M. Tanaka, A. Komiyama and S. Kurata. 2003. Prevention of antibiotic-associated diarrhea in children by *Clostridium butyricum* MIYAIRI. Pediatr Int. 45: 86–90.

Shornikova, A., I. Casas, E. Isolauri, H. Mykkanen and T. Vesikari. 1997a. *Lactobacillus reuteri* as a thrapeutic agent in acute diarrhea in young children. J. Pediatr Gastroenterol. Nutr. 24: 399–404.

Shornikova, A., I. Casas, H. Mykkanen, E. Salo and T. Vesikari. 1997b. Bacteriotherapy with *Lactobacillus reuteri* in rotavirus gastroenteritis. Pediatr Infect. Dis. J. 16: 1103–1107.

Sorokulova, I.B., D.L. Kirik and I.V. Pinchuk. 1997. Probiotics against *Campylobacter* pathogens. J. Travel Med. 4: 167–170.

Strus, M., A. Kucharska, G. Kukla, M. Brzychczy-Włoch, K. Maresz and P.B. Heczko. 2005. The *in vitro* activity of vaginal *Lactobacillus* with probiotic properties against *Candida*. Infect. Dis. Obstet Gynecol. 13: 69–75.

Szajewska, H. and A. Skorka. 2009. *Saccharomyces boulardii* for treating acute gastroenteritis in Children: updated meta-analysis of randomized controlled Trails. Aliment Pharmacol. Ther. 30: 955–963.

Szajewska, H., M. Ruszczynski and A. Radzikowski. 2006. Probiotics in the prevention of antibiotic-associated diarrhea in children: A meta-analysis of randomized controlled trials. J. Pediatr 149: 367–372.

Takahashi, M., H. Taguchi, H. Yamaguchi, T. Osaki, A. Komatsu and S. Kamiya. 2004. The effect of probiotic treatment with *Clostridium butyricum* on enterohemorrhagic *Escherichia coli* O157:H7 infection in mice. FEMS Immunol. Med. Microbiol. 41: 219–226.

Tojo, M., T. Oikawa, Y. Morikawa, N. Yamashita, S. Iwata, Y. Satoh, J. Hanada and R. Tanaka. 1987. The effect of *Bifidobacterium breve* administration on campylobacter enteritis. Acta Pediatr . 29: 160–167.

Uehara, S., K. Monden, K. Nomoto, Y. Seno, R. Kariyama and H. Kumon. 2006. A pilot study evaluating the safety and effectiveness of *Lactobacillus* vaginal suppositories in patients with recurrent urinary tract infection. Int. J. Antimicrob. Agents Suppl. 1: S30–34.

Viswanathan, V.K., M.J. Mallozzi and G. Vedantam. 2010. *Clostridium difficile* infection. An overview of the disease and its pathogenesis, epidemiology and interventions. Gut. Microbes 1: 234–242.

von Sonnenburg, F., N. Tornieporth, P. Waiyaki, B. Lowe, L.F. Peruski Jr., H.L. DuPont, J.I. Mathewson and R. Steffen. 2000. Risk and aetiology of diarrhea at various tourist destinations. Lancet 356: 133–134.

Waigankar, S.S. and V. Patel. 2011. Role of probiotics in urogenital healthcare. J. Midlife Health 2: 5–10.

Warmy, M., J. Pepin, A. Fang, G. Killgore, A. Thompson, J. Brazier, E. Frost and L.C. McDonald LC. 2005. Toxin production by an emerging strain of *Clostridium difficile* associated with outbreaks of severe disease in North America and Europe. Lancet 366: 1079–1084.

Wunderlich, P.F., L. Braun, I. Fumagalli, V. D'Apuzzo, F. Heim, M. Karly, R. Lodi, G. Politta, F. Vonbank and L. Zeltner. 1989. Double-blind report on the efficacy of lactic acid-producing *Enterococcus* SF68 in the prevention of antibiotic-associated diarrhoea and in the treatment of acute diarrhoea. J. Int. Med. Res. 17: 333–338.

Yasui, H., J. Kiyoshima, T. Hori and K. Shida. 1999. Protection against influenza virus infection of mice fed *Bifidobacterium breve* YIT4064. Clin. Diag. Lab. Immunol. 6: 186–92.

Zarate, G. and M.E. Nader-Macias. 2006. Influence of probiotic vaginal lactobacilli on *in vitro* adhesion of urogenital pathogens to vaginal epithelial cells. Lett. Appl. Microbiol. 43: 174–180.

Zhang, M.-M., J.-Q. Cheng, Y.-R. Lu, Z.-H. Yi, P. Yang and X.-T. Wu. 2010. Use of pre, pro- and synbiotics in patients with acute pancreatitis: A meta-analysis. World J. Gastroenterol. 16: 3970–3978.

18

Probiotics and Prebiotics in Cancer Prevention

Vijaya K. Gogineni[1] and *Lee E. Morrow*[2,*]

Introduction

The gut microbiota is a remarkable asset for human health. As a key element in the development and prevention of specific diseases, its study has yielded a new field of promising biotherapeutics. Homeostasis of the gut microbiota maintains various functions which are vital to the maintenance of human health. Disruption of the intestinal ecosystem equilibrium (gut dysbiosis) is associated with a plethora of human diseases, including autoimmune and allergic diseases, metabolic diseases, bacterial infections and altered cancer prevention. Beneficial modulation of the gut microbiota using agents such as prebiotics, probiotics, and antibiotic may favor health-promoting populations of bacteria and can be exploited to develop novel biotherapeutics. Evidence continues to emerge that the intestinal microbiota is intrinsically linked with overall health—including cancer risk.

[1] Division of Pulmonary, Critical Care and Sleep Medicine, Creighton University Medical Center, VA Medical Center, 4101 Woolworth Avenue, Room #9101, Omaha, Nebraska, USA, 68105.
Email: vijayagogineni@creighton.edu
[2] Associate Professor of Medicine and Pharmacy Practice, Division of Pulmonary, Critical Care and Sleep Medicine, Nebraska and Western Iowa Veterans' Affairs Hospital, Creighton University Medical Center, 601 North 30th Street, Suite #3820, Omaha, Nebraska, USA, 68131.
Email: lmorrow@creighton.edu
* Corresponding author

Role of Probiotics in Colorectal Cancer Prevention

Colorectal cancer represents a major public health problem accounting for over one million cases and about half a million deaths worldwide (Chau and Cunningham 2006). Five-year survival from colon cancer has been found to vary demographically: published estimates are 65% in North America, 54% in Western Europe, 34% in Eastern Europe, and 30% in India (Parkin et al. 2005). More than 80% of colorectal neoplasms occur sporadically, arising from adenomatous polyps via the long-term accumulation of mutations in genes including APC, K-ras and TP53 (Huycke and Gaskins 2004). Development of colon cancer represents a sequence of events that, although incompletely understood, occurs in definable steps. First is an initiating step, in which a carcinogen produces an alteration in the DNA. This step may be preceded by the metabolic activation of a precursor to produce the carcinogenic entity. The next clearly observable step is an overgrowth of colonic crypts, which can be seen morphologically as an aberrant crypt. Aberrant crypts (AC), which are considered pre-neoplastic structures, are enlarged and elevated relative to normal crypts and have a serpentine growth pattern. Aberrant crypts may occur singly or as groups of aberrant crypts within a single focus. A certain small but unknown fraction of these aberrant crypts will progress to polyps and eventually to tumors.

Dietary interventions and natural, bioactive supplements have been extensively evaluated for efficacy in reducing the risks of colon cancer. In the many existing animal and human studies of colon cancer, investigators have measured how diets or treatments affect specific predisposing factors such as increase in enzymes that activate carcinogens, increase pro-carcinogenic chemicals within the colon, or alter populations of certain bacterial genera or species. A number of studies have now shown that these predisposing factors are modified favorably by consumption of certain probiotics or prebiotics.

In vitro Studies

In vitro studies are of interest for ecological, metabolic and fermentation investigations. While such bench-top studies provide reproducible results, they inherently assume stability of the ecosystem and ignore numerous host factors. Existing *in vitro* studies have generally used fresh feces or colonic contents suspended in the buffer solution or culture media to overcome the limitation that only ~80% of stool bacteria are successfully cultivated.

Intestinal and lactic acid bacteria (Orrhage et al. 1994) *in vitro* can bind mutagenic compounds found in a Western diet—entities such as 3-amino-1-methyl-5H-pyrido[2,3-b]indole (Trp-P-2), 2-amino-1-methyl-6-phenylimidazo[4,5-b]pyridine (PhIP), 2-amino-3-methylimidazo[4,5-f] quinoline (IQ) and 2-amino-3,8-dimethylimidazo[4,5-f]quinoxaline (MeIQx). Binding of these compounds correlated well with the reduction of mutagenicity observed after exposure of the heterocyclic amines to bacterial strains. Such binding appears to be a physical phenomenon, mostly due to a cation exchange mechanism, and it has been suggested that cell wall peptidoglycans and polysaccharides are the two most important elements for binding to occur (Zhang and Ohta 1991). However, there is a danger in extrapolating these *in vitro* results to health claims for humans, as the reversibility of mutagen binding to cultures *in vivo* is unknown. Furthermore, the biologically significant levels of mutagens and of lactic acid bacteria in the human system are unknown.

In vivo Studies

Existing *in vivo* studies have been conducted using laboratory animals, gnobiotic animals (germ-free laboratory animals colonized with defined organisms), and healthy human volunteers. These studies provide metabolic, ecological, preclinical, clinical, host-bacterial and bacterial-bacterial interaction studies. The limitations of *in vivo* studies include differences between animals' and humans' gut microbiota composition, potential difference in host-bacterial interactions across species, and safety/ethical issues inherent with probiotic use in ill humans.

Early studies examined the effects of milk fermented with *Lactobacilli* and *Candida* species on tumor formation (Takano et al. 1985). Colon tumorgenesis induced by 1,2 dimethylhydrazine (DMH) was reduced in rats given the fermented milk product. Aberrant crypt (AC) formation in rats fed skim milk, skim milk fermented with *Bifidobacterium*, and the same bacteria incorporated into the diet, reduced the incidence of AC by 50% in probiotic treated animals (Abdelali et al. 1995). There was no difference in cecal lumenal pH, but the groups consuming the *Bifidobacteria* had decreased cecal β-glucuronidase activity. The measurement of β-glucuronidase represents an indirect indicator of risk because it is not clear whether it has a direct effect on the outcome of interest. *Lactobacillus casei* subspecies *rhamnosus GG* can interfere with the initiation or early promotional stages of DMH-induced intestinal tumorigenesis: this effect is most pronounced for animals fed a high-fat diet (Goldin et al. 1996).

Overnight cultures of *Lactobacillus acidophilus* have been shown to inhibit the formation of aberrant crypt foci which are thought to be precursor lesions of colon cancer induced by azoxymethane (Arimochi et al. 1997). Bolognani et al. (2001) reported that neither inulin, nor the probiotic *Lactobacillus acidophilus* had an effect on aberrant crypt foci formation when rats were fed a standard, low-fat diet. However, when rats consumed a high-fat diet, comparable to Western diets, both treatments significantly reduced aberrant crypt foci formation.

Epidemiological Studies

There are few epidemiological studies assessing the association between probiotics and colorectal cancer. An epidemiological study performed in Finland demonstrated that, despite a diet high in fat, colon cancer incidence was lower than in other countries presumably because of a relatively high consumption of milk, yoghurt, and other dairy products (Malhotra 1977, Maclennan and Jensen 1977). Separately, in two population-based case-control studies of colon cancer, an inverse relationship was seen with yoghurt (Peters et al. 1992) and cultured milk consumption (Young and Wolf 1988) after adjusting for potential confounding variables. In an epidemiologic study assessing the intestinal flora of populations with high risk for colon cancer, Moore and Moore (1995) were not able to definitively link high numbers of select bacterial species with a reduced risk of colon cancer. Fecal bacteria were compared in populations of polyp patients, Japanese-Hawaiians, North American Caucasians, rural native Japanese, and rural native Africans. The polyp patients and Japanese-Hawaiians were the subsets initially considered as the "high risk" groups. Fifteen bacterial subsets were associated with high risk of colon cancer (among these *Bacteroides* and *Bifidobacteria*) while five were associated with low risk of colon cancer (select *Lactobacilli* species and *Eubacterium aerofaciens*). This study does not clearly demonstrate cause-and-effect. Rather, the observed associations between bacterial species and risk of disease should be the starting point for further investigations.

Potential Mechanisms of Probiotic Action in Colorectal Cancer

The precise mechanisms by which probiotics may inhibit colon cancer are currently unknown and likely vary depending on the species. The postulated mechanisms by which the lactic acid bacteria might exert effects include: enhanced immune response; reduction of mutagenic, carcinogenic and genotoxic compounds; quantitative and/or qualitative alterations in the intestinal microflora; reduction of intestinal inflammation; and short chain fatty acid production.

Enhanced Immune Response: One mechanism by which probiotics may reduce cancer risk is via modulation of the mucosal and systemic immune responses. It has been demonstrated that lactic acid bacteria (LAB), particularly the cytoplasmic fraction of *Lactobacillus acidophilus* SNUL, *Lactobacillus casei* YIT9029, and *Bifidobacterium longum* HY8001 were able to significantly reduce tumor proliferation *in vitro*, to increase the survival of mice injected with tumor cells, and to promote anti-tumor activity via increased cellular immunity (Lee et al. 2004). Moreover, a recent study by Ghoneum et al. (2005) demonstrated that Caco-2 colonic adenocarcinoma cells underwent apoptosis *in vitro* upon phagocytosis of *Saccharomyces cerevisiae*. This was also observed in a breast cancer cell line (Ghoneum and Gollapudi 2004) suggesting that probiotic therapeutic interventions may not necessarily be restricted to cancers affecting the gastrointestinal system.

Probiotic alterations of cytokine profiles may also mediate host immunity. It has been demonstrated that *Bifidobacterium longum* and *Bifidobacterium animalis* promote the induction of inflammatory cytokines (IL-6, TNF-α) in mouse peritoneal cells (Sekine et al. 1994). Intrapleural administration of *Lactobacillus casei* strain Shirota into tumor-bearing mice induces the production of several cytokines (IFN-γ, IL-1 and TNF-α) in the thoracic cavity of mice, resulting in inhibited tumor growth and increased survival (Matsuzaki 1998). Sun and colleagues (Sun et al. 2005) have further demonstrated *in vivo* that peptidoglycan from a *Lactobacillus* species reduced the growth of CT26 colon cancer cells in BALB/c mice via an increased level of apoptosis in a dose-dependent manner. A protective effect of *Lactobacillus casei* against carcinogen-induced lesions in rat colon cells was observed when the probiotic was administered at a level of 1×10^{10} bacteria in 10 mL NaCl/kg body weight. Similarly, a single dose of living *Lactobacillus acidophilus*, *Lactobacillus gasseri*, *Lactobacillus confusus*, *Streptococcus thermophilus*, *Bacterium breve*, and *Bacterium longum* was found to prevent MNNG-induced DNA damage in the colon. However, a reduced bacterial dose of 50% or 90% resulted in the loss of the carcinogen protection effects (Wollowski et al. 2001). This may be attributed to a probiotic dose-dependent stimulation of gut immune cells to release inflammatory and regulatory cytokines such as IFN-γ, IL-12, IL-14 and IL -10 (Galdeano and Perdigon 2006).

These observations highlight that probiotic bacterial interactions with the gut—and systemic—immune system are extremely complex. Furthermore, they demonstrate that an immense amount of work remains in order to determine how individual agents affect overall gut health and development, which probiotics exert particular effects, and how probiotics can best be used to modulate gut immune homeostasis.

Reduction of Mutagenic, Carcinogenic and Genotoxic Compounds: A number of bacterial enzymes including β-glucuronidase and nitroreductase play an important role in cancer development as they hydrolyse carcinogenic compounds (de Moreno de LeBlanc and Perdigon 2005). Human studies have demonstrated that the capacity for probiotics to decrease the activity of these bacterial enzymes is strain-specific. *Lactobacillus plantarum* 299V, *Lactobacillus rhamnosus* DR20 and *Lactobacillus acidophilus* A1 were unable to decrease β-glucuronidase activity in healthy subjects (Goossens et al. 2003, Tannock et al. 2000, Marteau et al. 1990). However, *Lactobacillus casei* Shirota and *Lactobacillus acidophilus* significantly decreased enzymatic activity, indicating that they could reduce carcinogen production and, potentially, reduce the likelihood of colorectal cancer (Goldin et al. 1980, Spanhaak et al. 1998). Zhang and Ohta showed that freeze-dried lactic acid bacteria, intestinal bacteria and yeast cells significantly reduced the absorption of 3-amino-1,4-dimethyl-5H-pyrido[4,3-b]indole (Trp-P-1) from the small intestine in rats with subsequent decreased levels of this mutagen in the blood (Zhang and Ohta 1993). A similar study demonstrated reduced uptake of the mutagen Trp-P-2 and reduced deposition of its metabolites in various tissues of mice supplemented with dietary lactic acid bacteria (Orrhage et al. 2002). Lactobacilli have also been shown to degrade nitrosamines (Rowland and Grasso 1975). The consumption of lactobacilli by human volunteers has been shown to reduce the mutagenicity of urine and feces associated with the ingestion of carcinogens in cooked meat. *Lactobacillus plantarum* and *Bifidobacterium* Bb12 possess significant anti-genotoxic effects *in vitro*: both reduced fecal water genotoxicity towards HT-29 cells suggesting that these probiotics may be beneficial in preventing the early stages of colon cancer (O'Mahony et al. 2001). In aggregate these studies suggest that probiotic consumption may have utility in preventing colorectal cancers by reducing the levels of intraluminal carcinogenic compounds and reducing DNA damage.

Quantitative and/or Qualitative Alterations in the Intestinal Microflora: Consumption of probiotic organisms significantly reduces the number of fecal putrefactive bacteria such as coliforms while increasing the numbers of commensal bacteria such as Lactobacillus (O'Mahony et al. 2001) and Bifidobacteria (Gaudier et al. 2005). Such alterations have been associated with a reduced incidence of colonic adenocarcinoma in IL-10 knockout mice treated with *Lactobacillus salivarius* ssp. Salivarius UCC118 (O'Mahony et al. 2001). Dietary fat is considered to be a risk factor for colon cancer and it has been suggested that reductions in colon cancer may be mediated by increased levels of bile acids—mainly secondary bile acids produced by the action of bacterial 7α-dehydroxylase on primary bile acids—in the colon (Weisburger and Wynder 1987). A six-week course of *Lactobacillus acidophilus* fermented milk supplements to 14 colon cancer patients resulted in higher concentrations of soluble bile acids in feces (Lidbeck et al. 1991). Using azoxymethane-induced aberrant crypt foci in rats, Reddy et al. found that stimulated growth of Bifidobacteria in the colon inhibited colon carcinogenesis (Reddy et al. 1997). The authors suggested that the inhibition of aberrant crypt foci and crypt multiplicity was attributed to the pH-lowering effect of Bifidobacteria in the colon with subsequent inhibition of the growth of *Escherichia coli* and Clostridia (Kulkarni and Reddy 1994).

Reduction of Intestinal Inflammation: In pathological conditions like inflammatory bowel disease (IBD), immune homeostasis gives way to a chronic inflammatory state characterized by massive immune cell infiltration, immune-mediated tissue destruction, and attendant disruption of epithelial function and morphology. Intestinal inflammation has been linked to the development of colorectal cancer, with IBD increasing the likelihood of colon cancer development later in life (Collins et al. 2006). Probiotics have recently been shown to reduce intestinal inflammation in a number of animal models of IBD (Rachmilewitz et al. 2004) and in human IBD patients (Bibiloni et al. 2005). Indeed, epidemiological data suggest that up to 15% of human cancer incidence is associated with chronic inflammation (Mantovani et al. 2010). Evidence suggests that IBD develops, at least in part, as a response to changes in the normal microbiota (dysbiosis) rather than from pathogenic invasion (Mazmanian et al. 2008, Salzman and Bevins 2008). The dysbiosis model of IBD proposes that genetic or environmental changes alter gut homeostasis and shift the microbial balance away from symbiotic species (those with known health-promoting effects) and toward pathobiotic species (resident organisms such as Clostridium and Helicobacter that have pathogenic potential but are not normally pathogenic) (Round and Mazmanian 2009). This shift in turn leads to the induction of a chronic inflammatory state that significantly increases the risk for colorectal cancer.

Short Chain Fatty Acid Production: Short-chain fatty acids are the end products of carbohydrate fermentation—specifically resistant starches and dietary fiber—by anaerobic bacteria. Fecal concentrations of different short-chain fatty acids include acetate, propionate and butyrate in a molar ratio of ≈ 60:20:20 respectively. Butyrate has been regarded as the most important nutrient for colonocytes and has a major role in the regulation of cell proliferation and differentiation. Butyrate is produced mainly in the proximal colon where substrate availability is higher: accordingly this region of the colon has a lower lumenal pH. A number of studies have shown that butyrate inhibits human colon carcinoma cell proliferation and induces apoptosis in human colon carcinoma cells. Butyrate possesses possible anti-carcinogenic effects via suppression of COX-2 expression as has been demonstrated in HT-29 and Caco-2 cancer cell lines. Butyrate also induces expression of the host's glutathione-S-transferases and other stress response genes. Direct administration of butyrate in animal models of colon cancer has had variable results. While one study showed that luminal delivery of high-dose butyrate reduced aberrant crypt foci by 45% compared to untreated rats, other studies have shown butyrate to be ineffective (Salzman and Bevins 2008).

Clinical trials

There are limited data from clinical trials involving the use of probiotics for the prevention or treatment of colorectal cancer. To date, the results of *in vitro* studies have not been replicated in clinical trials given the inherent complexity of carcinogenesis, challenges with experimental design, variations in the tumor stages of the subjects, variation in the type of probiotic strains used, and difficulties in obtaining the adequate sample sizes. Table 1 summarizes existing randomized, controlled trials done on healthy human volunteers or cancer-free patients using probiotics to prevent colon cancer.

Role of Probiotics in Liver Cancer Prevention

Aflatoxins, produced by many species of fungus *Aspergillus*, have been implicated in liver cancer. Aflatoxin B1 (AFB1) is considered the most toxic of the aflatoxins and is produced by both *Aspergillus flavus* and *Aspergillus parasiticus*. After being metabolized in the liver, these toxins can bind to guanine in DNA, resulting in mutations at codon 249 of the TP53 tumor suppressor gene (Smela et al. 2001). Although aflatoxins are present in a typical Western diet, they are found at much higher levels in developing countries such as China. AFB1 toxin appears to bind to the bacterial surface of both *Lactobacillus rhamnosus* GG and *Lactobacillus rhamnosus* LC-705. A single viable bacterium is able to bind 10^7 AFB1 molecules, and binding appears to occur on the bacterial surface predominantly by hydrophobic interactions between the AFB1 molecules and the carbohydrate and protein components of the bacterial cell wall (Haskard et al. 2000). A randomized, placebo-controlled trial of 90 healthy men in Guangzhou, China demonstrated that dietary supplementation with *Lactobacillus rhamnosus* LC705 and *Propionibacterium freudenreichii* subsp. shermanii in healthy men exposed to dietary AFB1 resulted in reduced urinary excretion of one aflatoxin metabolite (AFB-N^7-guanine) known to be a biomarker for liver cancer risk (El-Nezami et al. 2006). The results of this probiotic intervention trial are encouraging for additional studies on the use of probiotics for unavoidable exposures to aflatoxins and other natural and/or environmental carcinogens.

Table 1. Randomized control trials of probiotic intervention for prevention or treatment of colon cancer.

Author	Study	Population	Intervention	Comments
Worthley et al. (2009)	Randomized, double-blind, placebo-controlled, crossover trial	Healthy human volunteers (n=20)	*Bifidobacterium lactis* alone or in combination with high-amylose maize starch and resistant starch	No significant effect on epithelial proliferation or crypt height. Fecal short chain fatty acid concentration was not different between groups. A greater proportion of patients harbored fecal *Lachnospiraceae* spp.
Hatakka et al. (2008)	Randomized, double-blind, placebo-controlled trial	Healthy male volunteers (n=38)	*Lactobacillus rhamnosus* LC705 with *Propionibacterium freudenreichii* ssp. shermanii JS	Increased fecal counts of Lactobacilli and Propionibacterium. Decreased β-glycosidase activity.
Rafter et al. (2007)	Randomized, double-blind, placebo-controlled trial	Patients with colon cancer (n=38) or removal of colon polyps (n=43)	Oligofructose-enriched inulin, *Lactobacillus rhamnosus* GG, and *Bifidobacterium lactis* Bb12	Reduced colorectal proliferation and improved epithelial barrier function. Increased secretion of IL-2 in patients with polyp removal. Increased IFN-γ in cancer patients.
Gianotti et al. (2010)	A randomized, double-blind trial	Colon cancer patients prior to resection (n=31)	*Bifidobacterium longum* and *Lactobacillus johnsonii.*	Lactobacillus modulated local immunity through adherence to mucosa and reduction in the concentration of pathogens.
Ishikawa et al. (2005)	A randomized, controlled trial	Colon cancer free patients (n=398)	Wheat bran and/or *L. casei*	*L. casei* prevented cellular atypia of colorectal tumors.

Role of Probiotics in Bladder Cancer Prevention

A clinical trial using probiotic *Lactobacillus casei* strain Shirota showed a significant decrease in superficial bladder cancer recurrence rates: 57% of probiotic-treated patients had a recurrence during one year of follow-up compared to 83% in the control group (Aso and Akaza 1992). A Japanese case-control study strongly suggested that habitual intake of lactic acid bacteria—widely consumed as fermented milk products—significantly reduced the risk of bladder cancer, supporting a potential anti-cancer role for probiotics (Ohashi et al. 2002).

Conclusions

The use of probiotics and prebiotics to prevent colon cancer has gained increasing attention due to positive outcomes from molecular and *in vivo* studies. Various mechanisms—enhanced immune response, reduction of mutagenic compounds, reducing intestinal inflammation, production of short chain fatty acids—have varying levels of supporting evidence. Animal model data suggest that probiotics with or without prebiotics have inhibitory effects on the development of precancerous lesions and malignant tumors. These effects are not entirely consistent and are small in some studies, but observed variations may represent dose and/or time effects. There is no convincing direct experimental evidence for cancer suppression in humans as a result of the consumption of probiotics, prebiotics or synbiotics. While certain combinations of probiotics, prebiotics or synbiotics have greater efficacy *in vivo* than either treatment alone, studies in humans have been less definitive in colorectal cancer. Accordingly, many researchers have pointed out the need for carefully designed human clinical trials with robust sample sizes. Furthermore, investigations are needed to identify the probiotic, prebiotic or synbiotic—alone or in combination—that will be most effective for human cancer prevention. However, there may not be one "ideal" treatment. Given inter-individual variability in host microbiota, the most effective pro- or prebiotic will likely be dependent upon the composition of each individual's existing intestinal microflora. Although the therapeutic application of probiotics is in its infancy, probiotics and prebiotics hold potential as a novel strategy for the prevention and/or treatment of select cancers.

References

Abdelali, H., P. Cassand, V. Soussotte, M. Daubeze, C. Bouley and J. Narbonne. 1995. Effect of dairy products on initiation of precursor lesions of colon cancer in rats. Nutr. Cancer. 24: 121–132.

Arimochi, H., T. Kinouchi, K. Kataoka, T. Kuwahara and Y. Ohnishi. 1997. Effect of intestinal bacteria on formation of azoxymethane-induced aberrant crypt foci in the rat colon. Biochem. Biophys. Res. Commun. 238: 753–757.

Aso, Y. and H. Akaza. 1992. Prophylactic effect of a *Lactobacillus casei* preparation on the recurrence of superficial bladder cancer. Urology Int. 49: 125–129.

Bibiloni, R., R.N. Fedorak, G.W. Tannock, K.L. Madsen, P. Gionchetti, M. Campieri, C. De Simone and R.B. Sartor. 2005. VSL#3 probiotic-mixture induces remission in patients with active ulcerative colitis. Am. J. Gastroenterol. 100: 1539–1546.

Bolognani, F., C.J. Rumney, B.L. Pool-Zobel and I.R. Rowland. 2001. Effect of lactobacilli, bifidobacterium and inulin on the formation of aberrant crypt foci in rats. Eur. J. Nutr. 40: 293–300.

Chau, I. and D. Cunningham. 2006. Adjuvant therapy in colon cancer—what, when and how? Annals of Oncology 17: 1347–1359.

Collins, P., C. Mpofu, A. Watson and J. Rhodes. 2006. Strategies for detecting colon cancer and/or dysplasia in patients with inflammatory bowel disease. Cochrane Database Syst. Rev. CD000279.

de Moreno de LeBlanc, A. and G. Perdigon. 2005. Reduction of beta-glucuronidase and nitroreductase activity by yoghurt in a murine colon cancer model. Biocell. 29: 15–24.

El-Nezami, H.S., N.N. Polychronaki, J. Ma, H. Zhu, W. Ling, E.K. Salminen, R.O. Juvonen, S.J. Salminen, T. Poussa and H.M. Mykkanen. 2006. Probiotic supplementation reduces a biomarker for increased risk of liver cancer in young men from Southern China. Am. J. Clin. Nutr. 83: 1199–1203.

Galdeano, M. and G. Perdigon. 2006. The probiotic bacterium *Lactobacillus casei* induces activation of the gut mucosal immune system through innate immunity. Clinical and Vaccine Immunology 13: 219–226.

Gaudier, E., C. Michel, J.P. Segain, C. Cherbut and C. Hoebler. 2005. The VSL# 3 probiotic mixturemodifies microflora but does not heal chronic dextran-sodium sulfate-induced colitis or reinforce the mucus barrier in mice. J. Nutr. 135: 2753–2761.

Ghoneum, M. and S. Gollapudi. 2004. Induction of apoptosis in breast cancer cells by *Saccharomyces cerevisiae*, the baker's yeast, *in vitro*. Anticancer Res. 24: 1455–1463.

Ghoneum, M., J. Hamilton, J. Brown and S. Gollapudi. 2005. Human squamous cell carcinoma of the tongue and colon undergoes apoptosis upon phagocytosis of *Saccharomyces cerevisiae*, the baker's yeast, *in vitro*. Anticancer Res. 25: 981–989.

Gianotti, L., L. Morelli, F. Galbiati, S. Rocchetti, S. Coppola, A. Beneduce, C. Gilardini, D. Zonenschain, A. Nespoli and M. Braga. 2010. A randomized double-blind trial on perioperative administration of probiotics in colo-rectal cancer patients. World J. Gastroenterol. 16: 167–175.

Goldin, B.R., L. Swenson, J. Dwyer, M. Sexton and S.L. Gorbach. 1980. Effect of diet and *Lactobacillus acidophilus* supplements on human fecal bacterial enzymes. J. Natl. Cancer Inst. 64: 255–261.

Goldin, B.R., L.J. Gualtieri and R.P. Moore. 1996. The effect of *Lactobacillus* GG on the initiation and promotion of DMH induced intestinal tumors in the rat. Nutr. Cancer 25: 197–204.

Goossens, D., D. Jonkers, M. Russel, E. Stobberingh, A. Van Den Bogaard and R. StockbrUgger. 2003. The effect of Lactobacillus plantarum 299v on the bacterial composition and metabolic activity in faeces of healthy volunteers: a placebo-controlled study on the onset and duration of effects. Ailment Pharmacol. Ther. 18: 495–505.

Haskard, C., C. Binnion and J. Ahokas. 2000. Factors affecting the sequestration of aflatoxin by *Lactobacillus rhamnosus* strain GG. Chem. Biol. Interact. 128: 39–49.

Hatakka, K., R. Holma, H. El-Nezami, T. Suomalainen, M. Kuisma, M. Saxelin, T. Poussa, H. Mykkanen and R. Korpela. 2008. The influence of *Lactobacillus rhamnosus* LC705 together with Propionibacterium freudenreichii ssp. shermanii JS on potentially carcinogenic bacterial activity in human colon. Int. J. Food Microbiol. 128: 406–410.

Huycke, M.M. and H.R. Gaskins. 2004. Commensal bacteria, redox stress, and colorectal cancer: mechanisms and models. Exp. Biol. Med. (Maywood) 229: 586–597.

Ishikawa, H., I. Akedo, T. Otani, T. Suzuki, T. Nakamura, I. Takeyama, S. Ishiguro, E. Miyaoka, T. Sobue and T. Kakizoe. 2005. Randomized trial of dietary fiber and *Lactobacillus casei* administration for prevention of colorectal tumors. Int. J. Cancer 116: 762–767.

Kulkarni, N. and B.S. Reddy. 1994. Inhibitory effect of *Bifidobacterium longum* cultures on the azoxymethane-induced aberrant crypt foci formation and fecal bacterial β-glucuronidase. Proceedings of the Society for Experimental Biology and Medicine 207: 278–283.

Lee, J.W, J.G. Shin, E.H. Kim, H.E. Kang, I.B. Yim, J.Y. Kim, H.G. Joo and H.J. Woo. 2004. Immunomodulatory and antitumor effects *in vivo* by the cytoplasmic fraction of *Lactobacillus casei* and *Bifidobacterium longum*. J. Vet. Sci. 5: 41–48.

Lidbeck, A., U. Geltner-Allinger and K.M. Orrhage. 1991. Impact of *Lactobacillus acidophilus* supplements on the faecal microflora and soluble faecal bile acids in colon cancer patients. Microbial Ecology in Health and Disease 4: 81–88.

Maclennan, R. and O.M. Jensen. 1977. Intestinal Microecology Group International Agency for Research on Cancer Dietary fibre, transit time, fecal bacteria, steroids, and colon cancer in two Scandinavian Populations. Lancet 2: 207–211.

Malhotra, S.L. 1977. Dietary factors in a study of colon cancer from cancer registry, with special reference to the role of saliva, milk, and fermented milk products and vegetable fibre. Med. Hypotheses 3: 122–134.

Mantovani, A., C. Garlanda and P. Allavena. 2010. Molecular pathways and targets in cancer-related inflammation. Ann. Med. 42: 161–170.

Marteau, P., P. Pochart, B. Flourie, P. Pellier, L. Santos, J.F. Desjeux and J.C. Rambaud. 1990. Effect of chronic ingestion of a fermented dairy product containing *Lactobacillus acidophilus* and *Bifidobacterium bifidum* on metabolic activities of the colonic flora in humans. Am. J. Clin. Nutr. 52: 685–688.

Matsuzaki, T. 1998. Immunomodulation by treatment with *Lactobacillus casei* strain Shirota. Int. J. Food Microbiol. 41: 133–140.

Mazmanian, S.K., J.L. Round and D.L. Kasper. 2008. A microbial symbiosis factor prevents intestinal inflammatory disease. Nature 453: 620–625.

Moore, W.E. and L.H. Moore. 1995. Intestinal floras of populations that have a high risk of colon cancer. Appl. Environ. Microbiol. 61: 3202–3207.

O'Mahony, L., M. Feeney, S. O'Halloran, L. Murphy, B. Kiely, J. Fitzgibbon, G. Lee, G. O'Sullivan, F. Shanahan and J.K. Collins. 2001. Probiotic impact on microbial flora, inflammation and tumor development in IL-10 knockout mice. Aliment Pharmacol. Ther. 15: 1219–1225.

Ohashi, Y., S. Nakai, T. Tsukamoto, N. Masumori, H. Akaza, N. Miyanaga, T. Kitamura and K. Kawabe. 2002. Habitual intake of lactic acid bacteria and risk reduction of bladder cancer. Urol. Int. 68: 273–280.

Orrhage, K., A. Annas and C.E. Nord. 2002. Effects of lactic acid bacteria on the uptake and distribution of the food mutagen Trp-P-2 in mice. Scandinavian Journal of Gastroenterology 37: 215–221.

Orrhage, K., E. Sillerstrom, J.A. Gustafsson, C.E. Nord and J. Rafter. 1994. Binding of mutagenic heterocyclic amines by intestinal and lactic acid bacteria. Mutat. Res. 311: 239–248.

Parkin, D.M., F. Bray, J. Ferlay and P. Pisani. 2005. Global cancer statistics 2002. CA Cancer Journal for Clinicians 55: 74–108.

Peters R.K., M.C. Pike, D. Garabrant and T.M. Mack. 1992. Diet and colon cancer in Los Angeles County, California. Cancer Causes Control. 3: 457–473.

Rachmilewitz, D., K. Katakura, F. Karmeli, T. Hayashi, C. Reinus, B. Rudensky, S. Akira, K. Takeda, J. Lee, K. Takabayashi and E. Raz. 2004. Toll-like receptor 9 signaling mediates the anti-inflammatory effects of probiotics in murine experimental colitis. Gastroenterology 126: 520–528.

Rafter, J., M. Bennett, G. Caderni, Y. Clune, R. Hughes, P.C. Karlsson, A. Klinder, M. O'Riordan, G.C. O'Sullivan, B. Pool-Zobel, G. Rechkem- mer, M. Roller, I. Rowland, M. Salvadori, H. Thijs, J. Van Loo, B. Watzl and J.K. Collins. 2007. Dietary synbiotics reduce cancer risk factors in polypectomized and colon cancer patients. Am. J. Clin. Nutr. 85: 488–496.

Reddy, B.S., R. Hamid and C.V. Rao. 1997. Effect of dietary oligofructose and inulin on colonic preneoplastic aberrant crypt foci inhibition. Carcinogenesis 18: 1371–1374.

Round, J.L. and S.K. Mazmanian. 2009. The gut microbiota shapes intestinal immune responses during health and disease. Nat. Rev. Immunol. 9: 313–323.

Rowland, I.R. and P. Grasso. 1975. Degradation of N-nitrosamines by intestinal bacteria. Appl. Microbiol. 29: 7–12.

Salzman, N.H. and C.L. Bevins. 2008. Negative interactions with the microbiota: IBD. Adv. Exp. Med. Biol. 635: 67–78.

Sekine, K., T. Kawashima and Y. Hashimoto. 1994. Comparison of the TNF-α levels induced by human-derived *Bifidobacterium longum* and rat-derived Bifidobacterium animalis in mouse peritoneal cells. Bifidobacteria Microflora 13: 79–89.

Smela, M.E., S.S. Currier, E.A. Bailey and J.M. Essigmann. 2001. The chemistry and biology of aflatoxin B(1): from mutational spectrometry to carcinogenesis. Carcinogenesis 22: 535– 45.

Spanhaak, S., R. Havenaar and G. Schaafsma. 1998. The effect of consumption of milk fermented by *Lactobacillus casei* strain Shirota on the intestinal microflora and immune parameters in humans. Eur. J. Clin. Nutr. 52: 899–907.

Sun, J., Y.H. Shi, G.W. Le and X.Y. Ma. 2005. Distinct immune response induced by peptidoglycan derived from Lactobacillus sp. World J. Gastroenterol. 11: 6330–6337.

Takano, T., K. Arai, I. Murota, K. Hayakawa, T. Mizutani and T. Mitsuoka. 1985. Effects of feeding sour milk on longevity and tumorigenesis in mice and rats. Bifid. Microflora 4: 31–37.

Tannock, G.W., K. Munro, H.J. Harmsen, G.W. Welling, J. Smart and P.K. Gopal. 2000. Analysis of the fecal microflora of human subjects consuming a probiotic product containing *Lactobacillus rhamnosus* DR20. Appl. Environ. Microbiol. 66: 2578–2588.

Weisburger, J.H. and E.L. Wynder. 1987. Etiology of colorectal cancer with emphasis on mechanism of action and prevention. *In*: De Vita, V.T., S. Hellman and S.A. Rosenberg (eds.). Important Advances in Oncology. Philadelphia: JB Lippincott, pp. 197–220.

Wollowski, I., G. Rechkemmer and B.L. Pool-Zobel. 2001. Protective role of probiotics and prebiotics in colon cancer. American Journal of Clinical Nutrition 73: 451S–455S.

Worthley, D.L., R.K. Le Leu, V.L. Whitehall, M. Conlon, C. Christophersen, D. Belobrajdic, K.A. Mallitt, Y. Hu, N. Irahara, S. Ogino, B.A. Leggett and G.P. Young. 2009. A human, double-blind, placebo-controlled, crossover trial of prebiotic, probiotic, and synbiotic supplementation: effects on luminal, inflammatory, epigenetic, and epithelial biomarkers of colorectal cancer. Am. J. Clin. Nutr. 90: 578–586.

Young, T.B. and D.A. Wolf. 1988. Case-control study of proximal and distal colon cancer and diet in Wisconsin. Int. J. Cancer 42: 167–175.

Zhang, X.B. and Y. Ohta. 1991. Binding of mutagens by fractions of the cell wall skeleton of lactic acid bacteria. J. Dairy Sci. 74: 1477–1481.

Zhang, X.B. and Y. Ohta. 1993. Microorganisms in the gastrointestinal tract of the rat prevent absorption of the mutagencarcinogen 3-amino-1, 4-dimethyl-5H-pyrido[4,3-b]indole. Can. J. Microbiol. 39: 841–845.

19

Probiotics and Prebiotics in Allergy and Asthma

Bengt Björkstén

Introduction

The prevalence of allergies, diabetes, Inflammatory Bowel Disease and other "immunologically mediated diseases of affluence" has increased progressively, particularly over the last 50 years. Over this time there has been growing recognition of the contributing role of declining microbial burden (the "hygiene hypothesis"). There has also been intense interest in the health benefits of dietary supplements (probiotic and prebiotic) that promote favourable colonisation. These two distinct, but rapidly converging areas of research have emphasised the need to understand, and ultimately to manipulate our physiological interactions with commensal microbiota.

An apparent decline in microbial exposure during early childhood is one of the most plausible causes of the escalating rates of allergic disease. Epidemiological support for this hypothesis has been progressively consolidated by a growing understanding of effects of microbial factors on immune development. While this story began with allergic disorders, autoimmune conditions (such as type I diabetes and Inflammatory Bowel Disease) are now increasingly included in these models. It is proposed that the underlying concepts of immune regulation by microbes are

Institute of Environmental Medicine, Karolinska Institutet Stockholm and Örebro University, Sweden.
Email: bengt.bjorksten@ki.se

similar for several immunologically mediated diseases, which have also been considered as "microbial deficiency syndromes" (Rook and Stanford 1998).

There is longstanding interest in the relationship between microbial exposure and allergic disease. It is common knowledge among clinicians and patients that respiratory infections can trigger and enhance allergic manifestations such as an asthma attack in sensitised individuals. It was therefore also assumed for many years that infections would enhance sensitisation and the development of allergies. Consequently, patients were advised to avoid exposure to infectious agents and parents of allergic children were advised to protect them from infections. While it is still certainly true that asthma symptoms are aggravated by respiratory tract infections, the exposure to microbial agents in infancy has other connotations as well. The story began in 1976, when the Canadian paediatrician John Gerrard suggested that infections in early childhood may actually protect against the development of allergies (Gerrard et al. 1976). It was not until 1989, however, that this notion raised general interest. That year, the British epidemiologist David Strachan suggested that infections early in life might prevent allergic rhinoconjunctivitis in adults (Strachan 1989). This formed the basis of the so called "hygiene hypothesis". The term is potentially misleading, as it would appear to question the enormous gains in Public Health caused by improved hygiene and vaccination programs. It also soon became obvious that a reduction in the number of respiratory infections in certain regions could not explain the global increase in allergic diseases.

Even if altered patterns of respiratory infections cannot explain the rise in allergic disease, changes in exposure to other microbial agents that stimulate the immune system early in life, seem to do so. The gut microbiota exert effects beyond the mucosal microenvironments and influences systemic precursor compartments, such as bone marrow and thymus (Hooper et al. 2012). The underlying mechanism(s) are likely to include stimulation of functional maturation of cells within the innate and adaptive immune systems during the early postnatal period. This process may ultimately determine the overall efficiency of immune/tolerance induction during early life, with major flow-on effects into adulthood. A full understanding of the underlying mechanisms may therefore open new venues for the prevention of allergies and other immunologically mediated diseases by modification of the gut microbiota. Thus, not only local disease manifestations, such as food allergy would be modified, but conceivably also diseases with manifestations at distant sites, such as respiratory allergies.

This chapter will review the current knowledge of the role of probiotics and prebiotics in the treatment and prevention of asthma and other manifestations of allergy. Initially, the immunological background and rationale for exploring a potential role for pro- and prebiotics in allergic disease will be briefly discussed. However, the main focus is on published evidence of the clinical effects of these products.

Immunological Background

The early T-cell responses are subject to a variety of postnatal regulatory mechanisms, which are driven by exposure of the infant immune system to environmental antigen. Cross-sectional and prospective studies indicate that in atopic children, consolidation of Th2-polarised immunity against inhalant allergens is initiated in early infancy and may be completed in the preschool years in children who do not develop clinically manifest allergy (Böttcher et al. 2006). However, a number of observations challenge the notion that allergy is the simple result of Th2 polarisation. For example, stronger Th1 responses to allergens that have also been noted in allergic children (Prescott et al. 1999, Ng et al. 2002, Thornton et al. 2004). The observations suggest that increased inappropriate reactivity to allergens is a result of failure of underlying regulatory pathways. There is also evidence that these pathways may be under environmental influence. For example, prospective studies from Estonia, with a low, and Sweden with a high prevalence of allergy, indicate that the regulatory mechanisms are established more rapidly in Estonia (Böttcher et al. 2006, Fagerås et al. 2011).

Studies investigating the relationship between early childhood infection and the risk of allergy and asthma have been inconsistent, or difficult to interpret. The immunological effects of microbial agents differ with the type of infectious agent and the site of infection. For example, only infections in the gastrointestinal tract appear to be protective (Matricardi et al. 2003). Furthermore, non-pathogenic colonizing organisms are likely to play a central role in immune development (Björkstén 2009). A large Danish national birth cohort study including more than 24,000 mother-child pairs found that respiratory infections early in life do not protect from atopic dermatitis (Benn et al. 2004). However, in that study we observed that other environmental factors, sometimes taken for indirect markers of microbial exposure (such as early day-care attendance, having three or more siblings, farm residence, and pet keeping) were all protective. It therefore seems that that these protective environmental factors are not due to fewer infections, but have other explanations. For example, the inverse relationship between number of older siblings and allergy risk may be due to altered maternal immunity as a consequence of repeated pregnancies and exposure to animals and could possibly be explained by high zone tolerance induction. Of relevance is also that there are observed differences in the composition of the gut microbiota between four-year old children with and without older siblings (Sjögren et al. 2009). This highlights the emerging concept that overall "microbial burden", rather than specific infections may be more relevant in early life. In this respect, gut microbiota are a more likely major source than the considerably less diverse microbial exposure in the respiratory tract.

Microbial Ecology

The intestinal tract performs many different functions. In addition to absorption and digestion, it is also the body's largest organ of host defence. Part of the intestinal mucosal barrier function is formed by a common mucosal immune system, which provides communication between the different mucosal surfaces of the body (Hooper et al. 2012). The total mucosal surface area of the adult human gastrointestinal tract is up to $300m^2$, making it by far the largest body area interacting with the environment. It is colonised with over 10^{14} micro-organisms, weighing over 1 kg and corresponding to more than ten times the total number of cells in the body (Turnbaugh et al. 2007).

Our gut microbiota can be pictured as a microbial organ placed within a host organ (Bäckhed et al. 2007). It is composed of different cell lineages with a capacity to communicate with one another and the host. The gut microbiome contains >100 times the number of genes in our genome and endows us with functional features that we have not had to evolve ourselves (Turnbaugh et al. 2007).

The gastrointestinal tract of the newborn baby is sterile. Soon after birth, however, it is colonised by numerous types of micro-organisms. Colonisation is complete after approximately one week, but the numbers and species of bacteria fluctuate markedly during the first months of life (Rautava et al. 2012). There is a continuous interaction between the microbial flora and the host, comprising a dynamic eco system that, once established, is surprisingly stable under normal conditions (Zoetendal et al. 1998). Environmental changes, e.g., a treatment period with antibiotics, only temporarily change the composition of the microbiota. A study of adult monozygotic twins living apart and their marital partners has emphasised either the potential dominance of host genotype over diet in determining microbial composition of the gut microbiota (Zoetendal et al. 2001), alternatively the significance of early life environment.

Microbial colonisation of the gastrointestinal tract, linked with lifestyle and/or geographic factors, may be important determinants of the heterogeneity in disease prevalence throughout the world (Björkstén 2009) and cohort studies have addressed this complex question. There are now numerous studies demonstrating differences in the composition of the gut microbiota between allergic and non-allergic individuals and between infants living in countries with a high and a low prevalence of allergy and between healthy and allergic infants (summarised in Björkstén 2005).

The clinical studies on microbial ecology published so far only indicate that there are geographic differences in the composition of gut microbiota and that there have been pronounced changes over the past 40–50 years in affluent countries with a market economy. However, very little is known about which changes are significant with regard to human health in general and immune regulation in particular. The reason is that until very recently, any ecological studies relied on rather crude, time consuming conventional isolation of bacteria on various media. Recent progress allowing the analysis of bacterial DNA and powerful statistical methods borrowed from analyses of gene expression and epigenetic analyses of relevant patient populations, will allow a better analysis and understanding of the complex microbial interactions in our gut, as well as microbe-host interactions.

Gut Microbiota and Immune Regulation

Epidemiological studies and experimental research suggest that the microbial environment and exposure to microbial products in infancy modify immune responses and enhance immune regulation and tolerance to ubiquitous antigens. The gut microbiota are the quantitatively most important source of microbial stimulation and may provide a primary signal for driving the postnatal maturation of the immune system and the development of a balanced immunity (Hooper et al. 2012). Thus, there is mounting evidence that commensal microbes acquired during the early postnatal period are required for the development of tolerance, not only to themselves, but also to other antigens. For example, Th2-mediated immune responses are not susceptible to oral tolerance induction in germ free mice (Sudo et al. 1997). Oral tolerance was only induced after the introduction of components of the normal microbiota.

Microbes activate the immune system through a range "pattern recognition receptors" system (Toll-like receptors, TLR). It is also recognised that interaction with the normal microbial flora of the gastrointestinal tract is the principal environmental signal for postnatal maturation of T-cell function (in particular the Th1 component) (Demengeot et al. 2006). Although TLR are found principally on cells of the innate immune system (including granulocytes, monocytes, and natural killer cells), they are also present on cells involved in programming and regulating "adaptive" immune responses (such as APC and regulatory T cells). It has been proposed that early microbial activation of both APC and regulatory T cells may promote Th1 maturation and play an important role in reducing the risk of Th2 mediated allergic responses (Wills-Karp et al. 2001). This is supported by animal studies, demonstrating that bacterial lipopolysaccharide (LPS) endotoxin exposure can prevent allergic sensitisation if given before allergic responses are established (Blumer et al. 2005). These effects may be of greater significance in genetically susceptible individuals who appear to have weaker Th1 responses in the perinatal period (Prescott et al. 1999). Genetic studies also support a role for the CD14/LPS (Baldini et al. 1999) and TLR (Eder et al. 2004) pathways in the development of allergic disease.

In animals, bacterial antigens (mycobacteria) have been used successfully to modify allergic inflammation in sensitised animals with evidence that effects are mediated by TGFβ and IL-10 producing regulatory T cells (Zuany-Amorim et al. 2002). Furthermore, supplementation with probiotic bacteria has also been shown to induce regulatory populations (Di Giacinto et al. 2005). There are also preliminary reports that bacteria may affect regulatory immune function in humans, with an increase in the *in vitro* production of regulatory cytokines (IL-10) after probiotic ingestion (Lammers et al. 2003). In addition, significant correlations between colonisation with bifidobacteria species have been observed in the first six months of life with the level of allergen-associated regulatory activity detected as FoxP3 expression in response to allergen stimulation (Martino et al. 2008, Martino and Prescott 2012).

A number of studies have suggested differences in colonization patterns between allergic and non-allergic children (reviewed in Björkstén 2009). Interestingly, prospective studies have shown that these differences were already apparent already at one week of age, i.e., well before the infants had developed any allergic manifestations, suggesting that early colonization can influence subsequent patterns of immune development. Observed differences include higher microbial counts of gram positive bacteria in neonates who do not develop allergic manifestations, less clostridia and a higher prevalence of bifidobacteria through the first year of life (Björkstén et al. 2001). These differences are present at least during the first five years of life (Sepp et al. 2005). Interestingly, similar differences were noted when comparing the gut microbiota in healthy one-year old infants living in two countries with a low and high prevalence of allergy (Estonia and Sweden) (Sepp et al. 1997). It was noted that the gut microbiota in Estonia in many respects was similar to that described in Western Europe in the early 1960's, before the emergence of the major difference in allergy prevalence between Eastern and Western Europe. As already mentioned, studies in germfree animals confirm that a microbial gut flora is essential for the development of oral tolerance and for the induction of normal immune regulation (Moreau et al. 1995, Björkstén 2005). The controversy regarding the role of gut bacteria in allergy development thus lies in the clinical consequences of these clinical and experimental findings and not as much to what extent they affect the immune system.

Probiotics and Allergy Treatment

Over the past 15 years several studies have explored the capacity of various probiotic strains to ameliorate allergic symptoms in patients suffering from allergic diseases. Several strains of probiotic bacteria have been tried both for treatment of clinical manifestations of allergy, such as eczema, asthma, hay fever and food allergy. The outcome of the studies varies considerably. In general terms, the results have been more encouraging in infant and young children than in older children and adults. The early clinical studies in infants were analysed in a Cochrane review 2007 (Osborn and Sinn 2007). The conclusion was that there was insufficient evidence to recommend the addition of probiotics to infant feeds for prevention of allergic disease or food hypersensitivity. Although there was a reduction in clinical eczema in infants, this effect was not consistent between studies and caution is advised in view of methodological concerns regarding included studies.The studies are limited to three species of lactobacilli, i.e., *Lactobacillus rhamnosus, fermentum* and *reuteri* and *Bifidobacterium lactis Bb-12*. In 2009, Boyle et al. reviewed clinical studies on eczema treatment, meeting defined quality criteria (Boyle et al. 2009). The conclusion of the meta-analysis of twelve trials was that there was no significant reduction in eczema symptoms with probiotic treatment compared with placebo (mean difference –0.90 points on a 20-point visual analogue scale; 95% confidence interval –2.84, 1.04). Meta-analysis of data from seven trials showed no significant difference in investigator rated eczema severity between probiotic and placebo treatments. Subgroup analysis by eczema severity or presence of atopy did not identify a specific population in which probiotic treatment was effective. There was significant heterogeneity between studies, however, the results of three studies that used the same probiotic strain were concordant.

The results of clinical studies limited to infants are slightly more encouraging (Table 1). Two small placebo-controlled studies from Finland were the first to report beneficial effects of probiotics on infantile eczema (Majamaa and Isolauri 1997, Isolauri et al. 2000). In the first study *Lactobacillus rhamnosus* GG was used and in the second study this strain was combined with *Bifidobacterium lactis Bb-12*. In addition to improved eczema scores, laboratory parameters were also affected. In a Danish study (Rosenfeldt et al. 2003), treatment with a combination of *Lactobacillus reuteri* and *L. rhamnosus* was associated with reduced extent of eczema, particularly in the subgroup of infants who also had a positive skin prick test. There was also a decrease in serum eosinophil cationic protein (ECP), but no significant changes in the production of the cytokines IL-2, IL-4, IL-10, or IFNγ.

In a study, a larger cohort comprising 230 infants (aged around 6 months) with atopic eczema/dermatitis syndrome (AEDS) and suspected cow's milk allergy were treated with the same probiotic strain (*Lactobacillus GG*), a

Table 1. Summary of placebo-controlled probiotic treatment studies for eczema with and without allergy in infants and children.

Investigators	Study population	Age	N=	Treatment Duration	Probiotics used	Outcomes and comments
Majamaa and Isolauri 1997	Infants with mild Eczema + CMA	<12 mon	27	1 mon	L GG	Improved SCORAD. Lower faecal α1-AT levels
Isolauri, Arvola et al. 2000	Breast fed babies with mild eczema	Mean 4.6 mon	27	2 mon	*L GG &* *B lactis, Bb-12*	Improved SCORAD by both probiotics. Reduced serum CD4 and urinary EPX
Rosenfeldt, Benfeldt et al. 2003	Eczema	1–13 yr	43	6 wl	*L rhamnosus & L.reuteri*	Reduced extent of eczema (p=0.02), particularly in sensitised children
Viljanen, Savilahti et al. 2005	Eczema and Suspected CMA	<12 mon	230	4 wk	*L GG* or Mixture of 4 strains	Reduced SCORAD by *L.GG* in IgE-sensitised infants
Weston, Halbert et al. 2005	Moderate or severe AD	6–18 mon	53	8 wk	*L fermentum*	Reduced SCORAD i(p=0.03) improvement (93% vs. 65%, p=0.01
Brouwer, Wolt-Plompen et al. 2006	Eczema	<5 mon	50	3 mon	*L GG* or *L rhamnosus*	No difference as compared with placebo
Sistek, Kelly et al. 2006	Eczema	1–10 yr	59	12 wk	*L GG &* *B lactis Bb-12* 2 x10^{10}*cfu*	Reduced SCORAD in food sensitive children
Giovannini, Agostoni et al. 2007	Asthma. RC	2–5 yr	156	12 mon	*L casei* 10^{8} *cfu*	No asthma prevention

CMA= Cow's milk allergy, RC = Rhinoconjuctivitis, SCORAD = SCORing index Atopic Dermatitis mixture of four probiotic strains, or placebo for four weeks (Viljanen et al. 2005). Beneficial clinical effects of the probiotics were seen in this study only in children with evidence of allergic sensitization and not in children with atopic dermatitis but no sensitization. This suggests that atopic dermatitis is a heterogeneous condition and that the effect of immune modifying agents, such as probiotics, will depend on the pattern of disease. Paired pre- and post-treatment plasma samples were analysed for concentrations of IL-2, IL-4, IL-6, IL-10, TNFα, IFNγ, soluble intercellular adhesion molecule 1, soluble E-selectin, TGF-β1, TGF-β2, and C-reactive protein (Viljanen et al. 2005). In infants with IgE-associated eczema, treatment with *Lactobacillus* GG (LGG) induced higher C-reactive protein levels than in the placebo group (P = .021). The IL-6 levels also increased after treatment and soluble E-selectin levels were higher after probiotic than after placebo treatment in infants with IgE-mediated CMA. Furthermore, faecal levels of α1-AT decreased in infants receiving lactobacilli, thus confirming the results of a previous study with the same micro-organism (Majamaa and Isolauri 1997).

In an Australian study, treatment with a strain of *Lactobacillus fermentum* (given at a dose of 10^9 colony forming units, cfu, twice daily) improved infantile eczema as assessed after 8 weeks (Weston et al. 2005). This study included infants with more severe eczema than any of the previously cited studies. The reduction in the SCORAD (SCORing index Atopic Dermatitis, a scoring system by which the extension and severity of eczema is quantitated) was significant in the probiotic, but not placebo group. Furthermore, significantly more children receiving probiotics had a SCORAD that was better than baseline at week 16 (93% vs. 63% in the placebo group, p=0.01). Interestingly, probiotic administration was associated with increased polyclonal Th1 IFNγ responses in the infants and the improvement in atopic dermatitis was directly proportional to the increase in IFNγ responses to Staphylococcus enterotoxin B (r=0.445, *P*=0.026) (Prescott et al. 2005). Increased IFNγ responses by probiotics have also recently been observed in infants treated for cow´s milk allergy (Pohjavuori et al. 2004).

In a study from New Zealand, the effect of two probiotics (*Lactobacillus rhamnosus GG* and *Bifidobacterium lactis*) given at a high dose ($2x10^{10}$) was studied in 59 children with established AEDS (Sistek et al. 2006). Although there was no significant difference between the probiotic and placebo groups after 12 weeks, a significant improvement was noted in the subgroup of food allergic children receiving probiotics.

A Dutch study could not document any beneficial effects of probiotics on infant eczema (Brouwer et al. 2006). After 4–6 weeks of baseline and double-blind, placebo-controlled challenges for diagnosis of cow's milk allergy (CMA), infants less than 5 months old with AD received a hydrolysed whey-based formula as placebo (n = 17), or supplemented with either *Lactobacillus rhamnosus* (n = 17) or *Lactobacillus GG* (n = 16) for

3 months. Before, during and after intervention, the clinical severity of AD was evaluated using SCORAD. Allergic sensitization was evaluated by measurement of total IgE and a panel of food-specific IgE antibodies, as well as skin prick testing for cow's milk. Inflammatory parameters were blood eosinophils, eosinophil protein X in urine, faecal alpha-1-antitrypsin and production of IL-4, IL-5 and IFNγ by peripheral blood mononuclear cells after polyclonal stimulation. There was no statistically significant effect of probiotic supplementation on SCORAD, sensitization, inflammatory parameters or cytokine production between groups. Only four infants were diagnosed with CMA, however, in this rather small study.

Thus, despite the conclusions of the meta-analysis (Osborn and Sinn 2007), most of the studies in infants in which probiotics were assessed for the treatment of eczema showed some beneficial effects, at least in subgroups of infants with documented allergy. However, the treatment was only associated with a small or modest reduction in symptoms. Furthermore, most of the studies were small, and mostly included only infants with mild eczema. In three of the studies there were also recorded effects on laboratory parameters, e.g., increased polyclonal IFNγ responses (Weston et al. 2005) or faecal chemokines (Majamaa and Isolauri 1997, Viljanen et al. 2005), lower serum ECP levels (Rosenfeldt et al. 2003), lower urinary eosinophil protein X and serum CD4 (Isolauri et al. 2000) and stabilised mucosal barrier function (Rosenfeldt et al. 2004).

There are also studies comprising older children and adults, in which probiotics have been tried as treatment of respiratory allergies. The results so far are not encouraging. In one study, 36 teenagers and young adults with pollen allergy were randomised to a 5.5-month treatment with *Lactobacillus rhamnosus* or placebo (Helin et al. 2002). The treatment had no effect on seasonal symptoms, or on the outcome of provocations. In contrast, in a study of 80 adults with perennial rhinitis a strain of *Lactobacillus paracasei* for 30 days was reported to slightly, although significantly, reduce frequency and severity of symptoms (Wang et al. 2004). There is also one large study in children, in which asthma and rhinoconjunctivitis was treated with a strain of *L. casei* (Giovannini et al. 2007). As for most studies in adults, the outcome was negative.

A strain of *Bifidobacterium longum* has been tried as treatment of Japanese Cedar pollinosis (JCP). In one randomized, double-blind trial, 44 patients received probiotic bacteria or placebo for 13 weeks during the pollen season (Xiao et al. 2006). The treatment was associated with decreases in rhinorrhea, nasal blockage and composite scores. The same authors also tried the same strain given in yoghurt for 14 weeks to patients with JCP (Wang et al. 2004). Slightly less eye symptoms were reported in the treatment as compared to the placebo group, but most differences did not reach statistical significance.

The group also reported similar results on ocular symptoms in a placebo-controlled study with cross-over design (Xiao et al. 2007).

Probiotics and Allergy Prevention

It is logical from an immunological standpoint to explore the benefits of probiotics very early in life when immune responses are still developing, and there are now a number of studies addressing the role of probiotics in primary allergy prevention (Table 2). The first study to assess the role of probiotics in this context administered *Lactobacillus rhamnosus* to mothers (starting 2–4 weeks before delivery) and to infants in the first 6 months of life. This was reported to reduce the incidence of eczema at 2 years by around 50% (Kalliomäki et al. 2001). Although the cumulative effect on eczema was still evident at 4 years, there was no reduction in respiratory allergy, IgE levels or allergic sensitisation (Kalliomäki et al. 2003). This was confirmed by a follow-up at seven years (Kalliomäki et al. 2007). Effects on underlying immune response were not reported and a number of methodological concerns have been raised about the study (Matricardi et al. 2003). A major concern was that many of the children (28 out of 64) included in the probiotic supplement group did not receive probiotics directly, as the supplement was given to the mother if babies were breastfed. These issues have made the results difficult to interpret.

There are at least ten other published studies performed in Australia, Germany, The Netherlands, New Zealand, Norway, Singapore, Sweden and United Kingdom, in which the potential to prevent the development of allergic disease by probiotics was tested (Table 2). Three strains of lactobacilli and four strains of bifidobacteria in doses ranging from 10^8 to 10^9 cfu daily were employed. An Australian study, using a *Lactobacillus acidophilus* strain, failed to show any reduction in allergic disease despite changes in colonization (Taylor et al. 2007). Rather, there was a concerning increase in sensitization and in IgE-associated atopic eczema. In this study, the treatment was started after birth, while in the all but two of the studies summarised in Table 2, the pregnant women were given the probiotic during the last month of gestation, in addition to the postnatal treatment of the babies.

In a Swedish study, *Lactobacillus reuteri* was given to pregnant mothers during the last four weeks and then daily to the infants for one year (Abrahamsson et al. 2007). The incidence of eczema and other potentially allergic manifestations were similar in the treatment and placebo groups. However, subgroup analyses showed that the probiotic treatment was associated with less IgE-associated atopic eczema during the second year of life. Furthermore, in the infants with atopic mothers there was a

Table 2. Summary of prevention studies in infants using strains of *Lactobacillus*, *Bifidobacterium* and/or *Proprionibacterium*.

		Study protocol				Outcomes			
Investigators, country (reference)	**n**	**organism(s) and dosage (Cfu)**	**Prenatal treatment**	**Postnatal (duration)**	**Follow-up**	**Less eczema**	**Less sensitisation**	**Reduction in other AD**	**Comment**
Kalliomäki et al. 2001, 2003, 2007	132	*LGG* 10^{10}	2–4 w	6 mon	7 yr	Yes	No	No	To mother if BF
Taylor et al. 2007; Jensen et al. 2012	189	*L. acidophillus* 3×10^8	No	6 mon	5 yr	No	No	No	More often sensitised
Abrahamsson et al. 2007	188	*L. reuteri* 10^8	2–4 w	12 mon	2 yr	IgE associated	subgroup with atopic mothers	subgroup with atopic mothers	
West et al. 2009	171	*L. paracasei*	No	From age 4 mon	13 mon	11% vs. 22% $p<0.05$	No	No	
Kopp et al. 2008	94	*LGG*10^{10}	4–6 w	6 mon	2 yr	No	-	No	To mother if BF
Niers et al. 2009	102	*B. bifidum*, *B. lactis*, *Lactococcus lactis*		12 mon	2 yr	Yes	-	-	
Wickens et al. 2012	425	*L. rhamnosus* $9x10^9$ or *B. animalis* $6x10^9$	4–6w	2 yr	4 yr	OR 0.57	-	OR 0.38 for RC	
Soh et al. 2009	245	*B longum* 10^7 & L rhamnosus 10^9	No	6 mon	12 mon	No	No	No	
Kim et al. 2010	68	*L. acidophilus*,*B. lactis*, *B. bifidum* 10^9	4–8 wk	6 mon	12 mon	Yes	No	No	To mother if BF
Dotterud et al. 2010	278	*LGG*, *B. lactis* $5x10^{10}$ *L. acidophilus* $5x10^9$	4 wk	3 mon	2 yr	Yes	No	No	To mother if BF

AD= Allergic Disease, BF= breast feeding, Cfu= Colony forming units, RC= Rhinoconjuctivitis LGG= *L. rhamnosus GG* reduction not only in IgE associated eczema and respiratory allergy, but also in the prevalence of allergen-specific IgE antibodies. In contrast, there was no effect of the treatment in infants who only had paternal allergy. This observation is interesting as the levels of IL-10 were higher and TGFβ lower in colostrum of mothers who had eaten probiotics during the last month of pregnancy (Böttcher et al. 2008). A follow-up at seven years of age did not confirm any preventive effect on respiratory allergy, however (Abrahamsson et al. 2013).

Despite all of the immunomodulatory effects described in experimental models, so far none of these studies has shown any clear effect preventive sensitization on any allergic disease other than eczema. Possible explanations for the varied results in the treatment and prevention studies include differences in the bacterial strains used, host factors that could influence microbial responsiveness and allergic propensity, and other environmental factors that could influence colonization or immune development. First, there are significant variations in the strains claimed to be probiotic. Second, it is of note that in five of the six studies suggesting at least some preventive effects, supplementation was started in pregnancy, whereas in the study that showed increased sensitisation (Taylor et al. 2007) probiotics were only given after birth. This may indicate that the supplementation to the mothers in late pregnancy is of particular importance. In light of this, it is of interest to note that in one study the levels of IL-10 was higher and TGFβ lower in colostrum of mothers receiving a probiotic as compared to placebo treated mothers (Böttcher et al. 2008). Third, there are differences in host susceptibility to microbial influence and to colonisation with a particular strain of bacteria. Functional genetic polymorphisms in microbial recognition pathways are well described (including TLR), and it is likely that this could result in individual variation in the effects of probiotics. Similarly there is some heterogeneity in the level of allergic risk in the study groups. Fourth, there are likely to be many environmental factors that influence both colonisation (such as maternal microbiota and other sources of microbial exposure, delivery method, antibiotics, and prebiotics in the diet and general microbial burden) and immune development. It is quite conceivable that administration of a certain strain may affect microbial ecology in one environment, but not in another. For example, in the Swedish study of allergy prevention, the probiotic *Lactobacillus reuteri* was isolated at least once in 12% of the mothers and the infants belonging to the placebo group. Thus the strain, which was originally isolated from breast milk, is a transient component of the gut microbiota, at least in Scandinavia. All of these factors are likely to make robust meta-analyses problematic to perform as more studies are completed.

Table 3. Summary of prevention studies in formula fed infants using mixtures of prebiotic oligosaccharides, alone or in combination with probiotics.

	Study protocol				Outcomes			
Reference	**Population Characteristics Study Group (n=)**	**Probiotics**	**Treatment**	**Follow-up, Age**	**Less eczema**	**Less sensitization**	**Reduction in other ARD**	**Comment**
(Arslanoglu et al. 2008)	FH allergy N=134	No	6 mo	2 yr	28 vs. 14% P<0.05	-	Urticaria 10 vs. 1.5%, p<0.05	Less wheeze
(Moro et al. 2006)	FH allergy N=206	No	6 mo	6 mon	10 vs. 23% P=0.01	-	No	Less wheeze
(Gruber et al. 2010)	No FH N=735	No	6 mo	12 mo	5.7% vs 9.7% P=0.04	No	No	
(Kukkonen et al. 2007, Kuitunen et al. 2009)	FH allergy N= 925	*L. rhamnosus GG & LC705 & B. breve & Proprionibact. freudenreichii* 2–5x10^9 cfu	(2–4 wk)* + 6 mon	5 yr	Yes at 2 yr No at 5 yr	No	No	Less ARD in CS group
(van der Aa et al. 2011)	Eczema infants N=75	*B. breve* 1.3x10^9 cfu	12 wk	8 mon	No	No	No	Less wheeze

Duration of prenatal treatment, ARD= Allergy Related Disease, CS= Born through Caesarean Delivery, FH= Family History

Prebiotics and Synbiotics

It would appear unlikely that supplementation with a single probiotic strain would be sufficient to have a major influence on the very diverse intestinal microbiota and the complex interaction between the gut bacteria and the host. This has led to an interest in dietary substrates that could have a more global effect on gut microbiota, namely prebiotics (non-digestible, fermentable oligosaccharides which stimulate the growth of allegedly "beneficial" bacteria, particularly of *Bifidobacterium* species). Altering the intake of foods containing these products could conceivably directly influence the composition and activity of intestinal microbiota. This could hypothetically explain some of the allergy protective effects of grains, cereals, citrus fruits and other food items that have been observed in some epidemiologic studies.

Whereas the probiotic approach adds only one or a few strains to a large spectrum of hundreds of species in the gut microbiota, the prebiotic approach aims at fertilization of the intestinal ecosystem. As various oligosaccharides are abundant in human milk and they seem to enhance the growth of *Bifidobacterium* species in infants and as these bacteria have been suggested to represent a "beneficial" gut flora, it would appear logical to employ a broader approach to administer oligosaccharides to bottle-fed babies, rather than merely supplementing with one or two strains of microbes. The concept of "beneficial bacteria" is poorly defined, however, and not clinically documented. It should also be noted that to date, as discussed previously, bacterial strains showing efficacy belong to the *Lactobacillus* family and not *Bifidobacterium*.

To date, there is limited support for the efficacy of prebiotics in relation to allergy management. As bifidobacteria are supposed to be particularly important in infancy and their growth is stimulated by breast milk which is a rich source of oligosaccharides, it is logical that the approach was studied by the same group of researchers testing the same oligosaccharide mixture (Moro et al. 2006, Arslanoglu et al. 2008, Gruber et al. 2010). The oligosaccharides tested were different from those present in human milk, though. In summary, the studies all reported a lower incidence of eczema among infants receiving the oligosaccharide-containing formula. Interestingly, in two of the studies, a lower incidence of wheezing associated with respiratory infections was also reported.

Combinations of oligosaccharides and probiotic bacteria ("symbiotic") have been tried in the treatment of eczema in infants (van der Aa et al. 2011, van der Aa et al. 2012) and as a possible means for prevention of allergic manifestations (Kukkonen et al. 2007, Kuitunen et al. 2009). Treatment of atopic dermatitis in infants with a symbiotic mixture of *Bifidobacterium breve* and galacto- and fructooligosacharides did not reduce the severity of eczema, except in the subgroup of infants with IgE associated eczema (van der Aa et al. 2010). The mixture did however reduce the likelihood of subsequent need for asthma medication (van der Aa et al. 2011), although no other immunomodulatory effects were noted.

The second study in which synbiotics were used showed a reduction in atopic eczema at age two, but no effects on sensitization or other allergic disease (Kukkonen et al. 2007). A mixture of four probiotic bacteria was given to the mothers during the last 2–4 weeks of pregnancy and then to the babies for six months. The statistical power was high in the study, since over 900 infants participated in the follow-up at two years. Probiotic treatment compared with placebo showed no effect on the cumulative incidence of allergic diseases but tended to reduce IgE-associated (atopic) diseases (odds ratio [OR], 0.71; 95% CI, 0.50–1.00; P = .052). The treatment also reduced eczema (OR, 0.74; 95% CI, 0.55–0.98; P = .035) and atopic eczema (OR, 0.66; 95% CI, 0.46–0.95; P = .025). The children were subsequently followed up at five years (Kuitunen et al. 2009). At that age the prevalence of eczema was similar in children receiving the symbiotic mixture and placebo. Interestingly, there was a significantly lower prevalence of IgE associated allergic manifestations in those who had received synbiotics during the first six months of life (24.3% vs. 40.5%, p=0.035).

Concluding Comments

While there is a sound theoretical basis for anticipating benefits of probiotic supplementation in allergic disease, there is currently insufficient data to recommend this as a part of standard therapy in allergic conditions in general, or for prevention. Although there has been promise in atopic dermatitis, it is generally accepted that more studies are needed to confirm this, and that any benefits are likely to be modest. However, faced with the stress and severe discomfort that can be associated with atopic dermatitis, many families are choosing to try probiotics in conjunction with their prescribed products.

So far, the most encouraging effects have been reported for the treatment and in particular prevention of infant eczema, the latter provided that not only the babies, but also the pregnant mother is treated. The fact that all beneficial effects are most pronounced in the young, rather than in adults are not surprising, given the role of the gut microbiota for the development of normal immune function and the fact that once established early in life the individual gut microbiota are surprisingly stable. Although probiotics are part of a normal gut microbiota and not associated with any serious adverse effects, further studies are needed to determine the significance of the increased rates of sensitisation associated with the use of probiotics in some prevention studies.

References

Abrahamsson, T.R., T. Jakobsson et al. 2007. Probiotics in prevention of IgE-associated eczema: a double-blind, randomized, placebo-controlled trial. J. Allergy Clin. Immunol. 119(5): 1174–1180.

Abrahamsson, T.R., T. Jakobsson, B. Bjõrkstén, G. Oldaeus and M.C. Yenmalin. 2013. No effect of probiotic on respiratory allergies: a seven-year follow up of a randomised control trial in infancy. Pediatric Allergy Immunol. 24: 556–561.

Arslanoglu, S., G.E. Moro et al. 2008. Early dietary intervention with a mixture of prebiotic oligosaccharides reduces the incidence of allergic manifestations and infections during the first two years of life. J. Nutr. 138(6): 1091–1095.

Baldini, M., I.C. Lohman et al. 1999. A polymorphism in the 5′- flanking region of the CD14 gene is associated with circulating soluble CD14 levels with total serum IgE. Am. J. Resp. Cell Mol._Biol. 20: 976–983.

Benn, C.S., M. Melbye et al. 2004. Cohort study of sibling effect, infectious diseases, and risk of atopic dermatitis during first 18 months of life. BMJ 328: 1223–1230.

Björkstén, B. 2005. Genetic and environmental risk factors for the development of food allergy. Curr. Opin. Allergy. Clin. Immunol. 5(3): 249–253.

Björkstén, B. 2009. Disease outcomes as a consequence of environmental influences on the development of the immune system. Curr. Opin. Allergy Clin. Immunol. 9: 185–189.

Björkstén, B., E. Sepp et al. 2001. Allergy development and the intestinal microflora during the first year of life. J. Allergy Clin. Immunol. 108(4): 516–520.

Blumer, N., U. Herz et al. 2005. Prenatal lipopolysaccharide-exposure prevents allergic sensitisation and airway inflammation, but not airway responsiveness in a murine model of experimental asthma. Clin. Exp. Allergy 35: 397–402.

Boyle, R.J., F.J. Bath-Hextall et al. 2009. Probiotics for the treatment of eczema: a systematic review. Clin. Exp. Allergy 39(8): 1117–1127.

Brouwer, M.L., S.A. Wolt-Plompen et al. 2006. No effects of probiotics on atopic dermatitis in infancy: a randomized placebo-controlled trial. Clin. Exp. Allergy 36: 899–906.

Bäckhed, F., R. Ley et al. 2007. Host-Bacterial Mutualism in the Human Intestine. Science 307: 1915–1920.

Böttcher, M.F., T.R. Abrahamsson et al. 2008. Low breast milk TGF-beta2 is induced by *Lactobacillus reuteri* supplementation and associates with reduced risk of sensitization during infancy. Pediatr Allergy Immunol. 19: 497–504.

Böttcher, M.F., M.C. Jenmalm et al. 2006. Cytokine responses to allergens during the first 2 years of life in Estonian and Swedish children. Exp. Allergy 36(5): 619–628.

Demengeot, J., S. Zelenay et al. 2006. Regulatory T cells in microbial infection. Springer Semin. Immunopathol. 28: 41–50.

Di Giacinto, C., M. Marinaro et al. 2005. Probiotics ameliorate recurrent Th1-mediated murine colitis by inducing IL-10 and IL-10-dependent TGF-beta-bearing regulatory cells. J. Immunol. 174: 3237–3246.

Eder, W., W. Klimecki et al. 2004. Toll-like receptor 2 as a major gene for asthma in children of European farmers. J. Allergy Clin. Immunol. 113: 482–488.

Fagerås, M., S. Tomicic et al. 2011. Slow salivary secretory IgA maturation may relate to low microbial pressure and allergic symptoms in sensitized children. Pediatr Res. 70: 572–577.

Gerrard, J., C. Geddes et al. 1976. Serum IgE levels in white and metis communities in Saskatchewan. Ann. Allergy 37: 91–100.

Giovannini, M., C. Agostoni et al. 2007. A randomized prospective double blind controlled trial on effects of long-term consumption of fermented milk containing *Lactobacillus casei* in pre-school children with allergic asthma and/or rhinitis. Pediatr Res. 62: 215–220.

Gruber, C., M. van Stuijvenberg et al. 2010. Reduced occurrence of early atopic dermatitis because of immunoactive prebiotics among low-atopy-risk infants. J. Allergy Clin. Immunol. 126(4): 791–797.
Helin, T., S. Haahtela et al. 2002. No effect of oral treatment with an intestinal bacterial strain, *Lactobacillus rhamnosus* (ATCC 53103), on birch-pollen allergy: a placebo-controlled double-blind study. Allergy 57(3): 243–246.
Hooper, L.V., D.R. Littman et al. 2012. Interactions between the microbiota and the immune system. Science 336(6086): 1268–1273.
Isolauri, E., T. Arvola et al. 2000. Probiotics in the management of atopic eczema. Clin. Exp. Allergy 30: 1604–1610.
Kalliomäki, M., S. Salminen et al. 2001. Probiotics in primary prevention of atopic disease: a randomised placebo-controlled trial. Lancet 357(9262): 1076–1079.
Kalliomäki, M., S. Salminen et al. 2003. Probiotics and prevention of atopic disease: 4-year follow-up of a randomised placebo-controlled trial. Lancet 361: 1869–1871.
Kalliomäki, M., S. Salminen et al. 2007. Probiotics during the first 7 years of life: a cumulative risk reduction of eczema in a randomized, placebo-controlled trial. J. Allergy Clin. Immunol. 119: 1019–1021.
Kuitunen, M., K. Kukkonen et al. 2009. Probiotics prevent IgE-associated allergy until age 5 years in cesarean-delivered children but not in the total cohort. J. Allergy Clin. Immunol. 123(2): 335–341.
Kukkonen, K., E. Savilahti et al. 2007. Probiotics and prebiotic galacto-oligosaccharides in the prevention of allergic diseases: a randomized, double-blind, placebo-controlled trial. J. Allergy Clin. Immunol. 119(1): 192–198.
Lammers, K.M., P. Brigidi et al. 2003. Immunomodulatory effects of probiotic bacteria DNA: IL-1 and IL-10 response in human peripheral blood mononuclear cells. FEMS Immunol. Med. Microbiol. 38(2): 165–172.
Majamaa, H. and E. Isolauri. 1997. Probiotics: a novel approach in the management of food allergy. J. Allergy Clin. Immunol. 99: 179–185.
Martino, D.J., H. Currie et al. 2008. Relationship between early intestinal colonization, mucosal immunoglobulin A production and systemic immune development. Clin. Exp. Allergy 38(1): 69–78.
Martino, D.J. and S.L. Prescott. 2012. Progress in Understanding the Epigenetic Basis for Immune Development, Immune Function, and the Rising Incidence of Allergic Disease. Curr. Allergy Asthma Rep.
Matricardi, P., B. Björkstén et al. 2003. Microbial products in allergy prevention and therapy. Allergy 58: 461–471.
Moreau, M.C., M. Coste et al. 1995. Oral tolerance to ovalbumin in mice: effect of some paramters on the induction and persistence of the suppression of systemic IgE and IgG antibody responses. Adv. Exp. Med. Biol. 371B: 1229–1234.
Moro, G., S. Arslanoglu et al. 2006. A mixture of prebiotic oligosaccharides reduces the incidence of atopic dermatitis during the first six months of age. Arch. Dis. Child 91(10): 814–819.
Ng, T., P. Holt et al. 2002. Cellular immune responses to ovalbumin and house dust mite in egg- allergic children. Allergy 57.
Osborn, D.A. and J.K. Sinn. 2007. Probiotics in infants for prevention of allergic disease and food hypersensitivity. Cochrane Database Syst. Rev. (4): CD006475.
Pohjavuori, E., M. Viljanen et al. 2004. Lactobacillus GG effect in increasing IFN-gamma production in infants with cow's milk allergy. J. Allergy Clin. Immunol. 114(1): 131–136.
Prescott, S.L., J.A. Dunstan et al. 2005. Clinical effects of probiotics are associated with increased interferon-gamma responses in very young children with atopic dermatitis. Clin. Exp. Allergy 35(12): 1557–1564.
Prescott, S.L., C. Macaubas et al. 1999. Development of allergen-specific T-cell memory in atopic and normal children. Lancet 353: 196–200.

Rautava, S., R. Luoto et al. 2012. Microbial contact during pregnancy, intestinal colonization and human disease. Nat. Rev. Gastroenterol. Hepatol. 9(10): 565–576.
Rook, G.A. and J.L. Stanford. 1998. Give us this day our daily germs. Immunol. Today 19: 113–116.
Rosenfeldt, V., E. Benfeldt et al. 2003. Effect of probiotic Lactobacillus strains in children with atopic dermatitis. J. Allergy Clin. Immunol. 111(2): 389–395.
Rosenfeldt, V., E. Benfeldt et al. 2004. Effect of probiotics on gastrointestinal symptoms and small intestinal permeability in children with atopic dermatitis. J. Pediatr 145(5): 612–616.
Sepp, E., K. Julge et al. 2005. Intestinal microbiota and immunoglobulin E responses in 5-year-old Estonian children. Clin. Exp. Allergy 35(9): 1141–1146.
Sepp, E., K. Julge et al. 1997. Intestinal microflora of Estonian and Swedish infants. Acta Paediatr 86(9): 956–961.
Sistek, D., R. Kelly et al. 2006. Is the effect of probiotics on atopic dermatitis confined to food sensitized children? Clin. Exp. Allergy 36: 629–633.
Sjögren, Y.M., M.C. Jenmalm et al. 2009. Altered early infant gut microbiota in children developing allergy up to 5 years of age. Clin. Exp. Allergy 39(4): 518–526.
Strachan, D. 1989. Hay fever, hygiene and household size. Brit. Med. J. 299: 1259–1256.
Sudo, N., S.-A. Sawamura et al. 1997. The requirement of intestinal bacterial flora for the development of an IgE production system fully susceptible to oral tolerance induction. J. Immunol. 159: 1739–1745.
Taylor, A.L., J.A. Dunstan et al. 2007. Probiotic supplementation for the first 6 months of life fails to reduce the risk of atopic dermatitis and increases the risk of allergen sensitization in high-risk children: a randomized controlled trial. J. Allergy Clin. Immunol. 119(1): 184–191.
Thornton, C.A., J.W. Upham et al. 2004. Functional maturation of CD4+CD25+CTLA4+CD45RA+ T regulatory cells in human neonatal T cell responses to environmental antigens/allergens. J. Immunol. 173(5): 3084–3092.
Turnbaugh, P.J., R.E. Ley et al. 2007. The human microbiome project. Nature 449: 804–810.
van der Aa, L.B., H.S. Heymans et al. 2010. Effect of a new synbiotic mixture on atopic dermatitis in infants: a randomized-controlled trial. Clin. Exp. Allergy 40: 795–804.
van der Aa, L.B., R. Lutter et al. 2012. No detectable beneficial systemic immunomodulatory effects of a specific synbiotic mixture in infants with atopic dermatitis. Clin. Exp. Allergy 42(4): 531–539.
van der Aa, L.B., W.M. van Aalderen et al. 2011. Synbiotics prevent asthma-like symptoms in infants with atopic dermatitis. Allergy 66(2): 170–177.
Wang, M.F., H.C. Lin et al. 2004. Treatment of perennial allergic rhinitis with lactic acid bacteria. Pediatr Allergy Immunol. 15(2): 152–158.
Weston, S., A. Halbert et al. 2005. Effects of probiotics on atopic dermatitis: a randomised controlled trial. Arch. Dis. Child 90(9): 892–897.
Viljanen, M., M. Kuitunen et al. 2005. Probiotic effects on faecal inflammatory markers and on faecal IgA in food allergic atopic eczema/dermatitis syndrome infants. Pediatr Allergy Immunol. 16(1): 65–71.
Viljanen, M., E. Pohjavuori et al. 2005. Induction of inflammation as a possible mechanism of probiotic effect in atopic eczema-dermatitis syndrome. J. Allergy Clin. Immunol. 115(6): 1254–1259.
Viljanen, M., E. Savilahti et al. 2005. Probiotics in the treatment of atopic eczema/dermatitis syndrome in infants: a double-blind placebo-controlled trial. Allergy 60: 494–500.
Wills-Karp, M., J. Santeliz et al. 2001. The germless theory of allergic disease: revisiting the hygiene hypothesis. Nat. Rev. Immunol. 1: 69–75.
Xiao, J.Z., S. Kondo et al. 2007. Clinical efficacy of probiotic Bifidobacterium longum for the treatment of symptoms of Japanese cedar pollen allergy in subjects evaluated in an environmental exposure unit. Allergol. Int. 56: 67–75.

Xiao, J.Z., S. Kondo et al. 2006. Probiotics in the treatment of Japanese cedar pollinosis: a double-blind placebo-controlled trial. Clin. Exp. Allergy 36: 1425–1435.
Zoetendal, E.G., A.D. Akkermans et al. 1998. Temperature gradient gel electrophoresis analysis of 16S rRNA from human fecal samples reveals stable and host-specific communities of active bacteria. Appl. Environ. Microbiol. 64(10): 3854–3859.
Zoetendal, E.G., K. Ben-Amor et al. 2001. DNA isolation protocols affect the detection limit of PCR approaches of bacteria in samples from the human gastrointestinal tract. Syst. Appl. Microbiol. 24(3): 405–410.
Zuany-Amorim, C., E. Sawicka et al. 2002. Suppression of airway eosinophilia by killed Mycobacterium vaccae-induced allergen-specific regulatory T-cells. Nat. Med. 8(6): 625–629.

20

Probiotics and Prebiotics in Crohn's Disease

Susan Mikolaitis, Prachi Chakradeo and *Ece A. Mutlu**

This chapter will discuss Crohn's disease (CD) and the role of probiotics and prebiotics in the treatment of it. Before we can discuss whether probiotics and prebiotics work in CD, we need to understand what CD is, and why probiotics or prebiotics may have a potential impact on the disease course. This introduction about the disease and its pathogenesis is neither exhaustive nor is meant to be the focus of this chapter (readers are referred to excellent reviews elsewhere for that), but should merely serve as the backdrop as CD relates to this topic. Secondly, we will discuss some of the potential mechanisms by which probiotics and prebiotics could affect CD pathogenesis. We will focus our attention on the clinical work done with prebiotics and probiotics and their results, and draw conclusions about the design of future work in this regard.

What is Crohn's Disease and why is there a Need for Probiotics and Prebiotics in the Treatment Armamentarium?

Crohn's is in the inflammatory bowel diseases (IBDs) family of chronic illnesses. Inflammatory bowel diseases are autoimmune diseases in which there is relapsing and remitting inflammatory insult to the gastrointestinal tract, resulting in mucosal and bowel wall damage with ulcerations. The

Rush University Medical Center, 1725 W. Harrison, Suite 206, Chicago, Illinois, 60612.
* Corresponding author

symptoms of IBD include abdominal pain, diarrhea, blood in the stool, fever, fatigue and malaise, and fecal incontinence, among others. Inflammatory bowel diseases have two main phenotypic forms: ulcerative colitis (UC) and Crohn's disease (CD). In UC, the inflammation typically involves only the mucosal layer of the colon; it starts at the rectum and extends proximally in a continuous fashion in the colon without any skip areas. In CD, the inflammation is patchy and may spread beyond the colon and may involve any part of the gastrointestinal tract from the mouth to the anus. Additionally, in CD, the inflammation is not limited to the mucosal layer and has greater depth by involving all layers of the bowel wall. This could lead to either narrowing of the intestinal lumen by stricturing over time, or extension of the inflammation through the wall of the bowel into surrounding tissues such as the abdominal wall and perianal skin, the bladder or the vagina by formation of abscesses and fistulae. There are also extraintestinal symptoms of IBDs, such as complications in the eye (e.g., uveitis) or the distant areas of the skin (e.g., pyoderma gangrenosum or erythema nodosum) or the joints (e.g., sacroiliitis or peripheral arthritis). As such, IBDs can be socially challenging diseases that could have a significant negative impact on quality of life.

The natural course of CD involves periods of symptoms (also termed active disease, relapse, or flare-up) that typically last weeks, to periods of feeling well when there are little or no symptoms (also termed inactive disease, or remission). The initial goal of treatment in CD is to relieve the patient of their symptoms (also termed induction of remission) and then to maintain the patient in an inactive state for extended periods of time (also termed maintenance of remission). With these goals in mind, the treatment typically involves immunosuppressive medications with potential serious side effects such as life threatening infections, bone marrow suppression, lung or liver fibrosis, lymphoproliferative diseases such as lymphoma, demyelinating diseases, development of psoriasis, osteoporosis, glaucoma, diabetes, weight gain, avascular bone necrosis to name a few. Over time, when disease complications such as strictures and fistulae occur, many patients with CD may also need surgery with bowel resections. The most common site of bowel resection in CD is an ileocolonic resection with removal of the ileocecal valve, which also could complicate the course of disease by predisposing the patient to additional long-term intestinal problems such as bile acid malabsorption and diarrhea, small intestinal bacterial overgrowth, as well as recurrence of the disease at the site of resection. Given the risks involved with immunosuppressive drugs as well as surgery, it is no surprise that a significant majority of patients with CD are exploring other therapies. Probiotics and prebiotics fill an unmet need for potentially safe, easily accessible and relatively cheap therapies for many patients (Mercer et al. 2012).

Current Thoughts on the Pathogenesis of Crohn's disease, and why and how Probiotics and Prebiotics could Impact the Disease

There are several lines of evidence that suggest CD is mostly of environmental origin. First and foremost of these is the fact that CD has been rising over the past 50 years: Population-based studies report that the incidence and prevalence of CD is highest in westernized nations but it is increasing in both the western and the developing world with industrialization (Molodecky et al. 2012, Zheng et al. 2010, Zvidi et al. 2009, Loftus et al. 2007, Mate-Jimenez et al. 1994, Munkholm et al. 1992, Brahme et al. 1975, Leong et al. 2004). Not only is CD prevalence increasing, there is also a rise especially in urban areas in both industrialized and westernized countries suggesting an environmental factor that is encountered more often in urban areas contributing to the development of CD (Soon et al. 2012, Hovde and Moum 2012, Lowe et al. 2009, Bernstein et al. 1999, Gilat et al. 1987, Klement et al. 2008, Ekbom et al. 1990, Radon et al. 2007). Secondly, when the concordance of CD is examined in monozygotic twins, it is noted to be about 50% or less (Tysk et al. 1988, Halfvarson 2011). This is yet another clue to the importance of environmental factors in IBD.

Among the environmental factors implicated for development of CD are excess sanitation that leads to an untrained immune system to recognize self; and microbial exposures, dietary and other lifestyle factors that may interact with immunoregulatory factors causing the development of disease in genetically susceptible individuals (Molodecky et al. 2012, Hanauer 2006, Mutlu and Gor 2008). In the western world and the developing world, fast paced life including increased travel, economic constraints to spend more time at work, and increased number of individuals working out of the home in a family unit, all come at the expense of time at home and time available to prepare meals. In turn, these contribute to the consumption of easily available, highly processed or restaurant-cooked meals on the run, and have become the replacement to preparing fresh and nutritionally superior foods at home. In addition to the convenience factor, in our experience, excessive rise in the cost of high quality fresh vegetables and fruits has made it very expensive to follow healthy eating guidelines especially for individuals on a fixed income in most western countries and urban areas. For example in the US, for every dollar spent on food by consumers, the farmer's share has dwindled from 41cents in 1952 to 15.8 cents in 2008, attesting that consumers knowingly or unknowingly pay primarily for the processing of their food, surpassing any investments in food quality and accessibility (Christian and Rashad 2009). The influence of these economic realities resulted in a major change in dietary habits over the past 50 years. Furthermore, a large scale systemic review of pre-illness diets and the development of IBD has found that high dietary intake of total fat, PUFAs, omega-6 fatty acids, and meat were associated with increased risks of CD and UC, whereas high fiber and fruit intake were associated with a decreased risk and high vegetable intake was associated with a decreased risk (Hou et al. 2011). The mechanisms by which diet can affect CD development could be direct, by altering the presence of luminal antigens, or indirect, by changing the resident microbiota in the gastrointestinal tract.

Resident microbiota is also thought to play an important role in the development of IBD. It is currently accepted that the aberrant immune attack in IBD occurs towards this resident microbiota in the setting of a decreased intestinal barrier. The clinical findings in CD patients that point toward a central role of enteric bacterial microbiota in IBD pathogenesis are as follows: CD typically involves areas of the bowel with the highest bacterial counts; namely the ileum, cecum and the rectum (Stenson 1999). Most fistulae in CD occur distally in the Gl tract, usually proximal to natural sphincters such as the ileocecal valve and the rectum (Sartor 1997, Balfour Sartor 1997). Placement of an ileostomy to divert colonic contents results in remission of CD and establishing continuity of the bowel with reversal of an ostomy restores exposure to the luminal material in the intestine and this causes recurrence (Rutgeerts et al. 1991, D'Haens et al. 1998). Also, installation of fecal material into normal undiseased loops of bowel in susceptible individuals can create inflammation with early features of CD (D'Haens et al. 1998). Colectomy and ileoanal pouch procedure (IAPP) results in remission of UC but recolonization of the pouch can lead to recurrence in the form of pouchitis (Farrell and LaMont 2002). Treatment with fecal enemas from healthy volunteers can cure severe recurrent UC as seen in a recent report of six patients with no evidence of active disease, with 1–13 years of follow up (Borody et al. 2003). Antibiotic treatments also show benefits in IBD therapy (Peppercorn 1997). For example, metronidazole therapy is efficacious for treatment of mild to moderate colonic and perianal CD (Jakobovits and Schuster 1984) and the combination of metronidazole and ciprofloxacin has been shown to have a comparable efficacy to steroids in active Crohn's colitis (Prantera et al. 1996). Modulation of the enteric flora with elemental diets can also be as useful as steroids in patients with IBD (Verma et al. 2000). Even one of the major genetic abnormalities found in patients with CD, alterations of the NOD2 gene in a small number of familial CD cases, points towards the importance of the microbiota. NOD2 protein has a leucine rich domain at its carboxy terminal, a feature of pattern recognition receptors that identify molecular patterns in microbial products (Hugot et al. 2001., Ogura et al. 2001). This protein is found in monocytes, epithelial cells, and in the Paneth cells of the ileum, which recognize and secrete defensins and other antimicrobial peptides into the lumen of the crypt base in an effort to protect epithelial cells from microbes (Ogura et al. 2003). Mutations of NOD2 in CD lead to a truncation of the 10th leucine-rich repeat and hence lead to a decreased recognition of muramyl dipeptide and bacterial lipopolypsaccharide and peptidoglycan (Inohara et al. 2001). Mutant NOD2 variants are then deficient in the activation of nuclear factor kappa B (NFKB), a key controller of the secretion of inflammatory cytokines such as TNF-α, via a serine threonine kinase, RICK. Eight percent of familial CD cases as opposed to 4% of controls have mutations in NOD2, whereas patients with UC do not have this alteration (Hugot et al. 2001, Ogura et al. 2001). Homozygocity for NOD2 mutations increases CD risk by 20 to 40 fold. While the exact link between NOD 2 mutations and the ultimate activation of NFKB is unknown, there certainly is altered bacterial pattern recognition in some patients with CD, and this could lead to overstimulation of the immune system through alternative activation pathways of NFKB. Furthermore, other CD susceptibility genes such as ATG16L1 and IRGM are involved in the microbial autophagy pathways that are felt to be down-stream of the NOD2-based NFKB signaling (Hampe et al. 2007, Prescott et al. 2007, Sehgal et al. 2012). Microbial autophagy pathways are postulated to be defective in CD potentially leading upto ineffective immune responses to microorganisms, especially intracellular bacteria. Microbial autophagy pathways have also been shown to be important in Paneth cell defenses and are especially important for development of ileal CD. Further studies are currently ongoing to determine the exact role of autophagy pathways in CD as they relate to bacterial composition (Frank et al. 2011). Besides these genetic abnormalities, antibodies against microbial antigens such as Anti-Saccharomyces Cerevisiae Antibodies (ASCA) and C-Bir found in the circulation in patients with CD are hallmarks of the disease and are used to help clinically differentiate CD from other illnesses of the GI tract (Quinton et al. 1998, Main et al. 1988, Lodes et al. 2004, Targan et al. 2005). Therefore, numerous clinical clues about the disease exemplify the importance of the interaction of the immune system and the microbiota in the pathogenesis of CD.

Evidence from animal models also points towards a central role of enteric bacterial microbiota in CD pathogenesis. Colonization with commensal bacteria is required in animal models of colitis regardless of the defect causing genetic susceptibility or barrier dysfunction in the animals. Examples of such models that do not develop disease in germ free environments are the indomethacin model in the rat, the carrageenan model in the guinea pig, HLA B27/human beta2 microglobulin transgenic rats, CD45 RB high SCID mice, IL-2 and IL-10 knock-out mice, TCR-α knockout mice (Farrell and LaMont 2002). Colitis can be transferred from animal to animal with T cells reactive against microbiota, e.g., in C3H-HeJBir mice and in stat-4 transgenic mice (Cong et al. 1998, Rath et al. 2001, Elson). Colitis can be attenuated or enhanced in animal models by the presence of certain bacteria (Sartor 1997, Sartor 2004). All of this data suggests that the microbiota could be the focus of the aberrant immune attack seen in CD.

Furthermore, in turn, the microbiota along the GI tract can also alter the aberrant immune response as well as gut barrier function, which are two other important factors in CD pathogenesis: Enteric flora influences the development and function of the normal mucosal immune system, termed the conditioning effect (Falk et al. 1998). Without the enteric flora, as seen in germ free animals, epithelial cell turnover is reduced and the colonic wall is thin (Gordon et al. 1997); and there are low levels of lymphocytes in the gut mucosa, small follicle structures as well as low immunoglobulin levels. As a result, germ free animals are highly susceptible to infections (Falk et al. 1998, Gordon et al. 1997). Exposure to commensal flora in these animals causes normalization of structure of the intestine and epithelial cell differentiation (Umesaki et al. 1993, Helgeland et al. 1996, Cebra et al. 1998, Klaasen et al. 1993, Umesaki et al. 1995, Jiang et al. 2001, Moreau and Gaboriau-Routhiau. 1996, Sudo et al. 1997, Frankel et al. 1994). The bacterial flora is well known to possess cytokine inducing molecules such as endotoxin via CD14/toll like receptors, and such as porins, lipid A associated proteins, superantigens, chaperonins, bacterial exotoxins, etc. via non-CD 14 mediated pathways (Henderson et al. 1996). Intestinal bacteria also possess the ability to depress inflammatory cytokines. For example, YopB from Yersinia inhibit TNF-α; and gapstatin like peptide from Actinomyces, a commensal in oral flora, can suppress IL-2, IL-4, and IL-5 (Henderson et al. 1996). In IBD, co-cultures of mucosal tissue explants of patients and probiotic bacteria such as Lactobacillus casei have been shown to reduce inflammatory cytokine release from the mucosa and decrease T cell activation (Carol et al. 2006). A strain specific effect of probiotic bacteria has been observed (Maassen et al. 2000). In animal models commensal bacteria such as Bacteroides thetaiotaomicron has been shown to depress NFKB activation, a central immune mechanism in CD (Kelly et al. 2004). Enteric flora can also affect gut barrier function positively through changes in enzymes that impact mucin production and even changes in gene expression of peptides found in desmosomes, as shown in the gnotobiotic mice colonized by the commensal Bacteroides thetaiotaomicron (Hooper et al. 2001). Probiotics such as *L. plantarum* and *L. rhamnosus* can induce mucin genes in intestinal epithelial cells, preventing adherence of enteropathogenic *E. coli* (Resta-Lenert and Barrett 2003). Similarly, probiotics such as Streptoccoccus thermophilus and Lactobacillus acidophilus have been shown to decrease the effects of enteroinvasive *E. coli* and to restore cytoskelatal and tight junctional protein phosphorylation, *in vitro* (Resta-Lenert and Barrett 2003)). A probiotic yeast, *Saccharomyces boulardii* also preserves barrier function against the effects of enteropathogenic *E. coli* (Czerucka et al. 2000). *In vitro* and *in vivo* studies also suggest that "beneficial bacteria" have the ability to change microbial composition in the GI tract or lie across the epithelium forming Gl tract biofilms that can prevent the localization and settlement of hostile bacteria that may induce immune activation, through induction of differing levels of bactericidal proteins and formation of complex biofilms above the epithelial surface (Resta-Lenert and Barrett 2003, Murphy et al. 2013). On the other hand, inflammatory effects of "pathogenic" bacteria in turn can also create vicious cycles of inflammation, oxidative injury, broken gut barrier, and increased permeability to other bacteria.

The resident microbiota in the GI tract is composed of billions of microorganisms, majority of which are bacteria. The highest concentrations of bacteria are in the colon, making up one the densest microbial environments in the body. In fact, it is estimated that the microbial cells in the body outnumber the body's own cells by ten to one, and that microbial metabolism in the GI tract is equivalent to that of the liver. Among single organisms implicated as the cause of CD are *Mycobacterium paratuberculosis*, measles virus, *Listeria monocytogenes*, *Pseudomonas multiphilia* and *florescens*, and non-pylori Helicobacter species. At least some of these are believed to either account for a small number of cases or be secondary colonizers of inflamed tissue. Helicobacter species however have been shown to either potentiate or attenuate experimental colitis depending on the type of bacteria and the model used rather than a consistent effect (Fox et al. 1999, Kullberg et al. 2002, Maggio-Price et al. 2002). *E. coli* species have been noted to increase in antibody titers for a greater variety of strains in CD patients compared to controls (Tabaqchali et al. 1978). Distinct adherent strains of *E. coli* have been described in ileum of CD patients (Darfeuille-Michaud et al. 2004) and have been implicated in the post-op recurrence of CD (Neut et al. 2002) but multiple different strains have also been detected in patients. Hence, a search for a single organism as the cause of IBD so far has been unsuccessful. In the past, it was difficult to study the microbiota associated with the GI tract: Methodology primarily involved anaerobic cultures and this required diligence and special training. Additionally many of the bacteria were unculturable due to lack of knowledge in their culture requirements, which has been postulated to include co-presence of one or more other bacterial taxa. Recent advents in genetic sequencing technology have provided insights into the gastrointestinal microbial world, which was difficult to study until the last half decade. Nowadays, the human microbiome project in the US (The Human Microbiome Project Consortium 2012a), and similar projects in Europe and Asia (Qin et al. 2010) are feverishly characterizing variations in the microbiota in health as well as disease states including CD. These studies have revealed that the bacterial microbiota in the GI tract are unique to a given individual suggesting a fingerprint like profile that varies from one person to another significantly (The Human Microbiome Project Consortium 2012b). Additionally, traditional demographic variables such as race or BMI did not explain majority of the variation in fecal microbiota. On the other hand, patients with CD and UC have been shown to have a dysbiotic bacterial composition that differs from normal, in multiple studies now (Qin et al. 2010). Nearly all studies have shown a decrease in the diversity of the GI tract microbiota in CD (Gillevet et al. 2010, Fujimoto et al. 2012, Hansen et al. 2012, Aomatsu et al. 2012, Manichanh et al. 2006, Ott et al. 2004). Some studies have also demonstrated associations of the bacterial microbiota with CD susceptibility genes (Frank et al. 2011). If a core set of dysbiotic bacteria can be found in a significant majority of CD patients, this could open the door to methods to change this "injurious and hostile" microbiota towards a "noninjurious and beneficial" one. Therefore, probiotics and prebiotics make a lot of sense, as specific treatments that could restore the intestinal bacterial composition back to a healthy state and thereby improve CD. In fact, numerous preclinical studies also point toward the potential utility of pro- and prebiotics as treatments for CD. For example, probiotic therapies in animal models of colitis, especially the IL-10 knock-out mice, have been very promising in preventing colitis; attenuating its severity; & even slowing the progression of inflammation to dysplasia and colon cancer (O'Mahony et al. 2001). A prebiotic, inulin, prevents colitis in the DSS model (Osman et al. 2006).

Do Probiotics Work for Crohn's Disease in the Clinical Setting?

Probiotics have been tested as a treatment for CD in multiple studies. The various clinical scenarios in which they have been tested include induction of remission, i.e., symptom abatement, as well as maintaining CD in remission, i.e., symptom prevention. Studies in active CD are summarized in Table 1 and are understandably relatively limited in number (Plein and Hotz 1993, Malchow 1997, Gupta et al. 2000, Fujimori et al. 2007, Steed et al. 2010). It is far more difficult to induce remission in CD than attempting to prevent inflammation. None of these studies have shown significant efficacy. There are several clinical trials that have examined probiotics as a potential therapy to prevent recurrence after surgically induced remission (Prantera et al. 2002, Marteau et al. 2006, Van Gossum et al. 2007, Chermesh et al. 2007): A Cochrane collaboration review of these studies have shown that four of these studies are methodologically reliable and are randomized, blinded and controlled studies (Doherty et al. 2009). These are shown in Table 2. None of these studies have identified a probiotic that prevented relapse after surgically-induced remission, and all were negative trials as shown in the table. When examined collectively, the clinical remission actually favored placebo with a relative risk of recurrence with any probiotic being 1.41. However, the confidence intervals were very wide (95% CI = 0.59–3.36) (Doherty et al. 2009). Endoscopic remission did not

Table 1. Clinical trials of probiotics in active CD.

Reference	*n*	**Trial design**	**Study groups**		**Duration (weeks)**	**Remission rate (%)**		*P*	**Notes**
			Intervention	**Control**		**Intervention**	**Control**		
Plein (1993)	17	R	S. boulardii (750 mg)	Placebo	9	NR	NR	0.05	Reduction in DAI
Malchow (1997)	28	R, OL	E. *coli* Nissle	Placebo	12	75	92	NS	Concomitant steroids used; no differences in relapse
Gupta et al. (2000)	4	OL	LGG (>10^{10} CFU)	-	24	100	-	-	Pediatric trial
Fujimori (2007)	10	OL	Synbiotic (Bifidobacteria + Lactobacilli + psyllium)	-	52		-	-	70% response rate; reduction in CDAI
Steed (2010)	35	R, DB	Synbiotic (*B. longum* + Synergy 1)	Placebo	24	62	45	NR, calculated p=0.68 (chi-sqr)	Reduction in mean CDAI, histology scores, and tissue TNF alpha expression

R=randomized
OL=open label
DB=double-blind
NR=not reported
NS= not significant
DAI=Disease activity index
CDAI=Crohn's disease activity index
P=p-value
n=number of cases
CFU= colony forming units
LGG= Lactobacillus GG

Table 2. Clinical trials of probiotics to prevent recurrence after surgically-induced remission of CD.

Reference	*n*	Trial design	Study groups		Duration (months)	Relapse rate (%)		*P*
			Intervention	Control		Intervention	Control	
Prantera et al. (2002)	45	R, DB	LGG (4·92 g)	Placebo	12	17	11	NS
Marteau et al. (2006)	98	R, DB	*L. johnsonii* (>10^9 CFU)	Placebo	6	49	64	NS
Van Gossum et al. (2007)	70	R, DB	*L. johnsonii* (>10^9 CFU)	Placebo	3	15	14	NS
Chermesh et al. (2007)	30	R, DB	Synbiotic 2000*	Placebo	24	25	20	NS

R=randomized; DB=double-blind; P=p-value; NS= not significant; CFU=colony forming unit; n=number of cases; LGG= Lactobacillus GG

favor either treatment with relative risk of endoscopic recurrence at 0.98 (95% CI =0.74–1.29) (Doherty et al. 2009). In summary, these treatments individually or collectively did not work in maintenance of remission of CD postoperatively. Maintenance trials after medically induced remission are shown in Table 3 and had variable outcomes and were also mostly negative (Malchow 1997, Fujimori et al. 2007, Guslandi et al. 2000, Schultz et al. 2004, Bousvaros et al. 2005, Garcia Vilela et al. 2008). A small trial found a decreased rate of relapse in CD patients in remission receiving *Saccharomyces boulardii* in addition to mesalamine (Guslandi et al. 2000). In general, as can be clearly seen from the tables, most of the studies are negative trials. However, the data to date has also many weaknesses. Notably, most of the reports pertaining to probiotics and CD tend to be vague in characterization of their subjects. In terms of outcomes, a significant majority did not use standardized definitions but employed surrogate markers of efficacy. Most studies were not controlled or blinded, and were usually single center trials with limited number of subjects. Some multicenter trials have been afflicted by concomitant use of medications that could have affected the outcome. The follow up periods are generally short and nearly none of the positive study results have been confirmed by additional larger studies.

Do Prebiotics Work for Crohn's Disease?

Prebiotics differ from probiotics as they stimulate *in situ* growth of beneficial resident colonic bacteria. There is increasing interest in the use of prebiotics in human health and disease. While prebiotic effects are demonstrated in healthy individuals, only recently have we begun to apply this knowledge to people who have chronic disease. Prebiotics, as strictly defined, include only two food ingredients: fructooligosaccharides and inulin, about which

Table 3. Clinical trials of probiotics to prevent recurrence after medically-induced remission of CD.

Reference	*n*	Trial design	Study groups		Duration (months)	Relapse rate (%)		*P*	Notes
			Intervention	Control		Intervention	Control		
Malchow (1997)*	28	R, OL	E. *coli* Nissle	Placebo	12	33	63	NS	Concomitant steroids used; no differences in relapse but numerically much lower
Guslandi et al. (2000)	32	R, OL	S. *boulardii* (1 g) + Mesalazine (2 g)	Mesalazine (3 g)	6	6	38	0·04	Probiotic better
Schultz et al. (2004)	11	R, DB	LGG (2X10^9 CFU)	Placebo	6	60	67	NS	Not significant to induce or maintain remission
Bousvaros et al. (2005)	75	R, DB	LGG (>10^9 CFU)	Placebo	24	31	17	NS	
Fujimori et al. (2007)	10	OL	Synbiotic with Bifidobacter and Lactobacillus	-	13	30	-	-	
Garcia Vilela et al. (2008)	34	R,OL	S. Boulardii	Placebo	3	NR	NR	NR	Improved intestinal permeability

R=randomized
OL=open label
DB=double-blind
NR=not reported
NS=not significant
P=p-value
n=number of cases
CFU=colony forming units
LGG=Lactobacillus GG

we have some data in CD (Gibson et al. 2004, Gibson and Roberfroid 1995). Prebiotics occur in a number of natural foods such as banana, wheat, chicory root, leeks, onions, artichokes particularly the Jerusalem variety, asparagus, and garlic. This discussion will revolve around the use of fructoologosaccharides, hereinafter referred to as "FOS", and inulin, in the treatment or prevention of CD flare-ups.

Potential mechanism of FOS in prevention of Crohn's disease relapse

It is widely accepted, in healthy individuals, that FOS selectively promotes the growth of *Bifidobacterium* spp. (Gibson and Wang 1994) and *Lactobacillus* spp. (Quigley 2012). These two bacteria are thought of as "health promoters" (Sartor 2004). The end metabolic products of *Bifidobacteria* are acetate and lactate, which lower the pH of the colon and thereby prevent growth of many pathogenic bacteria (Rasic 1983), which prefer a less acidic environment. Also, *Bifidobacteria* are able to excrete an end metabolic product that directly inhibits the growth of pathogens (Gibson and Wang 1994, Gibson and Wang 1994), and the exact mechanism by how this happens is unknown. *Bifidobacteria* are thought to offer several health promoting advantages to the human host (Gibson and Roberfroid 1995), but it is the inhibition of the growth of pathogens and the potential to act as immunomodulators that gives cause to promoting increased *Bifidobacteria* in the gut of the CD patient, a potential mechanism for prevention of flare-up and/or maintenance of remission. Further, the fermentation processes of *Bifidobacteria* on FOS provides a mixture of short-chain fatty acids (SCFAs) and lactate to the colon (Wang and Gibson 1993). SCFAs, in particular butyrate, are preferential substrates, nutrative to the colonocyte, and capable of modulating mucosal barrier function (Sartor 2004). Butyrate inhibits proinflammatory cytokine mRNA expression in the mucosa of the colon (Segain et al. 2000).

Human studies using prebiotics for the maintaning Crohn's disease in remission

Two human studies have found a decreasing trend for Bifidobacteria counts in patients with CD (Favier et al. 1997, Seksik et al. 2003). Further, metagenomic studies in CD indicate increased gram-negative bacteria, which produce molecules that are known to seduce pro-inflammatory cytokines. It intuitively makes sense that if one can increase Bifidobacteria counts, one should also decrease gram-negative bacteria via the mechanisms described above, thereby changing the colonic dysbiosis of CD to one that is a producer of butyrate, which seduces anti-inflammatory cytokine expression. Thus, human studies have begun to investigate the use of prebiotics as a means of production of butyrate in the colon, hence suppressing pro-inflammatory cytokines and upregulating anti-inflammatory cytokines in CD.

In the animal models of colitis, FOS feeding decreased disease activity, enhanced luminal bifidobacteria, inhibited nuclear factor kappa-B (NFkB), enhanced IL-10, and increased cecal secretory IgA levels compared to controls (Holma et al. 2002, Roller et al. 2004). Lindsay et al. (2006) performed an open-label, pilot study in ten patients with moderately active ileocolonic CD to determine the microbiological and immunological effects of FOS. Patients received 15 grams daily, for three weeks, of oligofructose and inulin in a single dose. These authors found FOS to be well tolerated in this group. Further, there was an increase in intestinal and fecal *Bifidobacteria* concentrations, an enhanced dendritic cell IL-10 production and toll-like receptor expression, and a significant reduction in disease activity as assessed with the Harvey-Bradshaw Index. Benjamin et al. conducted the first randomized, double-blind placebo-controlled trial to distinguish the impact of FOS on moderately active CD patients with respect to the clinical, immunological, and microbiological impact on the host (Benjamin et al. 2011). One-hundred and three patients were randomized to receive FOS or placebo for 4 weeks. Results showed no significant difference in achieving a clinical response between the groups, despite a significant improvement in quality of life scores in the FOS group over the placebo. The FOS group had reduced proportions of interleukin-6 in the lamina propria dendritic cells (DCs), an increase in IL-10 but no change in IL-12p40. There was augmentation of the adverse events of borborygmi, severity of flatulence, and abdominal pain in the FOS group over placebo, yet an improvement in quality of life scores of the FOS receiving group. Further, there was no significant difference in the fecal concentrations of *Bifidobacteria* at baseline or at the end of the trial. Therefore, this data indicate that the impact of FOS may not be primarily mediated by Bifidobacteria and suggest it is possibly mediated by increases in butyrate in the colon. The authors conclude that there is no clinical benefit to FOS supplementation for four weeks to patients with CD, but there is a significant improvement in quality of life scores as measured by the Inflammatory Bowel Disease Questionnaire. Joossens et al. (2005) undertook a similar trial: Sixty-seven inactive to moderately active CD patients enrolled to receive oligofructose-enriched inulin or placebo, 10 grams twice daily for 4 weeks. Similar to Benjamin et al. (2011), there were more drop-outs in the FOS than placebo group due to increased intensity of adverse events from baseline. Conversely, these authors noted a decrease in *Ruminococcus gnavus* and an increase in *Bifidobacteria longum*. A correlation between increases in *B. longum* and decreased disease activity was suggested as well.

There is scant evidence from randomized, double blind, placebo controlled studies administering prebiotics in CD for the induction and/or maintenance of remission, to allow definitive conclusions to be drawn. However, we now know that FOS administered in doses known to change human microbiota of healthy individuals, increases baseline adverse events reported by humans with CD. These include increased borborygmi, abdominal pain, distension, and diarrhea (Lindsay et al. 2006, Benjamin et al. 2011, de Vrese and Marteau 2007). Surely an increase in the severity of adverse events speaks to proceeding with caution in prescribing prebiotics to CD.

Our group at Rush University Medical Center has recently completed a pilot, randomized, double blind, placebo controlled trial of placebo vs. FOS vs. dietary intervention in 54 patients with inactive CD for maintenance of CD for 52 weeks. In this trial, we did titrate the FOS slowly over 8 weeks. The number of patients that withdrew before the end of the trial was not different across the three study groups; neither was the patient's self-report of adherence to the treatment assigned to them. This limited data suggests consideration be given to not only the total dose of prebiotics but also to titration of the dose incrementally, which is expected to reduce adverse events seen, compared to starting a bolus dose of prebiotics. The results of our study will be separately reported in the future.

Further research is needed to determine the mechanism by which prebiotics may exert their effects. These can be beyond simple induction of "beneficial" bacteria, and can be due to increases in sheer numbers of bacteria recognized by the host's immune cells; or due to the production of SCFAs and their effects on the colonic epithelial cells or on dendritic cell maturation and cytokine production; or even due to yet currently uncharacterized other functional changes in bacterial metabolism within the gastrointestinal tract (Hedin et al. 2007). Lastly, it is questioned whether a longer period of taking FOS can lead to a significant change in the total composition or function of the intestinal microbiota towards more closely resembling the microbiota of the healthy human and less like the dysbiosis seen in CD (Sartor 2004). These studies suggest that prebiotics may be more beneficial as an adjuvant therapy to maintaining CD in remission. However, at this time one cannot recommend FOS for improved clinical, immunological, or microbiological aspects of the gut until further scientifically conducted studies are completed and reported, using previous trials as a platform for future study designs.

Future Directions in Probiotic and Prebiotics for Crohn's Disease

In summary, the clinical results with probiotics and prebiotics have been mostly disappointing, despite the fact that they make perfect sense based on CD epidemiology and pathogenesis, and despite much preclinical data pointing towards a potentially beneficial effect in CD. There may be many reasons for this. One reason could be the fact that many of the tested organisms to date are actually minor components of the gut colonic flora that make up 2% or less of the total composition- they could be far outnumbered by those dysbiotic bacteria that may have formed a well-established community over extended periods of time before the development of CD in a particular patient. Perhaps this caveat could be overcome by designing probiotic preparations that contain multiple organisms that have been known to support the growth and survival of each other within a host. Secondly, the metabolic activity attributed to a particular probiotic bacterial phylotype in the preclinical setting could be altered by environmental factors within the host, leading to a different behavior within the host. As such, further preclinical studies are needed to characterize not only the dysbiotic microbiota composition in CD but also the alteration in functions of them. Only then, these functions could be understood and be able to be altered long-term. Thirdly, it should not be forgotten that CD to date has been clinically defined and merely represents a particular phenotype, however, the pathogenic events underlying it and the dominant pathways to disease onset/flares could be different from one patient to another. Definition of subgroups of patients within the CD phenotype could help delineate which groups of patients would be most likely to benefit from a particular type of probiotic or prebiotic targeting the dominant pathways to inflammation in that subgroup. Lastly, probiotic and prebiotic design efforts could take lessons from the large variability in microbiota composition in health. It is postulated that the diversity of bacteria in health creates redundancy and is also needed for horizontal transfer of genes across organisms. Future studies could envision designing organisms that have the potential to introduce new functional and controllable genes into the existing intestinal microbiota as an alternative strategy to the current approach which mainly attempts to populating the intestinal tract with new "beneficial" organisms.

References

Aomatsu, T. et al. 2012. Terminal restriction fragment length polymorphism analysis of the gut microbiota profiles of pediatric patients with inflammatory bowel disease. Digestion. 86(2): 129–135.

Balfour Sartor, R. 1997. Enteric Microflora in IBD: Pathogens or Commensals? Inflamm. Bowel Dis. 3(3): 230–235.

Benjamin, J.L. et al. 2011. Randomised, double-blind, placebo-controlled trial of fructo-oligosaccharides in active Crohn's disease. Gut. 60(7): 923–929.

Bernstein, C.N. et al. 1999. Epidemiology of Crohn's disease and ulcerative colitis in a central Canadian province: a population-based study. Am. J. Epidemiol. 149(10): 916–924.

Borody, T.J. et al. 2003. Treatment of ulcerative colitis using fecal bacteriotherapy. J. Clin. Gastroenterol. 37(1): 42–47.

Bousvaros, A. et al. 2005. A randomized, double-blind trial of Lactobacillus GG versus placebo in addition to standard maintenance therapy for children with Crohn's disease. Inflamm. Bowel Dis. 11(9): 833–839.
Brahme, F., C. Lindstrom and A. Wenckert. 1975. Crohn's disease in a defined population. An epidemiological study of incidence, prevalence, mortality, and secular trends in the city of Malmo, Sweden. Gastroenterology. 69(2): 342–351.
Carol, M. et al. 2006. Modulation of apoptosis in intestinal lymphocytes by a probiotic bacteria in Crohn's disease. J. Leukoc. Biol. 79(5): 917–922.
Cebra, J.J. et al. 1998. Development and maintenance of the gut-associated lymphoid tissue (GALT): the roles of enteric bacteria and viruses. Dev. Immunol. 6(1-2): 13–18.
Chermesh, I. et al. 2007. Failure of Synbiotic 2000 to prevent postoperative recurrence of Crohn's disease. Dig. Dis. Sci. 52(2): 385–389.
Christian, T. and I. Rashad. 2009. Trends in U.S. food prices, 1950–2007. Econ. Hum. Biol. 7(1): 113–120.
Cong, Y. et al. 1998. CD4+ T cells reactive to enteric bacterial antigens in spontaneously colitic C3H/HeJBir mice: increased T helper cell type 1 response and ability to transfer disease. J. Exp. Med. 187(6): 855–864.
Czerucka, D. et al. 2000. *Saccharomyces boulardii* preserves the barrier function and modulates the signal transduction pathway induced in enteropathogenic *Escherichia coli*-infected T84 cells. Infect. Immun. 68(10): 5998–6004.
D'Haens, G.R. et al. 1998. Early lesions of recurrent Crohn's disease caused by infusion of intestinal contents in excluded ileum. Gastroenterology. 114(2): 262–267.
Darfeuille-Michaud, A. et al. 2004. High prevalence of adherent-invasive *Escherichia coli* associated with ileal mucosa in Crohn's disease. Gastroenterology. 127(2): 412–421.
de Vrese, M. and P.R. Marteau. 2007. Probiotics and prebiotics: effects on diarrhea. J. Nutr. 137(3 Suppl. 2): 803S–811S.
Doherty, G. et al. 2009. Interventions for prevention of post-operative recurrence of Crohn's disease. Cochrane Database Syst. Rev. (4): p. CD006873.
Ekbom, A. et al. 1990. Perinatal risk factors for inflammatory bowel disease: a case-control study. Am. J. Epidemiol. 132(6): 1111–1119.
Elson, C.O. Experimental models of intestinal inflammation:New insights into mucosal mechanism of homeostasis, in Mucosal immunology, P. Ogra (ed.). Academic Press: San Diego.
Falk, P.G. et al. 1998. Creating and maintaining the gastrointestinal ecosystem: what we know and need to know from gnotobiology. Microbiol. Mol. Biol. Rev. 62(4): 1157–1170.
Farrell, R.J. and J.T. LaMont. 2002. Microbial factors in inflammatory bowel disease. Gastroenterol. Clin. North. Am. 31(1): 41–62.
Favier, C. et al. 1997. Fecal beta-D-galactosidase production and Bifidobacteria are decreased in Crohn's disease. Dig. Dis. Sci. 42(4): 817–822.
Fox, J.G. et al. 1999. A novel urease-negative Helicobacter species associated with colitis and typhlitis in IL-10-deficient mice. Infect. Immun. 67(4): 1757–1762.
Frank, D.N. et al. 2011. Disease phenotype and genotype are associated with shifts in intestinal-associated microbiota in inflammatory bowel diseases. Inflamm. Bowel Dis. 17(1): 179–184.
Frankel, W.L. et al. 1994. Mediation of the trophic effects of short-chain fatty acids on the rat jejunum and colon. Gastroenterology. 106(2): 375–380.
Fujimori, S. et al. 2007. High dose probiotic and prebiotic cotherapy for remission induction of active Crohn's disease. J. Gastroenterol. Hepatol. 22(8): 1199–1204.
Fujimoto, T. et al. 2013. Decreased abundance of *Faecalibacterium prausnitzii* in the gut microbiota of Crohn's disease. J. Gastroenterol. Hepatol. 28(4): 613–619.
Garcia Vilela, E. et al. 2008. Influence of *Saccharomyces boulardii* on the intestinal permeability of patients with Crohn's disease in remission. Scand. J. Gastroenterol. 43(7): 842–848.
Gibson, G.R. and M.B. Roberfroid. 1995. Dietary modulation of the human colonic microbiota: introducing the concept of prebiotics. J. Nutr. 125(6): 1401–1412.

Gibson, G.R. and X. Wang. 1994. Enrichment of bifidobacteria from human gut contents by oligofructose using continuous culture. FEMS Microbiol. Lett. 118(1-2): 121–127.
Gibson, G.R. and X. Wang. 1994. Regulatory effects of bifidobacteria on the growth of other colonic bacteria. J. Appl. Bacteriol. 77(4): 412–420.
Gibson, G.R. et al. 2004. Dietary modulation of the human colonic microbiota: updating the concept of prebiotics. Nutr. Res. Rev. 17(2): 259–275.
Gilat, T. et al. 1987. Childhood factors in ulcerative colitis and Crohn's disease. An international cooperative study. Scand. J. Gastroenterol. 22(8): 1009–1024.
Gillevet, P. et al. 2010. Quantitative assessment of the human gut microbiome using multitag pyrosequencing. Chem. Biodivers. 7(5): 1065–1075.
Gordon, J.I. et al. 1997. Epithelial cell growth and differentiation. III. Promoting diversity in the intestine: conversations between the microflora, epithelium, and diffuse GALT. Am. J. Physiol. 273(3 Pt 1): G565–G570.
Gupta, P. et al. 2000. Is lactobacillus GG helpful in children with Crohn's disease? Results of a preliminary, open-label study. J. Pediatr. Gastroenterol. Nutr. 31(4): 453–457.
Guslandi, M. et al. 2000. *Saccharomyces boulardii* in maintenance treatment of Crohn's disease. Dig. Dis. Sci. 45(7): 1462–1464.
Halfvarson, J. 2011. Genetics in twins with Crohn's disease: less pronounced than previously believed? Inflamm. Bowel Dis. 17(1): 6–12.
Hampe, J. et al. 2007. A genome-wide association scan of nonsynonymous SNPs identifies a susceptibility variant for Crohn disease in ATG16L1. Nat. Genet. 39(2): 207–211.
Hanauer, S.B. 2006. Inflammatory bowel disease: epidemiology, pathogenesis, and therapeutic opportunities. Inflamm. Bowel Dis. 12 Suppl. 1: S3–S9.
Hansen, R. et al. 2012. Microbiota of de-novo pediatric IBD: increased *Faecalibacterium prausnitzii* and reduced bacterial diversity in Crohn's but not in ulcerative colitis. Am. J. Gastroenterol. 107(12): 1913–1922.
Hedin, C., K. Whelan and J.O. Lindsay 2007. Evidence for the use of probiotics and prebiotics in inflammatory bowel disease: a review of clinical trials. Proc. Nutr. Soc. 66(3): 307–315.
Helgeland, L. et al. 1996. Microbial colonization influences composition and T-cell receptor V beta repertoire of intraepithelial lymphocytes in rat intestine. Immunology. 89(4): 494–501.
Henderson, B., S. Poole and M. Wilson. 1996. Microbial/host interactions in health and disease: who controls the cytokine network? Immunopharmacology. 35(1): 1–21.
Holma, R. et al. 2002. Galacto-oligosaccharides stimulate the growth of bifidobacteria but fail to attenuate inflammation in experimental colitis in rats. Scand. J. Gastroenterol. 37(9): 1042–1047.
Hooper, L.V. et al. 2001. Molecular analysis of commensal host-microbial relationships in the intestine. Science. 291(5505): 881–884.
Hou, J.K., B. Abraham and H. El-Serag. 2011. Dietary intake and risk of developing inflammatory bowel disease: a systematic review of the literature. Am. J. Gastroenterol. 106(4): 563–573.
Hovde, O. and B.A. Moum. 2012. Epidemiology and clinical course of Crohn's disease: results from observational studies. World J. Gastroenterol. 18(15): 1723–1731.
Hugot, J.P. et al. 2001. Association of NOD2 leucine-rich repeat variants with susceptibility to Crohn's disease. Nature. 411(6837): 599–603.
Inohara, N. et al. 2001. Human Nod1 confers responsiveness to bacterial lipopolysaccharides. J. Biol. Chem. 276(4): 2551–2554.
Jakobovits, J. and M.M. Schuster. 1984. Metronidazole therapy for Crohn's disease and associated fistulae. Am. J. Gastroenterol. 79(7): 533–540.
Jiang, H.Q., N.A. Bos and J.J. Cebra. 2001. Timing, localization, and persistence of colonization by segmented filamentous bacteria in the neonatal mouse gut depend on immune status of mothers and pups. Infect. Immun. 69(6): 3611–3617.
Joossens, S. et al. 2005. *Saccharomyces boulardii* in Crohn's disease: effect on anti-Saccharomyces cerevisiae antibodies and intestinal permeability. Inflamm. Bowel Dis. 11(9): 863–864.

Kelly, D. et al. 2004. Commensal anaerobic gut bacteria attenuate inflammation by regulating nuclear-cytoplasmic shuttling of PPAR-gamma and RelA. Nat. Immunol. 5(1): 104–112.

Klaasen, H.L. et al. 1993. Apathogenic, intestinal, segmented, filamentous bacteria stimulate the mucosal immune system of mice. Infect. Immun. 61(1): 303–306.

Klement, E. et al. 2008. Childhood hygiene is associated with the risk for inflammatory bowel disease: a population-based study. Am. J. Gastroenterol. 103(7): 1775–1782.

Kullberg, M.C. et al. 2002. Bacteria-triggered CD4 (+) T regulatory cells suppress Helicobacter hepaticus-induced colitis. J. Exp. Med. 196(4): 505–515.

Leong, R.W., J.Y. Lau and J.J. Sung. 2004. The epidemiology and phenotype of Crohn's disease in the Chinese population. Inflamm. Bowel Dis. 10(5): 646–651.

Lindsay, J.O. et al. 2006. Clinical, microbiological, and immunological effects of fructo-oligosaccharide in patients with Crohn's disease. Gut. 55(3): 348–355.

Lodes, M.J. et al. 2004. Bacterial flagellin is a dominant antigen in Crohn disease. J. Clin. Invest. 113(9): 1296–1306.

Loftus, C.G. et al. 2007. Update on the incidence and prevalence of Crohn's disease and ulcerative colitis in Olmsted County, Minnesota, 1940–2000. Inflamm. Bowel Dis. 13(3): 254–261.

Lowe, A.M. et al. 2009. Epidemiology of Crohn's disease in Quebec, Canada. Inflamm. Bowel Dis. 15(3): 429–435.

Maassen, C.B. et al. 2000. Strain-dependent induction of cytokine profiles in the gut by orally administered Lactobacillus strains. Vaccine. 18(23): 2613–2623.

Maggio-Price, L. et al. 2002. Helicobacter bilis infection accelerates and H. hepaticus infection delays the development of colitis in multiple drug resistance-deficient (mdr1a-/-) mice. Am. J. Pathol. 160(2): 739–751.

Main, J. et al. 1988. Antibody to Saccharomyces cerevisiae (bakers' yeast) in Crohn's disease. BMJ. 297(6656): 1105–1106.

Malchow, H.A. 1997. Crohn's disease and *Escherichia coli*. A new approach in therapy to maintain remission of colonic Crohn's disease? J. Clin. Gastroenterol. 25(4): 653–658.

Manichanh, C. et al. 2006. Reduced diversity of faecal microbiota in Crohn's disease revealed by a metagenomic approach. Gut. 55(2): 205–211.

Marteau, P. et al. 2006. Ineffectiveness of *Lactobacillus johnsonii* LA1 for prophylaxis of postoperative recurrence in Crohn's disease: a randomised, double blind, placebo controlled GETAID trial. Gut. 55(6): 842–847.

Mate-Jimenez, J. et al. 1994. Incidence and prevalence of ulcerative colitis and Crohn's disease in urban and rural areas of Spain from 1981 to 1988. J. Clin. Gastroenterol. 18(1): 27–31.

Mercer, M. et al. 2012. How patients view probiotics: findings from a multicenter study of patients with inflammatory bowel disease and irritable bowel syndrome. J. Clin. Gastroenterol. 46(2): 138–144.

Molodecky, N.A. et al. 2012. Increasing incidence and prevalence of the inflammatory bowel diseases with time, based on systematic review. Gastroenterology. 142(1): 46–54 e42; quiz e30.

Moreau, M.C. and V. Gaboriau-Routhiau. 1996. The absence of gut flora, the doses of antigen ingested and aging affect the long-term peripheral tolerance induced by ovalbumin feeding in mice. Res. Immunol. 147(1): 49–59.

Munkholm, P. et al. 1992. Incidence and prevalence of Crohn's disease in the county of Copenhagen, 1962–87: a sixfold increase in incidence. Scand. J. Gastroenterol. 27(7): 609–614.

Murphy, E.F. et al. 2013. Divergent metabolic outcomes arising from targeted manipulation of the gut microbiota in diet-induced obesity. Gut. 62(2): 220–226.

Mutlu, E.A. and N. Gor. 2008. To diet or not if you have inflammatory bowel disease. Expert Rev. Gastroenterol. Hepatol. 2(5): 613–616.

Neut, C. et al. 2002. Changes in the bacterial flora of the neoterminal ileum after ileocolonic resection for Crohn's disease. Am. J. Gastroenterol. 97(4): 939–946.

O'Mahony, L. et al. 2001. Probiotic impact on microbial flora, inflammation and tumour development in IL-10 knockout mice. Aliment. Pharmacol. Ther. 15(8): 1219–1225.

Ogura, Y. et al. 2001. A frameshift mutation in NOD2 associated with susceptibility to Crohn's disease. Nature. 411(6837): 603–606.

Ogura, Y. et al. 2003. Expression of NOD2 in Paneth cells: a possible link to Crohn's ileitis. Gut. 52(11): 1591–1597.

Osman, N. et al. 2006. Bifidobacterium infantis strains with and without a combination of oligofructose and inulin (OFI) attenuate inflammation in DSS-induced colitis in rats. BMC Gastroenterol. 6: 31.

Ott, S.J. et al. 2004. Reduction in diversity of the colonic mucosa associated bacterial microflora in patients with active inflammatory bowel disease. Gut. 53(5): 685–693.

Peppercorn, M.A. 1997. Antibiotics are effective therapy for Crohn's disease. Inflamm. Bowel Dis. 3(4): 318–319.

Plein, K. and J. Hotz. 1993. Therapeutic effects of *Saccharomyces boulardii* on mild residual symptoms in a stable phase of Crohn's disease with special respect to chronic diarrhea--a pilot study. Z. Gastroenterol. 31(2): 129–134.

Prantera, C. et al. 1996. An antibiotic regimen for the treatment of active Crohn's disease: a randomized, controlled clinical trial of metronidazole plus ciprofloxacin. Am. J. Gastroenterol. 91(2): 328–332.

Prantera, C. et al. 2002. Ineffectiveness of probiotics in preventing recurrence after curative resection for Crohn's disease: a randomised controlled trial with Lactobacillus GG. Gut. 51(3): 405–409.

Prescott, N.J. et al. 2007. A nonsynonymous SNP in ATG16L1 predisposes to ileal Crohn's disease and is independent of CARD15 and IBD5. Gastroenterology. 132(5): 1665–1671.

Qin, J. et al. 2010. A human gut microbial gene catalogue established by metagenomic sequencing. Nature. 464(7285): 59–65.

Quigley, E.M.M. 2012. Prebiotics and probitics: Their role in the management of gastrointestinal disorders in adults. Nutri. Clin. Pract. 195–200.

Quinton, J.F. et al. 1998. Anti-Saccharomyces cerevisiae mannan antibodies combined with antineutrophil cytoplasmic autoantibodies in inflammatory bowel disease: prevalence and diagnostic role. Gut. 42(6): 788–791.

Radon, K. et al. 2007. Contact with farm animals in early life and juvenile inflammatory bowel disease: a case-control study. Pediatrics. 120(2): 354–361.

Rasic, J.L. 1983. The role of dairy foods containing bifido and acidophilus bacteria in nutrition and health. N. Eur. Dairy J. 4: 80–88.

Rath, H.C. et al. 2001. Different subsets of enteric bacteria induce and perpetuate experimental colitis in rats and mice. Infect. Immun. 69(4): 2277–2285.

Resta-Lenert, S. and K.E. Barrett. 2003. Live probiotics protect intestinal epithelial cells from the effects of infection with enteroinvasive *Escherichia coli* (EIEC). Gut. 52(7): 988–997.

Roller, M., G. Rechkemmer and B. Watzl. 2004. Prebiotic inulin enriched with oligofructose in combination with the probiotics Lactobacillus rhamnosus and Bifidobacterium lactis modulates intestinal immune functions in rats. J. Nutr. 134(1): 153–156.

Rutgeerts, P. et al. 1991. Effect of faecal stream diversion on recurrence of Crohn's disease in the neoterminal ileum. Lancet. 338(8770): 771–774.

Sartor, R.B. 1997. Review article: Role of the enteric microflora in the pathogenesis of intestinal inflammation and arthritis. Aliment Pharmacol. Ther. 11 Suppl. 3: 17–22; discussion 22–3.

Sartor, R.B. 2004. Therapeutic manipulation of the enteric microflora in inflammatory bowel diseases: antibiotics, probiotics, and prebiotics. Gastroenterology. 126(6): 1620–1633.

Schultz, M. et al. 2004. Lactobacillus GG in inducing and maintaining remission of Crohn's disease. BMC Gastroenterol. 4: 5.

Segain, J.P. et al. 2000. Butyrate inhibits inflammatory responses through NFkappaB inhibition: implications for Crohn's disease. Gut. 47(3): 397–403.

Sehgal, R. et al. 2012. Mutations in IRGM are associated with more frequent need for surgery in patients with ileocolonic Crohn's disease. Dis. Colon Rectum. 55(2): 115–121.

Seksik, P. et al. 2003. Alterations of the dominant faecal bacterial groups in patients with Crohn's disease of the colon. Gut. 52(2): 237–242.

Soon, I.S. et al. 2012. The relationship between urban environment and the inflammatory bowel diseases: a systematic review and meta-analysis. BMC Gastroenterol. 12: 51.

Steed, H. et al. 2010. Clinical trial: the microbiological and immunological effects of synbiotic consumption - a randomized double-blind placebo-controlled study in active Crohn's disease. Aliment Pharmacol. Ther. 32(7): 872–883.

Stenson, W. 1999. Inflammatory Bowel Disease. Textbook of Gastroenterology. *In*: T. Yamada (ed.). Lippincott Williams & Wilkins: Philadelphia. 1775–1840.

Sudo, N. et al. 1997. The requirement of intestinal bacterial flora for the development of an IgE production system fully susceptible to oral tolerance induction. J. Immunol. 159(4): 1739–1745.

Tabaqchali, S., D.P. O'Donoghue and K.A. Bettelheim. 1978. *Escherichia coli* antibodies in patients with inflammatory bowel disease. Gut. 19(2): 108–113.

Targan, S.R. et al. 2005. Antibodies to CBir1 flagellin define a unique response that is associated independently with complicated Crohn's disease. Gastroenterology. 128(7): 2020–2028.

The Human Microbiome Project Consortium 2012a. A framework for human microbiome research. Nature. 486(7402): 215–221.

The Human Microbiome Project Consortium. 2012b. Structure, function and diversity of the healthy human microbiome. Nature. 486(7402): 207–214.

Tysk, C. et al. 1988. Ulcerative colitis and Crohn's disease in an unselected population of monozygotic and dizygotic twins. A study of heritability and the influence of smoking. Gut. 29(7): 990–996.

Umesaki, Y. et al. 1993. Expansion of alpha beta T-cell receptor-bearing intestinal intraepithelial lymphocytes after microbial colonization in germ-free mice and its independence from thymus. Immunology. 79(1): 32–37.

Umesaki, Y. et al. 1995. Segmented filamentous bacteria are indigenous intestinal bacteria that activate intraepithelial lymphocytes and induce MHC class II molecules and fucosyl asialo GM1 glycolipids on the small intestinal epithelial cells in the ex-germ-free mouse. Microbiol. Immunol. 39(8): 555–562.

Van Gossum, A. et al. 2007. Multicenter randomized-controlled clinical trial of probiotics (*Lactobacillus johnsonii*, LA1) on early endoscopic recurrence of Crohn's disease after lleo-caecal resection. Inflamm. Bowel Dis. 13(2): 135–142.

Verma, S. et al. 2000. Polymeric versus elemental diet as primary treatment in active Crohn's disease: a randomized, double-blind trial. Am. J. Gastroenterol. 95(3): 735–739.

Wang, X. and G.R. Gibson. 1993. Effects of the *in vitro* fermentation of oligofructose and inulin by bacteria growing in the human large intestine. J. Appl. Bacteriol. 75(4): 373–380.

Zheng, J.J. et al. 2010. Prevalence and incidence rates of Crohn's disease in mainland China: a meta-analysis of 55 years of research. J. Dig. Dis. 11(3): 161–166.

Zvidi, I. et al. 2009. The prevalence of Crohn's disease in Israel: a 20-year survey. Dig. Dis. Sci. 54(4): 848–852.

21

Probiotics and Prebiotics

New Hope for Genitourinary Health

Sarika Amdekar,[1] *Avnish Kumar,*[2] *Vivek Kumar Shrivastav,*[3] *Desh Deepak Singh*[4] and *Vinod Singh*[1,*]

Introduction

The Function of diet and/or nutrition is to supply nutrients to meet the hosts physiological requirements. As research behind diet and health/well being has evolved, the idea of 'functional foods' has become accepted among people. Functional foods, also known as nutriceuticals/biotherapeutics, are administered to obtain a specific result (Floch and Hong-Curtiss 2002). Foods which are 'functional' are those which exert certain positive properties or benefits over and above their normal nutritional value. Functional foods are universally popular but sometimes plagued by scarce research/claims; thus this concept is only commercially successful. Some organic and

[1] Department of Microbiology, Barkatullah University, Hoshangabad Road, Habibganj, Bhopal 462026, Madhya Pradesh, India.
Emails: drsarikaamdekar@gmail.com; vsingh3@rediff.com
[2] Department of Biotechnology, Dr B.R. Ambedkar University, Agra, 282004 Uttar Pradesh, India.
Email: avnishkumar81@gmail.com
[3] Department of Microbiology, College of Life Sciences, Cancer Hospital and Research Institute, Gwalior 474009, Madhya Pradesh, India
Email: vivek.shrivastav12@gmail.com
[4] Department of Virology, CSJM Medical University, Lucknow 286003, Uttar Pradesh, India.
Email: ddsbms@gmail.com
* Corresponding author

inorganic micronutrients, antioxidants, vitamins, dietary fibers, proteins like lactoferrin, certain bioactive peptides and polyunsaturated fatty acids are examples of functional foods. This concept has now changed to gastrointestinal function and maintaining the gut bacteria. The colon is a highly populated region of the gastrointestinal tract (GI tract) since normal microflora resides there, and due to this reason, GI tract is one of the most metabolically active organ of the body. Thus, the concept of modulating and/or improving the gut microbial function has a long history, as diet can have a major effect on gut microflora activities (Gibson and Roberfroid 1999). The microflora of the intestinal microenvironment works as a unit and has important defensive, metabolic and trophic roles (Canny and McCormick 2008).

Every day, human beings ingest a large number of microbes, mainly bacteria. Even though these organisms are naturally present in food and water, they can also be deliberately added during the processing of foods such as sausages, cheese, yoghurt and fermented milk products (Parvez et al. 2006). Increasing knowledge of nutritional needs, along with the growth and development in food technology, is transforming our concept of food from one of being necessary to give us energy and nutrients for our daily needs and to avoid the traditional nutrient deficiencies, to one of the reason(s) and mechanism(s) by which certain foods are associated with a reduced risk of chronic diseases. Emerging food revolution of probiotics and prebiotics over the last two or five years is another such example, which forms the focus of current debate and immense research.

Current Knowledge about Prebiotics and Probiotics

The word 'probiotic' is derived from the same Greek term 'biotikos' which may be literally translated as 'for life'. According to modern medical history; Metchnikoff established the importance of intestinal normal microflora (Metchnikoff 1907). Metchnikoff also found that ingestion of milk fermented with *Lactobacillus* reduces the pathogenic bacteria in gut and prolongs life. After his studies many researchers claimed the importance of *Lactobacillus* in gut protection (Kipeloff 1926, Rettger et al. 1935, Freter 1955). In 1965 the term "probiotics" was coined by Lilly and Stillwell (Lilly and Stillwell 1965). Fuller, for the first time, described probiotics as microbial supplements that benefits the host by maintaining the gut microbial balance (Fuller 1989). Probiotics have been used for many years in the animal feed industry, but they are now being increasingly made available in many forms and can be purchased from the market as freeze-dried preparations in health food (Mcfarlane and Cummings 1999, Tuohy et al. 2003).

In humans, *Lactobacillus* or *Bifidobacterium* are the most commonly used probiotics. They can be used as single specie or in mixed culture with other bacteria. Other probiotic candidates are *Escherichia, Bacillus, Enterococcus* and *Saccharomyces* sp. (Szajewska 2007). Probiotics have GRAS status (generally regarded as safe status) (Salimen et al. 1998). Not only this, their protective properties like antimicrobial, antioxidant, anti-diarrheal, anti-cancerous, anti-inflammatory and anti-lipidemic along with their lactose intolerance activities, are well documented. Probiotic organisms help in the treatment and prevention of diseases (John et al. 1997). Probiotics exert beneficial effects by various protective mechanisms like: maintaining the acidic pH, bacteriocins, production of hydrogen peroxide, prevention of colonization of pathogen, antimicrobial activity, degrading the toxins and stimulation of immunity of the host (Bernet et al. 1994, Singh et al. 2008, Amdekar et al. 2010). Probiotics have also shown their effect in alleviating symptoms of allergies (Yao et al. 2010, Kalliomaki et al. 2010), cancer (Kumar et al. 2010), AIDS (Trois et al. 2008), respiratory and urinary tract infections (Kaur et al. 2009, Amdekar et al. 2011).

Prebiotics are undigested nutrients that influence intestinal microbial flora (Boehm et al. 2004). The term 'Prebiotic' was introduced by Gibson and Roberfroid in 1995 (Gibson and Roberfroid 1995). It is a fiber found in some plants that reaches the colon undigested. Prebiotics beneficially affect the host by selectively stimulating the growth and/or activity of one or a limited number of bacteria in the colon and thereby improving host health.

Recently, scientists defined a dietary prebiotic as "a selectively fermented ingredient that results in specific changes in the composition and/or activity of the gastrointestinal microbiota, thus conferring benefit(s) upon host health. Prebiotics promote the growth and proliferation of useful bacteria in the digestive system. Edible sugar-like substances that are not broken down in the first part of the digestive tract and act as food for the good bacteria living in the colon. Prebiotics consist mainly of oligosaccharides, sugar molecules of three to six chains and soluble fibers. Prebiotic carbohydrates are found naturally in such fruits and vegetables as bananas, berries, asparagus, garlic, wheat, oatmeal, barley, etc. (Crittenden and Payne 2008). Oligosaccharides, lactulose, fructooligosaccharides (FOS), inulin, galactooligosaccharides (GOS), oligofructose (OF), isomalto-oligosaccharides, fiber gums, lactilol, lactosucrose, pyrodextrins, soligosaccharides, transgalacto-oligosaccharides and xylooligosaccharides are the various prebiotics mixed with probiotic to increase their efficiency (Sekhon and Jairath 2010). Prebiotics act as substrates for the beneficial bacteria. Hence they stimulate the growth of probiotic bacteria which are helpful for the host. The majority of the bacterial species found in the intestine are saccharolytic and ferment nutrients. All these lead to change in the gut ecology and ultimately improve the status of the host (Crittenden 1999).

Several studies have been conducted so far, on prebiotics, for their beneficial effect on hosts (Kullen et al. 2005). In a study, oligofructose and inulin stimulate the growth of *Bifidobacteria*, whereas they inhibit the growth of *E. coli* and *C. perfringens* (Cummings et al. 2001). Oral lactulose increases stool water content (Hebden et al. 1999) and accelerates colonic transit in healthy volunteers and increases stool frequency in constipation (Barrow et al. 1992). Symbiotic recipe of inulin plus oligofructose with *L. plantarum* plus *B. bifidum* increased the growth of *Bifidobacteria* and inhibited growth of the pathogenic strains of *Campylobacter jejuni*, *E. coli* and *Salmonella enteritidis* (Fooks and Gibson 2002).

Probiotics and prebiotics: they work synergistically and form "Synbiotics". Regular consumption of probiotics or prebiotics have health implications that include enhanced immune function, improved colonic integrity, decreased incidence and duration of intestinal infections, down-regulated allergic response, and improved digestion and elimination (Douglas and Sanders 2008). Probiotics add to an existing colony of bacteria, but prebiotics provide nutrients for existing flora, allowing the colony to grow naturally and flourish. Probiotics are mainly active in the small intestine and prebiotics are effective only in the large intestine, hence these both act synergistically (Gibson and Roberfroid 1995). Studies have shown that by harnessing both the benefits of these prebiotics and probiotics into synergy, the number of good bacteria may be increased many folds for the betterment of our health (Unne et al. 2001).

Genitourinary Disease: A Burning Problem

Genitourinary diseases/urogenital infections include the infection of bladder, kidneys, vagina, urethra, periurethra and cervix. It is a worldwide problem affecting >300 million females (Reid 2001). These infections are a common reason for women to visit a physician or urologist. Approximately 50–60% of women are probably affected with UTI in their lifetime (Rahn 2008). Increased prevalence and incidence of genitourinary infections in women is likely due to the result of several clinical factors including anatomic differences, hormonal effects and behavior patterns (Standiford et al. 2005). Recurrent UTI's may be managed by self-initiated therapies or prophylaxis, than by continuing to treat each case emergently.

Urinary Tract Infection (UTI) is an extremely common health problem, with an unpredictable history. Bacterial infection is the sole reason behind the genitourinary infection which includes *Escherichia coli*, *Proteus*, *Klebsiella* and *Staphylococcus saprophyticus*. Viruses, fungi and parasites can also cause UTIs. *E. coli* have been found to be the most common causative organism of UTI in many countries (Samra et al. 2005). Genitourinary infections in women are often characterized by an alteration in the local flora from predominance of lactobacilli to coliform uropathogens as a result of hormone deficiency, sexual activity, contraceptive measures and other factors (Forsum et al. 2005). In most of the cases, antibiotics are prescribed by the medical practitioners. Re-recurrence of UTI after antibiotic therapy is very common.

Doctors usually prescribe antibiotic drugs to kill the infection and a relief can be experienced, usually in a day or two. The biggest problem associated with a UTI is that it results in pyelonephritis which in turn, causes scarring and damage to the kidney and surrounding tissues. Persistent infections may lead to damage in the kidney's filter system. Antibiotic therapy is a good option against UTI. Typical antibiotics used against UTI's include trimethoprim-sulfamethoxamole, nitrofurantoin, ciprofloxacin, levofloxacin, or their chemical relatives, and certain penicillins such as amoxicillin. It may decrease the consequences of UTI to some extent but the spiraling cost of antibiotic therapy and appearance of multi drug resistant bacteria proves it to be an unsatisfactory therapeutic option. So far, no appropriate and successfully option for treating or preventing urinary tract infection is known (Borchert et al. 2008). The novel non-antibiotic therapy is preferred when compared to antibiotic therapy which is of limited use due to increase in drug resistance. Current social trends have changed from manmade chemicals towards the use of naturopathy and away from the chemotherapeutic regimens (Reid 1999). Alternative remedies against UTI are of interest to patients and their caregivers (Schmitt et al. 1992).

Modulating Urogenital Tract Microflora: Prebiotics and Probiotics

Pathogenic organisms are able to infect the vagina with bacterial vaginitis (BV) and yeast vaginitis (Reid and Bruce 2006). Microbiota of genital tract of a healthy woman comprises of approximately 50 species of organisms, which differ in composition according to the reproductive stages and exposure to several factors, including antibiotics and spermicides (Pascual et al. 2010). *Lactobacillus* is an important part of the normal flora; commonly found in the mouth cavity, gastrointestinal tract and female genitourinary tract (Kaewnopparat and Kaewnopparat 2009).

The concept of delivering *Lactobacillus* by oral route to repopulate the vagina is a new concept, which was firstly reported by Reid et al. in 2001. The concept of using probiotics was first considered by Bruce in 1973. He found that women who were not suffering from bladder infection showed vaginal *Lactobacillus* as the dominant microflora (Bruce et al. 1973). He hypothesized that *Lctobacillus* acts as a barrier to the ascension of uropathogens from the rectum to the bladder. Yet now, many strains have been shown to colonise the vagina and significantly reduce pathogen colonization and infection in the bladder and vagina. The important ones are *L. rhamnosus* GR-1, *L. reuteri* (previously *fermentum*) B-54 and *L. reuteri* (previously *acidophilus* and *fermentum*). RC-14 also reduced the infection in urinary bladder and vagina (Reid and Bruce 2001, Reid and Bruce 2003, Cadieux et al. 2002, Reid and Burton 2002, Reid and Bocking 2003).

In addition to these, Boris et al, found that *L. acidophilus* decreased the adherent of *Gardenella vaginalis* (Boris et al. 1998). A study by Asahara et al., suggested that *L. casei shirota* is a strain that is possibly useful for the prevention of UTI's (Asahara et al. 2001). Along with human beings, they have also been tested in animals. Patton et al., inserted one capsule of *L. crispatus* CTV-05 intravaginally into ten female animals and found that it colonized the vagina and thus protected the animals from UTI (Patton et al. 2003). Cadieux and coworkers proposed that *Lactobacillus* by products inhibit the growth and virulence of uropathogenic *E. coli* by inhibiting growth, inducing stress and down-regulating proteins critical for host attachment (Cadieux et al. 2009). In another study, it was found that vaginal *L. jensenii* KS119.1 and KS121.1 and *L. gasserii* KS120.1 and KS124.3 strains inhibited the adhesion and growth of uropathogenic *E. coli* IH11128 and 7372 strain (Atassi et al. 2006).

Results of various studies signify that the reoccurrence of UTI can be considerably reduced using one or two probiotic capsules vaginally per week for one year, with no side effects or yeast infections (Reid et al. 1995). The urogenital use of lactobacilli has been a major reason for expanding earlier definitions of "probiotics" from the intestine to: "live micro-organisms which when administered in adequate amounts confer a health benefit on the host" (FAO/WHO 2001).

Same as the probiotics, prebiotics also stimulate the native lactobacilli in the vaginal area. But this has been explored to a lesser extent. In a study by Reid et al., skimmed milk instillation into the vagina was found to significantly increase the lactobacilli counts (Reid et al. 1995, Reid et al. 1999). This was believed to aid in reduction of urinary tract infections through indigenous lactobacilli interference of pathogen ascension from the vagina into the bladder. Similarly, short-chain fructooligosaccharides (scFOS) were identified as Prebiotic for *Lactobacillus plantarum* WCFS1 through microarray tool (Saulnier et al. 2007). Effect of 10 oligosaccharides had shown a prebiotic function on the growth of *Lactobacillus* strain used as probiotics in chicken (Saminathan et al. 2011). This study has also suggested that, probiotic efficiency to utilize a specific prebiotic depends on both specific strain and substrate. Inulin and oligofructose act as prebiotic on *Lactobacillus rhamnosus* and *B. lactis* modulated the intestinal immune function in a rat model (Roller et al. 2004). In another study, synbiotic association of *L. helveticus*

M92 with inulin has shown the best effect on intestinal and fecal microflora and immune system of mice (Frce et al. 2009).

However, in many patients with recurrent urogenital infections, the lactobacilli count is low or absent, and therefore a prebiotic application would be less likely to function. Further *in vitro* studies have been performed to develop a new prebiotic formula that would allow lactobacilli to grow while not supporting growth of pathogens and not having any toxicity effects against the host. This task proved more difficult than was foreseen, especially as *Candida albicans* was able to grow in most media combinations. Nevertheless, the data suggested that some vitamins, mineral and other compounds had the potential to be prebiotics (Reid et al. 1999). The ultimate goal of this approach will be to test if such prebiotics override the local mucosal factors (nutrients, receptor sites) used by pathogens to colonise and infect the host and if a combination of prebiotic and probiotic complement each other in the gastrointestinal tract and would act at the urogenital site. Will prebiotics and probiotics act efficiently in urogenital tract of the host? Since synbiotics (combination of prebiotics and probiotics) work best in comparison of acting individually. Some more studies should be performed in this direction to prove the efficacy of prebiotics and probiotics in UTI's.

Mechanism of Action

Lactobacillus is naturally found in the healthy human vagina and urethra (Dong et al. 2011, Ravel et al. 2011). Urinary tract infection (UTI) is considered as a minor infection; however it can cause severe problems. UTI's have been usually studied during pregnancy, the postpartum period and after genital surgery. It is also common in postmenopausal and elderly women (Perez-Lopez et al. 2009). *Lactobacillus* is non pathogenic, non toxicogenic and it retains viability during storage. These make it a promising candidate to treat clinical manifestations. The mechanisms whereby lactobacilli function as anti-infective defenses are still not fully understood. Probiotic agents exert a helpful effect by using a wide array of actions. These generally resist pathogen by colonizing, production of antimicrobial compounds, inhibition of pathogen by colonization resistance by the direct suppression of harmful microorganisms and the stimulation of beneficial organisms (Fuller and Gibson 1997). These mechanisms vary according to the specific strain or combination of strains used, the presence of prebiotics (a non-digestible food ingredient which beneficially affects the host by selectively stimulating the growth and/or activity of one or a limited number of bacteria in the colon having the potential to improve host healths) (Gupta and Garg 2009) and the condition that is being treated in the patient (Devine and Marsh 2009).

Inhibitory Substances Produced by *Lactobacillus*

Lactic acid and acetic acid: Lactic acid and acetic acid is the important inhibitory substance produced by *Lactobacillus*. The accumulation of these acids results in the acidic pH and inhibition of gram positive and gram negative bacteria. It is responsible for the physiological acidification of the vagina (Hill et al. 1985). Lactic acid is harmful for *Nesseria gonorrhea* (Zhang et al. 1997). These acids diffuse through the membrane of the target organisms, in their hydrophobic undissociated form. This low pH results in inhibition of glycolysis, active transport hinderence and interferes with signal transduction. Anionic ions thus produced cannot diffuse freely through the cell wall and accumulate inside the bacterial cell. Thus, accumulation of anions leads to internal osmotic disorders for the bacteria (Kotikalapudi 2009).

Short chain fatty acids: Probiotics produce short chain fatty acids (SCFA) which lower the pH, favoring the growth of harmless microorganisms (Vandenbergh 1993). These include low molecular weight carboxylic acids with six to eight carbon atoms like acetate, propionate and butyrate (Cummings 1984). Their inhibition activity is high in associated form; since they have low pH and are responsible for the acidic nature. With this acidic pH they can easily penetrate the bacterial cell.

Hydrogen peroxide: Hydrogen peroxide is produced by many species of *Lactobacillus* as the antimicrobial agent (Dahiya and Speck. 1968). This process is oxygen dependent. As *Lactobacillus* does not produce catalase, therefore the hydrogen peroxide produced cannot be degraded and act as potent oxidants by forming the free radicals. These free radicals are harmful and cause irreversible damage to cell components such as enzymes, membrane constituents and DNA (Dalie et al. 2010).

Bacteriocins: Bacteriocins are proteinaceous antimicrobial substances. Sometimes they are also associated with lipids and carbohydrates. Bacteriocins have demonstrated inhibitory action on both gram positive and gram negative bacteria which affect the urogenital tract. Examples of bacteriocins are *Bacillus, Klebsiella, Pseudomonas, Proteus, Salmonella, Shigella, Staphylococcus, Vibrio* species and *E. coli* (Vila et al. 2010).

Effect on microflora: Probiotics modify the resident microflora. This mechanism has long been considered as an important mechanism of action of probiotics. Probiotic bacteria alter the physical environment so that the pathogenic bacteria cannot survive. Probiotics bacteria act by two modes. First they compete with pathogenic bacteria for food and energy source. Secondly, they produce an inhibitory substance which hinders the growth of the pathogen.

Competition for adhesion: The competition for space to adhere, between indigenous bacteria and exogenous pathogen, results in the competition exclusion of pathogenic bacteria (Ohashi and Ushida 2009). Lactobacilli interfere with genitourinary tract pathogens by various mechanisms. These include active exclusion of pathogens from the urinary tract epithelium, co-aggregation with certain pathogen, adherence to epithelial cells and biofilm formation which is auto-aggregation and surface hydrophobicity dependent (Dunne et al. 2001). Auto-aggregation is necessary for the adherence of the *Lactobacillus* to the wall of vaginal and urinary tract. This co-aggregation leads to the formation of a barrier which prevents pathogens from attaching to the wall and colonizing (Zhou et al. 2004). Surface hydrophobicity is an important factor for the adherence of *Lactobacillus* to the vaginal wall (Andreu et al. 1995). *Lactobacillus* cell wall contains lipoteichoic acid; which is responsible for the adhering properties of *Lactobacillus*. Steric hindrance physical phenomenon being the major factor in preventing uropathogens (Revolledo et al. 2006).

Probiotics: as immune enhancer: Advance knowledge in the field of immunology has proved that innate defense mechanisms play an important role in development of several diseases, and these 'danger-signals' (infectious as well as noninfectious) may trigger the body to protect against these diseases. Epithelial barrier consists of a thick mucus layer containing immunoglobulins mainly IgA and antimicrobial compounds. Their dynamic functional role is to regulate permeability between cells (Ohland and MacNaughton 2010). When this barrier function is interrupted due to chronic psychological stress, epithelial ion secretion and permeability is enhanced, binding of luminal bacteria to surface epithelia increases, the uptake of luminal antigens through follicle associated epithelium increases and mucosal inflammation initiates (Zareie et al. 2006).

Innate immune system is the first line of defense system against pathogenic microorganism (Kobayashi and Flavell 2004). Epithelial cells also produce different types of chemokines and cytokines, which are important for the activation of innate immune cells (Kayisli et al. 2002). Along with these, there are some another receptors known as Toll like receptors (TLRs) which are proteins, spanning and non catalytic entities. These TLRs recognized structurally conserved molecules derived from pathogens. Toll-like receptors are now regarded as the key molecules that alert the immune system in the presence of microbial infections. Toll-like receptors (TLRs) recognize antigens of microbial origin; which are referred as pathogen-associated molecular patterns (PAMPs). TLR's recognize various microbial products like lipopolysaccharide (LPS), peptidoglycan, lipoprotein and DNA. Stimulation of these receptors is responsible for the induction of acute inflammatory responses (Amdekar et al. 2011). These receptors are expressed on different immune cells like lymphocytes, macrophages and dendritic cells as well as also found in close proximity of epithelial cells. TLRs help the host to distinguish between pathogen associated molecular patterns (PAMP's) and self. TLR4 recognizes lipopolysaccharides (LPS) and is an important constituent in the cell wall of gram negative bacteria and a causative agent of endotoxin shock (Hoshino et al. 1999, Cario et al. 2002). *Lactobacillus* protects the urinary tract is still a very mysterious process because defending mechanisms are not fully understood till now. Down-regulation of pro-inflammatory cytokines, hydrogen peroxide and blocking the adherence of uropathogens are some of the mechanisms proposed by some researchers (Xia et al. 2006, Anukam et al. 2009, Velraeds et al. 2009). It has been observed that intestinal epithelium and uroepithelium contains several TLR's like TLR2, TLR3, TLR4 and TLR5 (Garcia-Lafuente 2001, Otte and Podolsky 2004). TLR2 and TLR4 are found on tubular cells. TLR4 actively participate in the clearance of uropathogens (mainly *E. coli*). Lipid A (a form of LPS) activates TLR4 in macrophages and this triggers the biosynthesis of diverse mediators of inflammation, in mononuclear and endothelial cells; lipid A also stimulates tissue factor production. Such events are desirable for clearing local infections. MyD88, a cytoplasmic adaptor molecule, is important for the signaling of IL-1R/TLR family. Ligand binding to IL-1R/TLR family results in the recruitment of MyD88 to Toll/IL-1 receptor domain, which bridges the signal to IL-1R-associated kinase. Eventually, the activation of transcription factor NF-kB occurs and it permits the transactivation of cytokine genes such as TNF-α and IL1-β and activates the production of co-stimulatory molecules required for the adaptive immune response.

Prebiotics beneficially affect the probiotics bacteria; as prebiotics increase the beneficial bacterial population and inhibit other bacteria and yeast infections.

Important Concerns about Probiotics and Prebiotics

One thing that is/are taken into account before using probiotics and prebiotics; viable cells are generally more effective at stimulating adaptive immunity, and the method used for cell killing should be considered if nonviable cells are used. While using cell supernatants, the most active compound should be purified and the physical and chemical properties of these compounds should be known. Doses of bacteria and growth phase at time of harvest are additional considerations in tandem with traditional methods of determining strain robustness or functional effects. Prebiotic should be an inert and harmless compound which should not react with the body. On the other side, genetic influences, site of action, the mucus and epithelial barrier, method of administration, diet and normal microflora of the host are also the important factors of using probiotics and prebiotics. In future trials, individual factors that potentially affect the efficacy of probiotics and prebiotics must be addressed. Continuing advancement in technologies, knowledge of the immune system, gut microbiota and improved biomarkers are essential to making human interventional studies with probiotics and prebiotics successful.

Summary and Conclusion

In summary, probiotic applications to areas of the body other than the intestine have a long history, likely starting with treatment of wounds and throat infections in paleolithic times. The resurgence of interest in this area is multidimensional, in part due to antibiotic failures and side effects, increasing infection rates despite current therapies, and a desire for more 'natural' therapies. In this era of increasing bacterial resistance to antibiotic therapy, progress in alternative harmless approaches is of utmost importance. Since infection is site-oriented probiotic therapy is one of the most encouraging therapeutics for the prevention of genitourinary tract infection in the post-antibiotic era. Probiotics and prebiotics have great potential, particularly today, with the increasing threat of antibiotic over-usage and prevalence of antibiotic resistant microorganisms. These are safe and effective. Their role will get a likely boost in the near future as evidence accrues from well documented studies on the efficacy of these agents when used in standardized and regulated formulations. The potential for prebiotics and probiotics to be used as an adjunct in the control of antibiotic resistance is particularly interesting. It is essential that the search be continued for useful prebiotic and most appropriate probiotic strains to be used in reducing urinary tract infection.

References

Amdekar, S., D. Dwivedi, P. Roy, S. Kushwah and V. Singh. 2010. Probiotics: multifarious oral vaccine against infectious traumas. FEMS Immunol. Med. Microbiol. 58(3): 1–8.

Amdekar, S., V. Singh and D.D. Singh. 2011. Probiotic therapy: Immunomodulating approach towards Urinary Tract Infection. Current Microbiol. 63(5): 484–490.

Andreu, A., A. Stapleton, C. Fennell, S. Hillier and E. Stamm. 1995. Hemagglutination, adherence and surface properties of vaginal *Lactobacillus* species. J. Infect. Dis. 171: 1237–1243.

Anukam, K.C., K. Hayes, K. Summers and G. Reid. 2009. Probiotic *Lactobacillus rhamnosus* GR-1 and *Lactobacillus reuteri* RC-14 may help downregulating TNF-α, Il-6, Il-8, Il-10 and IL-12 (p70) in the neurogenic bladder of spinal cord injured patient with urinary tract infection: A two case study. Adv. Urol. Article ID 680363.

Asahara, T., K. Nomoto, M. Watanuki and T. Yokokura. 2001. Antimicrobial activity of intrauretrally admisitered prebiotic *Lactobacillus casei* in a murine model of *Escherichia coli* urinary tract infection. Antimicrob. Agents Chemothera 45(6): 1751–1760.

Atassi, F., D. Brassart, P. Grob, F. Graf and A.L. Servin. 2006. Vaginal *Lactobacillus* isolates inhibit uropathogenic *Escherichia coli*. FEMS Microbiol. Lett. 257: 132–138.

Barrow, L., K.P. Steed, R.C. Spiller, P. J. Watts, C.D. Melia, M.C. Davies and C.G. Wilson. 1992. Scintigraphic demonstration of lactulose-induced accelerated proximal colon transit. Gastroenterol. 103: 1167–1173.

Bernet, M.F., D. Brassart, J.R. Neeser and A.L. Servin. 1994. *Lactobacillus acidophilus* LA 1 binds to human intestinal lines and inhibits cell attachment and cell invasion by enterovirulent bacteria. Gut. 35: 483–489.

Boehm, G., J. Jelinek, B. Stahl, K. van Laere, J. Knol, S. Fanaro, G. Moro and V. Vigi. 2004. Prebiotics in infant formulas. J. Clin. Gastroenterol. 38: S76–S79.

Borchert, D.L., A. Sheridan, A. Papatsoris, Z. Faruqu, J.M. Barua, I. Junaid et al. 2008. Prevention and treatment of urinary tract infection with probiotics: Review and Research perspective. Indian J. Urol. 24: 139–144.

Boris, S., J.E. Su´arez, F. V'azquez and C. Barb'es. 1998. Adherence of human vaginal lactobacilli to vaginal epithelial cells and interaction with uropathogens. Infection and Immunity, vol. 66(5): 1985–1989.

Bruce, A.W., P. Chadwick, A. Hassan and G.F. 1973. Recurrent urethritis in women. Can. Med. Assoc. J. 108 (8): 973–976.

Cadieux, P.A., J.P. Burton, E. Devillard and G. Reid. 2009. *Lactobacillus* byproducts inhibits the growth and virulence of uropathogenic *Escherichia coli*. J. Physiol. Pharmacol. 60(6): 13–18.

Cadieux, P., J. Burton, C.Y. Kang, G. Gardiner, I. Braunstein, A.W. Bruce and G. Reid. 2002. *Lactobacillus* strains and vaginal ecology. JAMA 287: 1940–1.

Canny, G.O. and B.A. McCormick. 2008. Minireview: Bacteria in the intestine, helpful residents or enemies from within? Inf. Immun. 76: 3360–3373.

Cario, E., D. Brown, M. McKee, K. Lynch-Devaney, G. Gerken and D.K. Podolsky. 2002. Commensal-associated molecular patterns induce selective toll-like receptor-trafficking from apical membrane to cytoplasmic compartments in polarized intestinal epithelium. Am. J. Pathol. 160: 165–173.

Crittenden, R. and M.J. Payne. 2008. Nutrition News. Facts and functions of prebiotics, probiotics and synbiotics., pp 1–2. Department of Human Nutrition, K-State Research and Extension, Kansas State University; Prebiotics. *In:* Y.K. Lee and S. Salminen (eds.). *Handbook of Probiotics and Prebiotics*. Wiley-Interscience, 2nd ed.; Chapter 4, pp. 535–582.

Crittenden, R.G. 1999. Prebiotics. *In: Probiotics: A Critical Review.* Edited by Tannock GW; Wymondham, UK; Horizon Scientific Press 141–156.

Cummings, J.H., G.T. Macfarlane and H.N. Englyst. 2001. Prebiotic digestion and fermentation. Am. J. Clin. Nutr. 73(suppl.): 415S–420S.

Cummings, J.H. 1984. Colonic absorption: The importance of SCFA in man. Scandinavian J. Gastroenterol. 19: 89–99.

Dahiya, R.S. and M.L. Speck. 1968. Hydrogen peroxide formation by lactobacilli and its effects on *Staphylococcus aureus*. J. Dairy Sci. 51(10): 1068–1072.

Dalie, D.K.D., A.M. Deschamps and F. Richard-Forget. 2010. Lactic acid bacteria—Potential for control of mould growth and mycotoxins: A review. Food Control. 21: 370–380.

Devine, D.A. and P.D. Marsh. 2009. Prospects for the development of probiotics and prebiotics for oral applications. J. Oral Microbiol. DOI: 10.3402/jom.v1i0.1949.

Dong, Q., D.E. Nelson, E. Toh, L. Diao, X. Gao, J.D. Fortenberry and B. Van der Pol. 2011. The microbial communities in male first catch urine are highly similar to those in paired urethral swab specimens. PLoS One 6:e19709.

Douglas, L.C. and M.E. Sanders. 2008. Probiotics and prebiotics in dietetics practice. J. Am. Diet Assoc. 108: 510–521.

Dunne, C., L. O'Mahony, L. Murphy, G. Thornton, D. Morrissey, S. O'Halloran, M. Feeney, S. Flynn, G. Fitzgerald, C. Daly, B. Kiely, G.C. CO'Sullivan, F. Shanahan and J.K. Collins. 2001. *In vitro* selection criteria for probiotic bacteria of human origin: Correlation with *in vivo* findings. Am. J. Clin. Nutr. 73: 386–392.

FAO/WHO. Food and Agriculture Organization of the United Nations and World Health Organization. 2001. (Evaluation of health and nutritional properties of powder milk and live lactic acid bacteria). Expert Consultation Report. FAO/WHO, Rome (http://www.fao.org/es/ ESN/Probio/probio.htm).

Floch, M.H. and J. Hong-Curtiss. 2002. Probiotics and Functional Foods in Gastrointestinal Disorders. Current Treatment Options in Gastroenterol. 5: 311–321.

Fooks, L.J. and G.R. Gibson. 2002. *In vitro* investigations of the effect of probiotics and prebiotics on selected human intestinal pathogens. FEMS Microbiol. Ecol. 39: 67–75.

Forsum, U., E. Holst and P.G. Larsson. 2005. Bacterial vaginosis a microbiological and immunological enigma. APIMS 113: 81–90.

Frece, J. B. Kos, I.K. Svetec, Z. Zgaga, J. Beganović, A. Lebos and J. Susković. 2009. Synbiotic effect of *Lactobacillus helveticus* M92 and prebiotics on the intestinal microflora and immune system of mice. J. Dairy Res. 76(1): 98–104.

Freter, R. 1955. The fatal enteric cholera infection in the guinea pig achieved by inhibition of normal enteric flora. J. Infect. Dis. 97: 57–64.

Fuller, R. and G.R. Gibson. 1997. Modification of the intestinal microflora using probiotics and prebiotics. Scand. J. Gastroenterol. 222 Suppl.: 28–31.

Fuller, R. 1989. Probiotics in man and animals. J. Appl. Bacteriol. 66: 365–378.

Garcia-Lafuente, A., M. Antolin, F. Guarner, E. Crespo and J.R. Malagelada. 2001. Modulation of colonic barrier function by the composition of the commensal flora in the rat. Gut. 48: 503–507.

Gibson, G.R. and M.B. Roberfroid. 1995. Dietary modulation of the human colonic microbiota. Introducing the concept of prebiotics. J. Nutr. 125: 1401–1412.

Gibson, G.R. and M.B. Roberfroid (eds.). 1999. Colonic Microbiota, Nutrition and Health. Kluwer Academic Publishers, Dodrecht.

Gupta, V. and R. Garg. 2009. Probiotics. Indian J. Med. Microbiol. 27(3): 202–209.

Hebden, J.M., P.J. Gilchrist, A.C. Perkins, C.G. Wilson and R.C. Spiller. 1999. Stool water content and colonic drug absorption: contrasting effects of lactulose and codeine. Pharm. Res. 16: 1254–9.

Hill, G.M, W.E. Neville, K.L. Jr. Richardson, P.R. Utley and R.L. Stewart. 1985. Castration method and progesterone-estradiol implant effects on growth rate of suckling calves. J. Dairy Sci. 68: 11: 3059–3061.

Hoshino, K. O. Takeuchi, T. Kawai, H. Sanjo, T. Ogawa, Y. Takeda, K. Takeda and S. Akira. 1999. Cutting edge: Toll-like receptor 4 (TLR4)-deficient mice are hyporesponsive to lipopolysaccharide: evidence for TLR4 as the Lps gene product. J. Immunol. 162: 3749–3752.

John, A., N.D. Catanzaro and Lisa Green. 1997. Microbial Ecology and Probiotics in Human Medicine (Part II). Alternative review medicine 2(4): 296–305.

Kaewnopparat, S. and N. Kaewnopparat. 2009. Formulation and Evaluation of vaginal Suppositories containing *Lactobacillus*. World Academy of Science, Engineering and Technology 55: 640–643.

Kalliomaki, M., J.M. Antoine, U. Herz, G.T. Rijkers, J.M. Wells and A. Mercenier. 2010. Guidance for substantiating the evidence for beneficial effects of probiotics: prevention and management of allergic diseases by probiotics. J. Nutr. 140: 713S–721S.

Kaur, I.P., A. Kuhad, A. Garg and K. Chopra. 2009. Probiotics: delineation of prophylactic and therapeutic benefits. J. Med. Food. 12: 219–235.

Kayisli, U.A., N.G. Mahutte and A. Arici. 2002. Uterine chemokines in reproductive physiology and pathology. Am. J. Reprod. Immuno.l 47: 213–221.

Kopeloff, N. 1926. *Lactobacillus Acidophilus*. Baltimore: Williams & Wilkins.

Kobayashi, K.S. and R.A. Flavell. 2004. Shielding the double-edged sword: negative regulation of the innate immune system. Leukocyte Biol. 75: 428–433.

Kotikalapudi, B.L. 2009. Characterization and encapsulation of probiotic bacteria using a pea-protein alginate matrix. Master thesis. University of Saskatchewan. Canada.

Kullen, M.J. and J. Bettler. 2005. The delivery of probiotics and prebiotics to infants. Curr. Pharm. Des. 11(1): 55–74.

Kumar, M., A. Kumar, R. Nagpal, D. Mohania, P. Behare, V. Verma, P. Kumar, D. Poddar, P.K. Aggarwal, C.J. Henry, S. Jain and H. Yadav. 2010. Cancer-preventing attributes of probiotics: an update. Int. J. Food. Sci. Nutr.

Lilly, D.M. and R.H. Stillwell. 1965. Probiotics: growth promoting factors produced by micro-organisms. Science 147: 747–748.

Saminathan, M., C.C. Sieo, R. Kalavathy, N. Abdullah and Y.W. Ho. 2011. Effect of prebiotic oligosaccharides on growth of *Lactobacillus* strains used as a probiotic for chickens. African J. Microbiol. Res. 5(1): 57–64.

Mcfarlane, G. and J.H. Cummings. 1999. Probiotics and prebiotics: Can regulating the activities of intestinal bacteria benefit health? BMJ 318: 999–1003.

Metchnikoff, E. 1907. The Prolongation of Life: Optimistic Studies. London: Butterworth-Heinemann.

Ohashi, Y. and U. Ushida. 2009. Health-beneficial effects of probiotics: its mode of action. Anim. Sci. J. 80: 361–371.

Ohland, C.L. and W.K. MacNaughton. 2010. Probiotic bacteria and intestinal epithelial barrier function. Am. J. Physiol. Gastrointest. Liver. Physiol. 298: G807–G819.

Otte, J.M. and D.K. Podolsky. 2004. Functional modulation of enterocytes by gram-positive and gram-negative microorganisms. Am. J. Physiol. Gastrointest. Liver Physiol. 286: G613–G626.

Parvez, S., K.A. Malik, S. Ah Kang and H.Y. Kim. 2006. Probiotics and their fermented food products are beneficial for health. J. Appl. Microbiol. 100(6): 1171–85.

Pascual, L.M., F. Ruiz, W. Giordano and L. Barberis. 2010. Vaginal colonization and activity of the probiotic bacterium *Lactobacillus fermentum* L23 in a murine model of vaginal tract infection. J. Med. Microbiol. 59: 360–364.

Patton, D.L., Y.T. Cosgrove Sweeney, M.A. Antonio, L.K. Rabe and S.L. Hillier. 2003. *Lactobacillus crispatus* capsules: single-use safety study in the Macaca nemestrina model. Sexually Transmiltted Diseases 30(7): 568–570.

Perez-Lopez, F.R., J. Hava and P. Chedraui. 2009. *Vaccinium macrocarpon*: An interesting option for women with recurrent tract infections and other health benefits. J. Obs. Gynaecol. Res. 35(4): 630–639.

Rahn, D.D. 2008. Urinary tract infections: contemporary management. Urol. Nurs. 28(5): 333–341.

Ravel, J., P. Gajer, Z. Abdo, G.M. Schneider, S.S. Koenig, S.L. McCulle, S. Karlebach, R. Gorle, J. Russell, C.O. Tacket et al. 2011. Vaginal microbiome of reproductive age women. Proc. Natl. Acad. Sci. USA 108(Suppl. 1): 4680–4687.

Reid, G. 1999. Potential preventive strategies and therapies in urinary tract infection. World J. Urol. 17: 359–363.

Reid, G. and A. Bocking. 2003. The potential for probiotics to prevent bacterial vaginosis and preterm labor. Am. J. Obs. Gynecol. 189(4): 1202–8.

Reid, G. and A.W. Bruce. 2006. Probiotics to prevent urinary tract infections: the rationale and evidence. World J. Urol. 24(1): 28–32.

Reid, G., A.W. Bruce, N. Fraser, C. Heinemann, J. Owen and B. Henning. 2001. Oral probiotics can resolve urogenital infections. FEMS Immunology & Medical Microbiology 30(1): 49–52.

Reid, G., A.W. Bruce and M. Taylor. 1995. Instillation of *Lactobacillus* and stimulation of indigenous organisms to prevent recurrence of urinary tract infections. Microecol. Ther. 23: 32–45.

Reid, G. and A.W. Bruce. 2001. Selection of *Lactobacillus* strains for urogenital probiotic applications. J. Infect. Dis. 183(S1): S77–80.

Reid, G. and A.W. Bruce. 2003. Urogenital infections in women –Can probiotics help? Postgrad. Med. J. 79: 429–32.

Reid, G. and J. Burton. 2002. Use of *Lactobacillus* to prevent infection by pathogenic bacteria. Microbes Infect. 4: 319–24.
Reid, G. 2001. Probiotic agents to protect the urogenital tract against infection. Am. J. Clin. Nutr. 73(suppl.): 437S–443S.
Rettger, L.F., M.N. Levy, L. Weinstein and J.E. Weiss. 1935. Lactobacillus Acidophilus and its Therapeutic Application. London: Yale University Press.
Revolledo, L., A.J.P. Ferreira and G.C. Mead. 2006. Prospects in *Salmonella* control: competitive exclusion, probiotics, and enhancement of avian intestinal immunity. J. Appl. Poult. Res. 15: 341–351.
Roller, M., G. Rechkemmer and B. Watzl. 2004. Prebiotic inulin enriched with oligofructose in combination with the probiotics *Lactobacillus rhamnosus* and *Bifidobacterium lactis* modulates intestinal immune functions in rats. J. Nutr. 134(1): 153–6.
Salimen, S. and A.C. Ouwehannd and E. Isolauri. 1998. Clinical application of probiotic bacteria. Int. Dairy J. 8: 563–572.
Samra, K.S.S., S. Cenk, G.M.G. Horu and O. Beril. 2005. Increasing antimicrobial resistance of *Escherichia coli* isolates from community acquired UTI during 1998–2003. Manisa Turkey Japan J. Infec. Dis. 58: 159–161.
Saulnier, D.M.A., D. Molenaar, W.M. de Vos, G.R. Gibson and S. Kolida. 2007. Identification of Prebiotic Fructooligosaccharide Metabolism in *Lactobacillus plantarum* WCFS1 through Microarrays. Appl. Environ. Microbiol. 73: 1753–1765.
Schmitt, C., J.D. Sobel and C. Meriwether. 1992. Bacterial vaginosis: treatment with clindamycin cream versus oral metronidazole. Obst. Gynecol. 79(6): 1020–1023.
Sekhon, B.S. and S. Jairath. 2010. Prebiotics, probiotics and synbiotics: an overview. J. Pharm. Edu. Res. 1(2): 13–36.
Singh, V., K. Singh, S. Amdekar, D.D. Singh, P. Tripathi, G.L. Sharma and H. Yadav. 2008. Innate and specifc gut-associated immunityand microbial Interference. FEMS Immunol. Med. Microbiol. 55(1): 6–12.
Standiford, C., L. Green and R. Van Harrison. 2005. Urinary Tract Infection Guideline Team. Taubman Medical Library 1–9.
Szajewska, H. 2007. Probiotics and prebiotics in pediatrics: where are we now? The Turkish Journal of Pediatrics 49: 231–244.
Trois, L., E.M. Cardoso and E. Miura. 2008. Use of probiotics in HIV-infected children: a randomized double-blind controlled study. J. Trop. Pediatr. 54: 19–24.
Tuohy, K.M., H.M. Probert, C.W. Smejkal and G.R. Gibson. 2003. Using probiotics and prebiotics to improve gut health. Drug Discov. Today 8: 692–700.
Unne, C., L. O'Mahony, L. Murphy et al. 2001. *In vitro* selection criteria for probiotic bacteria of human origin: correlation with *in vivo* findings. Am. J. Clin. Nutr. 73(suppl.): 386S–392S.
Vandenbergh, P.A. 1993. Lactic acid bacteria, their metabolic products and interference with microbial growth. FEMS Microbiol. Rev. 12: 221–238.
Velraeds, M.M., H.C. Van der Mei, G. Reid and H.J. Busscher. 2009. Inhibition of initial adhesion of uropathogenic *Enterococcus faeclis* by biosurfactants from *Lactobacillus* isolates. App. Environ. Microbiol. 62(6): 1958–1963.
Vila, B.E., Esteve- Garcia and J. Brufau. 2010. Probiotic micro-organism: 100 years of innovation and efficacy; modes of action. Word's Poult. Sci. J. 65: 369–380.
Xia, Y., K. Yamagata and T.L. Krukoff. 2006. Differential expression of the CD14/TLR4 complex and inflammatory signaling molecules following i.c.v. administration of LPS. Brain Res. 1095: 85–95.
Yao, T.C., C.J. Chang, Y.H. Hsu and J.L. Huang. 2010. Probiotics for allergic diseases: Realities and myths. Pediatr. Allergy Immunol. 21: 900–919.

Zareie, M., K. Johnson-Henry, J. Jury, P.C. Yang, B.Y. Ngan, D.M. McKay, J.D. Soderholm, M.H. Perdue and P.M. Sherman. 2006. Probiotics prevent bacterial translocation and improve intestinal barrier function in rats following chronic psychological stress. Gut. 55: 1553–1560.

Zhang, J., A.G. Thomas and E. Leybovich. 1997. Vaginal douching and adverse health effects: a meta-analysis. Am. J. Public Health 87: 1207–1211.

Zhou, X., S.J. Bent, M.G. Schneider, C.C. Davis, M.R. Islam and L.J. Forney. 2004. Characterization of vaginal microbial communities in adult healthy women as determined by sequencing clones and terminal restriction fragment length polymorphism (T-RFLP) analyses. Microbiol. 150: 2565–2573.

22

Genomics of Probiotics and Prebiotics

*Daniel J. O'Sullivan** and *Prakash M. Halami*[1]

Introduction

The scientific understanding of the probiotic concept has come a long way, since it was first conceived over a hundred years ago by Elie Metchnikoff in his insightful book, 'The Prolongation of Life' (Metchnikoff and Mitchell 1907). However, there is still a significant way to go before the association of specific probiotic health benefits with specific strains of bacteria can be fully scientifically proven. While many researchers believe this association is already scientifically established for a number of probiotic strains, the current failure of the European Food Safety Authority (EFSA) to approve any probiotic health claim for even the most highly studied strains suggests a better understanding of both the probiotic strain and the mechanism of action for its health benefit is needed. This is substantiated by the frequent use of the term 'inadequate strain characterization' by EFSA when rejecting many of these petitions. Genomics holds tremendous promise to address much of the scientific shortfall in probiotic understanding, especially characterization of probiotic strains, and many believe that it will provide the necessary scientific understanding to obtain probiotic strains with the highest efficacy for health claims such that regulatory approval will not be

Department of Food Science and Nutrition, Microbial and Plant Genomics Institute, University of Minnesota, 1500 Gortner Ave., St. Paul, MN 55108, USA.
[1] Current address: CSIR-Central Food Technological Research Institute, Mysore, India.
* Corresponding author: dosulliv@umn.edu

hindered. It is therefore reasonable to predict that regulatory authorities such as EFSA and FDA (Food and Drug Administration) will be satisfied with the scientific evidence associating specific probiotic attributes with specific strains in the not too distant future.

Genome Sequencing of Probiotic Bacteria

The genomics era in biological research truly began in 1995 following the publication of the first complete genome of a living organism, *Haemophilus influenza* (Fleishmann et al. 1995). This publication changed the approach to genome sequencing from a directed approach to a random shotgun approach and greatly increased the speed for generating genome sequences. Since then there have been further advances in sequencing technologies that have seen genome sequencing become an affordable and a routine practice today. The availability of genome sequences has revolutionized all aspects of biological research from human medicine to prokaryotic understanding and is currently furthering our understanding of probiotic bacteria and their association with the human gut.

The last decade has seen the publication of the genome sequences of strains representing all the major species associated with probiotics. These are listed in Table 1 and only include species that have a strong scientific basis for use in probiotics and are normal residents of the GI tract of most humans, with the possible exception of *B. animalis* subsp. *lactis*. There are many other organisms that are promoted and used commercially as probiotics, but there are currently too many unanswered questions about their use as probiotics to include them. For example, *E. coli* strain Nissle 1917 has attracted a lot of attention (Trebichavsky et al. 2010), but with the direct linkage between higher numbers of gram negative bacteria in the gut and endotoxin A in the blood stream, which is a potent stimulator of TNFα (pro-inflammatory cytokine), questions are raised about the use of any gram negative bacteria as a probiotic. Other organisms, such as the spore forming *Bacillus coagulans* and the yeast *Saccharomyces boulardii* are also promoted as probiotics but as they are not a normal part of GI flora in humans' questions arise about their functionality in the gut. Given metabolic functionality or being a normal resident of the GI flora are not necessarily prerequisites for a probiotic organism, these and other organisms may have probiotic functionality. However, further mechanistic understanding is required before their general inclusion as probiotics.

Table 1. *Lactobacillus* and *Bifidobacterium* species relevant to probiotic applications[1]

Lactobacillus	**Genome sequence (Acc No.)**	**Reference**	***Bifidobacterium***	**Genome sequence (Acc No.)**	**Reference**
L. acidophilus NCFM	CP000033	Altermann et al. 2005	*B. adolescentis* ATCC 15703	NC_008618	GB[2]
L. casei BD-11	CP002618	Ai et al. 2011	*B. animalis* subsp. *lactis* Bb-12	CP001853	Garrigues et al. 2010.
L. crispatus ST1	FN692037	Ojala et al. 2010	*B. breve* UCC203	CP000303	O,Connell-Motherway et al. 2011
L. fermentum CECT 5716	CP002033	Jiménez et al. 2010a	*B. bifidum* BGN4	CP001361	Yu et al. 2012
L. gasseri ATCC 3323	CP000413	Makarova et al. 2006	*B. longum* DJO10A	CP000605	Lee et al. 2008
L. helveticus ROO52	CP003799	Tompkins et al. 2012	*B. longum* subsp. *infantis* ATCC 15697	CP001095	Sela et al. 2008
L. johnsonii NCC 533	AE017198	Pridmore et al. 2004			
L. paracasei ATCC 25302	NA[3]	GB			
L. plantarum ST-III	CP002222	Wang et al. 2011			
L. reuteri JCM 1112	AP007281	Morita et al. 2008			
L. rhamnosus ATCC 53103	AP011548	Morita et al. 2009			
L. salivarius CECT 5713	CP002034	Jiménez et al. 2010b			

[1]Select strains representing each species are given; [2]only published in GenBank; [3]not applicable as only contig sequences deposited in GenBank

Genome Characteristics of Probiotic Lactobacilli and Bifidobacteria

Lactobacilli and bifidobacteria represent two distinct phyla of bacteria, the *Firmicutes* and *Actinobacteria* respectively. While they are genetically quite distinct, as illustrated in a phylogenetic tree based on their 16S rRNA gene sequences (Fig. 1), they have many phenotypic similarities, such as producers of lactic acid, and are important in fermented foods. For this

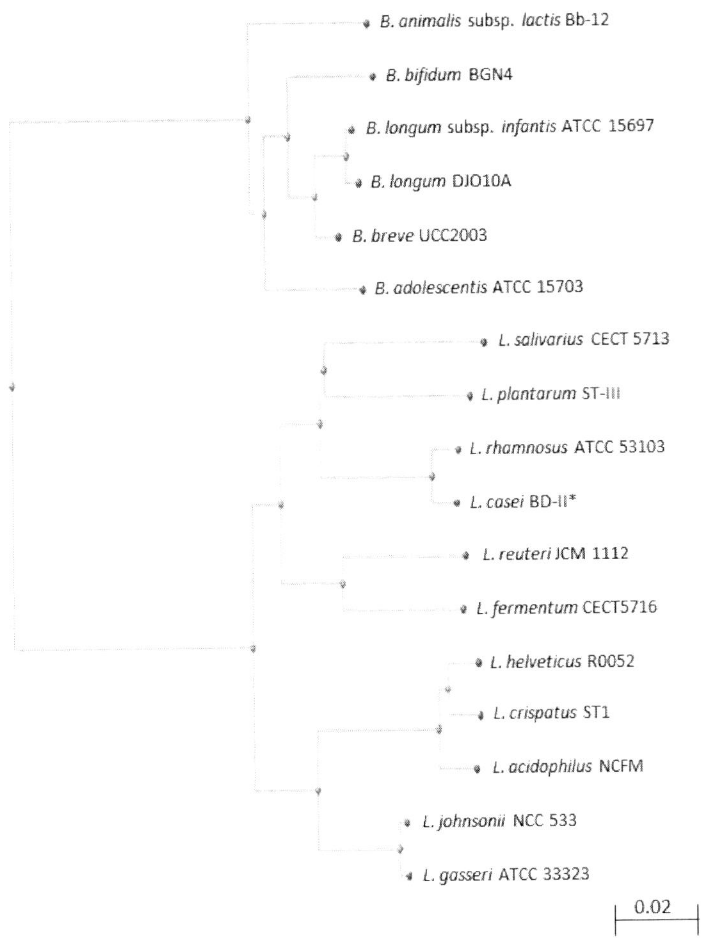

Fig. 1. Phylogenetic tree, generated using the BLAST Tree View software at the NCBI (National Center for Biotechnology Information), of the selected probiotic lactobacilli and bifidobacteria strains based on the 16S rRNA gene sequences deposited in GenBank. *, Indicates that the current genome sequence annotation data for this strain incorrectly annotates the 16S rRNA genes. The one chosen for this analysis was located within gene LCBD_0256, currently annotated as a hypothetical protein, at coordinates 256154–257727.

reason, bifidobacteria are often included in the Lactic Acid Bacteria (LAB), but since the LAB are a taxonomic group of genetically similar bacteria, the bifidobacteria cannot be included from a taxonomic point of view. Both are inhabitants of the GI flora of humans and animals, with the lactobacilli believed to be important for small intestine functionality and bifidobacteria for colon functionality (O'Sullivan 2005).

All the genomes of the strains representing the species of probiotic lactobacilli and bifidobacteria depicted in Table 2 consist of a single chromosome and in some cases one to three plasmids. It should be noted that plasmids are strain dependent and not all strains of these species contain plasmids. It is notable that plasmids are relatively common among lactobacilli compared to bifidobacteria and tend to be larger and encoding functional traits, such as bacteriocin production. In some cases mega plasmids are present, such as the 243 kb mega plasmid in *L. salivarius* CECT 5713 which encodes its salivaricin bacteriocin (Jiménez et al. 2010b). Plasmids are quite rare for bifidobacteria, typically only occurring in some strains of *B. longum* and are generally small and cryptic in nature (Lee and O'Sullivan 2006).

The genome sizes of the probiotic lactobacilli are quite diverse ranging from *L. gasseri* at 1.89 Mb to *L. plantarum* at 3.25 Mb (Table 2). The genome sizes of the probiotic bifidobacteria are more conserved at around 2 Mb, with *B. longum* subsp. *infantis* being notably the largest at 2.83 Mb. This increase in size over other bifidobacteria has been suggested to reflect its increased ability to utilize the diverse oligosaccharides present in human breast milk (Sela et al. 2008). Genome size is believed to be correlated with habitat, with more complex habitats requiring more features for successful existence, thus favoring larger genomes. The reverse is also the case, with bacteria exhibiting genome reduction when propagated over time in low complex and consistent habitats. This concept is best illustrated by the species of the symbiont *Buchnera*, which have reduced the size of their genomes dramatically during coevolution with their aphid hosts. This resulted in one of the smallest genome reported thus far at 450 kp, which is smaller than genomes for *Mycoplasma* which had previously been known as the smallest (Gill et al. 2002). Genome reduction has also been experimentally demonstrated for bifidobacteria during growth in laboratory pure culture media and constant environmental conditions. This was shown by growing *Bifidobacterium longum* DJO10A for ~ 1,000 generations in a laboratory medium and loosing > 50 kb from its genome size, specifically through two large deletions (Lee et al. 2008). These genome deletions were proposed to involve *IS*30 elements, given this insertion element was demonstrated to be active within this strain and also its association with the ends of one of the deleted regions. This therefore represents a significant challenge for the probiotic industry as retaining the integrity of a culture is critical for

Table 2. Genome characteristics of probiotic lactobacilli and bifidobacteria.

Bacterium	Genome		ORFs	Pro-phage	Bacteriocin	Number of R/M systems[1] and CRISPRs[2]
	Chr (Mb)	Plasmid no. (kb)				
L. acidophilus NCFM	1.99	0	1864	0	Helveticin J	CRISPR (1)
L. casei BD-II	3.07	1 (57)	3139	1	-	I (1), II (1), CRISPR (1)
L. crispatus ST-1	2.04	0	2024	0	Helveticin J	I (1), CRISPR (3)
L. fermentum CECT 5716	2.10	0	1109	0	-	CRISPR (2)
L. gasseri ATCC 3323	1.89	0	1810	0	Helveticin	I (1)
L. helveticus RO052	2.13	1 (6.4)	1989	0	Helveticin J	I (1), CRISPR (1)
L. johnsonii NCC 533	1.99	0	1821	2	Lactacin-F	CR?[3]
L. paracasei ATCC 253022[4]	2.88	-	-	-	-	-
L. plantarum ST-III	3.25	1 (54)	3013	3	Plantaricin	I (1)
L. reuteri JCM1112	2.03	0	1820	0	-	I (2)
L. rhamnosus ATCC 53103	3.00	0	2834	1	-	CRISPR (2)
L. salivarius CECT 5713	1.83	3 (45, 20 and 243)	1558	0	Salivaricin ABP-118	CRISPR (1)
B. adolescentis ATCC 15703	2.08	0	1630	0	-	I (1), III (1), CRISPR (2)
B. animalis subsp. *lactis* Bb-12	1.94	0	1642	0	-	II (1), CRISPR (1)
B. breve UCC2003	2.42	0	1854	1	-	CRISPR (3)
B. bifidum BGN4	2.22	0	1835	-	-	I (1)
B. longum DJO10A	2.38	2 (10, 3.6)	1990	1	Bisin	I (1), II (2), CRISPR (1)
B. longum subsp. *infantis* ATCC 15697	2.83	0	2498	3	Lactococcin 972	II (1), CR?

[1]Refers to type I, II or III restriction modifications systems; [2]CRISPRs were located using the online CRISPR finder from the Université de Paris Sud-11; [3]represents a partial CRISPR with one spacer and no *cas* genes; [4]only genome contigs deposited in GenBank. Chr, Chromosome; ORF, open reading frame

its optimum performance as a probiotic. Given that the majority of current cultures used as probiotics in foods today were isolated and laboratory maintained long before knowledge that genome reduction can occur over relative short evolutionary time frames, it is likely they have changed since their original isolation from the gut. These changes invariably involve genome regions that are not required for pure culture growth, but may be important for successful functionality in the gut. In the case of *B. longum* DJO10A, one of its genome deletions was characterized and it involved a deletion of the complete gene cluster involved in production of the broad spectrum lantibiotic, bisin. While lantibiotic production can be an important function for ecological competition in a gut environment, it is not needed during growth in pure culture and this loss of selective pressure can make it vulnerable for loss.

Antimicrobial production is an important feature for probiotic organisms to function in the gut. While production of lactate as well as acetate by bifidobacteria are important antimicrobials, bacteriocins can also provide a competitive edge for bacteria. The genomes of the probiotic lactobacilli and bifidobacteria do show this potential for some members, with seven lactobacilli genomes exhibiting bacteriocin genes and two bifidobacteria genomes (Table 2). In addition to this, the genome of *Lactobacillus reuteri* JCM 1112 exhibits the potential to produce reuterin, which is a very broad spectrum non-peptide antimicrobial believed to be important for its probiotic characteristics (Morita et al. 2008). Structurally, reuterin is 3-hydroxypropionaldehyde and is produced from anaerobic metabolism of glycerol (Talarico and Dobrogosz 1989). It is believed that *L. reuteri* acquired this ability via horizontal gene transfer, as reuterin production, together with cobalamin (vitamin B12) production, is encoded in a 58 gene island within its chromosome (Morita et al. 2008). While genomic islands are frequently associated with virulance functions in pathogenic bacteria, this is the first clear example of a genomic island in a probiotic bacterium associated with its proposed probiotic functionalities.

While bacteriocin production is a common feature among many of the LAB, it is not common in bifidobacteria. The characterization of lantibiotic production by *B. longum* DJO10A demonstrated a more complex transcriptional regulation controlling its production, such that no lantibiotic production occurs during growth in broth media (Lee et al. 2011). Given that traditional searches for bacteriocins involved growing cultures in broth media prior to conducting bioassays, it may explain the paucity of bacteriocins found for bifidobacteria thus far. This is not a common regulatory mechanism for bacteriocin production by the LAB or other bacteria, although the streptin lantibiotic produced by *Streptococcus pyogenes* is also not produced during growth in broth media (Wescombe and Tag 2003). Lantibiotics are a class of bacteriocins that undergo post translation modifications and are generally active against a wider range of bacteria than other types of bacteriocins. They are therefore of great interest as natural food preservatives and also for potential probiotic functionality. Todate, their production has only been detected for two phyla of bacteria, the *Firmicutes* and *Actinobacteria*, with the majority of those characterized coming from the *Firmicutes* (Li and O'Sullivan 2012). While lantibiotics can inhibit a wide range of bacteria, this range is limited to just gram positive bacteria, as the outer membrane of gram negative bacteria protects them from the lantibiotic, which needs to access the cell membrane to exert activity. A number of the lantibiotics from the *Actinobacteria* are showing novel post translational modifications that are expanding their antimicrobial range beyond the gram positive bacteria. For example microbisporicin, producted by *Microbispora coralline*, contains the novel amino acids dihydroxyproline and chlorotryptophan and has demonstrated activity against the gram negative bacteria, *Moraxella catarrhalis*, *Neisseria* spp., and *Haemophilus influenza*, but not the enterobacteria (Castiglione et al. 2008). This demonstrated the potential of lantibiotics to be able to overcome the outer membrane barrier of gram negative bacteria if they have the appropriate post translation modifications and gives hope that chacterization of more *Actinobacteria* lantibiotics will reveal more with expanded functionality. While the post translation modifications of bisin produced by *B. longum* DJO10A have not yet been characterized, it has been shown to also function against gram negative bacteria, including the enterobacteria (Lee et al. 2011). This may contribute to their functionality in the gut as high bifidobacteria numbers in feces have traditionally been associated with lower numbers of enterobacteria (O'Sullivan 2001).

There are many other features of interest to probiotic cultures present on these genomes. The presence of various restriction modification systems is quite evident as depicted in Table 2. It should be noted that these are often strain dependent and are more prevelant among different species. This has a direct impact on the ability to introduce plasmids via electroporation and explains why some species are much more difficult to introduce plasmids into than others. The prevalence of Clustered Regularly Interspaced Short Palindromic Repeats (CRISPRs) among the majority of the strains is also notable. These structures are composed of a series of evenly spaced direct repeats (as few as two or three to several hundred), with spacer regions that consist of DNA acquired from invading DNA such as bacteriophage or plasmids, that function along with genes encoding CAS proteins to give the bacterium immunity from future infection by the bacteriophage or plasmid (Horvath and Barrangou 2010). Hence they are analogous to a primitive immune system in bacteria.

Other features of interest to probiotic cultures include exopolysaccharide production and enzymes such as bile salt hydrolase (Fig. 2). Exopolysaccharide

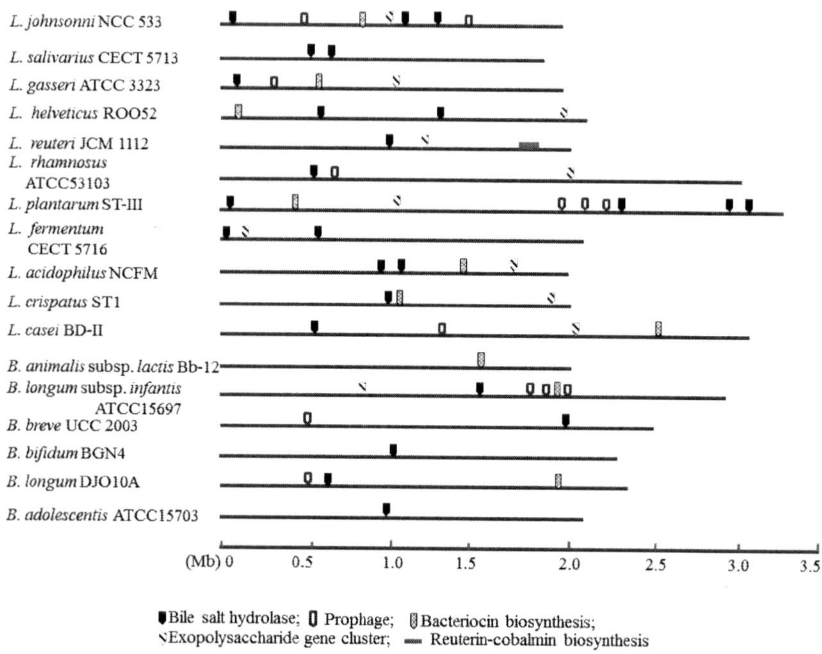

Fig. 2. Graphic representation of the genomes of the probiotic lactobacilli and bifidobacteria strains drawn approximately to scale together with the relative locations of different probiotic features.

production can offer protection for the bacterium and also it has functional properties in the gut, such as immuno stimulatory. It was also previously noted that ingesting high numbers of exopolysaccharide producing bifidobacteria can lubricate the colon, thus making it easier to pass feces (O'Sullivan 2001). This may be the underlying mechanism for yogurts containing bifidobacteria to facilitate colon regularity that are currently being marketed. Exopolysaccharide gene clusters are strain dependent and are present in ten of the eleven lactobacilli genomes in Fig. 2, but in only one (*B. longum* subsp. *infantis*) of the six bifidobacteria genomes. Bile salt hydrolase is considered an important trait as it enables survival in the small intestine from the bile released by the gall bladder to facilitate digestion. It is noteworthy that all the genomes in Fig. 2 contain bile salt hydrolase genes except for *B. animalis* subsp. *lactis* Bb-12. This may be due to gene loss from its evolution in the dairy environment, given it evolved significantly following its original isolation as *B. animalis* (Lee and O'Sullivan 2010). However, clinical studies have shown that it is viable in feces following ingestion, suggesting it may have alternative means of surviving bile in the gut (Palari et al. 2012).

Genome Safety Analysis of Probiotic Bacteria

When incorporating live and active bacteria in foods it is of paramount importance to ensure they are safe and are unlikely to contribute to the evolution of more virulaent pathogenic bacteria due to horizontal gene transfer. Traditional approaches to evaluate safety relied heavily on extended use in human foods without documented deleterious effects. Hence many of these bacterial cultures attained GRAS (generally regarded as safe) status from the FDA (Food and Drug Administration) in the USA. However, the existance of genes that may be associated with virulence functions or antibiotic resistance was always a concern, especially if they were associated with a mobile element, such as a plasmid, transposon or insertion element. The advent of genome sequences has now provided a tool to address this concern such that cultures with the potential to transfer unsavory genes can be recognized and not used in foods.

To conduct a thorough genome safety analysis, a functional assessment of the genome sequence should be generated. A common functional assessment method is provided by the COG (Clusters of Orthologous Groups) database at the NCBI. This can annotate the genome and predict functions due to motif similarities with known proteins. While many genome annotations deposited at GenBank do contain this type of information, others do not, indicating that relying solely on genome annotation deposited at GenBank may miss many functional hits. This is illustrated in Table 3, where no reference to antibiotic resistance is listed on the current genome annotation for *Lactobacillus johnsonii* NCC 533 deposited at GenBank. However, a COG analysis of its genome reveals several ABC-type multidrug transporters.

Given that the function of the majority of genes in a genome sequence is either totally unknown or weakly predicted, it is currently impossible to conduct a complete analysis. However, a genome analysis can readily uncover known virulence or potential antibiotic resistance genes that may be present. An analysis of the genome annotations for the probiotic lactobacilli and bifidobacteria deposited at GenBank doesn't show any obvious virulence genes that would be of a safety concern (Table 3). However, there are usually several genes predicted to be involved in antibiotic resistance in the majority of strains. These include ABC type multidrug transporters and genes predicted to encode resistance to tetracycline, bacitracin, bleomycin, daunorubicin, lincomycin, nitroimidazole, teicoplanin, methicillin or glycopeptide antibiotics. Their existance reveals a potential only, because they have not been functionally examined. In many cases, the potential gene may be truncated, thereby removing much of its concern, such as the *tetW1* and *tetW2* genes annotated in the genome of *B. longum* DJO10A. A potential antibiotic resistance gene would be of increased concern if it was located on

Table 3. Potential antibiotic resistance and hemolysin genes in the probiotic lactobacilli and bifidobacteria as reported in the genome annotations deposited in GenBank.

Bacterial genome	**Antibiotic resistance genes**	**Number of hemolysin genes**
L. jonhsonnii NCC 533	Not found[1]	2
L. salivarius CECT 5713	Putative genes for 3-multidrug resistance efflux pumps; 2-multidrug resistance ABC transporter ATP-binding and permease components; 4-multidrug resistance protein B; 8-transcriptional regulators, TetR family.	1
L. gasseri ATCC 3323	3-transcriptional regulators, TetR family; 19-ABC-type multidrug transport system ATPase and permease components; 1-putative gene for glycopeptide antibiotic resistance protein; 3-beta-lactamase class A; 3-transcriptional regulators, TetR family.	3
L. helveticus ROO52	1-multidrug resistance protein; 2-glycopeptide antibiotic resistance proteins; 3-TetR family transcriptional regulators.	4
L. reuteri JCM 1112	3-multidrug ABC transporter ATP-binding and permease components; 9-multidrug transport proteins, 1-putative drug efflux protein.	2
L. rhamnosus ATCC 53103	4-multidrug ABC transporter ATP-binding and permease components; 4-putative beta-lactamase; 4-beta-lactamase.	3
L. plantarum ST-III	1-multi-drug transporter; 1-EmrB/QacA family drug resistance transporter; 1-Na^+ driven multidrug efflux pump; 1-multidrug resistance ABC superfamily ATP binding cassette transporter.	3
L. fermentum CECT 5716	6-multidrug transport proteins; 1-bacitracin resistance protein; 1-bleomycin hydrolase.	0
L. acidophilus NCFM	A putative gene coding for bleomycin hydrolase; 1-daunorubicin resistance protein; 6-multidrug resistance proteins; 1-multidrug resistance ABC transporter ATP binding protein; 3-multidrug resistance efflux pumps; 1-lincomycin-resistance protein; 2-transcriptional regulators, TetR family.	2
L. crispatus ST1	1-EmrB/QacA family drug resistance transporter; 1-translation elongation factor homologous to tetracycline resistance protein.	2
L. casei BD-11	11-EmrB/QacA family drug resistance transporters; 1-daunorubicin resistance ATP-binding protein; 1-lincomycin resistance protein LmrB; 1-5-nitroimidazole antibiotic resistance protein.	2

Table 3. contd....

Table 3.contd.

Bacterial genome	Antibiotic resistance genes	Number of hemolysin genes
B. animalis subsp. *lactis* Bb-12	4-multidrug resistance protein B; 2-tetracycline resistance proteins; 1-nitroimidazole antibiotic resistance protein; 1-LmrB; 1-lincomycin resistance protein; 1-bacitracin resistance protein; 1-daunorubicin resistance ATP-binding protein; 3-multidrug resistance ABC transporter ATP-binding and permease components.	3
B. longum subsp. *infantis* ATCC 15697	4-methicillin resistance proteins; 4-glyoxalase/blomycin resistance proteins; 2-drug resistance transporters of the EmrB/QacA subfamily; 1-bacitracin resistance protein, 1-daunorubicin resistance ATP binding protein; 1-tetracycline resistance protein.	1
B. breve UCC2003	1-antibiotic resistance protein; 3-macrolide efflux proteins; 1-daunorubicin resistance DNA binding proteins; 2-multidrug resistance protein B; 1-putative bacitracin resistance protein.	0
B. bifidum BGN4	1-lincosamide resistance protein, LinA; 1-uncharacterized bacitracin resistance protein; 1-glycopeptide antibiotic resistance protein; 1-teicoplanin resistance protein; 1-nitroimmidazole resistance protein.	3
B. longum DJO10A	2-truncated tetracycline resistance genes (*tet*W1 and *tet*W2); 1-BacA hypothetical bacitracin resistance protein; 1-VanZ glycopeptide antibiotic resistance protein.	5
B. adolescentis ATCC 15703	1-EmrB/QacA putative drug resistance transporter; 1-putative drug resistance transporter; 1-putative Na^+-driven multidrug efflux pump; 1 BacA bacitracin resistance protein.	2

[1]While none are listed in its current genome annotation in GenBank, a COG analysis does show several ABC transporters, including multidrug transporters

a plasmid or a location of the chromosome that could be mobilized, such as a trasnposon or insertion element. This would increase the likelehood of horizontal transfer to other microbes in the gut. This would be a concern and should preclude use of the organism in foods, unless the gene was already ubiquous in nature.

While some LAB, such as many enterococci, encode genes predicted to encode cytolysins there are no such genes found in the annotations of the probiotic lactobacilli and bifidobacteria in Table 3. This would be a potential safety concern for including enterococci for probiotic applications. Genes annotated as hemolysin-like are quite common throughout both the probiotic lactobacilli and bifidobacteria, only being absent from the annotations of *L. fermentum* and *B. breve* (Table 3). While these genes suggest possible virulence, their function in the LAB or bifidobacteria has not been investigated and their ubiquous occurrence among these bacteria suggest they cannot preclude use of an organism solely for the presence of this gene.

Genome Insights into Prebiotic Utilization by Probiotic Lactobacilli and Bifidobacteria

Prebiotics include complex carbohydrates, such as oligosaccharides and sugar alcohols, that cannot be metabolized and absorbed in the small intestine and therefore can provide the gut microflora a nutrient source if they have the ability to metabolize them (Ziemer and Gibson 1998). Little is known about utilization of specific prebiotics by the majority of the gut microflora, but bifidobacteria in particular are known to be dominant users of oligosaccharides. This is especially true of human milk oligosaccharides (HMOs) and explains why breast fed infants exhibit a bifidobacteria dominant flora in their feces (Venema 2012). Genome analysis of probiotic bacteria can reveal their potential to utilize diverse prebiotic compounds.

Analysis of the genomes for the probiotic lactobacilli and bifidobacteria using the KEGG (Kyoto Encyclopedia of Genes and Genomes) database clearly shows a wider ability of bifidobacteria to metabolize complex carbohydrates than lactobacilli as they contain more genes predicted to encode enzymes needed to metabolize these substrates (Table 4). This suggests bifidobacteria are particularly better at metabolizing plant based carbohydrates, as they encode several enzymes to metabolize substrates, such as arabinofuran, arabinogalactan, xylan and glycan, while lactobacilli are mostly devoid of these features. It is also noteworthy that *B. longum* subsp. *infantis* is the only strain to encode both fucosidase and sialidase enzymes, which are enzymes involved in metabolizing HMOs. This clearly supports the association of this subspecies with infant guts. The decrease in numbers of the subspecies *infantis,* and increase in numbers of subspecies *longum* as infants are weaned, is also supported by its gene content as it encodes relatively few enzymes for metabolizing plant based oligosaccharides compared to the subspecies *longum.*

Future Contributions from Probiotic Genomics

Currently, genome sequences of probiotic bacteria are being used in conjunction with microarray and RNAseq technologies to determine gene expression levels in response to different stimuli. This has the potential

Table 4. Enzymes involved in prebiotic utilization for these probiotic lactobacilli and bifidobacteria.[1]

Enzyme	Substrate	Probiotic lactobacilli and bifidobacteria																
		La	Lca	Lcr	Lf	Lg	Lh	Lj	Lp	Lr	Lrh	Ls	Ba	Ban-l	Bbr	Bbf	Bl	Bl-i
Beta-Fructofuranosidase/inulinase	Fructofuran	2	1	3	1	2	1	1	1	0	1	1	1	1	1	0	1	1
Pullulanase/glycogen-debranching enzyme	Pullulan	2	1	0	0	0	0	1	0	0	1	0	2	3	1	0	2	2
Cyclomaltodextrinase /neopullulanase	Cyclodextrin	1	1	1	0	1	2	1	0	1	2	2	0	0	0	1	1	1
Galactofuranosyltransferase	Galactofuran	0	0	1	0	0	0	0	0	0	0	1	0	1	0	1	0	0
alpha-L-Arabinofuranosidase	Arabinofuran	0	0	1	0	0	0	0	0	0	0	0	2	1	0	0	6	1
Exo-alpha-L-arabinofuranosidase II	Arabinofuran	0	0	0	0	0	0	0	0	0	0	0	2	0	0	0	2	0
alpha-*N*-Arabinofuranosidase	Arabinofuran	0	0	0	0	0	0	0	0	0	0	0	0	0	0	2	1	0
Arabinogalactan endo-1,4-beta-galactosidase	Arabinogalactan	0	0	0	0	0	0	0	0	1	0	0	0	0	1	0	1	1
Arabinan endo-1,5-alpha-L-arabinosidase	Arabinan	0	0	0	0	0	0	0	0	0	0	0	0	0	0	0	2	0
Glycanase/glycogenase	Glycan	0	0	0	0	0	0	0	0	0	0	0	1	1	0	1	1	1
1,4-alpha-Glucan branching enzyme	Glucan	1	2	0	0	0	1	0	1	0	1	1	1	1	1	1	1	1
4-alpha-Glucanotransferase	Glucan	0	0	0	0	0	0	0	0	0	0	0	2	2	2	1	2	2
Glucan/glycogen phosphorylase	Glucan	1	0	0	0	0	0	0	1	0	1	0	1	1	2	1	1	1
beta-1,4-Endoglucanase/cellulase	Glucan	0	0	0	0	0	0	0	0	0	0	0	0	0	0	0	0	0
beta-1-3 exogluconase	Glucan	0	0	0	0	0	0	0	0	0	0	0	1	0	0	0	1	1

Table 4. contd....

Table 4. contd.

Enzyme	Substrate	Probiotic lactobacilli and bifidobacteria																
		La	**Lca**	**Lcr**	**Lf**	**Lg**	**Lh**	**Lj**	**Lp**	**Lr**	**Lrh**	**Ls**	**Ba**	**Ban-l**	**Bbr**	**Bbf**	**Bl**	**Bl-i**
Xylan esterase	Xylan	0	0	0	0	0	0	0	0	0	0	0	1	1	0	0	1	1
Endo-1,4-beta-xylanase	Xylan	0	0	0	0	0	0	0	0	0	0	1	0	1	0	2	2	0
Extracellular exoxylanase	Xylan	0	0	0	0	0	0	0	0	0	0	0	0	0	0	0	1	0
beta-Glucuronidase	Glycoside	0	0	0	0	0	0	0	0	0	1	0	0	0	0	0	1	0
alpha-1,4-Glucosidase	Glycoside	2	2	4	0	0	4	0	6	2	1	0	1	1	0	0	1	1
Oligo-1,6-glucosidase	Glycoside	2	2	0	1	1	0	1	3	0	1	1	1	1	2	0	1	1
beta-D-glucosidase	Glycoside	4	12	0	0	0	0	0	11	0	7	0	1	2	4	0	1	0
Thermostable beta-glucosidase B	Glycoside	0	0	6	0	0	4	0	0	0	0	0	1	1	0	0	1	1
Endo-beta-*N*-acetylglucosaminidase	Glycoprotein	0	0	0	0	0	0	0	1	0	0	0	0	0	0	0	1	0
N-Acyl-D-glucosamine 2-epimerase	Glycoprotein	2	0	1	3	0	2	0	0	0	0	0	0	1	0	0	1	1
Mannitol dehydrogenase	Mannitol	0	3	1	0	0	0	0	1	1	1	1	1	0	0	0	1	0
Sorbitol dehydrogenase	Sorbitol	0	2	0	0	0	0	0	0	0	2	1	1	0	0	0	1	0
beta-*N*-Acetylhexosaminidase	Glycan	0	0	1	0	0	1	0	0	0	1	1	1	1	1	2	2	2
Lacto-*N*-biose phosphorylase	Glycan	0	0	0	0	0	0	0	0	0	0	0	0	0	0	2	1	1
alpha-L-fucosidase	Glycan	0	2	0	0	0	0	0	0	0	3	0	0	0	0	2	0	3
Exo-alpha-sialidase	Glycan	0	0	0	0	0	0	0	0	0	0	0	0	0	0	0	0	1

[1]As determined by the KEGG database for these genomes; La, *Lactobacillus acidophilus* NCFM; Lca, *L. casei* BD-II; Lcr *L. crispatus* ST-1; Lf, *L. fermentum* CECT5716; Lg, *L. gasseri* ATCC 3323; Lh, *L. helveticus* RO052; Lj, *L. johnsonii* NCC 533; Lp, *L. plantarum* ST-III; Lr, *L. reuteri* JCM1112; Lrh, *L. rhamnosus* ATCC 53103; Ls, *L. salivarius* CECT5713; Ba, *B. adolescentis* ATCC15703; Ba-l, *B. animalis* subsp *lactis* Bb-12; Bbr, *B. breve* UCC2003; Bbf, *B. bifidum* BGN4; Bl, *B. longum* DJO10A; Bl-i, *B. longum* subsp *infantis* ATCC 15697

to understand how these organisms function both *in vivo* and *in situ* in the gut. Data from these types of approaches is currently furthering our understanding of how these bacteria respond to epithelial cells in the gut. For example, microarrays were recently used to show that *L. salvarius* can upregulate expression of its bacteriocin genes in response to interaction with a Caco-2 cell line and that this signal was mediated via a sortase anchored cell surface protein in *L. salvarius* (O'Callaghan et al. 2012). This type of functional analysis of genomes will greatly enlighten our knowledge base on mechanistic functioning of probiotic bacteria. This will lead to to a better scientific approach for the selection of strains for probiotic applications and will also strengthen the scientific rationale for using specific strains for specific probiotic functions. Once that occurs, regulatory authorities such as EFSA will be able to look favorably on petitions for probiotic claims.

References

Ai, L., C. Chen, F. Zhou, L. Wang, H. Zhang, W. Chen and B. Guo. 2011. Complete genome sequence of the probiotic strain *Lactobacillus casei* BD-II. J. Bacteriol. 193: 3160–3161.

Altermann, E., W.M. Russell, M.A. Azcarate-Peril, R. Barrangou, B.L. Buck, O. McAuliffe, N. Souther, A. Dobson, T. Duong, M. Callanan, S. Lick, A. Hamrick, R. Cano and T.R. Klaenhammer. 2005. Complete genome sequence of the probiotic lactic acid bacterium *Lactobacillus acidophilus* NCFM. Proc. Natl. Acad. Sci. USA 102: 3906–3912.

Castiglione, F., A. Lazzarini, L. Carrano, E. Corti, I. Ciciliato, L. Gastaldo, P. Candiani, D. Losi, F. Marinelli, E. Selva and F. Parenti. 2008. Determining the structure and mode of action of microbisporicin, a potent lantibiotic active against multiresistant pathogens. Chem. Biol. 15: 22–31.

Fleishmann, R.D., M.D. Adams, O. White, R.A. Clayton, E.F. Kirkness, A.R. Kerlavage, C.J Bult, J.F. Tomb, B.A. Dougherty, J.M. Merrick, K. McKenney, G. Sutton, W. FitzHugh, C. Fields, J.D. Gocayne, J. Scott, R. Shirley, L.-I. Liu, A. Glodek, J.M. Kelley, J.F. Weidman, C.A. Phillips, T. Spriggs, E. Hedblom, M.D. Cotton, T.R. Utterback, M.C. Hanna, D.T. Nguyen, D.M. Saudek, R.C. Brandon, L.D. Fine, J.L. Fritchman, J.L. Fuhrmann, N.S.M. Geoghagen, C. L. Gnehm, L.A. McDonald, K.V. Small, C.M. Fraser, H.O. Smith and J.C. Venter. 1995. Whole-genome random sequencing and assembly of *Haemophilus influenzae* Rd. Science 269: 496–512.

Garrigues, C., E. Johansen and M.B. Pedersen. 2010. Complete genome sequence of *Bifidobacterium animalis* subsp. *lactis* BB-12, a widely consumed probiotic strain. J. Bacteriol. 192: 2467–2468.

Gil, R., B. Sabater-Munoz, A. Latorre, F.J. Silva and A. Moya. 2002. Extreme genome reduction in *Buchnera* spp.: Toward the minimal genome needed for symbiotic life. Proc. Natl. Acad. Sci. USA 99: 4454–4458.

Horvath, P. and R. Barrangou. 2010. CRISPR/Cas, the Immune System of Bacteria and Archaea. Science 327: 167–170.

Jiménez, E., S. Langa, V. Martin, R. Arroyo, R. Martin, L. Fernandez and J.M. Rodriguez. 2010a. Complete genome sequence of *Lactobacillus fermentum* CECT 5716, a probiotic strain isolated from human milk. J. Bacteriol. 192: 4800.

Jiménez, E., R. Martin, A. Maldonado, V. Martin, A. Gomez de Segura, L. Fernandez and J.M. Rodriguez. 2010b. Complete genome sequence of *Lactobacillus salivarius* CECT 5713, a probiotic strain isolated from human milk and infant feces. J. Bacteriol. 192: 5266–5267.

Lee, J.-H. and D.J. O'Sullivan. 2006. Sequence analysis of two cryptic plasmids from *Bifidobacterium longum* DJO10A and construction of a shuttle cloning vector. Appl. Environ. Microbiol. 72: 527–535.
Lee, J.-H. and D.J. O'Sullivan. 2010. Genomic insights into bifidobacteria. Microbiol. Mol. Bio. Rev. 74: 378–416.
Lee, J.-H., X. Li and D.J. O'Sullivan. 2011. Transcription analysis of a lantibiotic gene cluster from *Bifidobacterium longum* DJO10A. Appl. Environ. Microbiol. 77: 5879–5887.
Lee, J.-H., V.N. Karamychev, S.A. Kozyavkin, D. Mills, A.R. Pavlov, N.V. Pavlova, N.N. Polouchine, P.M. Richardson, V.V. Shakhova, A.I. Slesarev, B. Weimer and D.J. O'Sullivan. 2008. Comparative genomic analysis of the gut bacterium *Bifidobacterium longum* reveals loci susceptible to deletion during pure culture growth. BMC Genomics 9: 247–262.
Li, X. and D.J. O'Sullivan. 2012. Contribution of the *Actinobacteria* to the growing diversity of lantibiotics. Biotechnol. Lett. 34: 2133–2145.
Makarova, K., A. Slesarev, Y. Wolf, A. Sorokin, B. Mirkin, E. Koonin, A. Pavlov, N. Pavlova, V. Karamychev, N. Polouchine, V. Shakhova, I. Grigoriev, Y. Lou, D. Rohksar, S. Lucas, K. Huang, D. M. Goodstein, T. Hawkins, V. Plengvidhya, D. Welker, J. Hughes, Y. Goh, A. Benson, K. Baldwin, J.-H. Lee, I. Díaz-Muñiz, B. Dosti, V. Smeianov, W. Wechter, R. Barabote, G. Lorca, E. Altermann, R. Barrangou, B. Ganesan, Y. Xie, H. Rawsthorne, D. Tamir, C. Parker, L. McKay, F. Breidt, J. Broadbent, R. Hutkins, D. O'Sullivan, J. Steele, G. Unlu, M. Saier, T. Klaenhammer, P. Richardson, S. Kozyavkin, B. Weimer and D. Mills. 2006. Comparative genomics of the lactic acid bacteria. Proc. Natl. Acad. Sci. USA 103: 15611–15616.
Metchnikoff, E. and P.C. Mitchell. 1907. The prolongation of life; optimistic studies. G. P. Putnam's sons, New York (Reprinted in 2004, Springer, NY).
Morita, H., H. To, S. Fukuda, H. Horikawa, K. Oshima, T. Suzuki, M. Murakami, Y. Kato, T. Takizawa, H. Fukuoka, T. Yoshimura, T. Masaoka, K. Itoh, D.J. O'Sullivan, L.L. McKay, J. Kikuchi and M. Hattori. 2008. Complete genomes of *Lactobacillus reuteri* and *Lactobacillus fermentum* reveal a genomic island for reuterin and cobalamin production. DNA Res. 15: 151–161.
Morita, H., H. Toh, K. Oshima, M. Murakami, T.D. Taylor, S. Igimi and M. Hattori. 2009. Complete genome sequence of the probiotic *Lactobacillus rhamnosus* ATCC 53103. J. Bacteriol. 191: 7630–7631.
O'Callaghan, J., L.F. Butto, J. MacSharry, K. Nally and P.W. O'Toole. 2012. Influence of adhesion and bacteriocin production by *Lactobacillus salivarius* on the intestinal epithelial cell transcriptional response. Appl. Environ. Microbiol. 78: 5196–5203.
O'Connell Motherway, M., A. Zomer, S.C. Leahy, J. Reunanen, F. Bottacini, M.J. Claesson, F. O'Brien, K. Flynn, P.G. Casey, J.A. Moreno Munoz, B. Kearney, A.M. Houston, C. O'Mahony, D.G. Higgins, F. Shanahan, A. Palva, W.M. de Vos, G.F. Fitzgerald, M. Ventura, P.W. O'Toole and D. van Sinderen. 2011. Functional genome analysis of *Bifidobacterium breve* UCC2003 reveals type IVb tight adherence (Tad) pili as an essential and conserved host-colonization factor. Proc. Natl. Acad. Sci. USA 108: 11217–11222.
O'Sullivan, D.J. 2001. Screening of intestinal microflora for effective probiotic bacteria. J. Ag. Food Chem. 49: 1751–1760.
O'Sullivan, D.J. 2005. Primary sources of probiotic cultures. *In*: I. Goktepe, V.K. Juneja and M. Ahmedna (eds.) Probiotics in Food Safety and Human Health. CRC Press, Taylor & Francis Group, Boca Raton, FL, pp. 89–105.
Ojala,T., V. Kuparinen, J.P. Koskinen, E. Alatalo, L. Holm, P. Auvinen, S. Edelman, B. Westerlund-Wikstrom, T.K. Korhonen, L. Paulin and M. Kankainen. 2010. Genome sequence of *Lactobacillus crispatus* ST1. J. Bacteriol. 192: 3547–3548.
Palaria, A., I. Johnson-Kandy and D.J. O'Sullivan. 2012. Effect of a synbiotic yogurt on levels of fecal bifidobacteria, clostridia, and enterobacteria. Appl. Environ. Microbiol. 78: 933–940.
Pridmore, R.D., B. Berger, F. Desiere, D. Vilanova, C. Barretto, A.C. Pittet, M.C. Zwahlen, M. Rouvet, E. Altermann, R. Barrangou, B. Mollet, A. Mercenier, T. Klaenhammer, F.

Arigoni and M.A. Schell. 2004. The genome sequence of the probiotic intestinal bacterium *Lactobacillus johnsonii* NCC 533. Proc. Natl. Acad. Sci. USA 101: 2512–2517.

Sela, D.A., J. Chapman, A. Adeuya, J.H. Kim, F. Chen, T.R. Whitehead, A. Lapidus, D.S. Rokhsar, C.B. Lebrilla, J.B. German, N.P. Price, P.M. Richardson and D.A. Mills. 2008. The genome sequence of *Bifidobacterium longum* subsp. *infantis* reveals adaptations for milk utilization within the infant microbiome. Proc. Natl. Acad. Sci. USA 105: 18964–18969.

Talarico, T.L. and W.J. Dobrogosz. 1989. Chemical characterization of an antimicrobial substance produced by *Lactobacillus reuteri*, Antimicrob. Agents Chemother. 33: 674–679.

Tompkins, T.A., G. Barreau and J.R. Broadbent. 2012. Complete genome sequence of *Lactobacillus helveticus* R0052, a commercial probiotic strain. J. Bacteriol. 194: 6349.

Trebichavsky, I., I. Splichal, V. Rada and A. Splichalov. 2010. Modulation of natural immunity in the gut by *Escherichia coli* strain Nissle 1917. Nutr. Rev. 68: 459–464.

Yu, D.S., H. Jeong, D.H. Lee, S.K. Kwon, J.Y. Song, B.K. Kim, M.S. Park, G.E. Ji, T.K. Oh and J.F. Kim. 2012. Complete genome sequence of the probiotic bacterium *Bifidobacterium bifidum* strain BGN4. J. Bacteriol. 194: 4757–4758.

Venema, K. 2012. Intestinal fermentation of lactose and prebiotic lactose derivatives, including human milk oligosaccharides. Int. Dairy J. 22: 123–140.

Wang, Y., C. Chen, L. Ai, F. Zhou, Z. Zhou, L. Wang, H. Zhang, W. Chen and B. Guo. Complete genome sequence of the probiotic *Lactobacillus plantarum* ST-III. J. Bacteriol. 193: 313–314.

Wescombe, P.A. and J.R. Tagg. 2003. Purification and characterization of streptin, a type A1 lantibiotic produced by *Streptococcus pyogenes*. Appl. Environ. Microbiol. 69: 2737–2747.

Ziemer, C.J. and G.R. Gibson. 1998. An overview of probiotics, prebiotics and synbiotics in the functional food concept: perspectives and future strategies. Int. Dairy J. 8: 473–479.

23

The Future of Prebiotics and Probiotics

Oscar Brunser,[1,*] *Javier A. Bravo*[2] and *Martin Gotteland*[3]

Introduction

Probiotics have been defined as "live microorganisms which, when administered in adequate amounts, confer a health benefit on the host" (FAO/WHO), and prebiotics as "non viable components of foods whose intake confers benefits to the host because they are associated with the modulation of his microbiota" (FAO). The simultaneous administration of prebiotics and probiotics (named symbiotics) may synergistically improve their health-promoting effects in the organism. A great number of bacterial strains, mainly belonging to different species of the genera *Lactobacillus* and *Bifidobacterium*, are currently considered as probiotics due to their ability to survive in the gastrointestinal tract and to exert, in a strain-specific manner, activities capable of inducing physiological responses and health-promoting effects in the consumer. Such probiotic activities have been widely described and some of these are listed in Table 1.

[1] Institute of Nutrition and Food Technology, Gastroenterology Unit and Biological Structure Laboratory, University of Chile, Santiago, Chile.
Email: brunser@entelchile.net
[2] Institute of Chemistry, Faculty of Sciences, Universidad Catoólica de Valparaiso, Valparaiso, Chile.
Email: javier.bravo@ucv.cl
[3] Department of Nutrition, Faculty of Medicine, University of Chile, Santiago, Chile.
Email: mgottela@med.uchile.cl
* Corresponding author

Table 1. Main activities of probiotics.

Probiotic Activity	**Mechanisms**	**Health effect**
Antibacterial	Release of organic acids, hydrogen peroxide, bacteriocins which modulate the growth or/and mucosal adhesion of pathogens or other bacterial populations of the IM	Prevention and/or management of gastrointestinal infections and diarrhea
Enzyme providing	ß-galactosidase activity α-galactosidase activity Esterase activity	Improvement of lactose digestion and tolerance in hypolactasic individuals Improvement of galactoside digestion and tolerance (raffinose, stachyose, melibiose from soy) Polyphenol deglycosilation
Immunomodulating and/or anti-inflammatory activities	Cellular components/ exopolisaccharide, lipoteichoic acid, etc.) capable of increasing the release of secretory IgA or defensins, stimulating toll-like receptors or natural killer activity of the cytotoxic T lymphocytes, stimulation of phagocytic activity of neutrophils or macrophages, stimulation or inhibition of the secretion of anti- or pro-inflammatory cytokines respectively, by host immunocompetent cells	Prevention and/or management of local and systemic bacterial and viral infections Decrease of symptomatology in allergic individuals Decrease of inflammation in subjects with inflammatory conditions (IBD, NEC, autoimmune pathologies, etc.)
Antitumoral activities	Inhibition of procarcinogenic enzymatic activities (ß-glucuronidase, nitroreductase, azoreductase) by bacterial populations of the IM Stimulation of apoptosis Antioxidant/anti-inflammatory activities Mutagen binding	Decrease of the risk of colorectal cancer
Trophic activity	Increase of polyamines (*Saccharomyces boulardii*)	Stimulates growth of the intestinal mucosa

The aim of this chapter is to describe recent emerging evidence suggesting that the administration of pro- and/or prebiotics may be of interest in areas traditionally not associated with their well-known health benefits, such as bone metabolism, renal function and the central nervous system (CNS). This also opens new insights into their future use in the management of a variety of health conditions.

Probiotics and Bone Mass Regulation

A number of studies have shown that the regulation of the intestinal microbiota (IM) by prebiotics or synbiotics may contribute to lower the risk of osteoporosis by improving calcium absorption and bone mineralization (Scholz-Ahrens et al. 2007). It has been proposed that this phenomenon is probably due to the production of short chain fatty acids (SCFAs) and the subsequent acidification of the colonic lumen, which favors calcium solubilization.

Beyond these effects, Sjögren et al. (2012) have recently proposed that the gut microbiota could act as a bone mass regulator through other, different mechanisms. They observed that the bone mass of germ-free mice was higher and that the number of osteoclasts per area of bone surface was lower than in conventional mice, indicating a decrease in osteoclast-mediated bone resorption in the former. In addition, germ-free animals also exhibited less CD_4^+ T cells in their blood and spleen, fewer CD_4^+ cells and osteoclast precursors in their bone marrow and lower bone contents of TNF-α, an osteolytic cytokine. The development of Th17 cells, an osteoclastogenic subset of CD_4^+ T cells linking T cell activation with bone resorption, as well as the ability of these cells to release IL-17 was also deficient in the microbiota-free animals. Most of these changes were normalized when the animals were conventionalized, i.e., when their gastrointestinal tract was colonized by a microbiota (Sjögren et al. 2012). It is noteworthy that compared with healthy subjects, patients with rheumatoid arthritis exhibit higher counts of Th17 cells which are involved in bone destruction during the active phase of the disease (Sarkar and Fox 2010); gut microbiota alterations have also been described in these patients (Vaahtovuo et al. 2008).

Interestingly, some probiotics have been shown to inhibit the IL-17 pro-inflammatory response in animal models of inflammation. For example, the oral administration of *L. gasseri* A5 has been shown to decrease IL-17 release in a mouse model of allergic asthma induced by *Dermatophagoides pteronyssinus* (Jan et al. 2012) and strains of *B. infantis* were able to inhibit IL-17 production by murine splenocytes and *ex vivo* dextran sulfate-treated colon, probably through the stimulation of IL-10 release (Tanabe et al. 2008). On the other hand, Kwon et al. (2010) also observed that the administration of probiotics slows the disease progression in Lewis rats with experimental rheumatoid arthritis, as reflected by the decrease of the clinical score, lymphocyte infiltration and inflammatory cytokine production (including IL-17); the authors suggested that this protective effect was probably due to the up-regulation of CD4(+) Foxp3(+) regulatory T (Treg) cells. Such effects seem to be strain-specific as, in opposition with these results, the exposure of human immune cells to other probiotic strains resulted in the strong induction of Th17 and Treg differentiation (Donkor et al. 2012). These observations suggest that some probiotic strains may be useful in the management of conditions affecting the bone and associated connective tissue, as in rheumatoid arthritis.

Probiotics, Prebiotics and Renal Function

Renal insufficiency is a highly prevalent condition generally resulting from long-standing atherosclerosis, hypertension, or diabetes. In end-stage renal disease (ESRD) most of the kidney function has been lost and the affected subjects require frequent hemodialysis (HD) or kidney transplantation. ESRD currently affects more than 2.5 million people worldwide and represents a heavy medical and economic burden. Impaired kidney function results in the progressive accumulation of toxic metabolites causing the uremic syndrome which impacts negatively on a number of physiological and biochemical functions of the patients (Meijers et al. 2010, Mutsaers et al. 2012, Sun et al. 2012). Some of these compounds, such as p-cresyl sulphate and indoxyl sulphate, circulate in the blood, bound to proteins, and are not eliminated efficiently by HD. They are responsible for the increased rate of complications and mortality in HD patients. In particular, increased p-cresyl sulphate levels are associated with the development of cardiovascular disease and associated mortality (Lin et al. 2012). p-cresol and indole (the precursors of p-cresyl- and indoxyl sulphate) are metabolites produced by intestinal bacteria from tyrosine and tryptophan, respectively (Evenepoel et al. 2009). Their production is increased in HD patients, probably due to the fact that their intestinal protein digestion is impaired (Bammens et al. 2004), resulting in more undigested protein reaching the colon where it is fermented by the microbiota (Davila et al. 2013). In addition, the gut microbiota of these patients is altered and they exhibit a prolonged colonic transit time and constipation, phenomena that favor protein fermentation and the production of metabolites (Smith and Macfarlane 1996, Wang et al. 2012). In order to confirm the role of the colon in the production of uremic solutes, Aronov et al. (2011) recently compared the plasma profiles in HD patients with intact colon with those from HD patients whose colon had been resected. A greater number of uremic solutes were detected in the plasma of the HD with an intact colon, compared with healthy subjects, with most of these solutes unidentified. A number of the uremic solutes detected in the patients with an intact colon were either absent or present in lower concentrations in those patients without colon; many of these solutes were not removed by the HD (Aronov et al. 2011).

In consequence, inhibiting the generation of uremic toxic precursors may be considered an interesting way to reduce their accumulation in HD patients. Probiotics and/or prebiotics have been proposed as tools for improving the characteristics of the intestinal environment and reducing undesirable intestinal nitrogenous metabolites in medical conditions such as hepatic encephalopathy (Shukla et al. 2011). Probiotics and/or prebiotics have also been shown to normalize and to regulate their bowel habits and constipation as well as their gut microbiota, which are altered in HD patients. In consequence, it may be postulated that prebiotics and/or probiotics could reduce the toxic effects of p-cresol in HD patients. Table 2 describes the results of human studies using prebiotics and/or probiotics with the aim of decreasing the urinary excretion of p-cresol and nitrogen (or NH_3) and increasing their fecal excretion; eight of these studies were carried out in healthy volunteers and six in patients with impaired renal function. Different prebiotics (trans-galacto-oligosaccharide, resistant starch, isomalt, lactulose, inuline, oligofructose-enriched inulin, arabinoxylooligosaccharides, gum Arabic) or probiotics (*S. boulardii*, *B. breve* Yakult + *L. casei* Shirota, *L. acidophilus* LC1, *B. longum* + *L. acidophilus* KB27 + *B. longum* KB31 + *S. thermophilus* KB19), alone or in combination, were used. In eight of these studies, the administration of prebiotics and/or probiotics stimulated bacterial proliferation in the colon and thereby the bacterial mass by stimulating the use of ammonia as nitrogen source for their growth. This phenomenon led to the increase of the fecal excretion of nitrogen and the correlative decrease of its urinary excretion. In six of these studies, p-cresol concentrations in stools, plasma or urine were decreased as a result of the reduced protein fermentation by the colonic microbiota. Additionally, in some of these studies improvements of the quality of life and of constipation and modifications of the fecal microbiota (increase of bifidobacteria) were also reported.

Although a limitation of these studies is the low number of subjects recruited (between 9 and 46), the results obtained support the need for the future development of clinical trials incorporating higher numbers of subjects to further confirm the usefulness of prebiotics and/or probiotics in the dietary management of patients with impaired renal function.

Probiotics and the Modulation of the Microbiota/gut/brain axis

There is strong evidence of functional communications between the gastrointestinal (GI) tract and the central nervous system (CNS) (Grenham et al. 2011, Clarke et al. 2012, Forsythe et al. 2010, Rhee et al. 2009, Bravo et al. 2012, Cryan and Dinan 2012). This traffic is bidirectional and involves both anatomical connections, like the vagus nerve (Forsythe et al. 2010), and humoral components including the immune system and the hypothalamus-pituitary-adrenal (HPA) axis (Forsythe et al. 2010, Bravo et al. 2012, Cryan and Dinan 2012, Dinan et al. 2006a). Recently new evidence has emerged suggesting the importance of another player in this interaction: the intestinal microbiota, leading to propose what is now recognized as the microbiota-

Table 2. Effects of pre-, pro- and synbiotics on urinary and fecal excretion of protein catabolites during hemodialysis.

Type of study	Subjects	Treatment	Results	Ref
Randomized crossover study	11 healthy subjects	Prebiotic: 5 or 39 g/d of resistant starch (RS) for 3 weeks	fecal nitrogen excretion and of total fecal phenols including p-cresol, without changes of urinary ammonia, urea, phenols, and total nitrogen	(Birkett et al. 1996)
Controlled crossover feeding trial	40 healthy subjects	Prebiotic: 7.5 or 15 g/d of TOS or placebo for3 weeks	fecal nitrogen density in the 15g TOS group. No effect on fecal concentrations of ammonia, indole, skatole, SCFAs and bile acids, nor the composition of the microbiota.	(Alles et al. 1999)
Randomized, placebo-controlled, crossover study	43 healthy volunteers	Prebiotic: 2x10g/d or 2x15g/d lactulose Probiotic: 2x250mg/d *S. boulardii* Synbiotic: 2x10g/d Lactulose + 2x250mg/d *S. boulardii* as a single dose or for a 4-week period	Single administration of lactulose dose-dependently urinary ^{15}N excretion. Long-term administration of lactulose urinary ^{15}N-excretion and the fecal ^{15}N-output (bacterial fraction) and *Bifidobacterium* population. *S. boulardii* alone had no significant effects.	(De Preter et al. 2006)
Double-blind, placebo-controlled, crossover design	19 healthy volunteers	Prebiotics: A controlled basal diet enriched with either 30 g isomalt or 30 g sucrose daily for 4 weeks	Isomalt diet fecal bifidobacterias and bacterial beta-glucosidase activity without effect on β-glucuronidase, sulfatase, nitroreductase and urease activities. Fecal SCFA, lactate, bile acids, neutral sterols, N, NH_3, phenol and p-cresol not affected by isomalt consumption.	(Gostner et al. 2006)
Pilot study	12 healthy volunteers	Prebiotics: Single administration of 5g inulin as test meal (pancake) Long term administration of 3 x 5 g inulin for 1 month + 1 week washout	The single dose of inulin faecal ^{15}N excretion and proportionally urinary ^{15}N excretion, probably reflecting an enhanced uptake of ammonia for bacterial biosynthesis; urinary p-cresol was also. A tendency towards decreased urinary excretion of p-cresol was noted after long term administration of inulin.	(Geboes et al. 2006)

Table 2. contd....

Table 2. contd.

Type of study	Subjects	Treatment	Results	Ref
Randomized, placebo-controlled, crossover study	20 healthy subjects	Prebiotic: 2x10g/d oligofructose-enriched inulin (OF-IN) Probiotic: $2x10^9$/d *B. breve* Yakult + $2x6.5.10^9$ /d *L. casei* Shirota Synbiotic: 2x10g/d OF-IN + $2x6.5.10^9$ /d *L. casei* Shirota for 3 days (short term) or 4 weeks (long term)	Both short- and long-term administration of OF-IN urinary p-cresol and ^{15}N content. Short-term (but not long term) OF-IN intake the ^{15}N content of the fecal bacterial fraction. Fecal bifidobacteria after long-term OF-IN intake. Long-term probiotics intake urinary excretion of ^{15}N and p-cresol.	(De Preter et al. 2007)
Acute feeding trial	12 healthy subjects	Prebiotic: Five test meals with 0.0, 0.2, 0.7, 2.2 or 4.9g of Arabinoxylooligosaccharides (AXOS)	2.2 and 4.9g of AXOS urinary nitrogen excretion and bacterial fermentation and fecal nitrogen excretion, without change in urinary p-cresol excretion	(Cloetens et al. 2008)
Randomized open study	14 healthy subjects	Prebiotics: 30 g fiber from potatoes (FPs, 12% resistant starch) or 30 g wrinkle pea starch (WPS, 70% resistant starch). Probiotics: 375 g of *L. acidophilus* LC1-containing yoghurt Each treatment for 10 days in random order.	^{15}N renal excretion was lower in the FP period than in the no-treatment period (p=0.034) and the LC1 period (p=0.001) and in the WPS period than in the LC1 (p=0.048). No change in the fecal ^{15}N-excretion was observed.	(Wutzke et al. 2010)

Prospective, randomized, single-blind, crossover design.	16 patients with chronic renal failure	*Prebiotics*: 2x25g/d of a highly fermentable fiber (gum arabic) or a placebo (1 g pectin/d) for 4 weeks, together with a low-protein diet	Serum urea nitrogen concentration of the subjects was >12% lower after 4 weeks of supplementation with gum arabic. Total fecal nitrogen content was 41 % higher during the gum arabic period than during the baseline or pectin periods. The bacterial fraction of feces accounted for 59% of the increase in total stool dry weight and total fecal nitrogen content.	(Bliss et al. 1996)
Randomized Controlled Trial	22 HD patients	*Probiotics*: *Bifidobacterium longum* in gastro-resistant seamless capsule (Bifina) or in powder formulation (LacB) for 5 weeks	The serum levels of indoxyl sulfate in the Bifina group after 5 weeks ($P < 0.005$), not in the Lac B group.	(Takayama et al. 2003)
Randomized Controlled Trial	9 patients with chronic renal failure	*Prebiotics*: 40 g/d of fermentable carbohydrate (FC) with a moderated restrictive protein diet (0.8 g/kg/d) for 5 weeks	FC fecal nitrogen excretion (+51%; $p<0.01$) and urinary nitrogen excretion (-12%; $p<0.01$), being unchanged the total amounts of nitrogen excreted by the two routes. The plasma urea concentration was (–23%; $p<0.05$) by FC.	(Younes et al. 2006)
Single centre, non-randomized, open-label phase I/II study	22 HD patients	*Prebiotics*: 10g/d OF-IN for 2 weeks followed by 2x10g/d for 2 weeks	p-cresyl sulfate generation and serum concentrations were by 20% at 4 weeks. In contrast, neither indoxyl sulfate generation rates nor serum concentrations were significantly changed	(Meijers et al. 2010)
Multicentric, randomized, double-blind, placebo controlled crossover trial	46 patients with chronic kidney disease (CKD) stages 3 and 4	*Probiotics*: $3 \times 3.0\times10^{10\,/d}$ of *L. acidophilus* KB27, *B. longum* KB31 and *S. thermophilus* KB19 for a total of a 3 month period.	BUN levels were by the probiotic treatment in 63% of the patients ($p<0.05$) while no effect was observed for creatinine and uric acid levels. 86% of the subjects ($p<0.05$) expressed a substantial improvement in their quality of life.	(Ranganathan et al. 2010)
Non randomized, non-controlled study	9 HD patients	*Synbiotic*: $3x1.10^8$ *L. casei* Shirota and *B.breve* Yakult/d + 3x1.7g GOS/d for 2 weeks after 2 weeks observation without symbiotic.	Synbiotic treatment serum p-cresol concentrations and improves constipation, normalizing stool quantity and consistency.	(Nakabayashi et al. 2011)

gut-brain axis (Grenham et al. 2011, Forsythe et al. 2010, Rhee et al. 2009, Bravo et al. 2012, Cryan and Dinan 2012, Bravo et al. 2011, Bercik et al. 2010, Collins et al. 2009, Cryan and O'Mahony 2010). The current evidence suggests that alterations of the regulation of this axis could be the underlying cause of a variety of functional bowel disorders including the irritable bowel syndrome (IBS).

The IBS is one of the most common functional gastrointestinal disorders in the Western world, with a prevalence of 10 to 20% in the United States, Europe and Asia (Madrid et al. 2005). It is a functional gastrointestinal condition, characterized by abdominal pain and changes in stool frequency and/or consistency (constipation, diarrhea, or alternation of both) (Cervero and Janig 1992, Mayer and Collins 2002, Mayer et al. 2006). Moreover, these symptoms arise in the absence of anatomical or biochemical markers that could serve as diagnostic tools for the condition (Clarke et al. 2009). Several strategies have been formulated for the pharmacological treatment of IBS, including the use of antispasmodic drugs, laxatives and anti-diarrheic compounds such as loperamide. 5-HT_4 partial agonists like tegaserod and 5-HT_3 antagonists have also been considered as treatment for IBS (Talley 2003; Dinan et al. 2006b). However, all of these strategies are effective only partially in reducing IBS symptoms.

Role of the intestinal microbiota

Over the last few years, the intestinal microbiota has emerged as a new target for the management of functional digestive alterations; there is a delicate balance between these bacteria, the small intestinal and colonic epithelia and the mucosa-associated lymphoid tissue that is important for gut homeostasis (Rhee et al. 2009, Shanahan 2010). The intestinal microbiota is currently considered a major participant in the modulation of multiple gastrointestinal functions including motility, secretion, blood flow, permeability, local immunity and visceral perception (Clarke et al. 2012). Alterations of this complex balance can lead to the appearance of pathologies, including IBS or inflammatory bowel diseases (IBD). It has been demonstrated, for example, that in germ-free mice the absence of an intestinal microbiota during early postnatal development results in altered anatomical and morphological features of the GI tract (e.g., an enlarged cecum, reduced intestinal surface area, increased enterochromaffin cell area, smaller Peyer's patches and reduced villous height) when compared to animals with a normal microbiota (Grenham et al. 2011, Shanahan 2002, Abrams et al. 1963, Gordon and Bruckner-Kardoss 1961). In addition, structural development and metabolic and protective functions are also affected by the changes occurring in germ-free animals (Grenham et al. 2011). Conventionalization of the germ-free animals with microbiota from conventionally reared animals is sufficient to restore the mucosal immune system (Umesaki et al. 1995). On the other hand, exposure to pathogens can lead to permanent changes in GI function in a manner that resembles the symptoms of IBS, and leading in some cases, to conditions resembling inflammatory bowel disease (IBD) (Navaneethan and Giannella 2011, Marshall et al. 2004, Thabane and Marshall 2009, Spiller and Garsed 2009). This evidence argues in favor of studying the role of the microbiota in the development of new strategies for the study and treatment of gastrointestinal disorders, as well as in its importance as a key to homeostasis.

The intestinal microbiota and the central nervous system

Functional gastrointestinal disorders, such as IBS, can be comorbid with stress-related psychiatric disorders, including major depression and anxiety (O'Mahony et al 2011). As a result, the search for therapies that target both the gut and the CNS is increasingly being explored. For example the treatment of hepatic encephalopathy, which seeks to reduce the mass of enteric bacteria that produce ammonia, involves the use of antibiotics (Forsythe et al. 2010, Bass 2007, Bercik et al. 2012) or lactulose, a nondigestible disaccharide that stimulates the growth of bacterial populations at the expense of the NH_3-producing bacteria. Moreover, alterations of the gut microbiota have been described in autism (Parracho et al. 2005, Finegold et al. 2002) in association with intestinal barrier disturbances, and where a possible treatment with antibiotics could be of some benefit (Posey et al. 2004, Sandler et al. 2000). Also, bacterial metabolites such as butyrate and propionate have been shown to affect behavior in rodents. Gundersen et al. (2009) showed that intraperitoneal injection of 100mg/kg of sodium butyrate to mice (3 times over a 24 h period) increases immobility in the forced swim test and the latency to consume food in a novel environment (Gundersen and Blendy 2009). These results suggest that acute treatment with butyrate produces depression and anxiety-like behaviors. However, this does not occur in mice treated with sodium butyrate for 21 days. In other studies, intracerebroventricular (ICV) injection of propionate to juvenile rats (a total of 1.04 μmoles per injection twice a day for 8 continuous days) increases locomotor activity in comparison to rats injected with phosphate buffer (Thomas et al. 2012), and it also increases repetitive behaviors while impairing social interactions (MacFabe et al. 2011). Together, these results do suggest that propionate and butyrate can affect animal behavior. However these findings have to be interpreted with some caution as the amount of butyrate and propionate necessary to promote changes in animal behavior are higher than those that can arise as a result of changes in the composition of the gut microbiota (Cryan and Dinan 2012). However, the effects of butyrate and propionate on other cell types, including nerve cells from the enteric nervous system cannot be ruled out. Another example comes from a study that elaborates on the fact that one of the side effects of the treatment of schizophrenia with atypical antipsychotic drugs is weight gain (Davey et al. 2012). In this study, Davey et al. (2012) observed that female rats treated with two doses of olanzapine (2 and 4 mg/kg for 21 days) had an increase in weight gain in comparison to those animals administered the vehicle only. Male rats treated with this drug had no change in weight gain compared to their respective controls. Moreover, and regardless of gender, olanzapine (4mg/kg for 21 days) increased the amount of visceral fat, and it also shifted the composition of the gut microbiota (a comparative larger number of *Firmicutes* and reduced number of *Bacteroidetes* in treated rats in comparison to their controls). It is interesting to note that the shift in the composition of the gut microbiota is similar to that observed in obesity studies (Ley et al. 2006). Therefore, it cannot be ruled out that pharmacological intervention on the CNS affects the composition of the gut microbiota, a change that can promote changes in metabolism that could lead to an obese phenotype.

These examples, although indirect, highlight the existence of crosstalk between the intestinal microbiota and the CNS and demonstrate how strategies directed to affect the gut microflora could promote benefits to mental health and the management of psychiatric illnesses.

On a more direct approach, emerging data has demonstrated that psychosocial stress in early stages of life has a considerable impact on gastrointestinal health in adulthood (O'Mahony et al. 2011, Drossman et al. 2011, Drossman et al. 1999, O'Mahony et al. 2009). For instance, early-life stress in rodents, which induces changes in brain neurochemistry, affecting, for example, the expression of corticotrophin releasing factor (CRF) and of both of its receptors, CRF_1 and CRF_2 (Bravo et al. 2010, O'Malley et al. 2010), also increases visceral sensitivity (O'Mahony et al. 2009). Moreover, early-life stress affects the composition of the gut microbiota (O'Mahony et al. 2009), which strongly suggests a "brain to gut" regulation of microbiota composition. On the other hand, there is evidence showing that alterations of the composition of the gut microbiota induce changes in animal behavior (Bravo et al. 2011, Goehler et al. 2008, Lyte et al. 1998, Lyte et al. 2006, Neufeld et al. 2011). For instance, the administration of the pathogenic bacteria *Campylobacter jejuni* or *Citrobacter rodentium* induces anxiety-like behavior in mice (Goehler et al. 2008, Lyte et al. 1998, Lyte et al. 2006). This altered behavior becomes evident hours after infection, suggesting that changes in the gut microbiota can rapidly induce biochemical changes in the CNS. This is further evidenced by the measuremnent of c-Fos as a marker for cell activation induced by *C. rodentium* infection in the parabrachial nucleus (Lyte et al. 2006). Furthermore, the stress induced by *C. rodentium* in mice may contribute to behavioral abnormalities (Garreau et al. 2011). This evidence demonstrates the bi-directionality of the microbiota-gut-brain axis, as the latter data strongly suggest the existence of an "intestinal microbiota to brain" modulation of behavior.

On the other hand, it has been shown that the absence of gut bacteria during development affects the hypothalamus-pituitary-adrenal (HPA) axis (Sudo et al. 2004). Germ-free mice produce higher levels of adrenocorticotropic hormone (ACTH) and corticosterone (CORT) in response to stress when compared to mice bearing a conventional microbiota (Sudo et al. 2004). In addition, expression of N-methyl D-aspartate (NMDA) receptor subunits NR-1 and NR-2a was reduced in the cortex and hippocampus of germ-free mice in comparison to their conventional controls. This effect was correlated with reduced levels of brain-derived neurotrophic factor (BDNF) in the same brain structures (Sudo et al. 2004). The altered HPA response in germ-free mice was partially prevented when the intestinal microbiota of germ-free mice was reconstituted with feces from control animals during the early stages of development (Sudo et al. 2004). In line with these findings, other studies have also shown similar alterations in HPA axis activity in germ-free mice (Neufeld et al. 2011) and moreover, the absence of gut microbiota during development reduces anxiety-like behavior in adult mice (Neufeld et al. 2011, Clarke et al. 2012) and increases locomotion in comparison to conventionally reared animals (Heijtz et al. 2011). Furthermore, other findings have been described in germ-free animals. For example, these animals have alterations in genes from at least four canonical pathways (citric acid cycle, long term potentiation, steroid hormone metabolism and cAMP signalling) in comparison to control mice (Heijtz et al. 2011), blunted immune responses towards components of the cell wall of gram negative bacteria, higher levels of 5-HT and its metabolite 5-hydroxyindoleacetic acid (5-HIAA) on the hippocampus, and increased levels of tryptophan in the plasma (Clarke et al. 2012). In addition, it is interesting to note that the alterations found by Clarke et al. (2012) in 5-HT and 5-HIAA levels were present in male germ-free rats but not in the females. This observation suggests that the absence of a gut microbiota during the early stages of development has a gender-dependent effect on the serotoninergic system. These results support the idea that the presence of the intestinal microbiota during early developmental stages is a key component for an adequate brain development, thus demonstrating the complex interactions between intestinal microbes, the gut and the CNS.

In other studies, it has been shown that colonization of germ-free BALB/c mice, that display a highly anxious phenotype (Browne et al. 2011) and NIH Swiss mice, that display a highly exploratory drive (Bercik et al. 2012), with the microbiota from each other, confers the behavioral profile of the donor. That is, germ-free BALB/c mice colonized with intestinal microbes from NIH Swiss animals have reduced latencies to step-down from an elevated platform, which can be interpreted as a reduced anxiety-like behavior while NIH Swiss germ-free mice colonized with BALB/c microbiota show increased step-down latencies (Bercik et al. 2011, Bercik et al. 2012), which suggests an increased anxiety-like phenotype. Therefore, alterations in the gut microbiota could underlie some of the behavioral traits associated with anxiety and depression.

The use of wide-spectrum non-absorbable antibiotics has also been used in order to gain insight on this microbiota-gut-brain axis. Intestinal dysbiosis induced in mice by the oral administration of neomycin, bacitracin and pimaricin, not only affected the composition of the gut microbiota, but it also increased their exploratory behaviour and BDNF levels in limbic regions of the brain (Bercik et al. 2011, Bercik et al. 2012). These effects were absent when the antimicrobials were administered intraperitoneally (Bercik et al. 2011, Bercik et al. 2012). Another approach to induce alterations of the composition of the gut microbiota is through dietary changes. For instance, including 50% of lean beef into normal mice chow significantly affects the bacterial composition in the feces in comparison to mice fed a regular chow diet (Li et al. 2009). Moreover, this change in diet and the concomitant change in microbiota the microbiota improve cognitive parameters and reduce anxiety-like behavior (Li et al. 2009). This further suggests that manipulation of the intestinal microbiota through modifications of the diet could promote beneficial changes to cognitive abilities. All this evidence further demonstrates the bidirectional communication between the microbiota, the gastrointestinal tract and the CNS. Therefore, alterations in this exchange of signals could represent not only a biological substrate to gastrointestinal function, but also for psychiatric conditions such as mood disorders (Forsythe et al. 2010).

Probiotics and the microbiota-gut-brain axis

Probiotics have come under the spotlight as a novel and safe tool for maintaining a healthy intestinal microflora. Moreover, probiotic bacteria are being studied for the treatment of IBS (Nikfar et al. 2008, Wilhelm et al. 2008, McFarland and Dublin 2008, Quigley and Flourie 2007, Whorwell et al. 2006). For instance, administration of *Bifidobacterium infantis* 35624 to IBS patients can effectively reduce abdominal pain, discomfort, distention/bloating, and difficult defecation, the cardinal symptoms of this condition (O'Mahony et al. 2005, Quigley 2008), thus demonstrating the usefulness of such strategies for the management of functional gastrointestinal disorders.

In terms of the mechanisms by which probiotics cause their effects, this encompasses a complex network of events, including displacement of pathogens (Collado et al. 2008), competition in metabolic interactions with hostile bacteria (Martin et al. 2010), production of bacteriocins (Corr et al. 2007), inhibition of bacterial translocation (Generoso et al. 2010), enhancement of mucosal barrier function (Liu et al. 2011), effects on calcium-dependent potassium channels in intestinal sensory neurons (Kunze et al. 2009), induction of opioid and cannabinoid receptors in intestinal epithelial cells (Rousseaux et al. 2007) and modulation of the immune system through signals from epithelial cells (Sanders 2011). Furthermore, public awareness has been raised through the media in terms of their health benefits, with some strains currently available in commercial food products and supplements (Guyonnet et al. 2007, Pereg et al. 2005, Hickson et al. 2007, Ortiz-Andrellucchi et al. 2008, Guerin-Danan et al. 1998). However, the benefits of probiotic bacteria seem to go beyond the realm of the gastrointestinal and immune systems. For example, it has been shown that administration of *Lactobacillus paracasei* NCC2461 was effective in reducing visceral pain in rodents (Verdu et al. 2006), one of the hallmark symptoms of IBS. More recently, and in relation to animal models of gastrointestinal functional disorders, Bercik et al. (2010) investigated the effects of probiotics in mice infected with the non-invasive parasite *Trichuris muris*; this caused inflammation and the appearance of anxiety-like behavior. These authors observed that anxiety-like behaviours were normalized after treatment with the probiotic *B. longum* NC3001 but not when *L. rhamnosus* NCC4007 was administered (Bercik et al. 2010). However, the heightened cytokine levels induced by the nematode infection were reduced neither by *B. longum* NC3001 nor *L. rhamnosus* NCC4007 treatment (Bercik et al. 2010), thus suggesting the specificity of the probiotic effects. In addition, the reduction in hippocampal BDNF expression induced by *T. muris* was also reversed by *B. longum* NC3001 treatment but not with *L. rhamnosus* NCC4007 (Bercik et al. 2010). This data shows the efficacy of probiotics in inducing changes in the CNS, and it also demonstrates the specificity of the probiotic effects. In addition, some of the behavioral aspects caused by chemically induced colitis can also be improved by the administration of *B. longum* NC3001, thus confirming the anxiolytic-like properties of this probiotic (Bercik et al. 2011a).

As mentioned earlier, rats exposed to early-life stress not only exhibit behavioral abnormalities that resemble mood disorders but they also have been demonstrated to serve as models for the study of brain-gut alterations (O'Mahony et al. 2011, O'Mahony et al. 2009). Thus, adult rats exposed to early-life stress being fed the probiotic *B. infantis* 35624 showed reduced signs of behavioral despair in the forced swim test; this microorganism also normalized the immune response and restored nor-adrenaline concentrations in the brain stem (Desbonnet et al. 2010). Moreover, *B. infantis* 35624 has the ability to reduce the visceral pain induced by colorectal distension in Sprague-Dawley (normosensitive) and Wistar-Kyoto (hypersensitive) rats (Gibney et al. 2010, McKernan et al. 2010) suggesting that probiotics can be of use for the treatment of psychiatric conditions in addition to functional gastrointestinal disorders.

Another example of a probiotic that can affect the gut-brain axis is *L. rhamnosus* JB-1. It has been demonstrated that this microorganism can modulate the immune system as it prevents the induction of interleukin (IL)-8 by tumor necrosis factor-(TNF)-α in human colon epithelial cell lines (T84 and HT-29) (Ma et al. 2004). This effect is mediated by live *Lactobacillus* only through the prevention of NF-kB nuclear translocation and the inhibition of IkB degradation (Ma et al. 2004). These findings suggest a complex interaction between the live bacteria and the epithelial cells, which is required to activate a very specific intracellular pathway in these latter. Furthermore, *L. reuteri* can increase the number of regulatory T cells that help improve the immune response towards different haptens in other organs in mice (Karimi et al. 2009). It also inhibits the cardio-autonomic response to colorectal distension (CRD) (Kamiya et al. 2006) and to gastric distention in rats (Duncker et al. 2011), a measure of visceral pain perception. Moreover, it has been shown that in the absence of inflammation, CRD increases the excitability of dorsal root ganglia (DRG) neurons in rats, but feeding *L. rhamnosus* JB-1 to these animals for 9 days prevents this phenomenon (Ma et al. 2009). This finding supports the idea that commensal bacteria can send signals to sensory neurons and reset their excitability status (Kamiya et al. 2006, Ma et al. 2009, Tanida et al. 2005). Additionally, *L. rhamnosus* JB-1 has been shown to decrease the contractile activity of mouse jejunum segments *ex vivo* (Wang et al. 2010). This effect was not observed when another probiotic *L. salivarius* was used, suggesting that the effects of *L. rhamnosus* JB-1 are strain-specific (Wang et al. 2010). More recently, Bravo et al. (2011b) demonstrated that feeding *L. rhamnosus* JB-1 to healthy adult BALB/c mice over a period of 4 weeks produced anxiolytic and antidepressant-like effects and improved responses in the cognitive as well as emotional abilities towards aversive stimuli (Bravo et al. 2011). In addition, *L. rhamnosus* JB-1 treatment prevented an exaggerated HPA axis response after a stressful situation and induced region-dependent changes in $GABA_{B1b}$, $GABA_{Aa2}$ and $GABA_{Aa1}$ receptor subunits in brain areas such as the prefrontal cortex, amygdala, *locus coeruleus* and hippocampus (Bravo et al. 2011). Moreover, Bravo et al. (2011) also demonstrated that the vagus nerve is responsible for the observed behavioral changes and brain gene expression as sectioning of the sub-diaphragmatic branch of the nerve prevented the effects of *L. rhamnosus* JB-1 on behavior and brain neurochemistry. This means that the vagus is a major anatomical pathway necessary for the communication between the enteric microbiota and the brain. More evidence in this respect comes from the observations by Bercik et al (2011) in which the effects of *B. longum* NC3001 in the model of chemically induced colitis is also vagally dependent (Bercik et al. 2011), which further emphasizes the role of the vagus in this functional axis. However, not all the effects of the microbiota are necessarily mediated by the vagus nerve. It has been shown that some of the alterations in mice behavior observed after oral treatment with wide spectrum antibiotics do occur in vagotomized animals (Bercik et al. 2011). This highlights that although the vagus is an important bridge between the gastrointestinal system and the CNS, there are probably other pathways of communication also, between the intestinal luminal space and the brain; this emphasizes the intricate nature of the connections of the microbiota-gut-brain axis.

As outlined above, most of the studies providing evidence about the connections and interactions between the gut microbiota and the CNS have been undertaken in rodents. There is as yet only a very limited number of studies in humans although it has been shown that the administration of a probiotic formulation containing *L. helveticus* R0052 and *B. longum* R0175 improves some markers of anxiety and depression as well as other psychological and cognitive parameters in healthy adult subjects (Messaoudi et al. 2011).

In summary, there is strong evidence suggesting that changes in gut microbiota could lead to changes in CNS function, including behavior (Bravo et al. 2011, Bercik et al. 2011, Bercik et al. 2010). Application of this knowledge could improve cognitive and emotional aspects of mental health and could lead to novel strategies for the treatment of psychiatric conditions such as mood disorders. Probiotic treatments could also increase the efficacy of psychopharmacology, as it has been shown that these microorganisms affect the expression of genes in the brain (Bravo et al. 2011); this may lead to the administration of lower doses of pharmacologically active compounds, thus improving their safety and reducing their toxic side effects, with considerable impact on drug safety. Lastly, selective alteration of the gut microbiota may by itself prove to have important beneficial outcomes for many systemic and gut conditions as a result of their actions on the central and enteric nervous systems.

Prebiotics

A large proportion of the indigestible molecules entering the large intestine are polysaccharides of plant origin whose constituent monosaccharides are linked forming chains of variable length and with varying degrees of branching. These polysaccharides are modified in the colon by the enzymatic activities of the microbiota during fermentation to produce short chain fatty acids (acetic, propionic and butyric at an approximate proportion of 1:1:3), gases (carbon dioxide, hydrogen, methane), and water (Crittenden and Playne 2009). This also induces changes of the physicochemical parameters of the colonic lumen, such as its pH, that are considered beneficial for the host's health.

These indigestible, fermentable carbohydrates were called prebiotics by Gibson and Roberfroid (1995) who defined them as "dietary components which have a specific fermentation pathway directed to the stimulation of populations of intestinal bacteria considered beneficial to health". As other investigators produced definitions that preferentially emphasized one or more different functional characteristics, and to further clarify and unify this concept, a meeting convened by FAO in Rome agreed on a definition that focused on what are considered the aspects fundamental for clinical, research and regulatory purposes: "Prebiotics are non viable components of foods whose intake confers benefits to the host because they are associated with the modulation of his microbiota" (Pineiro et al. 2008). This means that prebiotics allow manipulations of some colonic functions and especially of its microbiota.

Prebiotics serve as energy stores for the plants from which they are isolated; compounds such as chitin from the exoskeleton of crustaceans and insects and some of its derivates, as well as resistant starch and some synthetic molecules, also satisfy the criteria outlined in the FAO definition; some of these molecules are water soluble and may form viscous or non viscous solutions (Crittenden and Playne 2009). The prebiotics studied in greater detail are inulin (a polyfructan formed by multiple units of fructose linked by β 2-1 bonds with a terminal glucose unit), fructooligosaccharides (chains of fructose linked by β 2-1 bonds, FOS), synthetic galactooligosacharides (chains of galactose units with a terminal glucose unit, GOS), lactulose, a synthetic non-digestible sugar (4-O-β-D galactopyranosyl-β-D-fructofuranose) and polydextrose [randomly bonded glucose (with 1,6 glycosidic bonds predominating), some glucose, sorbitol (10%) and citric acid]. Other prebiotics have been less well studied: xylooligosaccharides, isomaltooligosaccharides, the oligosaccharides of soybeans (raffinose and stachyose), arabinoxylans, lactosucrose, lactitol, chitin and polysaccharides from fungi. Retrograded starch and some components of cereal fiber such as the hemicelluloses also function like prebiotics and are fermented by the resident colonic bacteria. Maternal milk contains a large number of oligosaccharides which probably additionally act as receptors and messengers for cells of the epithelial lining of the gut, stimuli for components of the immune system and for members of the resident microflora; human milk polysaccharides probably also protect the mucosal surface of the gastrointestinal tract from its colonization by enteropathogens. Traces of these molecules appear in the urine of breastfed infants (Bode 2009, Rudloff et al. 2012, Schwab and Ganzle 2011) indicating that they cross the intestinal epithelial layer. Of note, the nucleotides in human milk or those added to infant formulae have bifidogenic effects and may be considered in some sense to act as prebiotics.

In humans and in laboratory animals prebiotics induce not only changes in the intestinal microbiota and the mucosal surface of the colon but the transepithelial transport of the SCFA stimulates shifts of fluid to and from the lumen; furthermore, the transport of cationic minerals is stimulated by the lowered pH of the lumen. It has been shown recently that intraluminal colonic propionate induces the non-neuronal release of acetlylcholine (Ach) synthesized by the epithelial crypt cells to the serosal surface, especially in the distal colon and that this was associated with modifications of the electrical parameters of the mucosa and chloride excretion (Yajima et al. 2011). Increased calcium and magnesium retention and bone deposition have been observed in adolescents, in menopausal women and in laboratory animals consuming prebiotics. The increase of fecal bulk due to the transfer and retention of water in the lumen and the increase of the bacterial mass due to the proliferation of bifidobacteria and other species induce contractile responses of the intestinal smooth muscle and favor fecal evacuation. Gastrointestinal motility also increases because of electric signals by the pacemakers in the neural plexuses of the colonic wall and their propagation (Schwab and Ganzle 2011). This was supported recently by a study describing the effects of a GOS/FOS mixture in prematures (Mihatsch et al. 2006).

The increased transport of SCFA through the portal circulation modulates changes in systemic and hepatic lipid metabolism. At the systemic level decreases of blood cholesterol and triglycerides, have been demonstrated. The changes in triglyceride metabolism are the result of the algebraic sum of the effects of the absorbed acetate and of propionate and butyrate, the former stimulating and the latter inhibiting the synthesis of cholesterol and triglycerides in the liver, respectively. Butyric acid, the largest source of energy for the colonic epithelium, exerts antitumoral effects through the stimulation of colonocyte differentiation, the induction of apoptosis of damaged colonocytes, and decreased numbers of aberrant crypts in rats following the administration of 1,2-dimethyl hydrazine. The effects of prebiotics on the local and systemic immunity have been attributed to the stimulation of lymphocytes by SCFA. Some of these effects may occur in different segments of the colon depending on the length of the oligosaccharide chain as longer chains reach more distal portions and this shifts the fermentative processes and perhaps of their local effects (Milner 1999).

Cani and coworkers (2004, 2005) observed that rats fed a high-fat diet supplemented with oligofructose or long-chain inulin ingested significantly less energy than control animals associated with significant decreases of plasma and liver triglycerides and of the fat mass of the epididimal pads.

Higher levels of glucagon-like peptide (GLP)-1 (7–36) amide were detected in their portal vein as were increases in proglucagon mRNA in the mucosa of the proximal colon. Ghrelin, an orexigenic gastric peptide, remained low in plasma despite 8 hours of food deprivation. These authors postulated that short chain oligosaccharides that are fermented in the cecum and proximal colon modulate GLP-1 (7–36) amide and ghrelin production, with anorexigenic effects (Cani et al. 2005, Cani et al. 2004). In studies in human volunteers, oligofructose increased the perception of satiety after the morning and evening meals and reduced the hunger sensation and the prospective food intake in the evening. Interestingly, the intake of energy was also decreased for the lunch meals, which means that the anorexigenic effect may persist for a few hours. In volunteers on an *ad libitum* diet, oligofructose induced a negative energy balance equivalent to about 5% of the energy intake even though the fat content of the diet was high. This negative energy balance even in the presence of a high fat intake prebiotic may be useful in the management of obesity and/or diabetes (Cani et al. 2006). The metabolic effects of the products of fermentation of prebiotics are due to their detection by specialized cells in the colonic epithelium that generate signals that are transmitted either directly or via chemical (endocrine) or neural pathways to distant organs. In the proximal colon of rats prebiotics induce the proliferation of the entero-endocrine L-cell population that synthesizes glucagon–like peptides (GLP)-1 and 2. GLP-1 promotes insulin secretion and β-cell proliferation in the pancreas, controls the synthesis of glycogen in striated muscle fibers and promotes satiety (Cani et al. 2007).

The administration of 21g/day of short chain inulin to human volunteers decreased the intake of food and this was associated with lower body weight and fat mass. Peptide (P) YY plasma levels were increased and prolonged decreases of ghrelin levels were observed during tolerance tests (Parnell and Reimer 2009). Other studies have shown that increases of GLP-1, PYY and GIP occur in association with decreases in energy intake and decreased glycemic responses (Cani et al. 2009). This indicates that prebiotic intake and probably the modifications induced in the microbiota modulate the responses of the gut neuro-entero-endocrine and endocannabinoid systems potentiating the decreases of the adipose mass and of endotoxinemia related to the increased numbers of bifidobacteria and perhaps of other bacteria. It is not known whether modifications of the intestinal mucosal permeability, and of what magnitude may be induced (Weickert et al. 2006). Dewulf and coworkers (2011) demonstrated that inulin-type fructans may modulate the increase of white adipose tissue in animals fed a high fat diet through changes in the microbiota that favor the expansion of the bifidobacteria population at the expense of Roseburia spp. and Clostridium cluster XIVa. This is associated with blunting of the increases of receptors specific for the proliferation of subcutaneous adipose tissue (Dewulf et al. 2011). The energy sparing resulting from the fermentation of prebiotics may be compensated by increments of the metabolic activity of other bacteria whose biochemical processes increase primarily or secondarily as a result of the fermentation of oligosaccharides.

The mechanisms by which SCFAs act as chemical signals have been described rather recently. SCFA are detected by G protein coupled-receptors called Free Fatty Acid Receptor 2 (FFAR2, previously called GPR43 for G-protein Receptor 43) and FFAR3 (previously called GPR41) in numerous cell types (Stoddart et al. 2008). Acetate is preferentially linked *in vitro* by FFAR2, propionate interacts with both FFAR2 and FFAR3, and butyrate with FFAR3; there is evidence of the selective stimulation of specific receptors and cell types by the respective SCFAs. In laboratory animals butyrate and propionate suppress weight gain independently of the decrease of food intake (Lin et al. 2012). It is thought that chronic administration of butyrate and acetate to humans induces activation of adenosin 5′-monophosphate protein kinase (AMPK) and increased mitochondrial fatty acid oxidation while propionate inhibits food intake (Lin et al. 2012, Arora et al. 2011). In addition to their presence in L and other entero-endocrine cells, FFA receptors are present in smooth muscle fibers, adipocytes and lymphocytes and in other cell types and this may explain SCFA effects such as the modulation of mesenteric blood flow, changes in colonic motility and secretion of water, bicarbonate, chloride and potassium by the colonic mucosa (Yajima et al. 2011, Tazoe et al. 2009). However, the primary mechanism underlying the resistance to obesity remains obscure. GLP-1 also participates in the control of whole body glucose utilization and in the synthesis of glycogen through a neural pathway whose initial signals originate in enteric sensors of glucose that indicate to the striated muscle and the liver to prepare to metabolize glucose. When the flow of glucose to the circulation decreases, an opposite signal depresses and interrupts this synthetic pathway to avoid maintaining active a metabolic process that now lacks adequate amounts of substrate (Mithieux 2009).

In addition to contributing to weight loss, GLP-2 is a trophic factor for the intestine and it modifies intestinal epithelial permeability because it modulates the expression of tight junction proteins Zo-1 and occludin; this decreased permeability reduces the translocation of endotoxin across the junctional complex and the intensity of the low level inflammatory process associated with obesity (Cani et al. 2009).

GLP-1 is not the only enteric polypeptide with hormone activity whose synthesis is stimulated by prebiotics. Wheat fiber-enriched bread has been shown to be postprandially associated in females with blunted increases in circulating ghrelin and PYY, which does not occur when oat-fiber enriched bread is consumed. It has been postulated that the dissimilarities in content, composition and molecular weight of the hemicelluloses in their respective polysaccharides may induce differential endocrine responses not associated with differences in satiety ratings. In this case PYY and ghrelin levels in plasma do not seem to play a mayor role in the regulation of satiety (Weickert et al. 2006).

The demonstration of the transepithelial transfer of human milk oligosaccharides in infants suggests that some of these molecules may exert activities on remote tissues by acting on receptors in specialized cells or through modifications in the synthesis, delivery and/or sensitivity to cytokines (Ruddloff et al. 2012); this opens new perspectives to the functional and metabolic effects of these prebiotics (Wu et al. 2010, Froehlich et al. 2010).

Taking into account the precedent discussion, and although predicting future developments in biology and medicine is always risky, it is interesting to consider some aspects. In the first place, it is important to further advance in the identification and metabolic characterization of as many members of the resident microbiota of humans, both in normal circumstances or when affected by pathologies, as well as in defined strains of laboratory animals. This should extend to the modifications and adaptations these microorganisms experience in the presence of different prebiotics or other molecules in the colonic lumen. Another point of interest is the velocity with which these adaptations occur and their sequences and persistence as related to specific substrates. It is also interesting to characterize the responses of the different structures of the colonic mucosa: the epithelium, the neuro-entero-endocrine and neural components and their parameters of electrical activity in response to modifications in the conditions of the lumen. Another point of interest refers to possible responses of distant organs elicited by metabolites of different prebiotics.

Other aspects that deserve consideration refer to the functional changes induced by the less studied prebiotics: isomaltooligosaccharides, xylooligosaccharides, raffinose and stachyose from soybeans and chitin and acetylated chitin from fungi. It will also be of interest to explore the effects of mixtures of polysaccharides and changes in chain length and stereochemistry, including the spatial orientation of chemical bonds including the effects of different sequences of monosaccharides. Complex *in vitro* systems such as the "artificial colon" should allow a more detailed evaluation of the reactions occurring during complex fermentative processes.

The effects of individual human milk oligosaccharides and their specific local and systemic receptors merits detailed exploration, as well as their systemic affects. This is interesting because they are the result of hundreds of thousands of years of evolution to which complex tissue structure and metabolism of the gastrointestinal tract have adapted. The indigestible milk polysaccharides must play important roles in the maturation of digestive tract structure and functions, especially in newborn and young infants; the persistence of their effects deserves special attention in breastfed infants.

Finally, the function of the receptors that detect the metabolites of prebiotic fermentation and modulate the metabolism of the SCFA and induce effects such as anorexia and/or weight loss should be explored in depth. This could result in a set of useful tools for weight control. Within this context the effect of the multiple hormones of the entero-endocrine system should be explored and the possibility of synthetic forms with varying durations of their effects should also be the object of exploration.

The future and the labors of many specialists in different branches of medicine, biochemistry, physiology and associated sciences, will determine how many of these predictions were accurate.

References

Abrams, G.D., H. Bauer and H. Sprinz. 1963. Influence of the normal flora on mucosal morphology and cellular renewal in the ileum. A comparison of germ-free and conventional mice. Lab Invest. 12: 355–364.

Albrecht, S., H.A. Schols, E.G. van den Heuvel, A.G. Voragen and H. Gruppen. 2011. Occurrence of oligosaccharides in feces of breast-fed babies in their first six months of life and the corresponding breast milk. Carbohydr. Res. 346: 2540–2550.

Alles, M.S., R. Hartemink, S. Meyboom, J.L. Harryvan, K.M. Van Laere, F.M. Nagengast and J.G. Hautvast. 1999. Effect of transgalactooligosaccharides on the composition of the human intestinal microflora and on putative risk markers for colon cancer. Am. J. Clin. Nutr. 69: 980–991.

Aronov, P.A., F.J. Luo, N.S. Plummer, Z. Quan, S. Holmes, T.H. Hostetter and T.W. Meyer. 2011. Colonic contribution to uremic solutes. J. Am. Soc. Nephrol. 22: 1769–1776.

Arora, T., R. Sharma and G. Frost. 2011. Propionate. Anti-obesity and satiety enhancing factor? Appetite 56: 511–515.

Bammens, B., P. Evenepoel, K. Verbeke and Y. Vanrenterghem. 2004. Impairment of small intestinal protein assimilation in patients with end-stage renal disease: extending the malnutrition-inflammation-atherosclerosis concept. Am. J. Clin. Nutr. 80: 1536–1543.

Bass, N.M. 2007. Review article: the current pharmacological therapies for hepatic encephalopathy. Aliment Pharmacol. Ther. 25 Suppl. 1: 23–31.

Bercik, P., A.J. Park, D. Sinclair, A. Khoshdel, J. Lu, X. Huang, Y. Deng, P.A. Blennerhassett, M. Fahnestock, D. Moine, B. Berger, J.D. Huizinga, W. Kunze, P.G. McLean, G.E. Bergonzelli, S.M. Collins and E.F. Verdu. 2011a. The anxiolytic effect of Bifidobacterium longum NCC3001 involves vagal pathways for gut-brain communication. Neurogastroenterol. Motil 23: 1132–1139.

Bercik, P., E. Denou, J. Collins, W. Jackson, J. Lu, J. Jury, Y. Deng, P. Blennerhassett, J. Macri, K.D. McCoy, E.F. Verdu and S.M. Collins. 2011b. The intestinal microbiota affect central levels of brain-derived neurotropic factor and behavior in mice. Gastroenterology 141: 599–609..

Bercik, P., E.F. Verdu, J.A. Foster, J. Macri, M. Potter, X. Huang, P. Malinowski, W. Jackson, P. Blennerhassett, K.A. Neufeld, J. Lu, W.I. Khan, I. Corthesy-Theulaz, C. Cherbut, G.E. Bergonzelli and S.M. Collins. 2010. Chronic Gastrointestinal Inflammation Induces Anxiety-Like Behavior and Alters Central Nervous System Biochemistry in Mice. Gastroenterology 139: 2102-2112.e2101.

Bercik, P., S.M. Collins and E.F. Verdu. 2012. Microbes and the gut-brain axis. Neurogastroenterol. Motil 24: 405–413.

Birkett, A., J. Muir, J. Phillips, G. Jones and K. O'Dea. 1996. Resistant starch lowers fecal concentrations of ammonia and phenols in humans. Am. J. Clin. Nutr. 63: 766–772.

Bliss, D.Z., T.P. Stein, C.R. Schleifer and R.G. Settle. 1996. Supplementation with gum arabic fiber increases fecal nitrogen excretion and lowers serum urea nitrogen concentration in chronic renal failure patients consuming a low-protein diet. Am. J. Clin. Nutr. 63: 392–398.

Bode, L. 2009. Human milk oligosaccharides: prebiotics and beyond. Nutr. Rev. 67 Suppl. 2: S183–191.

Bravo, J.A., M. Julio-Pieper, P. Forsythe, W. Kunze, T.G. Dinan, J. Bienenstock and J.F. Cryan. 2012. Communication between gastrointestinal bacteria and the nervous system. Curr. Opin. Pharmacol. 12: 667–72.

Bravo, J.A., P. Forsythe, M.V. Chew, E. Escaravage, H.M. Savignac, T.G. Dinan, J. Bienenstock and J.F. Cryan. 2011. Ingestion of Lactobacillus strain regulates emotional behavior and central GABA receptor expression in a mouse via the vagus nerve. Proc. Natl. Acad. Sci. USA 108: 16050–16055.

Bravo, J.A., T.G. Dinan and J.F. Cryan. 2010. Alterations in the central CRF system of two different rat models of comorbid depression and functional gastrointestinal disorders. Int. J. Neuropsychopharmacol. 14: 666–683.

Browne, C.A., G. Clarke, T.G. Dinan and J.F. Cryan. 2011. Differential stress-induced alterations in tryptophan hydroxylase activity and serotonin turnover in two inbred mouse strains. Neuropharmacology 60: 683–691.

Cani, P. D., E. Joly, Y. Horsmans and N. M. Delzenne. 2006. Oligofructose promotes satiety in healthy human: a pilot study. Eur. J. Clin. Nutr. 60: 567–572.

Cani, P.D., A.M. Neyrinck, N. Maton and N.M. Delzenne. 2005. Oligofructose promotes satiety in rats fed a high-fat diet: involvement of glucagon-like Peptide-1. Obes. Res. 13: 1000–1007.

Cani, P.D., C. Dewever and N.M. Delzenne. 2004. Inulin-type fructans modulate gastrointestinal peptides involved in appetite regulation (glucagon-like peptide-1 and ghrelin) in rats. Br. J. Nutr. 92: 521–526.

Cani, P.D., S. Hoste, Y. Guiot and N.M. Delzenne. 2007. Dietary non-digestible carbohydrates promote L-cell differentiation in the proximal colon of rats. Br. J. Nutr. 98: 32–37.

Cani, P.D., S. Possemiers, T. Van de Wiele, Y. Guiot, A. Everard, O. Rottier, L. Geurts, D. Naslain, A. Neyrinck, D.M. Lambert, G.G. Muccioli and N.M. Delzenne. 2009. Changes in gut microbiota control inflammation in obese mice through a mechanism involving GLP-2-driven improvement of gut permeability. Gut. 58: 1091–1103.

Cervero, F. and W. Janig. 1992. Visceral nociceptors: a new world order? Trends Neurosci. 15: 374–378.

Clarke, G., E.M. Quigley, J.F. Cryan and T.G. Dinan. 2009. Irritable bowel syndrome: towards biomarker identification. Trends Mol. Med. 15: 478–489.

Clarke, G., J.F. Cryan, T.G. Dinan and E.M. Quigley. 2012. Review article: probiotics for the treatment of irritable bowel syndrome—focus on lactic acid bacteria. Aliment Pharmacol. Ther. 35: 403–413.

Clarke, G., S. Grenham, P. Scully, P. Fitzgerald, R.D. Moloney, F. Shanahan, T.G. Dinan and J.F. Cryan. 2012. The microbiome-gut-brain axis during early life regulates the hippocampal serotonergic system in a sex-dependent manner. Mol. Psychiatry 18: 666–673.

Cloetens, L., V. De Preter, K. Swennen, W.F. Broekaert, C.M. Courtin, J.A. Delcour, P. Rutgeerts and K. Verbeke. 2008. Dose-response effect of arabinoxylooligosaccharides on gastrointestinal motility and on colonic bacterial metabolism in healthy volunteers. J. Am. Coll. Nutr. 27: 512–518.

Collado, M.C., E. Isolauri and S. Salminen. 2008. Specific probiotic strains and their combinations counteract adhesion of Enterobacter sakazakii to intestinal mucus. FEMS Microbiol. Lett. 285: 58–64.

Collins, S.M., E. Denou, E.F. Verdu and P. Bercik. 2009. The putative role of the intestinal microbiota in the irritable bowel syndrome. Dig. Liver Dis. 41: 850–853.

Corr, S.C., Y. Li, C.U. Riedel, P.W. O'Toole, C. Hill and C.G. Gahan. 2007. Bacteriocin production as a mechanism for the antiinfective activity of Lactobacillus salivarius UCC118. Proc. Natl. Acad Sci. USA 104: 7617–7621.

Crittenden, R. and M.J. Playne. 2009. Prebiotics. *In*: Y.K. Lee and S. Salminen [eds.]. Handbook of probiotics and prebiotics. John Wiley & Sons., Hoboken, NJ, pp. 533–581.

Cryan, J. and S. O'Mahony. 2010. The Microbiome-Gut-Brain Axis: From Bowel to Behaviour. Neurogastroenterol. Motil 23: 187–192.

Cryan, J.F. and T.G. Dinan. 2012. Mind-altering microorganisms: the impact of the gut microbiota on brain and behaviour. Nat. Rev. Neurosci. 13: 701–712.

Davey, K.J., S.M. O'Mahony, H. Schellekens, O. O'Sullivan, J. Bienenstock, P.D. Cotter, T.G. Dinan and J.F. Cryan. 2012. Gender-dependent consequences of chronic olanzapine in the rat: effects on body weight, inflammatory, metabolic and microbiota parameters. Psychopharmacology (Berl) 221: 155–169.

Davila, A.M., F. Blachier, M. Gotteland, M. Andriamihaja, P.H. Benetti, Y. Sanz and D. Tomé. 2013. Intestinal luminal nitrogen metabolism: role of the gut microbiota and consequences for the host. Pharmacol. Res. 69: 114–126.

De Preter, V., T. Vanhoutte, G. Huys, J. Swings, L. De Vuyst, P. Rutgeerts and K. Verbeke. 2007. Effects of *Lactobacillus casei* Shirota, Bifidobacterium breve, and oligofructose-enriched inulin on colonic nitrogen-protein metabolism in healthy humans. Am. J. Physiol. Gastrointest Liver Physiol. 292: G358–368.

De Preter, V., T. Vanhoutte, G. Huys, J. Swings, P. Rutgeerts and K. Verbeke. 2006. Effect of lactulose and Saccharomyces boulardii administration on the colonic urea-nitrogen metabolism and the bifidobacteria concentration in healthy human subjects. Aliment Pharmacol. Ther. 23: 963–974.

Desbonnet, L., L. Garrett, G. Clarke, B. Kiely, J.F. Cryan and T.G. Dinan. 2010. Effects of the probiotic Bifidobacterium infantis in the maternal separation model of depression. Neuroscience.

Dewulf, E.M., P.D. Cani, A.M. Neyrinck, S. Possemiers, A. Van Holle, G.G. Muccioli, L. Deldicque, L.B. Bindels, B.D. Pachikian, F.M. Sohet, E. Mignolet, M. Francaux, Y. Larondelle and N.M. Delzenne. 2011. Inulin-type fructans with prebiotic properties counteract GPR43 overexpression and PPARgamma-related adipogenesis in the white adipose tissue of high-fat diet-fed mice. J. Nutr. Biochem. 22: 712–722.

Dinan, T.G., E.M. Quigley, S.M. Ahmed, P. Scully, S. O'Brien, L. O'Mahony, S. O'Mahony, F. Shanahan and P.W. Keeling. 2006a. Hypothalamic-pituitary-gut axis dysregulation in irritable bowel syndrome: plasma cytokines as a potential biomarker? Gastroenterology 130: 304–311.

Dinan, T., E. Quigley, L. O'Mahony, J. Cryan, C. Hill, P. Ross and F. Shanahan. 2006b. Brain-gut axis function in irritable bowel syndrome: focus on drug targets, gut flora and biomarkers. Alimentary Pharmabiotic Center, University College Cork.

Donkor, O.N., M. Ravikumar, O. Proudfoot, S.L. Day, V. Apostolopoulos, G. Paukovics, T. Vasiljevic, S.L. Nutt and H. Gill. 2012. Cytokine profile and induction of T helper type 17 and regulatory T cells by human peripheral mononuclear cells after microbial exposure. Clin. Exp. Immunol. 167: 282–295.

Drossman, D.A., F.H. Creed, K.W. Olden, J. Svedlund, B.B. Toner and W.E. Whitehead. 1999. Psychosocial aspects of the functional gastrointestinal disorders. Gut. 45 Suppl. 2: II25–30.

Drossman, D.A., L. Chang, N. Bellamy, H.E. Gallo-Torres, A. Lembo, F. Mearin, N.J. Norton and P. Whorwell. 2011. Severity in irritable bowel syndrome: a Rome Foundation Working Team report. Am. J. Gastroenterol. 106: 1749–1759.

Duncker, S.C., T. Kamiya, L. Wang, P. Yang and J. Bienenstock. 2011. Probiotic Lactobacillus reuteri alleviates the response to gastric distension in rats. J. Nutr. 141: 1813–1818.

Evenepoel, P., B.K. Meijers, B.R. Bammens and K. Verbeke. 2009. Uremic toxins originating from colonic microbial metabolism. Kidney Int. Suppl: S12–19.

Finegold, S.M., D. Molitoris, Y. Song, C. Liu, M.L. Vaisanen, E. Bolte, M. McTeague, R. Sandler, H. Wexler, E.M. Marlowe, M.D. Collins, P.A. Lawson, P. Summanen, M. Baysallar, T.J. Tomzynski, E. Read, E. Johnson, R. Rolfe, P. Nasir, H. Shah, D.A. Haake, P. Manning and A. Kaul. 2002. Gastrointestinal microflora studies in late-onset autism. Clin. Infect. Dis. 35: S6–S16.

Forsythe, P., N. Sudo, T. Dinan, V.H. Taylor and J. Bienenstock. 2010. Mood and gut feelings. Brain Behav. Immun. 24: 9–16.

Froehlich, J.W., E.D. Dodds, M. Barboza, E.L. McJimpey, R.R. Seipert, J. Francis, H.J. An, S. Freeman, J.B. German and C.B. Lebrilla. 2010. Glycoprotein expression in human milk during lactation. J. Agric. Food Chem. 58: 6440–6448.

Gareau, M.G., E. Wine, D.M. Rodrigues, J.H. Cho, M.T. Whary, D.J. Philpott, G. Macqueen and P.M. Sherman. 2011. Bacterial infection causes stress-induced memory dysfunction in mice. Gut. 60: 307–317.

Geboes, K.P., G. De Hertogh, V. De Preter, A. Luypaerts, B. Bammens, P. Evenepoel, Y. Ghoos, K. Geboes, P. Rutgeerts and K. Verbeke. 2006. The influence of inulin on the absorption of nitrogen and the production of metabolites of protein fermentation in the colon. Br. J. Nutr. 96: 1078–1086.

Generoso, S.V., M. Viana, R. Santos, F.S. Martins, J.A. Machado, R.M. Arantes, J.R. Nicoli, M.I. Correia and V.N. Cardoso. 2010. Saccharomyces cerevisiae strain UFMG 905 protects against bacterial translocation, preserves gut barrier integrity and stimulates the immune system in a murine intestinal obstruction model. Arch. Microbiol. 192: 477–484.

Gibney, S., R.D. Gosselin, T.G. Dinan and J.F. Cryan. 2010. Colorectal Distension-induced Prefrontal Cortex Activation in the Wistar Kyoto Rat: Implications for Irritable Bowel Syndrome (IBS). Neuroscience 165: 675–683.

Gibson, G.R. and M.B. Roberfroid. 1995. Dietary modulation of the human colonic microbiota: introducing the concept of prebiotics. J. Nutr. 125: 1401–1412.

Goehler, L.E., S.M. Park, N. Opitz, M. Lyte and R.P. Gaykema. 2008. Campylobacter jejuni infection increases anxiety-like behavior in the holeboard: possible anatomical substrates for viscerosensory modulation of exploratory behavior. Brain Behav. Immun. 22: 354–366.

Gordon, H.A. and E. Bruckner-Kardoss. 1961. Effect of normal microbial flora on intestinal surface area. Am. J. Physiol. 201: 175–178.

Gostner, A., M. Blaut, V. Schaffer, G. Kozianowski, S. Theis, M. Klingeberg, Y. Dombrowski, D. Martin, S. Ehrhardt, D. Taras, A. Schwiertz, B. Kleessen, H. Luhrs, J. Schauber, D. Dorbath, T. Menzel and W. Scheppach. 2006. Effect of isomalt consumption on faecal microflora and colonic metabolism in healthy volunteers. Br. J. Nutr. 95: 40–50.

Grenham, S., G. Clarke, J.F. Cryan and T.G. Dinan. 2011. Brain-gut-microbe communication in health and disease. Front Physiol. 2: 94.

Guerin-Danan, C., C. Chabanet, C. Pedone, F. Popot, P. Vaissade, C. Bouley, O. Szylit and C. Andrieux. 1998. Milk fermented with yogurt cultures and *Lactobacillus casei* compared with yogurt and gelled milk: influence on intestinal microflora in healthy infants. Am. J. Clin. Nutr. 67: 111–117.

Gundersen, B.B. and J.A. Blendy. 2009. Effects of the histone deacetylase inhibitor sodium butyrate in models of depression and anxiety. Neuropharmacology 57: 67–74.

Guyonnet, D., O. Chassany, P. Ducrotte, C. Picard, M. Mouret, C.H. Mercier and C. Matuchansky. 2007. Effect of a fermented milk containing Bifidobacterium animalis DN-173 010 on the health-related quality of life and symptoms in irritable bowel syndrome in adults in primary care: a multicentre, randomized, double-blind, controlled trial. Aliment Pharmacol. Ther. 26: 475–486.

Heijtz, R.D., S. Wang, F. Anuar, Y. Qian, B. Bjorkholm, A. Samuelsson, M.L. Hibberd, H. Forssberg and S. Pettersson. 2011. Normal gut microbiota modulates brain development and behavior. Proc. Natl. Acad. Sci. USA 108: 3047–3052.

Hickson, M., A.L. D'Souza, N. Muthu, T.R. Rogers, S. Want, C. Rajkumar and C.J. Bulpitt. 2007. Use of probiotic Lactobacillus preparation to prevent diarrhoea associated with antibiotics: randomised double blind placebo controlled trial. BMJ 335: 80.

Jan, R.L., K.C. Yeh, M.H. Hsieh, Y.L. Lin, H.F. Kao, P.H. Li, Y.S. Chang and J.Y. Wang. 2012. Lactobacillus gasseri suppresses Th17 pro-inflammatory response and attenuates allergen-induced airway inflammation in a mouse model of allergic asthma. Br. J. Nutr. 108: 130–139.

Kamiya, T., L. Wang, P. Forsythe, G. Goettsche, Y. Mao, Y. Wang, G. Tougas and J. Bienenstock. 2006. Inhibitory effects of Lactobacillus reuteri on visceral pain induced by colorectal distension in Sprague-Dawley rats. Gut. 55: 191–196.

Karimi, K., M.D. Inman, J. Bienenstock and P. Forsythe. 2009. Lactobacillus reuteri-induced regulatory T cells protect against an allergic airway response in mice. Am. J. Respir. Crit Care Med. 179: 186–93.

Kunze, W.A., Y.K. Mao, B. Wang, J.D. Huizinga, X. Ma, P. Forsythe and J. Bienenstock. 2009. Lactobacillus reuteri enhances excitability of colonic AH neurons by inhibiting calcium-dependent potassium channel opening. J. Cell Mol. Med. 13: 2261–2270.

Kwon, H.K., C.G. Lee, J.S. So, C.S. Chae, J.S. Hwang, A. Sahoo, J.H. Nam, J.H. Rhee, K.C. Hwang and S.H. Im. 2010. Generation of regulatory dendritic cells and CD4+Foxp3+ T cells by probiotics administration suppresses immune disorders. Proc. Natl. Acad. Sci. USA 107: 2159–2164.

Ley, R.E., P.J. Turnbaugh, S. Klein and J.I. Gordon. 2006. Microbial ecology: human gut microbes associated with obesity. Nature 444: 1022–1023.

Li, W., S.E. Dowd, B. Scurlock, V. Acosta-Martinez and M. Lyte. 2009. Memory and learning behavior in mice is temporally associated with diet-induced alterations in gut bacteria. Physiol. Behav. 96: 557–567.

Lin, C.J., H.L. Liu, C.F. Pan, C.K. Chuang, T. Jayakumar, T.J. Wang, H.H. Chen and C.J. Wu. 2012. Indoxyl sulfate predicts cardiovascular disease and renal function deterioration in advanced chronic kidney disease. Arch. Med. Res. 43: 451–456.

Lin, H.V., A. Frassetto, E.J. Kowalik, Jr., A.R. Nawrocki, M.M. Lu, J.R. Kosinski, J.A. Hubert, D. Szeto, X. Yao, G. Forrest and D.J. Marsh. 2012. Butyrate and propionate protect against diet-induced obesity and regulate gut hormones via free fatty acid receptor 3-independent mechanisms. PLoS One 7: e35240.

Liu, Z., Y. Ma and H. Qin. 2011. Potential prevention and treatment of intestinal barrier dysfunction using active components of Lactobacillus. Ann. Surg 254: 832–833; author reply 833.

Lyte, M., J.J. Varcoe and M.T. Bailey. 1998. Anxiogenic effect of subclinical bacterial infection in mice in the absence of overt immune activation. Physiol. Behav. 65: 63–68.

Lyte, M., W. Li, N. Opitz, R.P. Gaykema and L.E. Goehler. 2006. Induction of anxiety-like behavior in mice during the initial stages of infection with the agent of murine colonic hyperplasia Citrobacter rodentium. Physiol. Behav. 89: 350–357.

Ma, D., P. Forsythe and J. Bienenstock. 2004. Live Lactobacillus rhamnosus [corrected] is essential for the inhibitory effect on tumor necrosis factor alpha-induced interleukin-8 expression. Infect Immun. 72: 5308–5314.

Ma, X., Y. K. Mao, B. Wang, J.D. Huizinga, J. Bienenstock and W. Kunze. 2009. Lactobacillus reuteri ingestion prevents hyperexcitability of colonic DRG neurons induced by noxious stimuli. Am. J. Physiol. Gastrointest Liver Physiol. 296: G868–875.

MacFabe, D.F., N.E. Cain, F. Boon, K.P. Ossenkopp and D.P. Cain. 2011. Effects of the enteric bacterial metabolic product propionic acid on object-directed behavior, social behavior, cognition, and neuroinflammation in adolescent rats: Relevance to autism spectrum disorder. Behav. Brain Res. 217: 47–54.

Madrid, A.M., R. Quera, C. Defilippi, C. Defilippi, S. Olguín and S. Larraín. 2005. Prevalencia del Síndrome de Intestino Irritable en población chilena: estudio preliminar. Gastr Latinoam 16: 1.

Marshall, J.K., M. Thabane, A.X. Garg, W. Clark, J. Meddings and S.M. Collins. 2004. Intestinal permeability in patients with irritable bowel syndrome after a waterborne outbreak of acute gastroenteritis in Walkerton, Ontario. Aliment Pharmacol. Ther. 20: 1317–1322.
Martin, F.P., N. Sprenger, I. Montoliu, S. Rezzi, S. Kochhar and J.K. Nicholson. 2010. Dietary modulation of gut functional ecology studied by fecal metabonomics. J. Proteome Res. 9: 5284–5295.
Mayer, E.A. and S.M. Collins. 2002. Evolving pathophysiologic models of functional gastrointestinal disorders. Gastroenterology 122: 2032–2048.
Mayer, E.A., B.D. Naliboff and A.D. Craig. 2006. Neuroimaging of the brain-gut axis: from basic understanding to treatment of functional GI disorders. Gastroenterology 131: 1925–1942.
McFarland, L.V. and S. Dublin. 2008. Meta-analysis of probiotics for the treatment of irritable bowel syndrome. World J. Gastroenterol. 14: 2650–2661.
McKernan, D.P., P. Fitzgerald, T.G. Dinan and J.F. Cryan. 2010. The probiotic Bifidobacterium infantis 35624 displays visceral antinociceptive effects in the rat. Neurogastroenterol. Motil 22: 1029–1035, e1268.
Meijers, B.K., V. De Preter, K. Verbeke, Y. Vanrenterghem and P. Evenepoel. 2010. p-Cresyl sulfate serum concentrations in haemodialysis patients are reduced by the prebiotic oligofructose-enriched inulin. Nephrol. Dial Transplant 25: 219–224.
Messaoudi, M., R. Lalonde, N. Violle, H. Javelot, D. Desor, A. Nejdi, J.F. Bisson, C. Rougeot, M. Pichelin, M. Cazaubiel and J.M. Cazaubiel. 2011. Assessment of psychotropic-like properties of a probiotic formulation (*Lactobacillus helveticus* R0052 and *Bifidobacterium* longum R0175) in rats and human subjects. Br. J. Nutr. 105: 755–764.
Mihatsch, W.A., J. Hoegel and F. Pohlandt. 2006. Prebiotic oligosaccharides reduce stool viscosity and accelerate gastrointestinal transport in preterm infants. Acta Paediatr 95: 843–848.
Milner, J.A. 1999. Functional foods and health promotion. J. Nutr. 129: 1395S–1397S.
Mithieux, G. 2009. A novel function of intestinal gluconeogenesis: central signaling in glucose and energy homeostasis. Nutrition 25: 881–884.
Mutsaers, H.A., M.J. Wilmer, D. Reijnders, J. Jansen, P.H. van den Broek, M. Forkink, E. Schepers, G. Glorieux, R. Vanholder, L.P. van den Heuvel, J.G. Hoenderop and R. Masereeuw. 2012. Uremic toxins inhibit renal metabolic capacity through interference with glucuronidation and mitochondrial respiration. Biochim. Biophys. Acta 1832: 142–150.
Nakabayashi, I., M. Nakamura, K. Kawakami, T. Ohta, I. Kato, K. Uchida and M. Yoshida. 2011. Effects of synbiotic treatment on serum level of p-cresol in haemodialysis patients: a preliminary study. Nephrol. Dial. Transplant 26: 1094–1098.
Navaneethan, U. and R.A. Giannella. 2011. Infectious colitis. Curr. Opin. Gastroenterol. 27: 66–71.
Neufeld, K.M., N. Kang, J. Bienenstock and J.A. Foster. 2011. Reduced anxiety-like behavior and central neurochemical change in germ-free mice. Neurogastroenterol. Motil 23: 255–264.
Nikfar, S., R. Rahimi, F. Rahimi, S. Derakhshani and M. Abdollahi. 2008. Efficacy of probiotics in irritable bowel syndrome: a meta-analysis of randomized, controlled trials. Dis. Colon Rectum 51: 1775–1780.
O'Mahony, L., J. McCarthy, P. Kelly, G. Hurley, F. Luo, K. Chen, G.C. O'Sullivan, B. Kiely, J.K. Collins, F. Shanahan and E.M. Quigley. 2005. Lactobacillus and bifidobacterium in irritable bowel syndrome: symptom responses and relationship to cytokine profiles. Gastroenterology 128: 541–551.
O'Mahony, S.M., J.R. Marchesi, P. Scully, C. Codling, A.M. Ceolho, E.M. Quigley, J.F. Cryan and T.G. Dinan. 2009. Early life stress alters behavior, immunity, and microbiota in rats: implications for irritable bowel syndrome and psychiatric illnesses. Biol. Psychiatry 65: 263–267.
O'Mahony, S.M., N.P. Hyland, T.G. Dinan and J.F. Cryan. 2011. Maternal separation as a model of brain-gut axis dysfunction. Psychopharmacology (Berl) 214: 71–88.

O'Malley, D., T.G. Dinan and J.F. Cryan. 2010. Neonatal maternal separation in the rat impacts on the stress responsivity of central corticotropin-releasing factor receptors in adulthood. Psychopharmacology (Berl). 214: 221–9.

Ortiz-Andrellucchi, A., A. Sanchez-Villegas, C. Rodriguez-Gallego, A. Lemes, T. Molero, A. Soria, L. Pena-Quintana, M. Santana, O. Ramirez, J. Garcia, F. Cabrera, J. Cobo and L. Serra-Majem. 2008. Immunomodulatory effects of the intake of fermented milk with *Lactobacillus casei* DN114001 in lactating mothers and their children. Br. J. Nutr 100: 834–845.

Parnell, J.A. and R.A. Reimer. 2009. Weight loss during oligofructose supplementation is associated with decreased ghrelin and increased peptide YY in overweight and obese adults. Am. J. Clin. Nutr. 89: 1751–1759.

Parracho, H.M., M.O. Bingham, G.R. Gibson and A.L. McCartney. 2005. Differences between the gut microflora of children with autistic spectrum disorders and that of healthy children. J. Med. Microbiol. 54: 987–991.

Pereg, D., O. Kimhi, A. Tirosh, N. Orr, R. Kayouf and M. Lishner. 2005. The effect of fermented yogurt on the prevention of diarrhea in a healthy adult population. Am. J. Infect. Control. 33: 122–125.

Pineiro, M., N.G. Asp, G. Reid, S. Macfarlane, L. Morelli, O. Brunser and K. Tuohy. 2008. FAO Technical meeting on prebiotics. J. Clin. Gastroenterol. 42 Suppl 3 Pt 2: S156–9.

Posey, D.J., D.L. Kem, N.B. Swiezy, T.L. Sweeten, R.E. Wiegand and C.J. McDougle. 2004. A pilot study of D-cycloserine in subjects with autistic disorder. Am. J. Psychiatry 161: 2115–2117.

Quigley, E.M. 2008. Probiotics in functional gastrointestinal disorders: what are the facts? Curr. Opin. Pharmacol. 8: 704–708.

Quigley, E.M. and B. Flourie. 2007. Probiotics and irritable bowel syndrome: a rationale for their use and an assessment of the evidence to date. Neurogastroenterol. Motil 19: 166–172.

Ranganathan, N., P. Ranganathan, E.A. Friedman, A. Joseph, B. Delano, D.S. Goldfarb, P. Tam, A. V. Rao, E. Anteyi and C.G. Musso. 2010. Pilot study of probiotic dietary supplementation for promoting healthy kidney function in patients with chronic kidney disease. Adv. Ther. 27: 634–647.

Rhee, S.H., C. Pothoulakis and E.A. Mayer. 2009. Principles and clinical implications of the brain-gut-enteric microbiota axis. Nat. Rev. Gastroenterol. Hepatol. 6: 306–314.

Rousseaux, C., X. Thuru, A. Gelot, N. Barnich, C. Neut, L. Dubuquoy, C. Dubuquoy, E. Merour, K. Geboes, M. Chamaillard, A. Ouwehand, G. Leyer, D. Carcano, J.F. Colombel, D. Ardid and P. Desreumaux. 2007. Lactobacillus acidophilus modulates intestinal pain and induces opioid and cannabinoid receptors. Nat. Med. 13: 35–37.

Rudloff, S., G. Pohlentz, C. Borsch, M.J. Lentze and C. Kunz. 2012. Urinary excretion of *in vivo* 13C-labelled milk oligosaccharides in breastfed infants. Br. J. Nutr. 107: 957–963.

Sanders, M.E. 2011. Impact of probiotics on colonizing microbiota of the gut. J. Clin. Gastroenterol. 45 Suppl.: S115–119.

Sandler, R.H., S.M. Finegold, E.R. Bolte, C.P. Buchanan, A.P. Maxwell, M.L. Vaisanen, M.N. Nelson and H.M. Wexler. 2000. Short-term benefit from oral vancomycin treatment of regressive-onset autism. J. Child Neurol. 15: 429–435.

Sarkar, S. and D.A. Fox. 2010. Targeting IL-17 and Th17 cells in rheumatoid arthritis. Rheum. Dis. Clin. North Am. 36: 345–366.

Scholz-Ahrens, K.E., P. Ade, B. Marten, P. Weber, W. Timm, Y. Acil, C.C. Gluer and J. Schrezenmeir. 2007. Prebiotics, probiotics, and synbiotics affect mineral absorption, bone mineral content, and bone structure. J. Nutr. 137: 838S–846S.

Schwab, C. and M. Ganzle. 2011. Lactic acid bacteria fermentation of human milk oligosaccharide components, human milk oligosaccharides and galactooligosaccharides. FEMS Microbiol. Lett. 315: 141–148.

Shanahan, F. 2002. The host-microbe interface within the gut. Best Pract. Res. Clin. Gastroenterol. 16: 915–931.

Shanahan, F. 2010. Probiotics in perspective. Gastroenterology 139: 1808–1812.

Shukla, S., A. Shukla, S. Mehboob and S. Guha. 2011. Meta-analysis: the effects of gut flora modulation using prebiotics, probiotics and synbiotics on minimal hepatic encephalopathy. Aliment Pharmacol. Ther. 33: 662–671.

Sjogren, K., C. Engdahl, P. Henning, U.H. Lerner, V. Tremaroli, M.K. Lagerquist, F. Backhed and C. Ohlsson. 2012. The gut microbiota regulates bone mass in mice. J. Bone Miner Res. 27: 1357–1367.

Smith, E.A. and G.T. Macfarlane. 1996. Enumeration of human colonic bacteria producing phenolic and indolic compounds: effects of pH, carbohydrate availability and retention time on dissimilatory aromatic amino acid metabolism. J. Appl. Bacteriol. 81: 288–302.

Spiller, R. and K. Garsed. 2009. Postinfectious irritable bowel syndrome. Gastroenterology 136: 1979–1988.

Stoddart, L.A., N.J. Smith and G. Milligan. 2008. International Union of Pharmacology. LXXI. Free fatty acid receptors FFA1, -2, and -3: pharmacology and pathophysiological functions. Pharmacol. Rev. 60: 405–417.

Sudo, N., Y. Chida, Y. Aiba, J. Sonoda, N. Oyama, X.N. Yu, C. Kubo and Y. Koga. 2004. Postnatal microbial colonization programs the hypothalamic-pituitary-adrenal system for stress response in mice. J. Physiol. 558: 263–275.

Sun, C.Y., S.C. Chang and M.S. Wu. 2012. Uremic toxins induce kidney fibrosis by activating intrarenal renin-angiotensin-aldosterone system associated epithelial-to-mesenchymal transition. PLoS One 7: e34026.

Takayama, F., K. Taki and T. Niwa. 2003. Bifidobacterium in gastro-resistant seamless capsule reduces serum levels of indoxyl sulfate in patients on hemodialysis. Am. J. Kidney Dis. 41: S142–145.

Talley, N.J. 2003. Evaluation of drug treatment in irritable bowel syndrome. Br. J. Clin. Pharmacol. 56: 362–369.

Tanabe, S., Y. Kinuta and Y. Saito. 2008. Bifidobacterium infantis suppresses proinflammatory interleukin-17 production in murine splenocytes and dextran sodium sulfate-induced intestinal inflammation. Int. J. Mol. Med. 22: 181–185.

Tanida, M., T. Yamano, K. Maeda, N. Okumura, Y. Fukushima and K. Nagai. 2005. Effects of intraduodenal injection of *Lactobacillus johnsonii* La1 on renal sympathetic nerve activity and blood pressure in urethane-anesthetized rats. Neurosci. Lett. 389: 109–114.

Tazoe, H., Y. Otomo, S. Karaki, I. Kato, Y. Fukami, M. Terasaki and A Kuwahara. 2009. Expression of short-chain fatty acid receptor GPR41 in the human colon. Biomed. Res. 30: 149–156.

Thabane, M. and J.K. Marshall. 2009. Post-infectious irritable bowel syndrome. World J. Gastroenterol. 15: 3591–3596.

Thomas, R.H., M.M. Meeking, J.R. Mepham, L. Tichenoff, F. Possmayer, S. Liu and D.F. MacFabe. 2012. The enteric bacterial metabolite propionic acid alters brain and plasma phospholipid molecular species: further development of a rodent model of autism spectrum disorders. J. Neuroinflammation 9: 153.

Umesaki, Y., Y. Okada, S. Matsumoto, A. Imaoka and H. Setoyama. 1995. Segmented filamentous bacteria are indigenous intestinal bacteria that activate intraepithelial lymphocytes and induce MHC class II molecules and fucosyl asialo GM1 glycolipids on the small intestinal epithelial cells in the ex-germ-free mouse. Microbiol. Immunol. 39: 555–562.

Vaahtovuo, J., E. Munukka, M. Korkeamaki, R. Luukkainen and P. Toivanen. 2008. Fecal microbiota in early rheumatoid arthritis. J. Rheumatol. 35: 1500–1505.

Verdu, E.F., P. Bercik, M. Verma-Gandhu, X.X. Huang, P. Blennerhassett, W. Jackson, Y. Mao, L. Wang, F. Rochat and S.M. Collins. 2006. Specific probiotic therapy attenuates antibiotic induced visceral hypersensitivity in mice. Gut. 55: 182–190.

Wang, B., Y.K. Mao, C. Diorio, M. Pasyk, R.Y. Wu, J. Bienenstock and W.A. Kunze. 2010. Luminal administration ex vivo of a live Lactobacillus species moderates mouse jejunal motility within minutes. Faseb J. 24: 4078–4088.

Wang, I.K., H.C. Lai, C.J. Yu, C.C. Liang, C.T. Chang, H.L. Kuo, Y.F. Yang, C.C. Lin, H.H. Lin, Y.L. Liu, Y.C. Chang, Y.Y. Wu, C.H. Chen, C.Y. Li, F.R. Chuang, C.C. Huang, C.H. Lin and H.C. Lin. 2012. Real-time PCR analysis of the intestinal microbiotas in peritoneal dialysis patients. Appl. Environ. Microbiol. 78: 1107–1112.

Weickert, M.O., J. Spranger, J.J. Holst, B. Otto, C. Koebnick, M. Mohlig and A.F. Pfeiffer. 2006. Wheat-fibre-induced changes of postprandial peptide YY and ghrelin responses are not associated with acute alterations of satiety. Br. J. Nutr. 96: 795–798.

Whorwell, P.J., L. Altringer, J. Morel, Y. Bond, D. Charbonneau, L. O'Mahony, B. Kiely, F. Shanahan and E.M. Quigley. 2006. Efficacy of an encapsulated probiotic Bifidobacterium infantis 35624 in women with irritable bowel syndrome. Am. J. Gastroenterol. 101: 1581–1590.

Wilhelm, S.M., C.M. Brubaker, E.A. Varcak and P.B. Kale-Pradhan. 2008. Effectiveness of probiotics in the treatment of irritable bowel syndrome. Pharmacotherapy 28: 496–505.

Wu, S., N. Tao, J.B. German, R. Grimm and C.B. Lebrilla. 2010. Development of an annotated library of neutral human milk oligosaccharides. J. Proteome Res. 9: 4138–4151.

Wutzke, K.D., M. Lotz and C. Zipprich. 2010. The effect of pre- and probiotics on the colonic ammonia metabolism in humans as measured by lactose-[$^{15}N_2$]ureide. Eur. J. Clin. Nutr. 64: 1215–1221.

Yajima, T., R. Inoue, M. Matsumoto and M. Yajima. 2011. Non-neuronal release of Ach plays a key role in secretory response to luminal propionate in rat colon. J. Physiol. 589 (Part 4): 953–962.

Yajima, T., R. Inoue , M. YAjima, T. Tsuruta, S. Karaki, T. Hira and A. Kuwahara. 2001. The G-protein in cholesterol-rich membrane microdomains mediates mucosal sensing of short-chain fatty acids and secretory response in rat colon. Acta Physiol. (Oxf.) 203: 381–389.

Younes, H., N. Egret, M. Hadj-Abdelkader, C. Remesy, C. Demigne, C. Gueret, P. Deteix and J.C. Alphonse. 2006. Fermentable carbohydrate supplementation alters nitrogen excretion in chronic renal failure. J. Ren. Nutr. 16: 67–74.

Index

F

G

H

I

L

M

N

Color Plate Section

Chapter 3

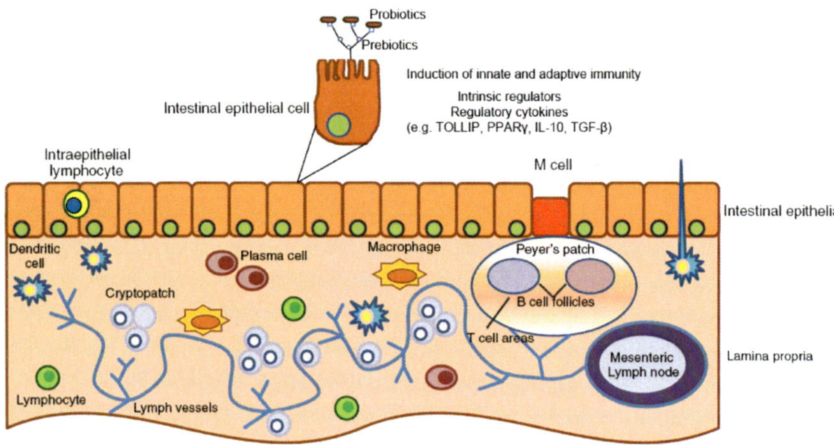

Fig. 1. Interactions of prebiotics and immune system in the intestinal mucosa, which display immunomodulatory functions (Modified from Choque Delgado et al. 2011).

Chapter 5

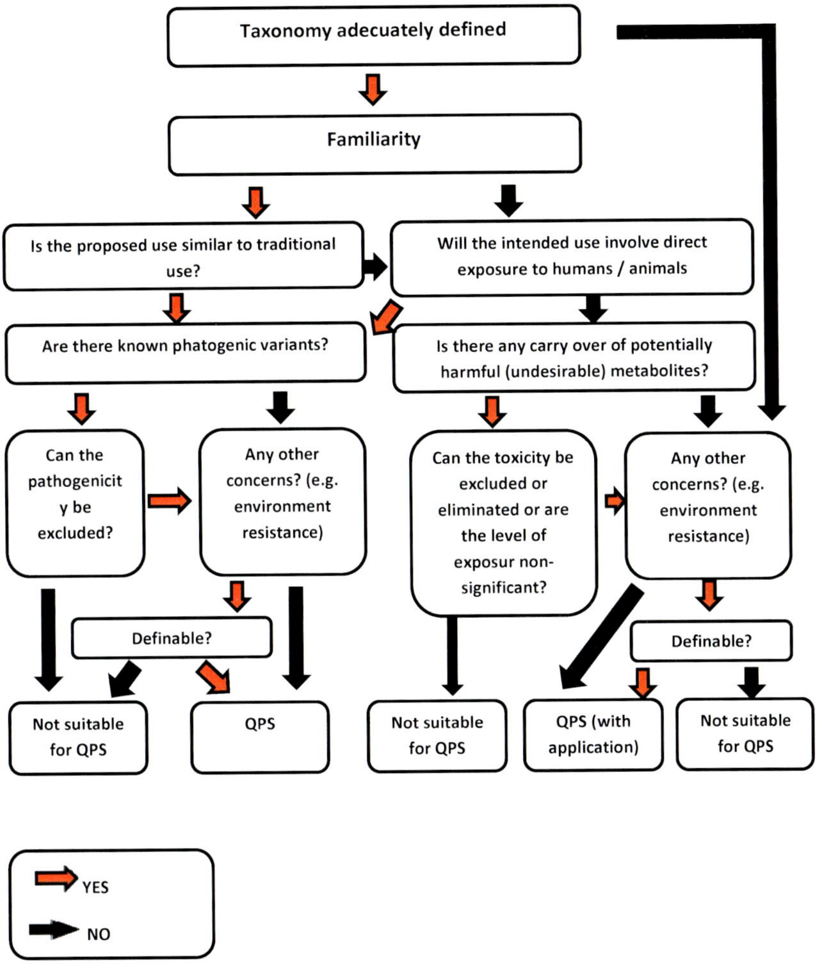

Fig. 1. Decision tree approach for the acceptance of a QPS micro-organism.

Chapter 9

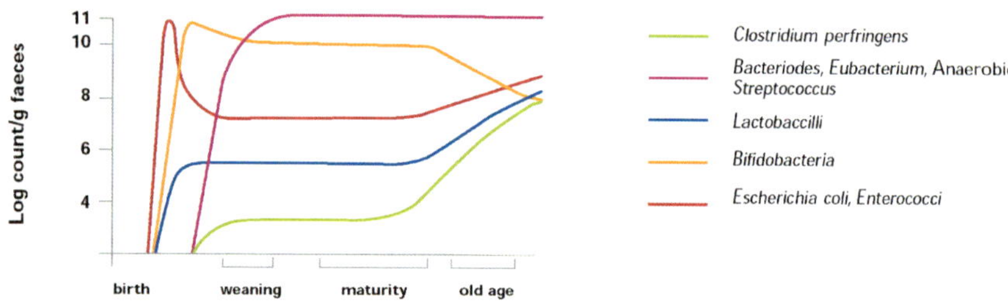

Fig. 1. Changes in gut microbiota groups with ageing in humans (adapted from Mitsuoka 1992).

Chapter 10

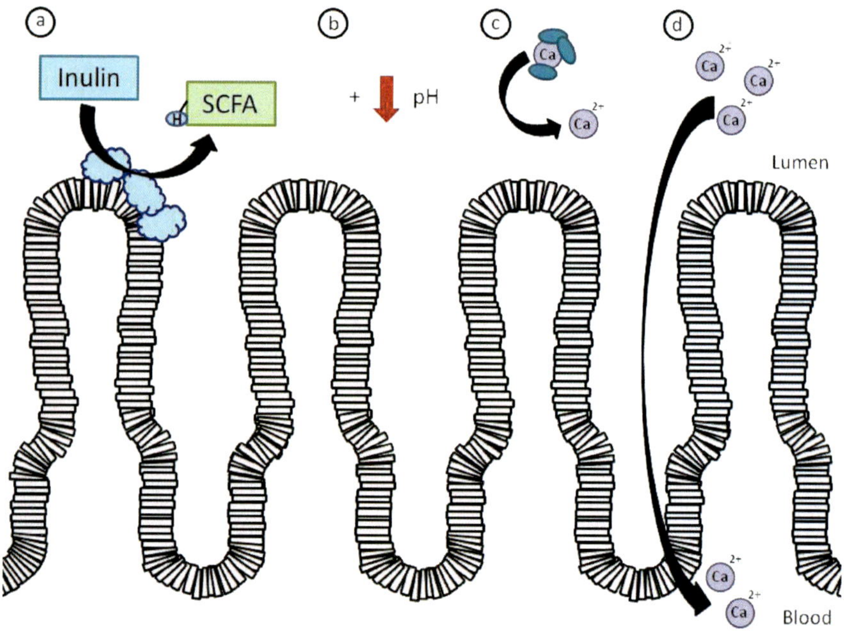

Fig. 3. Fermentation of prebiotics enhances calcium uptake at the enterocyte through the production of short chain fatty acids (SCFAs). a) Prebiotics such as inulin are fermented by saccharolytic bacteria residing in the colon to form SCFAs such as butyrate and acetate b) Accumulated SCFAs decreases the luminal pH c) This newly acidic environment ionizes calcium, freeing it from compounds to which it is bound d) Ionized calcium is absorbed more easily across the intestinal wall.

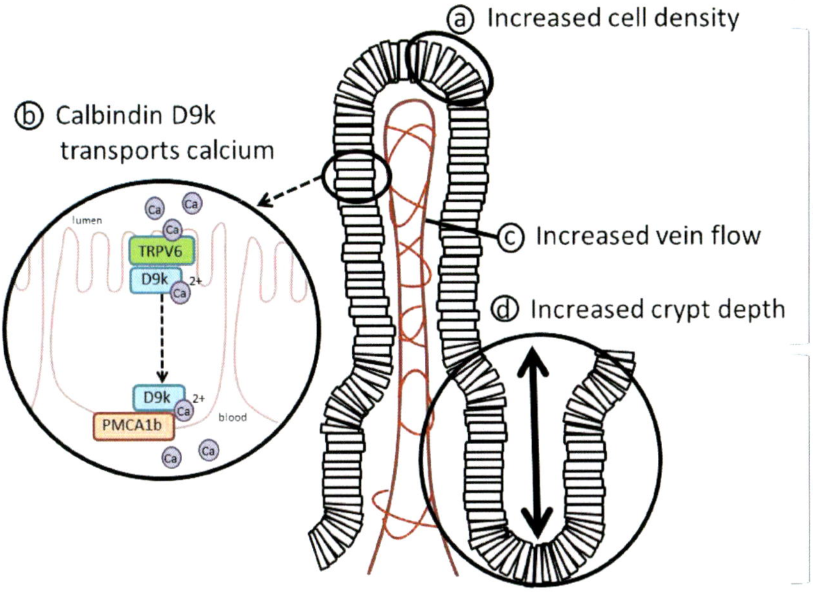

Fig. 4. Prebiotic-induced mineral absorption is mediated at the cellular level. a) epithelial cell density increases, b) calbindin D9k increases the active transport of calcium across the enterocyte, c) cecal vein flow increases resulting in increased surface area which allows for greater absorption, and d) crypt depth increases.

Chapter 15

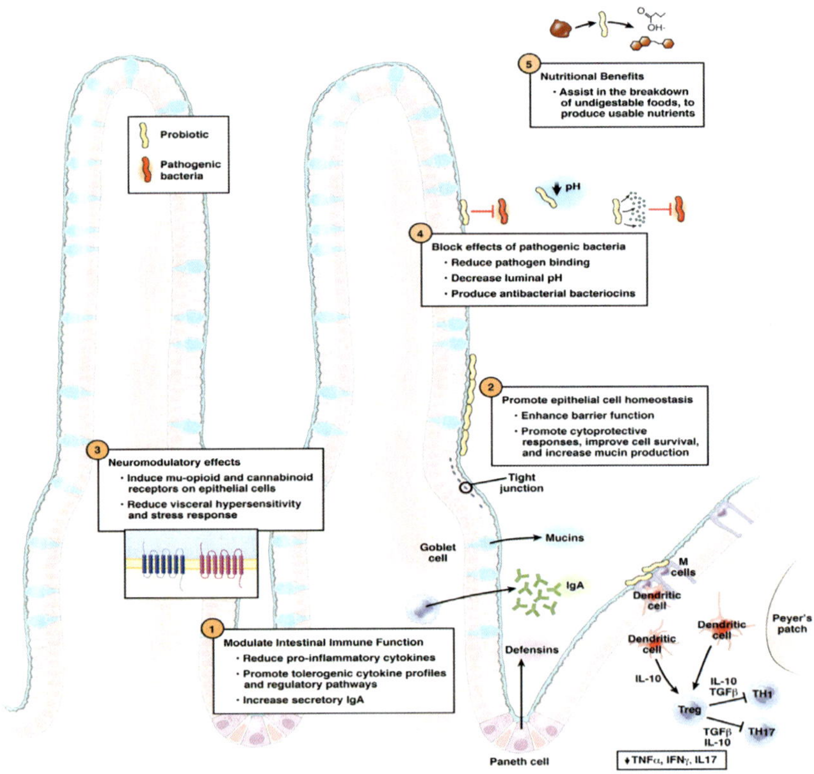

Fig. 1. Mechanisms of action of probiotics in the gastrointestinal tract (Ciorba 2012).